Simulation and Optimization Methods in Machining and Structure/Material Design

Simulation and Optimization Methods in Machining and Structure/Material Design

Guest Editors

Wuyi Ming
Xiaoke Li
Wenbin He

MDPI

Basel • Beijing • Wuhan • Barcelona • Belgrade • Novi Sad • Cluj • Manchester

Guest Editors
Wuyi Ming
College of Mechanical and Electrical Engineering
Zhengzhou University of Light Industry
Zhengzhou
China

Xiaoke Li
College of Mechanical and Electrical Engineering
Zhengzhou University of Light Industry
Zhengzhou
China

Wenbin He
College of Mechanical and Electrical Engineering
Zhengzhou University of Light Industry
Zhengzhou
China

Editorial Office
MDPI AG
Grosspeteranlage 5
4052 Basel, Switzerland

This is a reprint of the Special Issue, published open access by the journal *Metals* (ISSN 2075-4701), freely accessible at: https://www.mdpi.com/journal/metals/special_issues/metals_metallic_glasses.

For citation purposes, cite each article independently as indicated on the article page online and as indicated below:

Lastname, A.A.; Lastname, B.B. Article Title. *Journal Name* **Year**, *Volume Number*, Page Range.

ISBN 978-3-7258-7478-1 (Hbk)
ISBN 978-3-7258-7479-8 (PDF)
https://doi.org/10.3390/books978-3-7258-7479-8

Contents

About the Editors

Wuyi Ming

Wuyi Ming was born in Hubei, China, in 1981. He received his bachelor's degree from Huazhong Agricultural University in 2002, master's degree from Zhengzhou University in 2010, and Ph.D. from Huazhong University of Science and Technology in 2014, all in Engineering. He conducts theoretical and applied research in non-traditional machining and deep learning. He currently works at Zhengzhou University of Light Industry. He has published nearly 110 SCI-indexed articles (more than 2500 citations, H30) in journals such as the *Journal of Cleaner Production*, *Applied Surface Science*, *Journal of Materials Processing Technology*, *Measurement*, and *Metals*. His main research interests include artificial intelligence and green manufacturing. He won second prize in the Henan (China) Science and Technology Progress Award in 2018.

Xiaoke Li

Xiaoke Li received his bachelor's degree from Henan University of Science and Technology in 2009 and his Ph.D. in Mechanical Manufacturing and Automation from Huazhong University of Science and Technology in 2017. His research focuses on complex product design and manufacturing, structural reliability analysis, surrogate modeling and sequential sampling, and process simulation and parameter optimization. He has led several national- and provincial-level research projects, including those funded by the National Natural Science Foundation of China. He has published more than 60 academic papers, with over 40 indexed in SCI journals, and holds over 20 invention patents. He is a youth member of the Mechanical Design Division of the Chinese Mechanical Engineering Society and serves as a reviewer for SCI journals such as RESS and IJAMT.

Wenbin He

Wenbin He received his bachelor's degree in Mechanical Engineering from Henan Polytechnic University in 1991, master's degree in Mechanical Engineering from Tsinghua University in 2002, and Ph.D. in Mechanical Engineering from Tsinghua University in 2014. He is currently the Director of the Henan Key Laboratory for Intelligent Manufacturing of Mechanical Equipment. His research focuses on the digital design and manufacturing of complex equipment, vehicle dynamics, and intelligent control. He has led over ten major provincial projects and participated in three National Natural Science Foundation projects, in addition to coordinating eight enterprise-funded research programs. He has received two first prizes and three second prizes in the Henan Science and Technology Progress Award. He holds over 30 invention patents and has published more than 80 academic papers, with over 50 indexed in SCI/EI. He also serves as a council member of the Henan Artificial Intelligence Society.

metals

Editorial

Simulation and Optimization Methods in Machining and Structure/Material Design

Wuyi Ming *, Xiaoke Li * and Wenbin He

Department of Electromechanical Science and Engineering, Zhengzhou University of Light Industry, Zhengzhou 450000, China; hwb@zzuli.edu.cn
* Correspondence: mingwuyi@gmail.com (W.M.); lixiaoke@zzuli.edu.cn (X.L.)

1. Introduction and Scope

Machining and structural/material design are fundamental pillars of modern manufacturing systems. As industries continue to demand higher performance, higher efficiency, and tighter tolerances, the design and control of manufacturing processes have become more intricate than ever. These involve a wide array of complex physical interactions that are inherently nonlinear and time-dependent. Factors such as process parameters, thermo-mechanical coupling [1], phase transformation, and microstructural evolution [2,3] are deeply interconnected, collectively influencing product performance, manufacturing efficiency, and cost. The simultaneous occurrence of heat transfer, stress development, material flow, solidification, and microstructural change presents significant challenges for both real-time process control and predictive design [4].

Understanding and mastering these phenomena require more than empirical experience—it demands a scientific, physics-based modeling approach. To this end, numerical modeling and multi-scale simulation techniques have emerged as indispensable tools in both academic and industrial domains [5,6]. These tools enable engineers and scientists to visualize and quantify internal states and transitions that are otherwise difficult to observe directly, such as grain boundary movement, dislocation activity, residual stress evolution, and thermal field distribution [7,8]. By capturing behaviors across multiple length and time scales, from atomic arrangements to component-level responses, multi-scale modeling provides a comprehensive framework for linking processing conditions to material outcomes [9,10]. In recent years, the integration of simulation technologies with advanced optimization algorithms, such as machine learning, evolutionary computing, and surrogate modeling, has further enhanced the ability to explore large design spaces efficiently [11,12]. These intelligent methods, when combined with systematic experimental validation, help to minimize dependence on costly and time-consuming trial-and-error experimentation. Moreover, the adoption of digital twins, physics-informed neural networks (PINNs), and real-time data assimilation into simulation environments is further narrowing the gap between modeling and practical application [13–15]. These developments are accelerating the convergence of virtual prototyping and smart manufacturing. As a result, product development cycles can be shortened, manufacturing processes can be fine-tuned in silico, and overall production costs can be significantly reduced without compromising performance. These advances also open new opportunities for material-by-design strategies, where optimal microstructures and component geometries can be predicted before fabrication, significantly reducing waste and resource consumption.

This Special Issue presents a curated collection of recent advances in the field of machining and structural/material design, with particular emphasis on the simulation and optimization of complex processing phenomena. The selected papers represent a broad and balanced mix of fundamental theory, computational methodology, and application-oriented research. Together, they showcase how simulation-driven innovation continues to reshape the understanding and advancement of modern manufacturing systems. One of the key thematic focuses of this collection is multi-physics modeling, a critical approach for capturing the coupled behavior of different physical fields in real processing environments. The papers delve into various modeling strategies, including coupled thermal–mechanical simulations that enable more accurate predictions of temperature fields, stress distribution, and deformation during manufacturing processes such as welding, forming, and additive manufacturing. Several contributions further investigate the quantitative relationship between microstructure evolution and mechanical properties, providing insight into the process–structure–property triad and enabling performance-driven material design. Another important dimension addressed by this Special Issue is real-time process modeling and feedback control. Some papers explore how in situ monitoring data can be integrated into simulation frameworks to support adaptive control, defect prevention, and quality assurance in advanced manufacturing settings. This reflects a growing trend toward digital twin technologies, which fuse real-time sensor data with predictive models to optimize performance dynamically. In addition, the integration of simulation tools with material characterization techniques and optimization algorithms is a recurring highlight throughout the collection. The use of artificial intelligence, surrogate modeling, and metaheuristic algorithms to guide process design and parameter tuning is gaining momentum, especially in high-dimensional or computationally expensive scenarios. These approaches not only enhance predictive accuracy but also significantly reduce development costs and time. Collectively, this Special Issue reflects a strong and growing commitment within the research community to build robust, accurate, and scalable simulation and optimization frameworks. These frameworks are essential for enabling the next generation of intelligent, data-driven, and sustainable manufacturing solutions, supporting the transformation toward smart factories, autonomous decision-making, and agile production paradigms.

2. Contributions

In this Special Issue, twenty high-quality papers contributed by distinguished researchers from diverse academic and industrial backgrounds have been published, reflecting the current forefront of research and innovation in the areas of machining and structure/material design. These contributions collectively highlight the growing maturity and versatility of simulation and optimization methods as powerful tools for addressing complex, multidisciplinary challenges inherent to modern manufacturing systems. As manufacturing transitions toward higher performance, digitalization, and sustainability, the need for predictive modeling, data-driven decision-making, and physics-informed optimization has become more pronounced than ever. The scope of the published works ranges from fundamental investigations into material behavior and process–structure–property relationships to application-driven innovations in process control, energy efficiency, and structural performance enhancement. Several studies explore advanced numerical techniques such as finite element modeling, phase-field simulation, and molecular dynamics to capture multi-scale and multi-physics interactions in real manufacturing environments. Others integrate intelligent algorithms—ranging from machine learning to evolutionary optimization—with experimental validation to accelerate process development and enhance design reliability. The material systems covered include conventional steels, high-strength alloys, titanium-based materials, and complex composites, with applications that extend across automotive, aerospace, energy, and biomedical sectors. Collectively, these twenty papers demonstrate how the synergy between simulation, optimization, and experimentation is driving a new paradigm in materials and process engineering. By enabling more accurate predictions, rapid iteration, and integrated evaluation, these approaches contribute significantly to reducing development time, improving product performance, and ensuring more robust and adaptive manufacturing practices. This Special Issue not only showcases recent technical progress but also outlines a promising path forward for the continued evolution of intelligent, efficient, and sustainable design and production methodologies in advanced manufacturing.

Among the contributions, five papers concentrated on material forming processes and the quantitative relationships between processing conditions, microstructure evolution, and resulting mechanical properties. These studies represent comprehensive efforts to understand and control material response under complex thermal–mechanical environments, where the accurate modeling of heat flow, plastic deformation, and microstructural transitions is crucial for ensuring component integrity and performance. Specifically, one paper investigated how varying tool rotation speeds affected the degree of dynamic recrystallization, grain size distribution, and mechanical strength in a high-copper-content aluminum alloy subjected to friction stir welding, offering insights into the optimization of welding parameters for strength–ductility balance (contribution 1). Another work constructed forming limit diagrams (FLDs) for 22MnB5 hot-stamped steel through a combination of experiments and finite element simulations, thereby providing practical references for predicting fracture onset and improving die design in automotive hot forming (contribution 2). A third study focused on the deformation mechanisms of 42CrMo high-strength steel during multi-pass warm upsetting, revealing the influence of temperature, strain rate, and pass schedule on strain accumulation and property uniformity, which is critical for large-diameter shaft or gear component manufacturing (contribution 3). The fourth paper evaluated how laser power, scanning speed, and hatch spacing influenced porosity, melt pool stability, and microstructure in TA15 titanium alloy fabricated via selective laser melting (SLM), and correlated these parameters with tensile strength and elongation, supporting quality control in additive manufacturing of high-performance Ti alloys (contribution 4). Lastly, the microstructural evolution and texture changes in an Al-Mg-Si-Cu alloy under different welding speeds in friction stir welding were examined, demonstrating how controlled welding conditions could refine equiaxed grains, reduce anisotropy, and improve performance stability across the weld zone (contribution 5).

In addition, two papers focused on cutting and machining behavior modeling, which are critical for understanding thermal load effects, tool wear mechanisms, and energy efficiency in precision machining. The first study developed an analytical–numerical hybrid model for predicting transient temperature profiles in cutting tools under varying turning parameters, considering heat generation, conduction, and convection at the tool–chip–workpiece interface, and validated the model with embedded thermocouple measurements (contribution 6). The second study explored how drilling energy consumption relates to changes in subsurface microhardness, residual strain, and plastic deformation in the material beneath the machined surface, highlighting how energy efficiency metrics can serve as indirect indicators of machining-induced damage and surface integrity degradation (contribution 7).

Three additional papers emphasized microstructure-level simulations and the interplay of multiple physical fields during advanced material processing. These contributions reflect the increasing use of high-fidelity computational models to simulate fine-scale interactions that are difficult to observe experimentally. One study employed a phase-field method to simulate dendritic growth in a WC particle-reinforced Ni-based alloy during plasma cladding, incorporating residual stress fields into the model and demonstrating how mechanical loading alters solute diffusion and interface morphology, thus guiding parameter control for composite coatings (contribution 8). Another used molecular dynamics (MD) to investigate the structure and thermal response of $Ni_{60}Nb_{40}$ metallic glass in particle and thin film forms, revealing that reduced dimensions lower the glass transition temperature and affect atomic packing efficiency, which has implications for nanoscale amorphous material design and thermal stability assessment (contribution 9). The final paper in this group developed a finite element model that links plastic deformation in 35CrMoA steel to ultrasonic nonlinear responses, specifically second harmonic generation. By coupling dislocation density evolution with acoustic wave propagation modeling, this work advanced our ability to noninvasively detect early-stage plastic damage, contributing to nondestructive evaluation (NDE) techniques in safety-critical components (contribution 10).

One notable paper in this collection delivered an innovative contribution to the field of architected materials and structural mechanics. The authors proposed and experimentally verified a family of metallic re-entrant auxetic structures with negative Poisson's ratio behavior, focusing on the influence of geometric parameters on mechanical responses such as energy absorption, elastic recovery, and cyclic stability. This work provides valuable insights into the design of lightweight, flexible metamaterials that are applicable to fields ranging from aerospace to biomedical devices, where both strength and adaptability are required (contribution 11).

Five additional papers focused on optimization strategies in processing and structural/material design. These contributions demonstrate the growing integration of intelligent algorithms and computational tools to enhance material performance, structural efficiency, and manufacturing sustainability. One study optimized the microstructure and mechanical properties of alloys through multiple remelting and heat treatment cycles, improving the overall performance and reliability of the materials (contribution 12). Another employed an Extreme Learning Machine (ELM) surrogate model combined with a Particle Swarm Optimization (PSO) algorithm to optimize the dynamics of multiblade centrifugal fans, enhancing aerodynamic performance and operational stability (contribution 13). A third work carried out a multi-objective lightweight design of vehicle aluminum subframes, achieving a balance between strength, stiffness, and weight, which supports the development of fuel-efficient vehicles (contribution 14). The fourth paper optimized hopper structure using an E-SVR surrogate model to improve material flow behavior and device efficiency, contributing to smarter material handling system design (contribution 15). The final study in this group focused on improving the energy efficiency of large-scale automotive glass molding processes, successfully reducing energy consumption and carbon emissions, and advancing green manufacturing practices (contribution 16).

Finally, four comprehensive review papers examined simulation and optimization methods in machining and structural/material design, covering several cutting-edge technologies and emerging application directions. These contributions highlight the growing role of intelligent computation and advanced modeling in enhancing manufacturing precision, efficiency, and innovation. One review focused on the application of artificial intelligence in the design and analysis of steel structures, emphasizing how intelligent algorithms significantly improve design efficiency and accuracy, while enabling optimization under increasingly complex conditions (contribution 17). Another provided an overview of recent advances in the finite element modeling of electrical discharge machining (EDM), analyzing how this method effectively simulates material removal and temperature field distribution, thereby offering a theoretical basis for process parameter optimization (contribution 18). A third paper surveyed the use of swarm intelligence algorithms to optimize EDM parameters, with notable achievements in enhancing machining accuracy, reducing surface damage, and improving overall efficiency—demonstrating the strong potential of swarm intelligence in advanced manufacturing (contribution 19). The final review explored ultrasound-assisted wire-arc additive manufacturing, outlining its key role in improving forming precision, shortening production cycles, and reducing costs, which supports innovation in metal rapid prototyping technologies (contribution 20). Collectively, these studies offer innovative optimization strategies and engineering insights that support material processing, mechanical design, and industrial technology upgrading.

As Guest Editors of this Special Issue, we hope that the papers included in this collection will be beneficial to scientists and engineers in advancing their research and development endeavors.

3. Conclusions and Outlook

In the fields of machining and structure/material design, simulation and optimization methods have been a hot topic in academic research and industrial applications, and have been evolving with the progress of science and technology. With the continuous integration of multidisciplinary technologies, especially the combination of computer science [16], materials science, and engineering technology [17], numerical models and algorithms for simulating machining processes and structural responses have been making significant progress. Currently, the conditions for accurate simulation are continuously improving, enabling the more accurate reflection of various physical phenomena in complex machining environments, such as stress distribution, thermal field changes, and the evolution of material properties [18].

In terms of optimization decisions, technological developments have greatly expanded the range of factors that can be taken into account. In the past, optimization methods usually focused on the adjustment of a single process parameter, but currently, with the proposal of multi-objective optimization, researchers have begun to focus on a more complex decision space, such as the collaborative optimization of multiple factors, including material properties, production costs, environmental impacts, process stability, and so on. This not only improves the overall performance of products but also promotes the exploration of more sustainable manufacturing models [19]. In particular, with the combination of advanced physical models, intelligent optimization algorithms, and large-scale material property databases, researchers are able to accurately simulate the deformation and phase transition processes of multiphase materials under different processing conditions, as well as their interactions with environmental factors. For example, with applications in industrial processes such as machining, mold making, and composite molding, these simulation tools provide more reliable theoretical support for the production process, helping to reduce unnecessary test costs, improve productivity, and reduce energy consumption [20].

However, despite the advances in new numerical techniques, machine learning methods, and computational power, the fields of machining and structure/material design still face many challenges in modeling and optimization. For complex multiphase materials, it is still a considerable challenge to accurately simulate their behavioral changes under different machining environments. Moreover, with the continuous upgrading of industrial demands, existing optimization algorithms often display inefficiency and lack of accuracy when dealing with large-scale, multi-objective complex systems. Therefore, how to design efficient optimization algorithms to adapt to the changing industrial demands is an important trend for future research.

Acknowledgments: As Guest Editors, we appreciate the invaluable contributions of our distinguished contributing authors, and the time and effort of the reviewers, editors, and editorial staff of *Metals*.

Conflicts of Interest: The authors declare no conflicts of interest.

List of Contributions:

1. Li, W.; Luo, Z.; Sun, Y.; Liu, X. Effect of Tool Speed on Microstructure Evolution and Mechanical Properties of Friction Stir Welded Joints of Al-Mg-Si Alloy with High Cu Content. *Metals* **2024**, *14*, 758. https://doi.org/10.3390/met14070758.
2. He, W.; Yang, B.; Zhang, X.; Li, M.; Sun, S.; Wang, B.; Ma, Q. Investigation into the Hot-Forming Limit for 22MnB5 Hot-Forming Steel under a Stamping Process. *Metals* **2024**, *14*, 561. https://doi.org/10.3390/met14050561.
3. Xiao, Z.; Wang, H.; Liu, J.; Jiang, J.; Yu, L.; Zhang, Y. Study on Deformation Behavior and Mechanical Properties of 42CrMo High-Strength Steel with Multi-Station Warm Upsetting. *Metals* **2024**, *14*, 135. https://doi.org/10.3390/met14020135.
4. Jiang, J.; Liang, C.; Chen, Y.; Wang, Y.; Cui, H.; Xu, J.; Zhou, F.; Wang, P.; Zhang, D.Z. The Influence of Process Parameters on the Density, Microstructure, and Mechanical Properties of TA15 Titanium Alloy Fabricated by Selective Laser Melting. *Metals* **2025**, *15*, 233. https://doi.org/10.3390/met15030233.
5. Luo, Z.; Sun, Y.; Li, W.; He, J.; Luo, G.; Liu, H. Evolution of Microstructures, Texture and Mechanical Properties of Al-Mg-Si-Cu Alloy under Different Welding Speeds during Friction Stir Welding. *Metals* **2023**, *13*, 1120. https://doi.org/10.3390/met13061120.
6. Liu, H.; Meurer, M.; Bergs, T. Modeling and Monitoring of the Tool Temperature During Continuous and Interrupted Turning with Cutting Fluid. *Metals* **2024**, *14*, 1292. https://doi.org/10.3390/met14111292.
7. Storchak, M.; Hlembotska, L.; Melnyk, O.; Baranivska, N. Interaction of Mechanical Characteristics in Workpiece Subsurface Layers with Drilling Process Energy Characteristics. *Metals* **2024**, *14*, 683. https://doi.org/10.3390/met14060683.
8. Wei, D.; Chen, M.; Zhang, C.; Ai, X.; Xie, Z. Simulation of Localized Stress Impact on Solidification Pattern during Plasma Cladding of WC Particles in Nickel-Based Alloys by Phase-Field Method. *Metals* **2024**, *14*, 1022. https://doi.org/10.3390/met14091022.
9. Zhang, W.; Ma, Y. Study of Size Effect on Ni60Nb40 Amorphous Particles and Thin Films by Molecular Dynamic Simula-tions. *Metals* **2024**, *14*, 835. https://doi.org/10.3390/met14070835.
10. Yu, S.; Hu, L.; Yang, X.; Ji, X. Finite Element Modeling of Acoustic Nonlinearity Derived from Plastic Deformation of 35CrMoA Steel. *Metals* **2025**, *15*, 343. https://doi.org/10.3390/met15040343.
11. Plewa, J.; Płońska, M.; Junak, G. Metallic Metamaterials with Auxetic Properties: Re-Entrant Structures. *Metals* **2024**, *14*, 1272. https://doi.org/10.3390/met14111272.
12. Tang, P.; Pang, R.; Chen, H.; Ren, Y.; Tan, J. Optimization of Microstructure and Mechanical Properties in Al-Zn-Mg-Cu Alloys Through Multiple Remelting and Heat Treatment Cycles. *Metals* **2025**, *15*, 234. https://doi.org/10.3390/met15030234.
13. Meng, F.; Wang, L.; Ming, W.; Zhang, H. Aerodynamics Optimization of Multi-Blade Centrifugal Fan Based on Extreme Learning Machine Surrogate Model and Particle Swarm Optimization Algorithm. *Metals* **2023**, *13*, 1222. https://doi.org/10.3390/met13071222.
14. Meng, X.; Sun, Y.; He, J.; Li, W.; Zhou, Z. Multi-Objective Lightweight Optimization Design of the Aluminium Alloy Front Subframe of a Vehicle. *Metals* **2023**, *13*, 705. https://doi.org/10.3390/met13040705.
15. Li, X.; Jiang, Q.; Long, Y.; Chen, Z.; Zhao, W.; Ming, W.; Cao, Y.; Ma, J. Design Optimization of Chute Structure Based on E-SVR Surrogate Model. *Metals* **2023**, *13*, 635. https://doi.org/10.3390/met13030635.
16. Chen, Y.; Zhang, S.; Hu, S.; Zhao, Y.; Zhang, G.; Cao, Y.; Ming, W. Study of Heat Transfer Strategy of Metal Heat-ing/Conduction Plates for Energy Efficiency of Large-Sized Automotive Glass Molding Process. *Metals* **2023**, *13*, 1218. https://doi.org/10.3390/met13071218.

17. Sarfarazi, S.; Mascolo, I.; Modano, M.; Guarracino, F. Application of Artificial Intelligence to Support Design and Analysis of Steel Structures. *Metals* **2025**, *15*, 408. https://doi.org/10.3390/met15040408.
18. Li, L.; Sun, S.; Xing, W.; Zhang, Y.; Wu, Y.; Xu, Y.; Wang, H.; Zhang, G.; Luo, G. Progress in Simulation Modeling Based on the Finite Element Method for Electrical Discharge Machining. *Metals* **2024**, *14*, 14. https://doi.org/10.3390/met14010014.
19. Chen, Y.; Hu, S.; Li, A.; Cao, Y.; Zhao, Y.; Ming, W. Parameters Optimization of Electrical Discharge Machining Process Using Swarm Intelligence: A Review. *Metals* **2023**, *13*, 839. https://doi.org/10.3390/met13050839.
20. Cao, Y.; Zhang, Y.; Ming, W.; He, W.; Ma, J. Review: The Metal Additive-Manufacturing Technology of the Ultrasonic-Assisted Wire-and-Arc Additive-Manufacturing Process. *Metals* **2023**, *13*, 398. https://doi.org/10.3390/met13020398.

References

1. Wang, B.; Meng, S.; Gao, B.; Wang, K.; Xu, C. Ultra-high temperature mechanical behavior and microstructural evolution of needle-punched carbon/carbon composites under time-varying thermo-mechanical coupling conditions. *Compos. Struct.* **2025**, *365*, 119192. [CrossRef]
2. Zhao, S.; Xiao, H.; Li, Y.; Zhang, Z.; Wang, Y.; Huang, Q.; Cao, L.; Gao, F.; Tracy, C.L.; Ewing, R.C.; et al. Multi-stage phase transformation pathways in MAX phases. *Nat. Commun.* **2025**, *16*, 1554. [CrossRef]
3. Wang, Z.J.; Zheng, Z.; Fu, M.W. Aluminum matrix composites: Structural design and microstructure evolution in the deformation process. *J. Mater. Res. Technol.* **2024**, *30*, 3724–3754. [CrossRef]
4. Zhang, Z.; Yu, C.; Ren, G.; Shen, S.; Yi, H. Laser Additive Manufacturing of Three-Dimensional Porous Structures: Structural Design, Microstructure, Mechanical Properties and Applications. *J. Mater. Res Technol.* **2025**, *36*, 3684–3725. [CrossRef]
5. Zong, J.; Zhang, L.; Liang, X.; Zhang, Y.; Zhang, Y.W.; Nie, Z.; Zhao, J. Numerical simulation of plate-fin heat exchanger and design of novel turbulator structures using a multi-scale simulation approach. *Int. Commun. Heat Mass Transf.* **2025**, *164*, 108897. [CrossRef]
6. Shi, G.; Hou, L.; Zhao, H. Numerical study on the seismic behaviour of high-strength steel moment-resisting frames: Multi-scale modelling and validation. *Thin-Walled Struct.* **2025**, *208*, 112807. [CrossRef]
7. Yan, H.; Zeng, X.; Cui, Y.; Zou, D. Numerical and experimental study of residual stress in multi-pass laser welded 5A06 alloy ultra-thick plate. *J. Mater. Res. Technol.* **2024**, *28*, 4116–4130. [CrossRef]
8. Mao, S.; Yang, B.; Liu, G.; Liu, G.; Zhang, Z. Temperature distribution and residual stress evolution at the interface of CuCrZr/316 L multi-material by laser powder bed fusion. *Opt. Laser Technol.* **2023**, *163*, 109355. [CrossRef]
9. Jiang, D.; Xie, L.; Wang, L. Current application status of multi-scale simulation and machine learning in research on high-entropy alloys. *J. Mater. Res. Technol.* **2023**, *26*, 1341–1374. [CrossRef]
10. Ji, H.; Song, Q.; Liu, Z. Micro-milling machinability prediction for crystalline materials via numerical-analytical hybrid modelling and strain rate-dependent grain-scale simulation. *J. Manuf. Process.* **2024**, *124*, 972–984. [CrossRef]
11. Fang, T.; Xie, L.; Niu, Y.; Hong, D.; Zhu, Y.; Wang, Z.; Zheng, X. Thermal shock cycle finite element simulation and optimization design of $Yb_2Si_2O_7$-based high temperature abradable sealing coatings based on machine learning. *Ceram. Int.* **2025**, *51*, 14160–14173. [CrossRef]
12. Zhou, W.; Xie, Z. Sealing rubber ring design based on machine learning algorithm combined progressive optimization method. *Tribol. Int.* **2025**, *201*, 110173. [CrossRef]
13. Fang, X.; Song, Q.; Wang, X.; Li, Z.; Ma, H.; Liu, Z. An intelligent tool wear monitoring model based on knowledge-data-driven physical-informed neural network for digital twin milling. *Mech. Syst. Signal Process.* **2025**, *232*, 112736. [CrossRef]
14. Zhang, S.; Dinavahi, V.; Liang, T. Towards hydrogen-powered electric aircraft: Physics-informed machine learning based multi-domain modeling and real-time digital twin emulation on FPGA. *Energy* **2025**, *322*, 135451. [CrossRef]
15. Li, X.; Zhu, H.; Chen, Z.; Ming, W.; Cao, Y.; He, W.; Ma, J. Limit state Kriging modeling for reliability-based design optimization through classification uncertainty quantification. *Reliab. Eng. Syst. Saf.* **2022**, *224*, 108539. [CrossRef]
16. Ming, W.; Sun, P.; Zhang, Z.; Qiu, W.; Du, J.; Li, X.; Zhang, Y.; Zhang, G.; Liu, K.; Wang, Y.; et al. A Systematic Review of Machine Learning Methods Applied to Fuel Cells in Performance Evaluation, Durability Prediction, and Application Monitoring. *Int. J. Hydrogen Energy* **2023**, *48*, 5197–5228. [CrossRef]
17. Chang, H.; Zhang, Z.; Wang, P.; Zhang, G. A Novel Composite Flm with Superhydrophobic Graphene for Anti-icing/deicing Via Chemical-assisted Magnetically Controllable Picosecond Laser Writing. *Mater. Today Phys.* **2025**, *54*, 101726. [CrossRef]
18. Wang, P.; Zhang, Z.; Wei, S.; Hao, B.; Chang, H.; Huang, Y.; Zhang, G. Study on Plasma Behaviour, Ablation Mechanism, and Surface Morphology of CFRP by Underwater Laser-induced Plasma Micro-machining. *J. Mater. Process. Technol.* **2025**, *338*, 118757. [CrossRef]

19. Li, X.; Wang, P.; Zhao, M.; Su, X.; Tan, Y.; Ding, J. Customizable Anisotropic Microlattices for Additive Manufacturing: Machine Learning Accelerated Design, Mechanical Properties and Structural-property Relationships. *Addit. Manuf.* **2024**, *89*, 104248. [CrossRef]
20. Arvanitis, K.; Nikolakopoulos, P.; Pavlou, D.; Farmanbar, M. Machine Learning-based Fatigue Lifetime Prediction of Structural Steels. *Alex. Eng. J.* **2025**, *125*, 55–66. [CrossRef]

metals

MDPI

Article

Finite Element Modeling of Acoustic Nonlinearity Derived from Plastic Deformation of 35CrMoA Steel

Shumin Yu [1], Lei Hu [2,*], Xingbin Yang [1] and Xiangyu Ji [2]

1 Zhengzhou Guodian Mechanical Design Institute Co., Ltd., Zhengzhou 450046, China; yusm@chec.com.cn (S.Y.); yangxb@chec.com.cn (X.Y.)
2 School of Materials Science and Engineering, Anhui University of Technology, Ma'anshan 243032, China; xiangyujiahut@sina.com
* Correspondence: hulei828@ahut.edu.cn

Abstract: Acoustic nonlinearity derived from microstructural evolution of metallic materials during plastic deformation has been found to be a promising nondestructive technique to identify early stage plastic damage in metallic structural components. In the current investigation, the propagation of longitudinal ultrasonic waves in plastically deformed 35CrMoA steel plates was simulated using finite element (FE) methods based on the theory of dislocation-induced acoustic nonlinearity to establish the relationship between acoustic nonlinearity parameters and plastic strain. Experiments were conducted to validate the numerical model. Both simulated and experimental results demonstrate a monotonic increase in the acoustic nonlinearity parameter with applied plastic strain. The simulated ultrasonic nonlinear parameters deviate from experimental measurements in a two-stage pattern. In the low-strain regime (plastic strain < 8.5%), FE predictions underestimate experimental values, possibly due to dislocation entanglement in high-density regions that restricts dislocation mobility and suppresses acoustic nonlinearity. The FE model overestimates the parameters when plastic strain exceeds about 8.5%. This reversal is related to the formation of dislocation cells and walls with enhanced acoustic nonlinearity.

Keywords: 35CrMoA steel; ultrasonic nonlinearity; second-harmonic generation; plastic damage; finite element simulation

1. Introduction

Plastic deformation is a fundamental phenomenon in the manufacture and service of metallic materials. Plastic damage is generated with the accumulation of plastic deformation resulting from changes in microstructure, such as dislocation multiplication, annihilation, and subgrain initiation.

Failure occurs when the level of plastic damage exceeds a certain threshold [1–3]. Therefore, the quantitative detection of plastic damage is of great significance in predicting the residual service life of metallic structures. Nonlinear ultrasonics is considered to be one of the effective methods for assessing early-stage damage of materials, such as plastic damage, fatigue damage, and creep damage [4–6].

The ultrasonic nonlinearity arises from the nonlinear stress–strain relationship resulting from the evolution of microstructures, such as crystal lattices and dislocations. The magnitude of damage in metal materials can be characterized by cumulative plasticity resulting from changes in microstructure, such as dislocation density. The magnitude of ultrasonic nonlinearity exhibits distinct differences between intact and damaged material states. This discrepancy enables the evaluation of material degradation through ultrasonic nonlinearity measurements, which are quantified by analyzing harmonic amplitudes in propagated ultrasonic waves [7–9]. To quantitatively describe the relationship between ultrasonic nonlinearity and material microstructure, a series of theoretical models have been established, including the dislocation string model, dislocation dipole model, and cumulative plastic strain model [10–13]. Hikata et al. [14] modeled the dislocation motion contribution to acoustic nonlinearity based on the dislocation string vibration model of Granato and Lücke [15] and experimentally verified in a high purity single crystal aluminum. The Hikata et al. model was later extended by Cash and Cai [16] to take into consideration orientation-dependent line energy and verified with dislocation dynamics simulations. Zhang et al. [17] improved the Hikata et al. model by taking into account the nonlinearity arising from screw and edge dislocations. However, it is still a challenge to capture the integrating quantitative microstructural contributions (e.g., dislocation density, precipitation, etc.) by experimental observation to establish the damage to constitutive relationships. Numerical simulations have been increasingly applied in ultrasonic nonlinearity studies due to their high efficiency and low cost. Some numerical models have been proposed to simulate the ultrasonic nonlinearity of media and/or damage, such as the finite element method (FEM) [18], finite difference time domain method (FDTD) [19], and local interaction simulation approach (LISA) [20]. Zhu et al. [21] simulated the longitudinal wave propagation in plastically deformed 30Cr2Ni4MoV martensite stainless steel and a monotonically increasing tendency of the acoustic nonlinearity parameter as a function of plastic strain was found. Wang et al. [22] modeled the effect of porosity on ultrasonic nonlinearity. The results showed that the relative nonlinear coefficient increases with the increasing pore length and decreases with the increasing pore width. Kim et al. [23] presents a non-destructive method for quantitative assessment of fatigue damage of materials with linear elastic properties. The semi-analytical finite element (SAFE) technique was used to obtain the phase-matching modes of the specimens. Li et al. [24] proposed a 2D hybrid spectral/finite element scheme for numerically resolving crack-induced contact acoustic nonlinearity in solid structures. Cao et al. [25] proposed a dedicated FE model to scrutinize the modulation mechanism of various modalities of pitting damage on the probing ultrasonic waves. Chen et al. [26] simulated the nonlinear mixed-frequency ultrasonic wave transmission mechanism from the cracks with different depths and the excited frequency. However, it is still a challenge to directly model the microscopic microstructure (e.g., dislocations, precipitates, etc.) of a macroscopic engineering structure due to the huge computational cost.

In the current investigation, a finite element model based on the dislocation-induced ultrasonic nonlinearity theory was established to simulate the ultrasonic nonlinearity of plastically deformed material. The Kocks–Mecking model was used to simulate the evolution of dislocation density. The ultrasonic nonlinear coefficients of material at different plastic damages were simulated to obtain the relationship between plastic damage, dislocation evolution and ultrasonic nonlinearity of the material. Subsequently, experimental validation was carried out to verify the accuracy of this model. The simulated results provide a theoretical foundation for the ultrasonic nonlinear evaluation of plastic damage of metal material.

2. Experimental Procedure

The material used in this study is 35CrMoA steel with the following chemical composition (wt. %): C 0.357, Cr 1.025, Mo 0.188, Si 0.212, Mn 0.688, P 0.015, S 0.0048, and Fe as the balance. The dimensions of the specimens are shown in Figure 1. Initially, the specimens were stretched to a predefined strain at room temperature using a universal testing machine with a controlled strain rate of 2.0 mm/min. A total of six groups of specimens were prepared, each containing two specimens in each group. Group A0 represented the original specimen used as a reference while group A5 was stretched until it fractured to determine the maximum elongation. Groups A1 to A4 were stretched to 20%, 40%, . . . , 80% of the maximum elongation, respectively, resulting in the corresponding strain values of 0.4%, 3.5%, 6.8% and 11.2%, respectively.

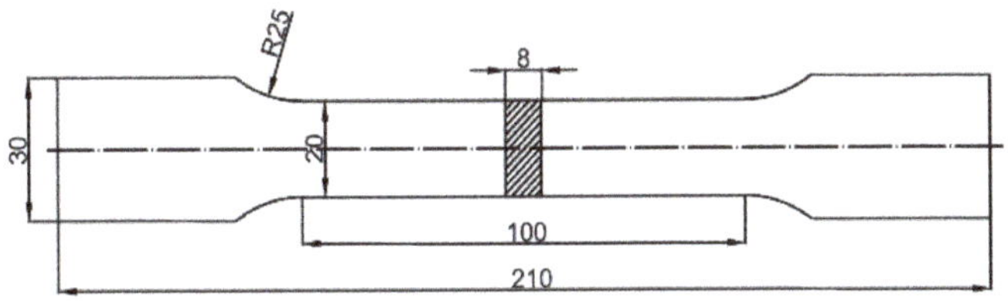

Figure 1. Dimensions of the specimen (Unit: mm).

Subsequently, the acoustic nonlinearity coefficients of both the original and plastically deformed specimens were measured using the RITEC SNAP RAM-5000 nonlinear ultrasonic testing system (RITEC Inc., Warwick, RI, USA). The experimental setup for nonlinear ultrasonic measurements is shown in Figure 2. A calibration step was conducted using a pulse-echo setup before the nonlinearity measurements to obtain the calibration function. The calibration function converts the electrical output of the receiving transducer into displacement amplitude to compensate for the coupling effect of the receiving transducer [27]. It is important to maintain the same contact condition between the specimen and the transducers in all tests. For this purpose, a specially designed fixture was used to secure the transducers, ensuring that a constant pressure of 3 kg × 9.8 = 29.40 N was applied in all tests.

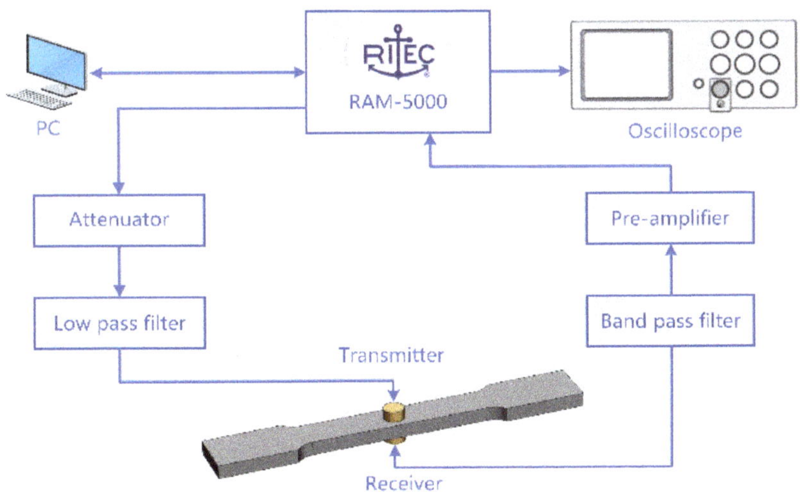

Figure 2. Experimental setup for nonlinear ultrasonic measurements.

To generate and measure the longitudinal wave, two commercial piezoelectric transducers with a diameter of $\Phi = 8$ mm were positioned at the center of the top and bottom surfaces of the specimens. A narrow-band longitudinal piezoelectric transducer was used to transmit a high-power fundamental wave while the receive transducer operated in pulse-echo mode and was excited with a broadband signal. The central frequencies of the transmitter and receiver are 5 MHz and 10 MHz, receptively, to the second harmonic components of longitudinal wave. The cutoff frequencies of the low- and high-pass filters are 1 and 20 MHz, respectively. The transmitter is operated with a voltage of 450 V. A high-voltage tone-burst signal consisting of 5 cycles, with a spatial length of approximately 8.0 mm, is modulated using a Hanning window and then fed into the transmitter. The transmitted signal waveform is shown in Figure 3.

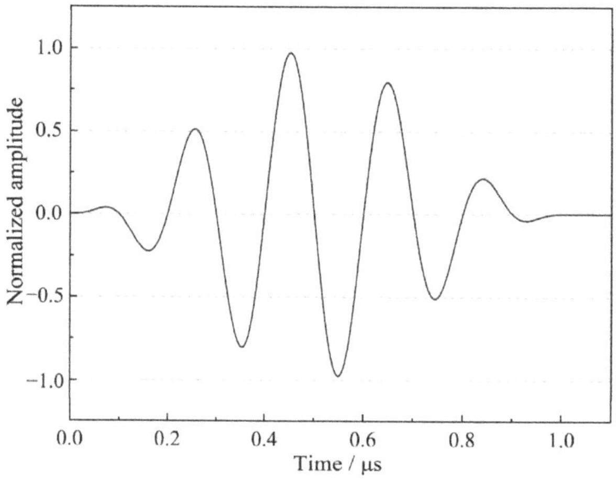

Figure 3. Waveform of the transmitted signal.

The one-dimensional nonlinear constitutive equation in an isotropic solid medium can be described as [28]:

$$\sigma_{xx} = E\varepsilon_x(1 + \beta\varepsilon_x + \ldots) \quad (1)$$

where, E is the elastic modulus, σ_{xx} is stress, ε_x is the strain, x is the material coordinate, and β is the second-order nonlinear coefficient. When the strain is small, the relationship between strain and displacement is given by:

$$\varepsilon_x = \frac{\partial u}{\partial x} \quad (2)$$

where, u is displacement. Ignoring the effects of scattering and attenuation, the wave equation of one-dimensional longitudinal wave in the solid medium is:

$$\rho\frac{\partial^2 u}{\partial t^2} = \frac{\partial \sigma_{xx}}{\partial x} \quad (3)$$

where ρ is the density of the medium, and t is time. By substituting Equations (1) and (2) into Equation (3) and ignoring the high order terms, the one-dimensional longitudinal wave equation in nonlinear media can be derived as:

$$\rho\frac{\partial^2 u}{\partial t^2} = E\frac{\partial^2 u}{\partial x^2} + 2\beta E\frac{\partial u}{\partial x} \cdot \frac{\partial^2 u}{\partial x^2} \quad (4)$$

Consider a single-frequency sinusoidal incident wave, namely $u = A_1\sin(\omega t)$, where amplitude of fundamental wave is A_1 and ω is angular velocity. According to the perturbation theory, the solution of the wave equation is [29]:

$$u = A_1\sin(kx - \omega t) - \frac{1}{8}\beta k^2 A_1^2 x\cos 2(kx - \omega t) \quad (5)$$

where, k is the wave number of the propagating wave. According to Equation (5), amplitude A_2 of the second harmonic is:

$$A_2 = \frac{1}{8}\beta k^2 A_1^2 x \tag{6}$$

Then, the second-order nonlinear coefficient of the material can be expressed as:

$$\beta = \frac{8}{k^2 x}\frac{A_2}{A_1^2} \tag{7}$$

According to Equation (7), the second order acoustics nonlinear coefficient of the material can be obtained through the Fast Fourier transform (FFT) transformation of the experimentally measured time domain received signal. Subsequently, the amplitudes A_1 and A_2 of the fundamental wave and the second harmonic in the received signal can be determined. To minimize the influence of ultrasonic near-field effects, the secondary reflected echo from both the lower and upper surfaces of the specimen was selected for measuring the second order acoustics nonlinear coefficient. The frequency scan range of the received signals begins at 1 MHz and extends up to 20 MHz. Each received signal file comprises approximately 1200 data points, providing detailed information at various frequencies. Additionally, a post-averaging process was applied to the harmonic signals to enhance stability. This process ensures that the measured values accurately reflect the overall level of the nonlinear effect present in each plastically deformed specimen. Through post-averaging, the values are smoothed and become more consistent, thus improving the reliability of the results. To reduce the testing error, each sample was measured three times. Figure 4 illustrates one of the frequency spectrum curves of the received signal of the Group A4 specimen, highlighting the features of received signal.

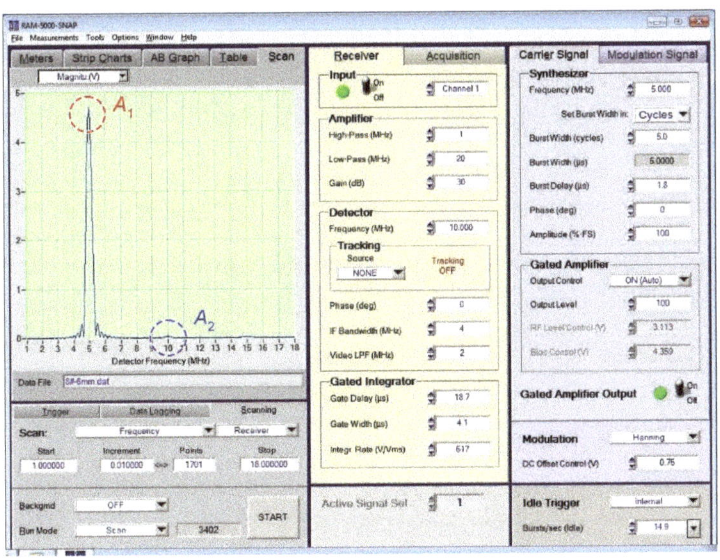

Figure 4. Frequency spectrum curve of the received signal of group A4 specimen.

From Figure 4, both the 5 MHz fundamental wave (A_1) and the 10 MHz second harmonic wave (A_2) were captured by the receiver. The amplitude of fundamental wave (A_1) is nearly 4.5 V, whereas the amplitude of the second harmonic wave (A_2) is less than 0.1 V. This indicates that the acoustics nonlinearity of plastically deformed 35CrMoA steel is relatively weak, and generating an observable second harmonic wave requires a high-power fundamental wave.

3. Mathematical and Finite Element Models

3.1. Theory of Dislocation-Induced Acoustics Nonlinearity

The plastic damage in metal materials is closely related to the evolution of dislocation structure. When subjected to an external load, dislocations within the material nucleate, accumulate and become entangled, resulting in changes to ultrasonic nonlinear parameters. Gliding dislocations are impeded by impurity atoms and precipitation phase particles within the material. Under stress, the dislocation line between two adjacent pinning points deforms into a bow-like string. When a periodic stress wave is applied to this dislocation string, it undergoes forced vibration, leading to waveform distortion of the incident wave and the generation of high-order harmonics.

The propagation of ultrasonic waves generates a small oscillatory stress with an amplitude of $\Delta\sigma$ in addition to the applied stress and results in an additional stran $\Delta\varepsilon$. The relation between the additional stress $\Delta\sigma$ and the additional strain $\Delta\varepsilon$ caused by elastic wave in crystal can be expressed as [30]:

$$\begin{aligned} \Delta\sigma &= \tfrac{\partial\sigma}{\partial\varepsilon}(\Delta\varepsilon) + \tfrac{1}{2}\tfrac{\partial^2\sigma}{\partial\varepsilon^2}(\Delta\varepsilon)^2 + \ldots \\ &= A(\Delta\varepsilon) + \tfrac{1}{2}B(\Delta\varepsilon)^2 + \ldots \end{aligned} \tag{8}$$

where

$$\begin{aligned} A &= \left[\tfrac{1}{A_2^H} + \tfrac{4(1-v)}{3}\cdot\tfrac{\Omega\Lambda L^2 R}{\mu}(1+vf_s-2vf_e)^{-1}\right]^{-1} \\ B &= \frac{-2\left[-\tfrac{1}{2}\tfrac{A_3^H}{\left(A_2^H\right)^3} + \tfrac{12}{5}(1-v)^3\cdot\tfrac{\Omega\Lambda L^4 R^3}{\mu^3 b^2}\cdot\sigma_b\cdot(1+vf_s-2vf_e)^{-3}\right]}{\left[\tfrac{1}{A_2^H} + \tfrac{4(1-v)}{3}\cdot\tfrac{\Omega\Lambda L^2 R}{\mu}(1+vf_s-2vf_e)^{-1}\right]^3} \end{aligned} \tag{9}$$

where μ is the shear modulus, b is the modulus of the Burgers vector, ν is Poisson's ratio, Λ is the dislocation density, L is half of the length of the dislocation string, σ_b is the back stress, Ω and R are the conversion factors of shear strain and shear stress to longitudinal strain and longitudinal stress respectively. A_2^H and A_3^H are the second-order and third-order Huang coefficients of the lattice respectively, and f_e and f_s are the fractions of edge and screw dislocations in total dislocations, respectively. Ultrasonic nonlinear coefficient β can be expressed as:

$$\beta = -\frac{B}{A} = \frac{-\tfrac{A_3^H}{\left(A_2^H\right)^3} + \tfrac{24}{5}(1-v)^3\cdot\tfrac{\Omega\Lambda L^4 R^3}{\mu^3 b^2}\cdot\sigma_b\cdot(1+vf_s-2vf_e)^{-3}}{\left[\tfrac{1}{A_2^H} + \tfrac{4(1-v)}{3}\cdot\tfrac{\Omega\Lambda L^2 R}{\mu}(1+vf_s-2vf_e)^{-1}\right]^2} \tag{10}$$

In most cases, $4(1-v)\Omega\Lambda L^2 R(1+vf_s-2vf_e)^{-1}/3\mu$ is much smaller than $1/A_2^H$ and can be neglected. Therefore, the ultrasonic nonlinear coefficient β can be further simplified as

$$\beta = -\frac{A_3^H}{A_2^H} + \frac{192(1-v)^3}{5}\frac{\Lambda L^4\Omega R^3\left(A_2^H\right)^2}{\mu^3 b^2}\times\sigma_b(2-v+3v\cos 2\theta)^{-3} \tag{11}$$

where θ is the angle between the dislocation line and the Burgers vector.

3.2. Dislocation Density Evolution Model

The plastic deformation of metal materials is related to the dislocation motion. According to the Kocks–Mecking model [31], the flow stress σ_s is solely related to dislocation density Λ, which can be expressed as:

$$\sigma_s = \sigma_0 + \alpha\mu b\sqrt{\Lambda}\,(1) \tag{12}$$

where α is a constant, which is 0.5 for most metals. During plastic deformation, the change rate of dislocation density is given as

$$\frac{d\Lambda}{d\varepsilon} = k_1\sqrt{\Lambda} - k_2\Lambda\,(2) \tag{13}$$

where k_1 and k_2 are dislocation formation and annihilation coefficients, respectively. Dislocation annihilation is the result of dynamic recovery process, and its effect can be neglected in room temperature tensile. The coefficient k_1 can be obtained by fitting the flow curve of 35CrMoA steel.

3.3. Finite Element Simulation

In the current investigation, the commercial finite element software COMSOL Multiphysics (V6.2, COMSOL Inc., Stockholm, Sweden) was used to model the constitutive equation of acoustics nonlinearity and simulate the propagation of ultrasonic nonlinearity in a plastically deformed 35CrMoA steel specimen. In homogeneous isotropic media, ultrasonic nonlinearity arises from material nonlinearity, geometric nonlinearity, and plastic deformation. It is generally believed that the ultrasonic nonlinearity caused by plastic damage mainly results from the evolution of dislocations in materials. The finite element model is shown in Figure 5.

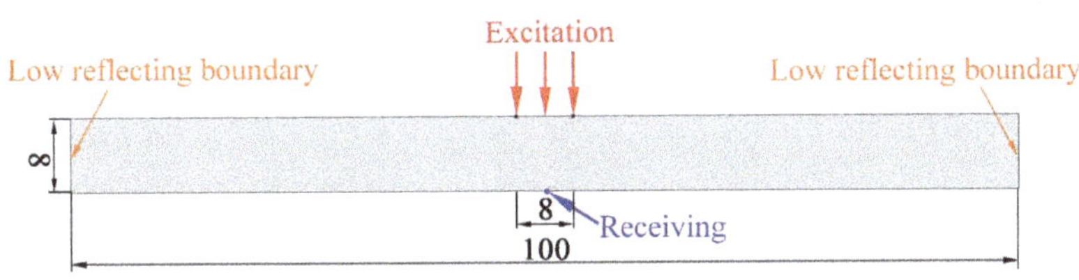

Figure 5. Schematic diagram of the simulation model (Unit: mm).

The top and bottom sides, except for the area where the excitation signal is applied, are stress-free and non-reflecting boundary conditions are prescribed for the lateral sides. The fundamental wave was excited by imposing longitudinal stress loadings with a 5-cycle Hanning-windowed sinusoidal tone burst at 5.0 MHz. The width of the excited area is 8 mm (identical to the diameter of the transducers) and the amplitude of the fundamental wave is 20 MPa. The received signal was obtained by extracting the longitudinal displacement at the center of the bottom edge. Subsequently, a Fast Fourier transformation (FFT) was performed on the extracted time domain signal to get the amplitudes of the fundamental and second harmonic signals. Finally, the acoustic nonlinearity parameters can be calculated using Equation (7).

The fourth-order Lagrange element was used in the FE model. To ensure the convergence of calculation, each wavelength should contain at least approximately 1.5 grid elements [32]. The wavelength of an ultrasonic longitudinal wave in steel is approximately 1.2 mm, so the grid size of finite element model elements is selected as 0.8 mm. According to the Fourier or von Neumann stability analysis [33], the time step, Δt, should be smaller than the time required for the longitudinal wave to travel across the element length, which is expressed as:

$$\Delta t \leq \frac{\Delta y}{c_L} \tag{14}$$

where c_L is the longitudinal wave velocity in steel and Δy is the length of element. The material parameters used in the calculation were $A_2^H = -3A_3^H = 195$ GPa, $\mu = 82$ GPa, $\upsilon = 0.3$, $b = 2.5$ nm, $L = 23$ nm, and the conversion factors Ω and R were chosen as 0.33 [34]. The edge dislocation and the screw dislocation were assumed to each make up half of the total dislocations. Figure 6 shows the propagation of the ultrasonic wave in the specimen when the plastic strain $\varepsilon = 0.112$.

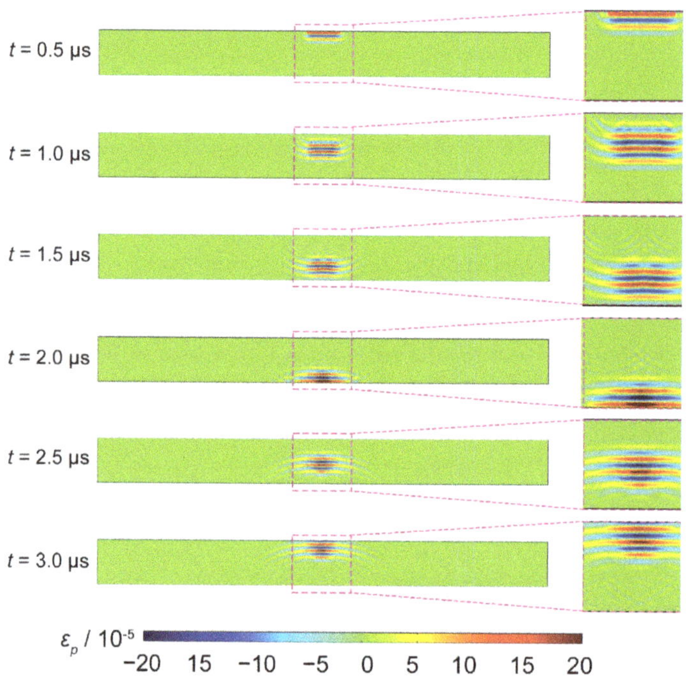

Figure 6. Propagation of elastic waves in the specimen.

It can be seen that the ultrasonic waves propagate to the bottom edge of the specimen after about t = 1.4 μs, and reflect several times on the bottom and top edges. For the simulated specimens in this investigation, the ultrasonic nonlinearity mainly arises from the material nonlinearity. The received signal in the time and frequency domain of the received signal in Figure 6 is shown in Figure 7.

The first wave is the signal that the transmitted signal directly reaches the receiving probe on the bottom edge, while the second wave is the signal that the transmitted signal reaches the receiving probe after being reflected once each from the bottom and top edges of the specimen, respectively. The Hanning window is used to intercept the second wave (t = 4–5.5 μs) of the received signal in Figure 7a, and then the Fast Fourier transform (FFT) was performed on this signal to obtain the spectrum distribution of the received signal. It can be seen that due to the nonlinear stress–strain relationship of the material, the second harmonic at $2f_0$ frequency is generated in the received signal.

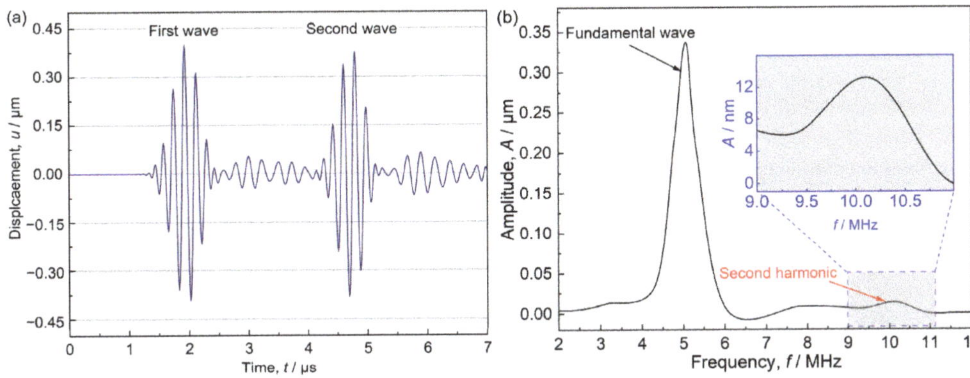

Figure 7. Received signal in time (**a**) and frequency (**b**) domain.

4. Discussion

The normalized relative nonlinearity parameters of the plastically deformed specimens with respect to the ultrasonic nonlinear parameters β_0 of the original unstretched specimens were defined as a function of the plastic strain. The results from experimental measurement and finite element simulation are shown in Figure 8.

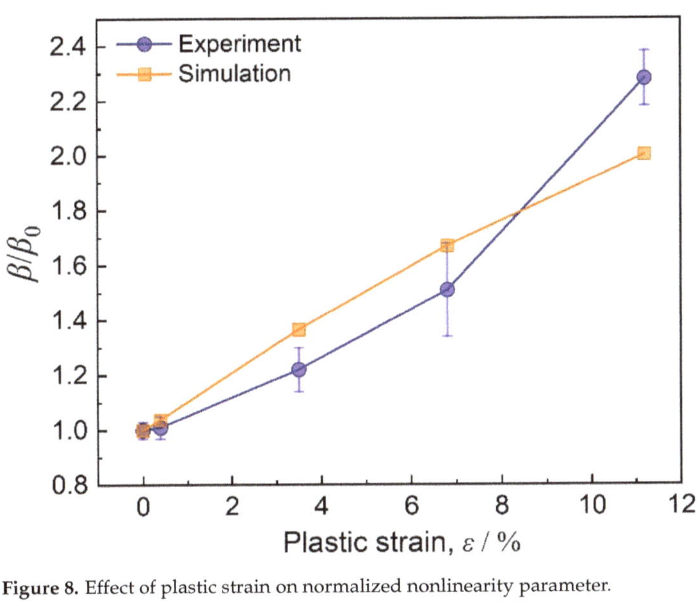

Figure 8. Effect of plastic strain on normalized nonlinearity parameter.

From Figure 8, it is evident that the ultrasonic nonlinear coefficient increases with the increase in plastic deformation. In the early stage of plastic deformation, the rise in the ultrasonic nonlinear coefficient can be mainly attributed to the increase in dislocation density. However, in the late stage of plastic deformation, the increase of the ultrasonic nonlinear coefficient is primarily associated with the formation of dislocation walls and dislocation cells [29].

The simulated ultrasonic nonlinear parameters exhibit a two-stage deviation pattern compared to experimental measurements. In the low-strain regime (plastic strain < 8.5%), the finite element (FE) predictions systematically underestimate experimental values. This deviation likely originates from dislocation entanglement mechanisms in high-density regions, where mutually trapped dislocations restrict their mobility and thereby suppress dislocation-induced ultrasonic nonlinearity [10]. In contrast, a crossover phenomenon emerges at plastic strains exceeding 8.5%, with the FE model overestimating the experimental ultrasonic nonlinear parameters. This reversal correlates with microstructural evolution during plastic deformation: Progressive dislocation accumulation leads to self-organization into cellular structures (dislocation walls/cells) that demonstrate enhanced acoustic nonlinearity compared to planar dislocation configurations [29]. Notably, the current modeling framework exclusively accounts for planar dislocation dynamics, failing to capture the diminished nonlinear response associated with these three-dimensional dislocation structures. This modeling limitation explains the increasing discrepancy at advanced deformation stages.

5. Conclusions

Investigations into the propagation of longitudinal waves and the ultrasonic nonlinear response of plastically deformed 35CrMoA steel were conducted through experiments and finite element simulations, ultilizing the dislocation-induced acoustic nonlinearity theory as a basis. The following conclusions have been drawn:

(1) Simulated and experimental results show a strong correlation between the acoustic nonlinearity parameter and plastic strain. Both simulated and experimental results demonstrate a monotonic increase in the nonlinear acoustic parameter with applied plastic strain.

(2) The simulation results for the ultrasonic nonlinear coefficient are lower than the experimental results when the plastic strain is lower than about 8.5%. This difference may be attributed to the dislocation entanglement in the areas with high dislocation density during the early stage of plastic deformation, which hinders dislocation movement and consequently reduces the nonlinear ultrasonic response of dislocations.

(3) When the plastic strains exceed about 8.5%, as dislocation density increases, the dislocations pile up with each other, forming dislocation cells and dislocation walls. The ultrasonic nonlinear response induced by dislocation cells and dislocation walls is higher than that of plane dislocations. Therefore, the calculated results for ultrasonic nonlinear coefficients in the later stage of plastic deformation are lower than the experimental results.

Author Contributions: Conceptualization, X.Y.; Methodology, S.Y.; Software, X.J.; Validation, L.H.; Investigation, S.Y. and X.J.; Resources, L.H.; Writing—original draft, S.Y.; Writing—review & editing, L.H.; Supervision, X.Y. All authors have read and agreed to the published version of the manuscript.

Funding: This research received no external funding.

Data Availability Statement: The original contributions presented in this study are included in the article. Further inquiries can be directed to the corresponding author.

Conflicts of Interest: Author Shumin Yu, Xinbin Yang are employed by the company Zhengzhou Guodian Mechanical Design Institute Co., Ltd. The remaining authors declare that the research was conducted in the absence of any commercial or financial relationships that could be construed as a potential conflict of interest.

References

1. Zeng, J.; Liu, J.; Jia, Y.; Zhao, G. Study on the Deformation and Fracture Mechanisms of Plastic Metals Considering Void Damage. *Metals* **2023**, *13*, 1556. [CrossRef]
2. Liu, Y.; Liu, J.; He, Y. Formation Mechanism and Evolution of Plastic Damage in Billet during Reduction Pretreatment. *Metals* **2023**, *13*, 747. [CrossRef]

3. Wang, X.; Chen, J.; Su, G.; Li, H.; Wang, C. Plastic damage evolution in structural steel and its non-destructive evaluation. *J. Mater. Res. Technol.* **2020**, *9*, 1189–1199. [CrossRef]
4. Sampath, S.; Sohn, H. Detection and localization of fatigue crack using nonlinear ultrasonic three-wave mixing technique. *Int. J. Fatigue* **2022**, *155*, 106582. [CrossRef]
5. Williams, C.L.; Lear, M.H.; Shokouhi, P. On the utility of linear and nonlinear ultrasound for evaluating microstructure-property relationships in laser powder bed fusion 316 L stainless steel. *Addit. Manuf.* **2023**, *73*, 103653. [CrossRef]
6. Yun, H.; Rayhana, R.; Pant, S.; Genest, M.; Liu, Z. Nonlinear ultrasonic testing and data analytics for damage characterization: A review. *Meas. J. Int. Meas. Confed.* **2021**, *186*, 110155. [CrossRef]
7. Bang, S.; Song, D.; Jhang, K. Comparisons of second- and third-order ultrasonic nonlinearity parameters measured using through-transmission and pulse-echo methods. *NdtE Int.* **2023**, *133*, 102757. [CrossRef]
8. Ryu, S.; Park, S.; Jhang, K. Plastic properties estimation of aluminum alloys using machine learning of ultrasonic and eddy current data. *NdtE Int.* **2023**, *137*, 102857. [CrossRef]
9. Nilsson, M.; Ulriksen, P.; Rydén, N. Nonlinear ultrasonic characteristics of a corroded steel plate. *Nondestruct. Test. Eval.* **2023**, *38*, 456–479. [CrossRef]
10. Rauter, N.; Lammering, R. A constitutive model for the analysis of second harmonic Lamb waves in unidirectional composites. *Int. J. Solids Struct.* **2018**, *135*, 184–196. [CrossRef]
11. Deng, M. Second-harmonic generation of ultrasonic guided wave propagation in an anisotropic solid plate. *Appl. Phys. Lett.* **2008**, *92*, 111910. [CrossRef]
12. Masurkar, F.; Tse, P.; Yelve, N.P. Evaluation of inherent and dislocation induced material nonlinearity in metallic plates using Lamb waves. *Appl. Acoust.* **2018**, *136*, 76–85. [CrossRef]
13. Xiang, Y.; Deng, M.; Liu, C.; Xuan, F. Contribution of mixed dislocations to the acoustic nonlinearity in plastically deformed materials. *J. Appl. Phys.* **2015**, *117*, 214903.
14. Hikata, A.; Chick, B.B.; Elbaum, C. Dislocation Contribution to the Second Harmonic Generation of Ultrasonic Waves. *J. Appl. Phys.* **1965**, *36*, 229–236.
15. Granato, A.; Lücke, K. Theory of mechanical damping due to dislocations. *J. Appl. Phys.* **1956**, *27*, 583–593.
16. Cash, W.D.; Cai, W. Dislocation contribution to acoustic nonlinearity: The effect of orientation-dependent line energy. *J. Appl. Phys.* **2011**, *109*, 014915.
17. Zhang, J.; Xuan, F.Z.; Xiang, Y. Dislocation characterization in cold rolled stainless steel using nonlinear ultrasonic techniques: A comprehensive model. *Europhys. Lett.* **2013**, *103*, 68003.
18. Hong, M.; Su, Z.; Wang, Q.; Cheng, L.; Qing, X. Modeling nonlinearities of ultrasonic waves for fatigue damage characterization: Theory, simulation, and experimental validation. *Ultrasonics* **2014**, *54*, 770–778. [CrossRef]
19. Matsuda, N.; Biwa, S. A Finite-Difference Time-Domain Technique for Nonlinear Elastic Media and Its Application to Nonlinear Lamb Wave Propagation. *Jpn. J. Appl. Phys.* **2012**, *51*, 07GB14.
20. Lee, B.C.; Staszewski, W.J. Lamb wave propagation modelling for damage detection: I. Two-dimensional analysis. *Smart Mater. Struct.* **2007**, *16*, 249–259. [CrossRef]
21. Zhu, W.; Deng, M.; Xiang, Y.; Xuan, F.; Liu, C.; Wang, Y. Modeling of ultrasonic nonlinearities for dislocation evolution in plastically deformed materials: Simulation and experimental validation. *Ultrasonics* **2016**, *68*, 134–141. [CrossRef] [PubMed]
22. Wang, X.; He, C.; He, H.; Xie, W. Simulation and experimental research on nonlinear ultrasonic testing of composite material porosity. *Appl. Acoust.* **2022**, *188*, 108528. [CrossRef]
23. Kim, J.; Park, J.; Zhu, B.; Cho, Y. Nonlinear Ultrasonic Guided Wave Method Using Semi-Analytical Finite Element (SAFE) Technique on a Damaged SWO-V Spring Coil. *Metals* **2021**, *11*, 752. [CrossRef]
24. Li, F.; Zou, F. A hybrid spectral/finite element method for accurate and efficient modelling of crack-induced contact acoustic nonlinearity. *J. Sound Vib.* **2021**, *508*, 116198. [CrossRef]
25. Cao, W.; Xu, L.; Su, Z.; Pang, B.; Chi, R.; Wang, L.; Wang, X. Modeling of pitting damage-induced ultrasonic nonlinearity in AL-Whipple shields of spacecraft: Theory, simulation, and experimental validation. *Int. J. Mech. Sci.* **2021**, *207*, 106659. [CrossRef]
26. Chen, H.; Li, S. Collinear Nonlinear Mixed-Frequency Ultrasound with FEM and Experimental Method for Structural Health Prognosis. *Processes* **2022**, *10*, 656. [CrossRef]
27. Kim, J.; Song, D.; Jhang, K. A method to estimate the absolute ultrasonic nonlinearity parameter from relative measurements. *Ultrasonics* **2017**, *77*, 197–202. [CrossRef] [PubMed]
28. Li, J.; Rokhlin, S.I. Propagation and scattering of ultrasonic waves in polycrystals with arbitrary crystallite and macroscopic texture symmetries. *Wave Motion* **2015**, *58*, 145–164. [CrossRef]
29. Matlack, K.H.; Kim, J.Y.; Jacobs, L.J., Qu, J. Review of Second Harmonic Generation Measurement Techniques for Material State Determination in Metals. *J. Nondestruct. Eval.* **2015**, *34*, 273. [CrossRef]

30. Zhou, W.J.; Ichchou, M.N. Wave propagation in mechanical waveguide with curved members using wave finite element solution. *Comput. Method Appl. Mech. Eng.* **2010**, *199*, 2099–2109. [CrossRef]
31. Meckings, H.; Kocks, U.F. Kinetics of flow and strain-hardening. *Acta Metall.* **1981**, *29*, 1865–1875. [CrossRef]
32. Xu, B.; Shen, Z.; Ni, X.; Lu, J. Numerical simulation of laser-generated ultrasound by the finite element method. *J. Appl. Phys.* **2004**, *95*, 2116–2122. [CrossRef]
33. Sewell, G. *The Numerical Solution of Ordinary and Partial Differential Equations*, 2nd ed.; John Wiley & Sons: Hoboken, NJ, USA, 1989.
34. Zhang, J.; Xuan, F.Z. Fatigue damage evaluation of austenitic stainless steel using nonlinear ultrasonic waves in low cycle regime. *J. Appl. Phys.* **2014**, *115*, 113. [CrossRef]

metals

MDPI

Article

Optimization of Microstructure and Mechanical Properties in Al-Zn-Mg-Cu Alloys Through Multiple Remelting and Heat Treatment Cycles

Peng Tang [1,2], Ronghui Pang [1], Hanbing Chen [1,*], Yulong Ren [1] and Jun Tan [3,*]

1 School of Resources, Environment and Materials, Guangxi University, Nanning 530004, China; tp@gxu.edu.cn (P.T.); 2139200226@st.gxu.edu.cn (R.P.); 2415301056@st.gxu.edu.cn (Y.R.)
2 School of Mechanical and Aerospace Engineering, Nanyang Technological University, 50 Nanyang Avenue, Singapore 639798, Singapore
3 College of Materials Science and Engineering, Chongqing University, Chongqing 400044, China
* Correspondence: 2315391006@st.gxu.edu.cn (H.C.); j.tan@cqu.edu.cn (J.T.)

Abstract: This study explores the influence of multiple remelting and repeated T6 heat treatment on the microstructure and mechanical properties of Al-Zn-Mg-Cu alloys. With the increase in remelting cycles, the alloy experiences structural alterations due to recovery, recrystallization, and grain growth during remelting, which enhances its mechanical properties. However, the continuous transformation of alloy phases during remelting leads to the deterioration of the morphology and size of α-Al dendrites and $MgZn_2$ phase, causing a decline in mechanical properties. The results indicate that after three remelting cycles, the grain size of the alloy is significantly reduced, and the α-Al dendrites and $MgZn_2$ phase maintain favorable morphology and size, thereby achieving the effects of grain refinement and strength enhancement. These findings offer a theoretical foundation for optimizing the properties of Al-Zn-Mg-Cu alloys.

Keywords: Al-Zn-Mg-Cu alloy; multiple remelting; T6 heat treatment; microstructure; mechanical properties

1. Introduction

Al-Zn-Mg-Cu alloys are widely used in structural fields because of their high strength–density ratio and good fracture toughness [1,2]; as the structural material in the aerospace field, the Al-Zn-Mg-Cu alloy is a high-toughness aluminum alloy, due to its high hardness, strength, wear resistance, and other excellent mechanical properties [3]. The Al-Zn-Mg-Cu alloy is one of the most widely used aluminum alloy types at present, with very high strength, but its corrosion performance is poor, will affect its service life. In the reprocessing process of aluminum alloy, remelting will bring some problems, such as the increase of internal stress, which will lead to warping or distortion of processed parts and even cracking. After reaching a certain degree, cracks or even fractures will occur. The processes of remelting and heat treatment are known to induce alterations in the micro-structural characteristics and mechanical properties of metallic alloys. These alterations necessitate further investigation to elucidate the underlying mechanisms and their implications on the alloy's performance.

Compared with industrial primary aluminum, waste aluminum alloy has lower melting temperature, shorter melting time, lower metal burn loss, and lower energy consumption [4]. More important, secondary aluminum production energy consumption is only

3–5% of the original aluminum production, which can not only reduce the production cost but also help reduce the emission of carbon dioxide and sulfur dioxide pollutants in the industrial process [5,6], playing an important role in energy conservation and environmental protection [7]. Zhou et al. [6] studied the melting, extrusion, and heat treatment processes of the secondary Al-Zn-Mg-Cu alloy, and the results showed that the composition distribution of secondary Al-Zn-Mg-Cu alloy ingots was more uniform, and the microstructure was more optimized. Lin et al. [8] prepared four kinds of secondary Al-Zn-Mg-Cu alloy profiles by melting, composition adjustment, refining, and casting the secondary AA7075 ingot, and finally by extrusion and T6/T73 heat treatment. The mechanical properties of all the secondary profiles were superior to the requirements of ASTM B209-14 and B221M-13. The utilization of scrap aluminum products in the production of secondary aluminum not only facilitates resource conservation but also significantly contributes to the promotion of green manufacturing and sustainable development [9,10]. At the end of the product life cycle, parts made from Al-Zn-Mg-Cu alloys need to be replaced or secondary. Therefore, the recovery or remelting of Al-Zn-Mg-Cu alloys is essential due to the huge use, demand, and technological advances. The remelting process of other alloys can be used for reference in the study of Al-Zn-Mg-Cu alloy remelting [11,12], so it needs to be further explored.

The T6 heat treatment is a pivotal process for optimizing the mechanical attributes of 7xxx series aluminum alloys, which are celebrated for their superior strength in critical industries, such as aerospace and transportation. This treatment excels in inducing the precipitation of strengthening phases, refining grain structures, and directing microstructural development [13]. J.-Y. Choi et al. [14] studied the effects of multiple heat treatment on the microstructure and mechanical properties of the alloy and found that repeated heat treatment would affect the precipitation of the second phase in the alloy, thus causing changes in the microstructure and mechanical properties. Zhao et al. [15] reported significant alterations in the microstructure of TB15 titanium alloy with an increasing number of solution-aging treatment cycles. They observed the coalescence and growth of the α phase, along with a thickening of the grain boundaries of the original β grains. Concurrently, there was a noted deterioration in the mechanical properties of the alloy as the frequency of the solution aging treatments escalated.

In comparison, research on multiple remelting is still relatively limited. Existing studies have shown that the remelting process can alter the grain structure, precipitate phase behavior, and mechanical properties of aluminum alloys [16]. Barot and Sutaria [17] investigated the multiple remelting behavior of Al-Si-Cu alloy and found that with the increase in the number of remelting cycles, the grain size first decreased and then increased, while the precipitation-strengthening effect gradually weakened. However, there has not yet been a systematic study on the effects of multiple remelting + T6 treatment on the micro-structure and mechanical properties of Al-Zn-Mg-Cu alloys.

Although many researchers confirm the differences in microstructure between heat treatment aluminum and raw aluminum remelting, few researchers have systematically explored the mechanism of microstructure and properties of heat-treated alloys after repeated remelting. Therefore, in this study, the changes in microstructure and mechanical features of Al-Zn-Mg-Cu alloys after multiple remelting and repeated heat treatment were conducted to investigate the melting process, utilizing optical microscopy (OM), scanning electron microscopy (SEM), and X-ray diffraction (XRD) with times as the independent variable. This study reveals a strategy for simultaneously enhancing strength and ductility without changing the manufacturing process and provides the possibility of producing high-performance Al-Zn-Mg-Cu alloys for industrial applications.

2. Materials and Methods

The Al-Zn-Mg-Cu alloy used in this study was provided by ALG Aluminium Inc. (Nanning, China), with its chemical composition listed in Table 1. To ensure consistent composition across different remelting cycles, inductively coupled plasma optical emission spectrometry (ICP-OES) was used after each remelting to adjust alloying elements accordingly.

Table 1. Chemical composition of the experiment Al-Zn-Mg-Cu alloy.

Alloys	Elements (wt.%)							
	Al	Zn	Mg	Cu	Cr	Fe	Si	Mn
A1	89.66	5.58	2.61	1.40	0.03	0.18	0.08	0.15
A2	89.36	5.87	2.48	1.50	0.02	0.18	0.09	0.13
B1	89.37	5.85	2.48	1.56	0.02	0.18	0.09	0.13
B2	89.06	6.12	2.43	1.66	0.02	0.20	0.09	0.13
C1	90.04	5.87	1.83	1.49	0.02	0.18	0.08	0.14
C2	89.94	6.01	1.78	1.55	0.02	0.20	0.09	0.14
D1	90.40	5.73	1.70	1.46	0.02	0.18	0.08	0.14
D2	90.67	5.93	1.20	1.48	0.02	0.21	0.09	0.14

The experimental flowchart is shown in Figure 1. In the process of remelting experiment, the samples were subjected to T6 heat treatment before each remelting. The aging treatment was carried out in a DG450C drying oven. Before melting, the experimental alloys were preheated to 200 °C. The crucible with raw materials is put into the resistance furnace, the temperature is raised to 720~740 °C, and the heat is kept for a certain time until the alloy is completely melted, followed by the covering agent and the refining agent, respectively. Then, the heat is raised to 800 °C, and the casting is finished after a certain time. These metal melts were picked at 800 °C cast and cooled to room temperature with an iron mold.

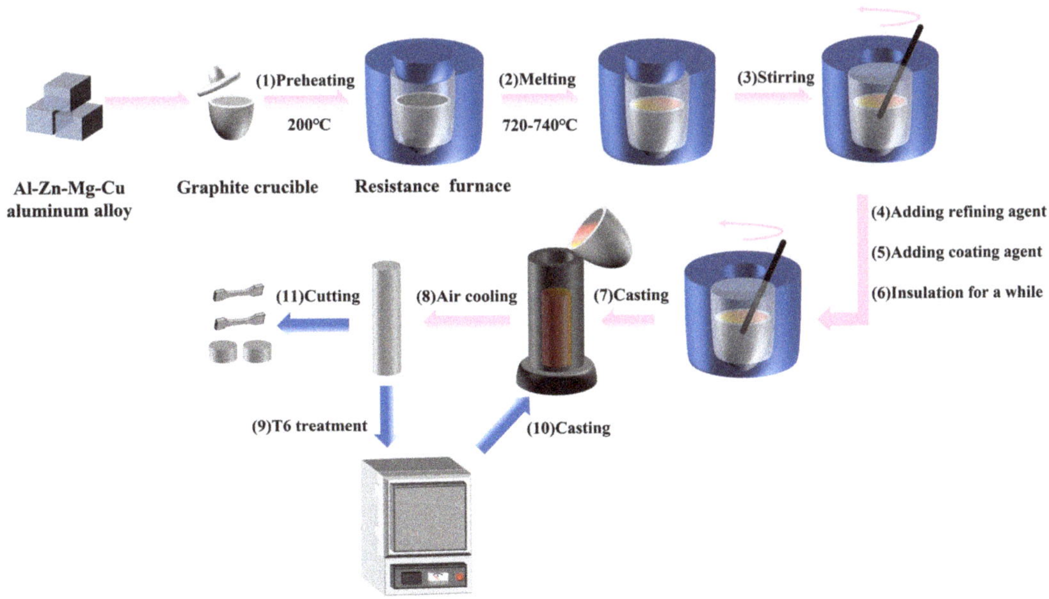

Figure 1. The experimental process flowchart.

The processing steps of alloy samples are as follows: the rolled Al-Zn-Mg-Cu (T6) aluminum alloy is sample A, and sample a is obtained after remelting and T6 heat treatment, and so on. Detailed sample information is shown in Table 2.

Table 2. The multiple remelting and heat treatment procedures parameters of the Al-Zn-Mg-Cu alloy.

Step	T6 Heat Treatment	Multiple Remelting + Repeated T6	Remelting Cycles
A1	470 °C × 1 h + 120 °C × 24 h	1st T6 (Rolled condition)	0
A2	470 °C × 1 h + 120 °C × 24 h	Sample A (T6 + remelting)	1
B1	470 °C × 1 h + 120 °C × 24 h	Sample A2 (T6)	1
B2	470 °C × 1 h + 120 °C × 24 h	Sample B (T6 + remelting)	2
C1	470 °C × 1 h + 120 °C × 24 h	Sample B2 (T6)	2
C2	470 °C × 1 h + 120 °C × 24 h	Sample C (T6 + remelting)	3
D1	470 °C × 1 h + 120 °C × 24 h	Sample C2 (T6)	3
D2	470 °C × 1 h + 120 °C × 24 h	Sample D (T6 + remelting)	4

Initial rolled sample (A), first remelting + T6 treated sample (a), second remelting + T6 treated sample (b), and so forth up to the fourth remelting (d). The metallographic sample was uniformly taken at the bottom 20 mm of the bar sample (φ25 mm × 130 mm), polished step by step with abrasive paper according to the standard, and then polished with a diamond-polishing paste, corroded with Keller's reagent. The microstructures were characterized using optical metallographic microscope (OM, Zeiss, Oberkochen, Germany) and desktop scanning electron microscope (SEM, TM400PLUS, Hitachi, Japan) equipped with an energy-dispersive spectrometer (EDS, IXRF5500, Tokyo, Japan). Samples for microstructure observation were polished and etched by Keller's reagent for a certain time. The metallographic analysis was carried out on an optical microscope (OM, Zeiss/Observer. A1, Oberkochen, Germany). In order to analyze and compare phase morphology more accurately, ten OM images of different annealed samples were taken at 100× magnification. In SEM and OM tests, a section of 1 cm height is cut from a cylindrical ingot as a test sample. The samples of XRD and thermal analysis are a small fraction of the samples clipped after tensile test. And the mechanical properties were characterized using a universal testing machine (8801, INSTRON, Buckinghamshire, UK). Tensile testing was conducted on a computerized testing machine (Shimadzu/AG-X, Kyoto, Japan) at room temperature. Parameters of yield strength (YS, $\sigma_{0.2}$), ultimate tensile strength (UTS), and ductility were obtained at a strain rate of 1 mm/min. According to ASTM E8M-04 standard [18] (sub-size sample), 21 totally different samples (3/per ingot) were machined out for tensile properties assessment. The Vickers hardness was characterized using an automatic turret microhardness tester (HV, AVH-5L, Shanghai Hengyi Precision Instruments, Shanghai, China). The electron backscatter diffraction (EBSD) was characterized via a NORDLYS electron backscatter diffraction microscope.

3. Results and Discussion

3.1. Grain Structure Evolution

Figure 2 show the metallographic images of the Al-Zn-Mg-Cu alloy after electrolytic polishing. Figure 2A–D depict the metallographic images before remelting, while Figure 2a–d show the corresponding images after remelting. Prior to remelting, Figure 2A reveals significant grain size variation, along with the presence of small dark brown particles. As remelting progresses, Figure 2B–D illustrate a gradual transformation in grain morphology, with the phase distribution within the matrix becoming increasingly homogeneous. The grains evolve from irregular, fractured polygons to more complete, predominantly polygonal shapes, accompanied by a denser distribution of the α-Al phase.

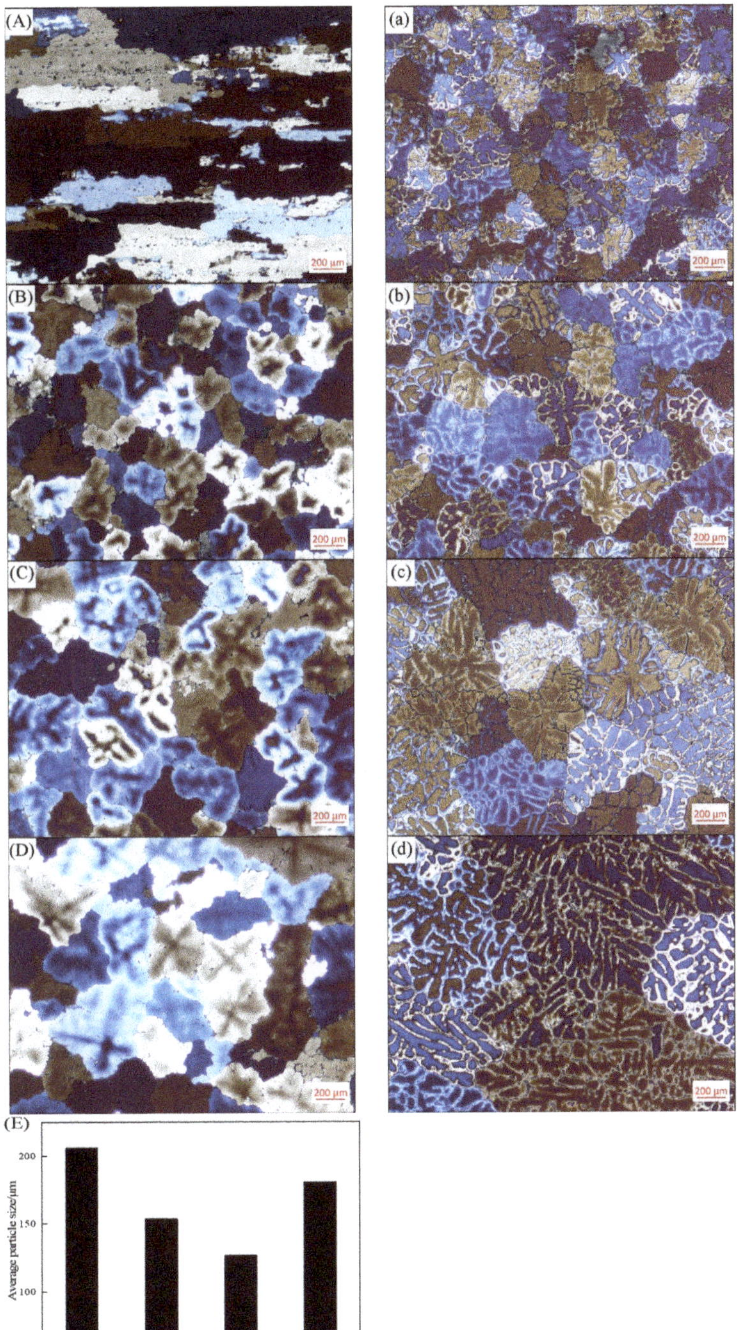

Figure 2. Structure and average grain size of the Al-Zn-Mg-Cu alloy after electrolytic polishing; light microscope, subfigures (**A**) and (**a**) correspond to Alloys A1 and A2, (**B**) and (**b**) to Alloys B1 and B2, (**C**) and (**c**) to Alloys C1 and C2, and (**D**) and (**d**) to Alloys D1 and D2, as detailed in Table 1. (**E**) shows the average grain size of some alloys.

Under identical experimental conditions, the remelting process (Figure 2a–c) leads to a progressive reduction in grain sizes, with more pronounced dendrite structures, indicating effective grain refinement and an enhancement in the alloy's strength. However, after the fourth remelting cycle (Figure 2d), grain coarsening occurs, characterized by an abrupt increase in grain size and the formation of intermittent grain boundaries.

At this stage, the alloy's performance begins to deteriorate. To investigate the effects of remelting iterations, an in-depth analysis was conducted on Al-Zn-Mg-Cu alloy specimens, as shown in Figure 3. This study focused on examining the microstructure and grain orientation distribution under different remelting conditions. Crystal diffraction data of the polished samples were obtained using a scanning electron microscope (SEM) equipped with electron backscatter diffraction (EBSD) accessories. The results were processed with Channel 5 software to analyze grain orientation and microscopic texture variations in the Al-Zn-Mg-Cu alloy across different remelting iterations. Figure 3a represents the alloy after a single remelting, while Figure 3b corresponds to three remelting cycles. As observed in Figure 3a, although the grain size is relatively small, dendrite growth is insufficient, and the grain orientation appears disordered without a clear pattern. The texture is distributed across different crystallographic directions (X_0, Y_0, and Z_0). In contrast, Figure 3b shows that after three remelting cycles, the grains are larger, but the dendrites are fully developed, exhibiting a strong preferred orientation. Most textures converge toward the <001> crystal direction along X_0, indicating a more defined structural alignment.

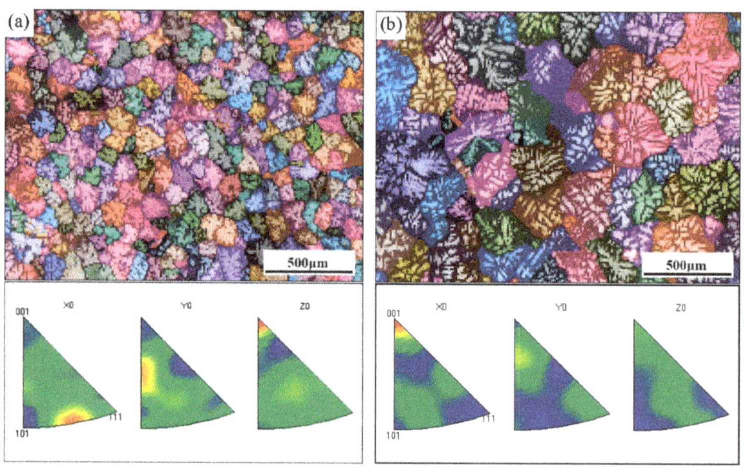

Figure 3. (**a**) and (**b**) present the microstructure images of the alloy after the first and third remelting cycles, respectively (where (**a**) corresponds to alloy A2 in Table 1, and (**b**) corresponds to alloy C2 in Table 1).

An in-depth analysis was conducted on the Al-Zn-Mg-Cu alloy specimens subjected to varying remelting iterations, as depicted in Figure 4.

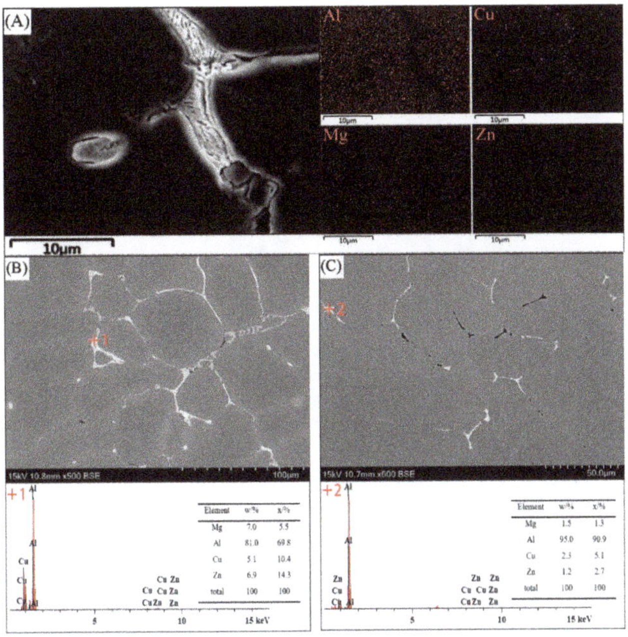

Figure 4. SEM-EDS analysis of A, B, and C on the first T6 remelted surface scan, the fourth T6 remelted surface, and the non-remelted surface (where (**A**), (**B**) and (**C**) correspond to alloy A2, alloy D1, and alloy D2 in Table 1, respectively).

3.2. *Impact on Mechanical Properties*

Figure 5 shows the effects of different remelting times and repeated T6 heat treatment times on the mechanical properties of the alloy. In the absence of remelting, the alloy exhibits a trend in initial decline followed by a subsequent increase in fracture strength, yield strength, and hardness. Conversely, under the state of remelting, the alloy demonstrates an initial increase followed by a decrease in both fracture strength and yield strength. Comparative analysis reveals that, relative to the original sample, the ultimate tensile strength undergoes a significant enhancement of 88.7%, rising from 115.4 MPa to 217.7 MPa. Similarly, the yield strength increases by 84.5%, from 115.1 MPa to 212.5 MPa. However, at the fourth remelting iteration, both tensile and yield strengths undergo a notable decrease, with ultimate tensile strength dropping to 147.2 MPa and yield strength to 147.5 MPa.

A concurrent examination of the metallographic diagram in Figure 2c reveals that the grain distribution attains a more uniform state at the third remelting cycle, coinciding with the optimal fracture strength and ultimate tensile strength observed at this point. As the number of remelting cycles increases, α-Al dendrites and $MgZn_2$ phase maintain their morphological and dimensional integrity, thereby preserving high mechanical properties. However, the continuous phase transformation during remelting eventually leads to the degradation of the morphology and size of α-Al dendrites and $MgZn_2$ phase, ultimately resulting in a decrement in the mechanical properties of the alloy.

Upon analysis of Figure 6a–c, it becomes evident that with increasing remelting iterations, the fracture surface gradually exhibits cleavage features, concurrent with an enhancement in tensile properties. Specifically, upon the third remelting, the cleavage steps become more pronounced, marking the optimal tensile performance. However, subsequent to the fourth remelting, excessive remelting leads to repeated recrystallization of grains, resulting in the appearance of fine cracks within the alloy. This ultimately leads to a deterioration in the overall properties of the alloy.

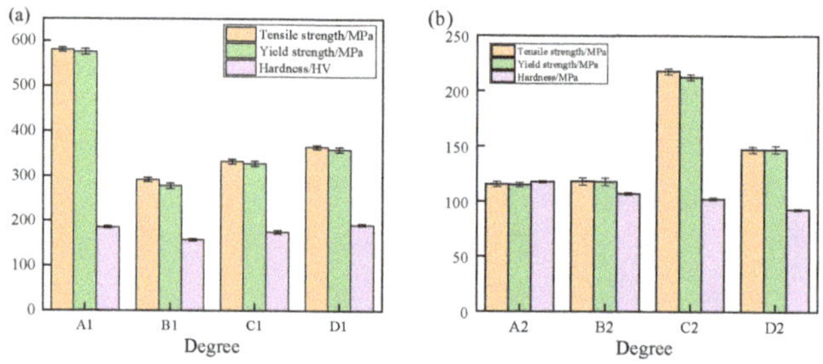

Figure 5. Effect of remelting cycles and repeated T6 heat treatments on alloy properties. Subgraph (**a**) presents the mechanical properties of alloys A1, B1, C1, and D1, and subgraph (**b**) corresponds to the mechanical properties of alloys A2, B2, C2, and D2.

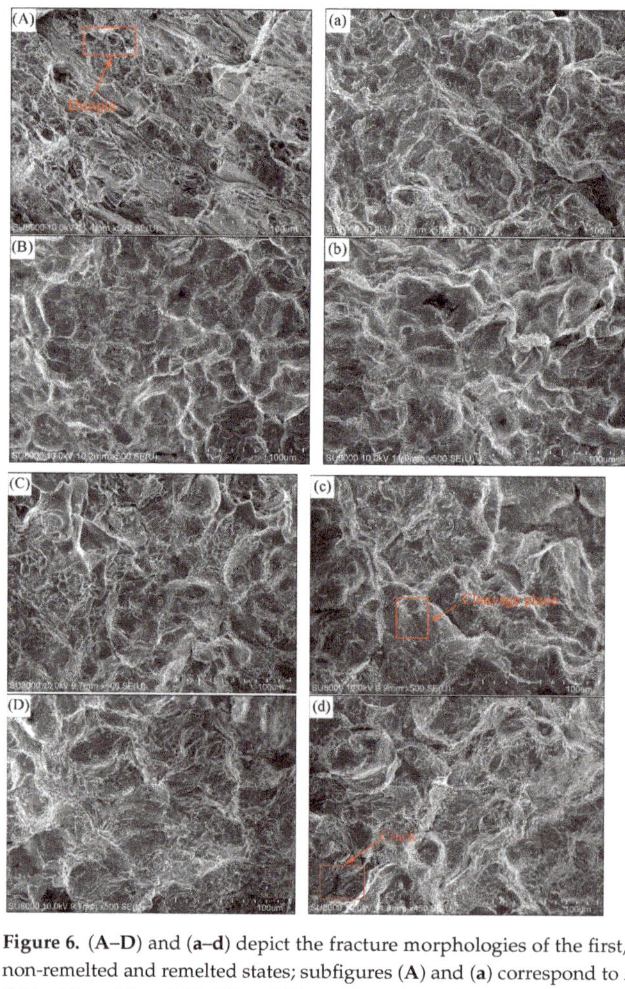

Figure 6. (**A–D**) and (**a–d**) depict the fracture morphologies of the first, second, third, and fourth non-remelted and remelted states; subfigures (**A**) and (**a**) correspond to Alloys A1 and A2, (**B**) and (**b**) to Alloys B1 and B2, (**C**) and (**c**) to Alloys C1 and C2, and (**D**) and (**d**) to Alloys D1 and D2, as detailed in Table 1.

A comparative analysis of the fracture morphology between non-remelted and remelted alloys, under identical remelting conditions, reveals intriguing differences. As illustrated in Figure 6C,c, in the absence of remelting, a cleavage surface emerges at the fracture, indicating a correlation with the structural morphology and tensile properties. Upon introducing remelting and with subsequent increases in remelting iterations, the alloy's fracture surface gradually transitions towards a smoother and flatter appearance. However, upon reaching the fourth remelting cycle, small cracks begin to manifest at the alloy's fracture, indicating a transition towards a typical cleavage fracture mechanism. Based on these observations, it can be concluded that the alloy attains its optimal performance at the third remelting iteration.

3.3. Microstructure and Precipitation Phases

Based on the SEM mapping image in Figure 4, following the initial remelting iteration, the Al element is predominantly concentrated within the crystalline structure, while a small amount of Mg is present in the second phase. In contrast, the Cu element is more concentrated in the second phase, with only trace amounts detectable in the matrix. This observation suggests that the black particles correspond to the Al_2Cu phase. As the number of remelting iterations increases, the grain size undergoes a gradual refinement process. This phenomenon can be attributed to the reduction in size and enhancement in density of the precipitated phases within the alloy during the remelting process, which collectively contribute to additional strengthening effects [19]. Additionally, subsequent heat treatments result in the formation of numerous small, high-density η′ ($MgZn_2$) phase, necessitating a higher consumption of Mg and Zn elements.

Consequently, the coarse second-phase particles that persist after solid solution treatment undergo partial dissolution. As evident in Figure 2a–d, the coarse second-phase particles in the alloy undergo a significant reduction following the third remelting iteration. It is apparent that the primary precipitated phase in the heat-treated alloy is the η′ phase. Given the finer and more uniform grain boundary distribution observed in the alloy after the third remelting iteration, its hardness and strength are superior to other iterations. Furthermore, the small size and high density of the η′ phase at this stage contribute to the alloy's peak hardness and strength.

In face-centered cubic crystalline materials, the cube texture is typically aligned close to the <001> direction and resides within the soft orientation region. Brass and Goss textures occupy the mid-orientation region, whereas the copper texture, proximate to the <111> direction, resides within the hard orientation region [20]. Grains situated in the soft orientation region are predisposed to slipping. As the Schmidt factor value increases, the Young's modulus decreases, facilitating material yield and consequently reducing its strength. In general, in aluminum alloys, the cube texture is usually arranged near the <001> direction, within the soft orientation region. The brass and Gaussian textures are located in the medium-oriented region, while the copper textures near the <111> direction are located in the hard-oriented region. In the third remelting, due to the preferred orientation of dendrites, their texture gathers in the soft orientation region <001>, resulting in grain slip easily. With the increase in Schmidt coefficient, Young's modulus decreases, effectively reducing the internal stress of the material and improving the tensile property of the material. This also explains the phenomenon that fractures in Figure 6a–c gradually show cleavage characteristics.

The influence of varying remelting times on the strength and hardness of the Al-Zn-Mg-Cu aluminum alloy exhibits a distinct trend in initial decrease followed by a subsequent increase. Initially, prior to any remelting, the microstructure assumes an equiaxial grain structure, characterized by a preponderance of bulk grains and a near absence of subgrain boundaries.

During the first remelting cycle, a synergistic effect of material recrystallization and precipitated phase comes into play. This process potentially negates prior work-hardening phenomena, yet the emergence of regenerated grain boundaries enhances the pinning effect of the strengthening phase within the material. Consequently, a renewed work-hardening phenomenon manifests, leading to an elevation in the alloy's strength and hardness.

However, the recrystallization rate decreased with the increase in the number of remelting cycles. At the same time, the diffusion ability of atoms and the migration rate of grain boundaries are enhanced, which leads to the rapid growth of grains [21]. It is worth noting that the strength of the alloy decreases with the increase in the number of remelts. Apparently, at the fourth remelting, the alloy showed signs of softening, the most obvious manifestation being a decrease in its strength and hardness. Recrystallization particularly facilitates the reformation of deformed grains into novel equiaxed crystals, thereby mitigating residual deformation structures. Furthermore, the purification of grain boundaries during recrystallization reduces the presence of impurity particles, while accelerated grain growth further diminishes deformation structures and residual stress, ultimately enhancing the overall performance of the alloy material.

3.4. Phase Composition and Grain Structure

Figure 7 presents the XRD pattern of the alloy, exhibiting distinct variations based on the varying remelting durations. To quantitatively assess the $MgZn_2$ content within the Al-Zn-Mg-Cu alloy, following the third and fourth remelting processes, the adiabatic method was employed. This methodology ensures precise and reliable estimations.

To ensure the accuracy of measurements, the X'PERT HighScore Plus 3.0 software was utilized to meticulously determine the areas of the most prominent peaks associated with Al and $MgZn_2$. The software's precision and reliability are paramount in ensuring the integrity of our data. Additionally, the research referenced the PDF card for accurate K values. The RIR values for Al and $MgZn_2$ at the third and fourth peaks are as follows: RIRAl(3) = 4.29, RIRMgZn$_2$(3) = 3.67, RIRAl(4) = 4.61, and RIRMgZn$_2$(4) = 3.67, respectively. These values are crucial in the subsequent calculations, as they directly influence the estimation of $MgZn_2$ content.

Subsequently, all these values were substituted into Formula (1) [6] to calculate the $MgZn_2$ content. This approach ensures that calculations are rigorous, accurate, and in line with established scientific methodologies. The results obtained from this analysis provide valuable insights into the compositional changes in the alloy during the remelting process.

$$W_{MgZn_2} = \frac{I_{MgZn_2}}{K_{Al}^{MgZn_2}\left(\frac{I_{Al}}{K_{Al}^{Al}} + \frac{I_{MgZn_2}}{K_{Al}^{MgZn_2}}\right)} \quad (1)$$

W_{MgZn_2} is the mass fraction of $MgZn_2$, I_{MgZn_2} is the strongest peak area of $MgZn_2$, $K_{Al}^{MgZn_2}$ is the ratio of K_{MgZn_2}, and K_{Al}, K_{Al}^{Al} is the ratio of K_{Al} and K_{Al}.

According to the calculation, the contents of $MgZn_2$ in the Al-Zn-Mg-Cu alloy at the third and fourth remelting are 2.32% and 2.18%, respectively. The results show that the semi-fused η′ phase is transformed into non-fused η phase in the remelted state. Therefore, based on the above analysis, the distribution of a large number of small dispersed GP region and η′ phase in the grain makes it have a high strength at the third remelting.

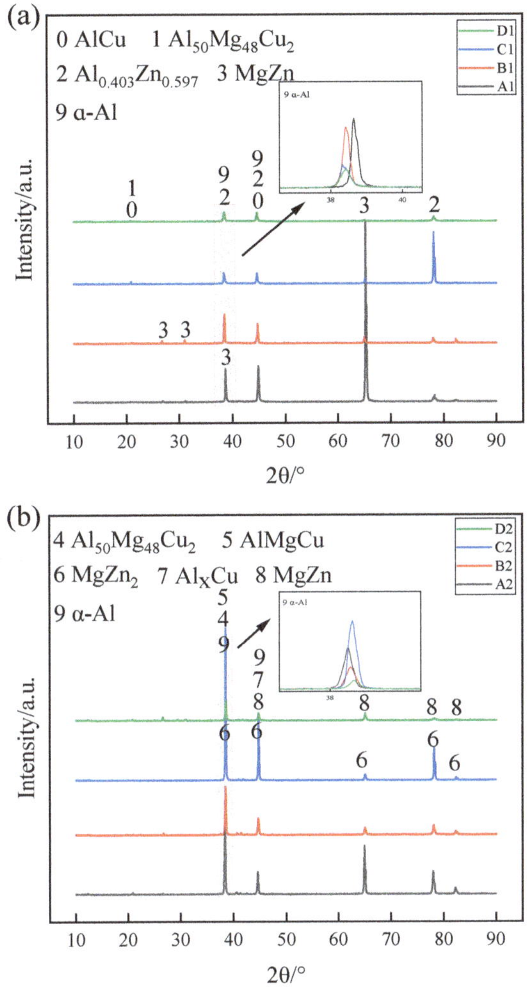

Figure 7. The XRD patterns of the alloy with different remelted times. Subfigure (**a**) shows the XRD patterns of alloys A1, B1, C1 and D1. Subfigure (**b**) shows the XRD patterns of alloys A2, B2, C2 and D2.

From Figure 7a, it is evident that following the initial T6 treatment, the diffraction peak of MgZn is particularly prominent. However, subsequent multiple T6 treatments combined with remelting result in a gradual decrease in the diffraction peak intensity of MgZn, indicating a reduction in the MgZn phase. These observations suggest that the remelting process has a significant impact on the phase composition of the alloy. Additionally, it is conceivable that a portion of Mg and Zn within the matrix may react with other elements, leading to the formation of Al_2Cu and $S(Al_2CuMg)$ phases. This reaction could contribute to the depletion of Mg and Zn in the matrix.

Figure 7b reveals that during the third remelting, the diffraction peak intensity of $MgZn_2$ attains its peak value. This phenomenon can be attributed to the fine-grained structure of $MgZn_2$, which is intimately interlocked with the matrix. The smaller grain size of $MgZn_2$ enhances its strengthening effect. Alternatively, it is possible that with continued remelting, MgZn and Al_xCu phases within the alloy become more prevalent than $MgZn_2$, resulting in a morphological shift. When the number of remelting exceeds a certain threshold, the $MgZn_2$ within the alloy transforms entirely into MgZn.

Figure 8 illustrates the impact of remelting cycles on the grain structure evolution of the Al-Zn-Mg-Cu aluminum alloy. Following a T6 heat treatment, the ordered atomic groups and dispersed phases within the melt serve as nucleation sites during alloy casting [22]. The characteristics of these nucleation sites, including their structure and size, significantly influence the microstructure of the alloy upon solidification. When the die-casting of the alloy results in a fine grain structure, it is indicative of a higher number of crystal nuclei and grain boundaries within the material.

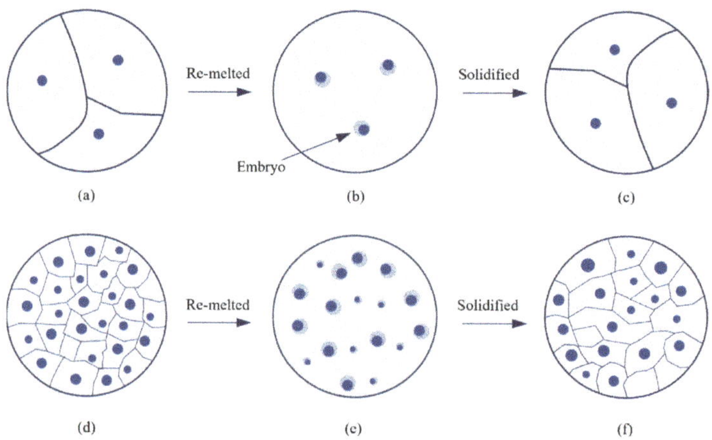

Figure 8. **(a–f)** Evolution process of Al-Zn-Mg-Cu aluminum alloy grain size affected by remelted times.

During the remelting process, the atomic groups within the melt become more ordered, leading to an increased number of potential nucleation embryos. Consequently, even after remelting, the grain structure remains relatively fine. Conversely, if the initial casting exhibits a coarser grain structure, it suggests the presence of fewer ordered atomic groups within the aluminum alloy melt following remelting. This, in turn, results in a reduced number of nucleation embryos, leading to a persistence of the coarser grain structure even after the remelting process. These observations highlight the intricate relationship between the thermal treatment, melt composition, and the resulting grain structure of 7075 aluminum alloy. Understanding and controlling these variables is crucial for optimizing the mechanical properties and performance of the alloy.

It is evident from observations that as the number of remelting iterations increases, the atomic clusters within the alloy's microstructure undergo gradual melting, particularly those of smaller sizes. Consequently, there is a decrease in the overall count of atomic clusters with each subsequent remelting process. This decrease, in turn, translates to a reduced nucleation substrate during solidification, leading to a smaller quantity of grains formed. It is noteworthy that each atomic cluster possesses the capability to serve as a germination site for novel grains. Therefore, a decrement in the number of crystal nuclei during solidification necessitates an elevation in grain size. As the grain size enlarges, there is a corresponding decrement in the grain boundary area and, consequently, a reduction in grain boundary defects. However, this enlargement in grain size is indicative of a gradual decline in the microstructural properties of the alloy. Excessive remelting results in grain coarsening and a progressive deterioration of microstructural properties.

Dendritic textures, along with some unmelted particles and portions of the matrix surface, can serve as nucleation sites for crystal formation, resulting in a fine and uniform microstructure in the alloy after the third remelting. In general, the function of yield strength and grain size can be expressed by the Hall–Petch equation [23]:

$$\sigma = \sigma_0 + Kd^{-1/2} \quad (2)$$

where σ is the required yield strength, σ_0 is the yield stress of a single crystal, and K is a constant, representing the influence of grain boundaries on strength; d is the average size of each grain. Upon undergoing three iterations of remelting, the grain size of the alloy is significantly reduced. This refinement in grain size leads to an enhancement in both the yield stress and toughness of the alloy, thereby achieving the desired effects of grain refinement and strength augmentation.

4. Conclusions

In this study, the effects of multiple remelting and repeated heat treatment on Al-Zn-Mg-Cu alloys were investigated to analyze property changes during the remelting process and determine the optimal number of remelting cycles. The key conclusions are as follows:

(1) The properties of the alloy were studied after multiple heat treatment and four times remelting. It can be seen from the metallographic structure that under the same conditions, the grain size becomes smaller and smaller and the dendrites are developed during remelting, that is, the grains are refined during the remelting process. With the increase in the number of remelting times, the alloy underwent different structural changes due to the recovery, recrystallization and grain growth during the remelting process, and the mechanical properties of the alloy were improved. Under the state of remelting, the alloy demonstrates an initial increase followed by a decrease in both fracture strength and yield strength. Comparative analysis reveals that, relative to the A2 sample, the ultimate tensile strength undergoes a significant enhancement of 88.7%, rising from 115.4 MPa to 217.7 MPa. Similarly, the yield strength increases by 84.5%, from 115.1 MPa to 212.5 MPa. However, at the fourth remelting iteration, both tensile and yield strengths undergo a notable decrease, with ultimate tensile strength dropping to 147.2 MPa and yield strength to 147.5 MPa. The yield strength (YS, $\sigma 0.2$) and ultimate tensile strength (UTS) are the parameters obtained when the strain rate is 1 mm/min.

(2) Under the same conditions, it is found that the fracture strength and yield strength of the alloy without remelting are better than those in the melting state. With the increase in remelting times, α-Al dendrites and $MgZn_2$ phase continue to maintain favorable grain shape and size, and their mechanical properties are still high. However, due to the continuous transformation of the alloy phase during remelting, the morphology and size of α-Al dendrites and $MgZn_2$ phase begin to deteriorate, leading to the decrease in mechanical properties.

(3) Upon successive remelting, the $MgZn_2$ phase within the alloy undergoes a transformation to MgZn, which, along with Al_xCu, exhibits a more favorable morphology relative to $MgZn_2$. With each iteration of heat treatment, the $MgZn_2$ phase that precipitates become increasingly coherent with the aluminum matrix, with a diminutive size correlating positively with strengthening effects.

(4) In the context of the remelting process, there is a progressive ordering of atomic groups within the melt, which augments the population of potential nucleation embryos. This ordering leads to a refinement of the grain structure, which is maintained even

post-remelting. In contrast, an initial coarser grain structure in the casting implies a scarcity of ordered atomic groups within the aluminum alloy melt subsequent to remelting. This deficiency in ordered atomic groups correlates with a diminished number of nucleation embryos, perpetuating the coarser grain structure through the remelting process. These findings underscore the complex interplay between thermal treatments, melt composition, and the resultant grain structure in the 7075 aluminum alloy, emphasizing the significance of these factors in dictating material properties.

Author Contributions: Research idea design, P.T., H.C. and J.T.; research acquisition, analysis, and interpretation, R.P., Y.R. and H.C.; writing—manuscript preparation, writing—review and editing, P.T. and H.C.; data management, P.T., R.P. and Y.R.; software, J.T.; validation, formal analysis and investigation, H.C., Y.R., and R.P.; resources, P.T.; supervision, P.T. and J.T.; project management, P.T. All authors have read and agreed to the published version of the manuscript.

Funding: This research was supported by the Guangxi Natural Science Foundation (2023GXNSFAA026374, 2024JJA160370, 2023AB38012), the National Natural Science Foundation of China (52261024), the China Scholarship Council Program (202306660028), and the Guangxi University Students' Innovation and Entrepreneurship Training Program (S202310593342).

Data Availability Statement: The original contributions presented in this study are included in the article. Further inquiries can be directed to the corresponding authors.

Conflicts of Interest: The authors declare no conflicts of interest.

References

1. Chen, H.; Xie, Y.; Tang, P.; Tang, Z.; Liao, C.; Pang, R. Demystifying the Influence of Scrap to Raw Material Ratio on Microstructural and Mechanical Properties of the Secondary 7075 Aluminum Alloy. *Mater. Lett.* **2025**, *382*, 137847. [CrossRef]
2. Alatorre, N.; Ambriz, R.R.; Noureddine, B.; Amrouche, A.; Talha, A.; Jaramillo, D. Tensile Properties and Fusion Zone Hardening for GMAW and MIEA Welds of a 7075-T651 Aluminum Alloy. *Acta Metall. Sin. (Engl. Lett.)* **2014**, *27*, 694–704. [CrossRef]
3. Khalid, M.Y.; Umer, R.; Khan, K.A. Review of recent trends and developments in aluminium 7075 alloy and its metal matrix composites (MMCs) for aircraft applications. *Results Eng.* **2023**, *20*, 100372. [CrossRef]
4. Zhou, B.; Liu, B.; Zhang, S.; Lin, R.; Jiang, Y.; Lan, X. Microstructure evolution of Secondary 7075 aluminum alloy and its mechanical and corrosion properties. *J. Alloys Compd.* **2021**, *879*, 160379. [CrossRef]
5. Miller, W.S.; Zhuang, L.; Bottema, J.; Wittebrood, A.J.; De Smet, P.; Haszler, A.; Vieregge, A. Recent development in aluminium alloys for the automotive industry. *Mater. Sci. Eng. A* **2000**, *280*, 37–49. [CrossRef]
6. Damgaard, A.; Larsen, A.W.; Christensen, T.H. Recycling of metals: Accounting of greenhouse gases and global warming contributions. *Waste Manag. Res.* **2009**, *27*, 773–780. [CrossRef] [PubMed]
7. Muñiz-Lerma, J.A.; Paliwal, M.; Jung, I.-H.; Brochu, M. Fractional Crystallization Model of Multicomponent Aluminum Alloys: A Case Study of Aircraft Recycling. *Metall. Mater. Trans. B* **2017**, *48*, 1024–1034. [CrossRef]
8. Lin, R.; Liu, B.; Zhang, J.; Zhang, S. Microstructure evolution and properties of 7075 aluminum alloy Secondary from scrap aircraft aluminum alloys. *J. Mater. Res. Technol.* **2022**, *19*, 354–367. [CrossRef]
9. Wang, P.; Lei, H.; Zhu, X.; Chen, H.; Fang, D. Investigation on the mechanical properties of epoxy resin with void defects using digital image correlation and image-based finite element method. *Polym. Test.* **2018**, *72*, 223–231. [CrossRef]
10. Barot, R.P.; Sutaria, M.P. Effect of multiple remelting on behaviour of AlSi5Cu3 Aluminium alloy. *Mater. Today Proc.* **2022**, *62*, 4046–4051. [CrossRef]
11. Liu, T.; Hu, M.; Li, X.; Piao, J. Synthesis of Al-5Ti-1B-1Ce alloy from remelted chips and its refinement effect. *Mater. Lett.* **2023**, *349*, 134798. [CrossRef]
12. Matejka, M.; Bolibruchová, D.; Kuriš, M. Crystallization of the Structural Components of Multiple Remelted AlSi9Cu3 Alloy. *Arch. Foundry Eng.* **2020**, *20*, 41–45. [CrossRef]
13. Rometsch, P.A.; Zhang, Y.; Knight, S. Heat treatment of 7xxx series aluminium alloys—Some recent developments. *Trans. Nonferrous Met. Soc. China* **2014**, *24*, 2003–2017. [CrossRef]
14. Choi, J.-Y.; Moon, J.; Jang, J.-H.; Lee, T.-H.; Hong, H.-U.; Kim, H.C.; Kang, N.H.; Shin, J.-H.; Lee, C.-H. Enhancement of mechanical properties by repeated heat treatment in reduced activation ferritic/martensitic steel with Ta and Ti. *J. Nucl. Mater.* **2021**, *557*, 153321. [CrossRef]

15. Zhao, A.; Hu, S.; Chen, S.; Zhang, Y.; Zhao, H.; Zhang, Y. Effect of repeated solution aging treatment on microstructure and mechanical properties of TB15 titanium alloy. *Heat Treat. Met.* **2022**, *47*, 91–94. [CrossRef]
16. Yang, Y.; Xie, Y.; Tang, P.; Lu, J. Mechanism of Fe-rich phase supercooling induced by CDS process to regulating microstructure and strengthen properties of Al-Si-Fe alloy. *J. Alloys Compd.* **2024**, *1007*, 176462. [CrossRef]
17. Barot, R.P.; Sutaria, M.P. Microstructural and mechanical characteristics of AlSi5Cu3 aluminum alloy cast in metal mold and sand mold. *Mater. Today Proc.* **2023**. [CrossRef]
18. *ASTM E8M-2004*; Standard Test Methods for Tension Testing of Metallic Materials, American Society for Materials and Testing. ASTM: West Conshohocken, PA, USA, 2004.
19. Oh, J.; Park, H.D.; Gwak, M.; Lee, J.; Son, S.; Amanov, A.; Kim, H.S.; Seol, J.B.; Sung, H.; Kim, J.G. Mechanical property enhancement in gradient structured aluminum alloy by ultrasonic nanocrystalline surface modification. *Mater. Sci. Eng. A* **2021**, *812*, 141101. [CrossRef]
20. Liao, W.; Zhang, C.; Qiang, H.; Song, W.; Ren, H. Effect and Mechanism of Solidified Microstructure on Deformation Behavior, Mechanical Properties, and Residual Stress of Cu-Ni-Si Alloy. *Materials* **2022**, *15*, 8724. [CrossRef] [PubMed]
21. Tian, Y.Q.; Li, W.; Zheng, X.P.; Song, J.Y.; Wei, Y.L.; Chen, L.S. Study on Carbide precipitation Behavior and Microstructure Properties of two-phase zone Annealing hot rolled Medium Manganese Steel. *Mater. Rev.* **2019**, *33*, 2765–2770. [CrossRef]
22. Wang, X.; Rong, Q.; Shi, Z.; Li, Y.; Cao, J.; Chen, B.; Lin, J. Investigation of stress effect on creep, precipitation and dislocation evolution of Al–Li alloy during creep age forming. *Mater. Sci. Eng. A* **2022**, *836*, 142723. [CrossRef]
23. Andani, M.T.; Lakshmanan, A.; Sundararaghavan, V.; Allison, J.; Misra, A. Estimation of micro-Hall-Petch coefficients for prismatic slip system in Mg-4Al as a function of grain boundary parameters. *Acta Mater.* **2022**, *226*, 117956. [CrossRef]

metals

MDPI

Article

The Influence of Process Parameters on the Density, Microstructure, and Mechanical Properties of TA15 Titanium Alloy Fabricated by Selective Laser Melting

Junjie Jiang [1], Chuang Liang [1], Yuanchao Chen [2], Yongbiao Wang [1,*], Hongyang Cui [1], Jianlin Xu [1], Fang Zhou [1], Pengpeng Wang [1] and David Z. Zhang [3,*]

[1] Henan Provincial Key Laboratory of Intelligent Manufacturing of Mechanical Equipment, Zhengzhou University of Light Industry, Zhengzhou 450002, China; violet5257@163.com (J.J.); liangc2025@163.com (C.L.); cuihongyang92@163.com (H.C.); 15603823281@163.com (J.X.); 2020021@zzuli.edu.cn (F.Z.); wpeng121@126.com (P.W.)
[2] Zhengzhou RuiFei Biotechnology Co., Ltd., Zhengzhou 450001, China; zhuoenduo@126.com
[3] College of Engineering, Mathematics and Physical Sciences, University of Exeter, North Park Road, Exeter EX4 4QF, UK
* Correspondence: wsbiaoyongwang@163.com (Y.W.); zhangzw@cqu.edu.cn (D.Z.Z.)

Abstract: With superior manufacturing freedom capability, Selective Laser Melting (SLM) technology is capable of fabricating high-strength Ti-6Al-2Zr-1Mo-1V (TA15) complex titanium alloy parts, thereby finding extensive applications in the aerospace sector. This paper primarily investigates the influence of process parameters on the relative density, microstructure, and mechanical properties of SLMed TA15 under conditions of similar laser linear energy density. The results indicate that the laser linear energy density significantly affects the single-track morphology of SLMed TA15; excessive energy density leads to keyhole defects, while insufficient energy density causes balling phenomena, resulting in discontinuous clad tracks. When the laser linear energy density is appropriate, the scanning spacing affects the forming density of the parts, with both excessively large and small spacings having adverse effects. With a fixed scanning spacing of 100 μm, high-density samples can be produced within a suitable range of linear energy density. However, when the laser linear energy density is comparable, a lower scanning speed leads to heat accumulation, causing in situ decomposition of the α' martensite and the formation of coarser $\alpha + \beta$ phases, which reduces strength and hardness but improves plasticity. At a laser power of 90 W, a scanning speed of 400 mm/s, and a scanning spacing of 100 μm, the specimen exhibits a tensile strength of 1233 MPa and an elongation of 8.4%, achieving relatively excellent comprehensive properties.

Keywords: selective laser melting; TA15; linear energy density; single-track; microstructure; mechanical properties

1. Introduction

Titanium alloys are extensively applied in aerospace, automotive, chemical, medical, and other fields as they have low density, high specific strength, high thermal strength, extraordinary corrosion resistance, and excellent biocompatibility. Capable of sustained operation at 500 °C for up to 3000 h, the TA15 titanium alloy is a near-α titanium alloy with a maximum short-term operating temperature of 800 °C for up to 5 min. Its primary applications include large integral parts, load-bearing structures, and welded components for aircraft, missiles, launch vehicles, and satellites [1,2]. The traditional processing methods for TA15 structural components mainly include casting, forging, and mechanical processing [1,3,4]. Substantial challenges in the processing and manufacturing of the TA15 alloy arise from its high reactivity, low thermal conductivity, and considerable deformation resistance [5]. One of the primary methods for metal additive manufacturing is SLM technology, which overcomes the drawbacks of traditional technologies by fully melting the metal powder with excellent forming precision and manufacturing flexibility.

In recent years, several studies have begun to investigate the SLMed TA15 alloy. Wu et al. [6] prepared TA15 samples using SLM and found that the room-temperature and high-temperature tensile properties of the TA15 samples were visibly higher than those of TA15 fabricated by conventional manufacturing techniques and of other near-α titanium alloys, due to the effect of refined α' martensitic grains and nanoscale twins. The author conducted preliminary research on the mechanical properties and microstructure evolution of the SLMed TA15 alloy before and after annealing and found that the annealed TA15 sample inherited a martensitic hierarchical structure and precipitated nanoscale-sized β phase. In addition, the samples before and after annealing exhibited weak textures, resulting in inconspicuous anisotropy of mechanical properties [7]. Huang et al. [8] also studied the influence of heat treatment on SLMed TA15 and found that, with the increase in heat-treatment temperature, the martensitic structure of the morphologically formed TA15 sample gradually transformed into an $\alpha + \beta$ structure, and the tensile properties decreased while the plasticity increased. Cai et al. [9] studied the influence of changes in laser power and scanning speed on the microstructure and mechanical properties of TA15 formed by SLM under fixed scanning spacing. The research results showed that different process parameters would affect the relative density of the TA15 alloy, thereby affecting its mechanical properties. In addition, the author conducted a preliminary study on the influence of process parameters and scanning strategies on the surface quality of SLMed TA15 parts [10]. Yin et al. [11] investigated the effects of dual shot peening on the residual stress and microstructure of SLMed TA15 alloys and analyzed the thermal relaxation behavior post-peening. Wei et al. [12,13] studied the differences in the performance of TA15 formed by SLM and Laser Metal Deposition (LMD). The study found that the microstructure of the SLMed sample was α' martensite, which is isotropic, while the microstructure of the LMDed sample was $\alpha + \beta$ basket-weave microstructure and anisotropic. At the same time, the SLMed sample presented higher tensile strength and lower plasticity.

Evident from the studies above is the fact that current research on SLMed TA15 predominantly focuses on the microstructural and mechanical properties following forming and post-heat treatment, with limited investigations into the effects of process parameter variations on TA15 properties. It is well known that modifying process parameters to increase the forming density of titanium alloys is a mainly effective strategy, based on studies of SLMed titanium alloys [14–18]. For instance, Wei et al. [19] analyzed the quality of single-track formations and, considering forming efficiency, selected continuous clad tracks free from balling, warping, and defects. They then achieved a Ti-5Al-2Sn bulk with a density of 99.95% by choosing an appropriate scanning spacing. Zhou et al. [14] fixed the scanning spacing and varied the laser power and scanning speed to obtain Ti-13Nb-13Zr titanium alloy parts with the highest density. Similar methodologies have been reported in references [14,18]. Additionally, Sun et al. [15] and Khorasani et al. [16] employed the Taguchi method to determine the SLM process parameters for Ti6Al4V that yield the optimal density. These studies reveal that, under the premise of consistent single-track formation quality, dense entities can be formed by matching the appropriate scanning spacing. However, the influence of the laser power, scanning speed, and scanning spacing on formability is not consistent. Regrettably, scant research has focused on the impact of changing process parameters on the microstructure and properties of SLMed specimens under consistent single-track formation quality. This study concentrates its focus precisely on this aspect.

Therefore, this study aims to analyze the influence of SLMed process parameters on the single-track morphology and the characteristic size of the TA15 alloy, reveal the single-track-forming mechanism, and determine the optimization process interval for single-track-forming. The focus is on studying the effects of changes in scanning spacing and laser power scanning speed on the microstructure and mechanical properties of SLMed TA15 alloys under similar single-track-forming quality to achieve real-time control of the microstructure and properties of SLMed TA15 alloys during the processing by changing the process parameters.

2. Experimental Details

2.1. Powder Material

TA15 titanium alloy powder was provided by Beijing Zhonghang Maite Powder Metallurgy Technology Co., Ltd. (Beijing, China), and its chemical composition is listed in Table 1. The surface morphology and particle size distribution of the powder are shown in Figure 1. It can be observed that the TA15 powder exhibits good overall sphericity with smooth surfaces, although there are a small number of irregularly shaped powder particles and satellite particles present. It can be found from Figure 1b that the powder particle size presents a normal distribution, and the powder particle size is from 11.2 μm to 76 μm; the average size is 34.3 μm. Additionally, D_{10} and D_{90} are 20.4 μm and 55.7 μm, respectively.

Table 1. Chemical composition of TA15 powder, wt. %.

Element	Ti	Al	V	Zr	Mo	Si	Fe	C	O	N
As-received powders	Bal.	6.37	2.17	2.16	1.3	0.017	0.072	0.029	0.1	0.02
Standard [20]	Bal.	5.5–7.0	0.8–2.5	1.5–2.5	0.5–2.0	≤0.15	≤0.25	≤0.1	≤0.15	≤0.05

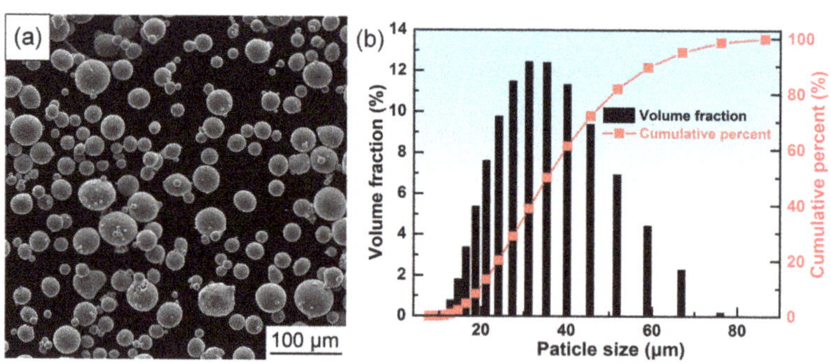

Figure 1. (**a**) Morphology and (**b**) particle size distribution of TA15 powders.

The apparent density of the TA15 powder measured by the Hall flowmeter was 2.365 g/cm^3, and the density of the TA15 alloy was 4.45 g/cm^3 [20]; the loose specific weight calculated to 53.13%. The powder fluidity test showed that the average time for 50 g TA15 powder to flow through the standard funnel (aperture 2.5 mm) was 40.87 s. It is usually accepted that a liquidity evaluation is better if the repose angle is less than 30°, and the TA15 powder's was 27°. It may be said that the experiment's TA15 powder had high spreadability and fluidity. Figure 2 displays the result of the experiment.

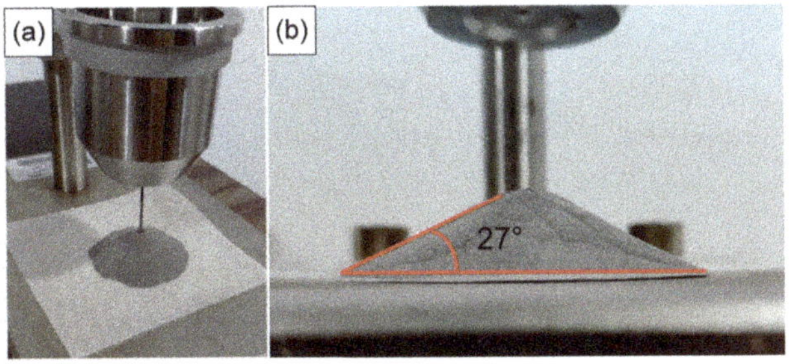

Figure 2. (**a**) Flowability and (**b**) spreading property of TA15 powders.

2.2. SLM Experimental Platform

The SLM experimental platform used in this apparatus was EOSINT M280 (EOS GmbH, Krailling, Germany).

2.3. Experimental Characterization Method

2.3.1. Surface Morphology Observation

A digital microscope (VHX-1000, Keyence (China) Co., Ltd., Shanghai, China) was used to examine the surface morphology of SLMed samples.

2.3.2. Single-Track Morphology Observation and Characteristic Size Testing

The schematic diagram of the SLMed single-track is shown in Figure 3a. The results of the single-track formation are depicted in Figure 3c. The single-track was formed on a TC4 titanium alloy baseplate with dimensions of 100 × 50 × 5 mm, and the length of each single-track was 10 mm. The macro morphology of a single-track was observed and photographed by VHX-1000. Then, a single longitudinal section was cut along the vertical direction, and the morphology of the molten pool was observed and photographed under the microscope. Digimizer was used to measure the width, height, depth, and other single-track characteristic sizes of the molten pool, as shown in Figure 3b. The characteristics of the melt pool are depicted in Figure 3d, wherein the boundary of the melt pool represents the microstructural demarcation between the TA15 single-track and the TC4 baseplate. To circumvent the impact of errors, three clad tracks were fabricated for each set of process parameters for the purpose of statistical measurement.

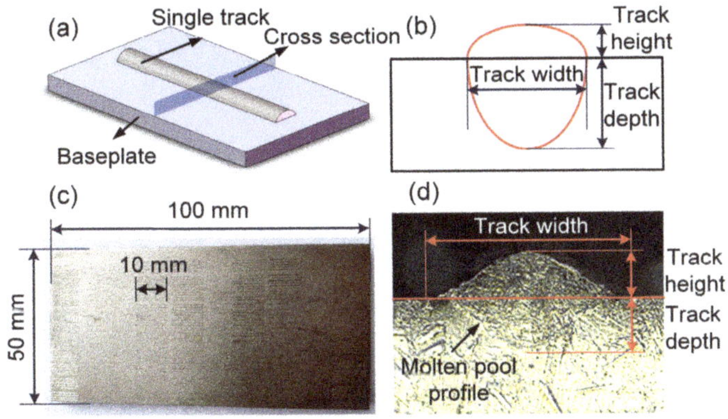

Figure 3. (**a**) Schematic diagram of SLM forming single-track, (**b**) characteristic size, (**c**) single-track-forming result, and (**d**) characteristics of the melt pool.

2.3.3. Relative Density Testing

We measured the density of SLMed parts using the Archimedes method. The measuring instrument was an electronic analytical balance with a density measurement component, model 2204, manufactured by Shanghai Zhuojing Electronic Technology Co., Ltd. (Shanghai, China). The calculation formula for sample density ρs is as follows:

$$\rho_s = \frac{m_1 \times \rho_w}{m_1 - m_2} \quad (1)$$

where m_1 is the mass of the sample in air, m_2 is the mass of the sample in distilled water, and ρ_w is the density of distilled water in the measurement environment. The formula for calculating the relative density of the sample, θ, is as follows:

$$\theta = \frac{\rho_s}{\rho_{mth}} \quad (2)$$

where ρ_{mth} represents the standard density of the sample. The standard density used in this study is the density of the rod used in the production of TA15 powder (4.451 g/cm^3).

2.3.4. Microstructure Characterization Methods

1. Observation of microstructure

Following electropolishing, the metallographic structure was corroded with Kroll's reagent, which is composed of 3% HF + 6% HNO_3 + 91% H_2O. It takes 30 to 40 s for rust to occur. During the process, a magnetic stirrer was used to stir the corrosion solution. We used the VHX-1000 to examine and take pictures of the corroded sample.

2. EBSD analysis

The EBSD experiment was carried out using a field emission scanning electron microscope (FESEM) (JSM-7800F, JEOL Co., Ltd., Tokyo, Japan). The acceleration voltage was set at 20 kV with a scanning area of approximately 40 × 40 μm^2 and a scanning step size of 0.2 μm. Using HKL Channel 5, we proceeded with data analysis after completing data collection.

2.3.5. Mechanical Performance Characterization Methods

1. Hardness test

The dimensions of the hardness test specimen were 10 × 10 × 5 mm, as shown in Figure 4a. The surface for hardness testing was the polished top face of the specimen. The hardness test was performed on a Vickers hardness tester (310HVS-5, Laizhou Huayin Testing Instrument Co., Ltd., Laizhou, China). Every sample was tested at least ten times, with a test load of 9.807 N and a loading time of 15 s. The average value was obtained by subtracting the greatest and lowest values.

2. Tensile test

Figure 4 displays the finished photograph of the tensile specimen and its dimensions. The dimensions of the tensile components used in this experiment were designed by the standard GB/T 228.1-2010 [21]. The tensile specimens were polished to a smooth finish prior to tensile testing to avoid the influence of surface defects. The tensile test adopted an electronic universal testing machine (SANS CMT5105, MTS Co., Ltd., Eden Prairie, MN, USA), with a tensile rate of 1 mm/min. Each sample underwent three tensile tests. The observation of the fracture morphology of the tensile specimen was performed on a ESEM (MIRA3, TESCAN Co., Ltd., Brno, Czech Republic).

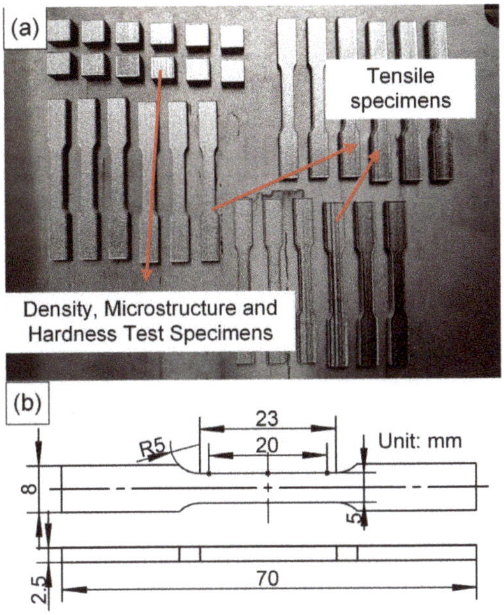

Figure 4. (**a**) SLMed TA15 specimens and (**b**) schematic diagram of the tensile specimen dimensions.

3. Results and Analysis

3.1. SLM Forming TA15 Single-Track Experiment

The formability of SLM is largely determined by single-track-forming, which is the basis of SLM forming [22]. To investigate the impact of process parameters on single-track characteristics, single-track-forming experiments were first conducted on a TA15 alloy. Under stable equipment parameters such as smoke extraction, oxygen content, laser, and fixed powder layer thickness, single-track-forming is mainly affected by the laser power P (W) and laser scanning speed V (mm/s). Set different laser powers (50–195 W) and scanning speeds (200–2000 mm/s) to form the TA15 single-track.

The type statistics of every single-track morphology are displayed in Figure 5. The horizontal axis in the figure represents the laser scanning speed, and the vertical axis represents the laser power. Laser power and scanning speed significantly affect the quality of a single-track, as shown in Figure 6. Increasing the scanning speed will progressively result in cladding track distortion, balling phenomena, and even discontinuity while the laser power is fixed. The cladding's quality will gradually decline while the scanning speed stays the same and the laser power decreases. Divided into three types, corresponding to the three zones in the figure, is the single-track morphology by observation and analysis.

Zone A: Stable forming zone. The cladding track is uniformly wide, straight, and successive. The regular-shaped "fish scale" ripples that were left over when the melt solidified are apparent.

Zone B: Unstable forming zone. Despite being continuous, the cladding track's varied width causes distortion and even balling phenomena. As the laser power decreases or the laser scanning speed increases, the distortion and balling phenomena of the cladding track become more pronounced.

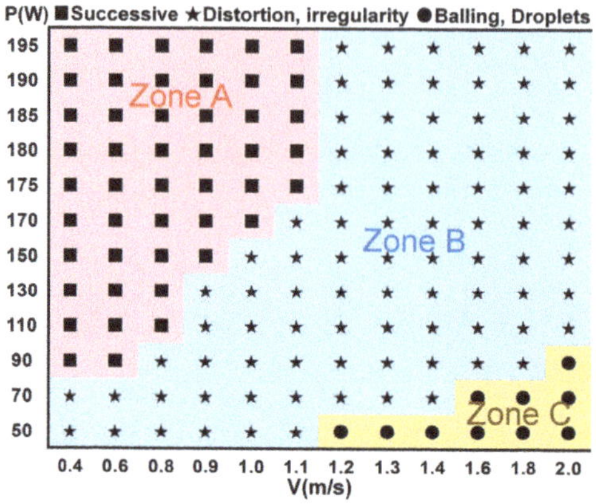

Figure 5. Statistics of single-track morphology.

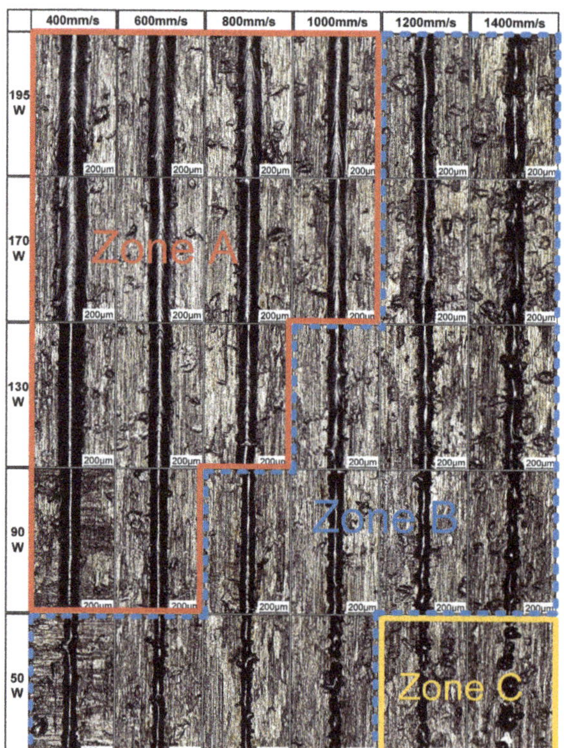

Figure 6. Macroscopic morphologies of TA15 single-tracks with different laser powers and scanning speeds.

Zone C: Discontinuous zone. The balling phenomenon of the cladding track is severe, presenting a discontinuous "droplet like" appearance.

The single-track width exhibits a diminishing trend as the laser power decreases and laser scanning speed increases, as shown in Figure 7, reduced to about 50 μm from a maximum of 250 μm.

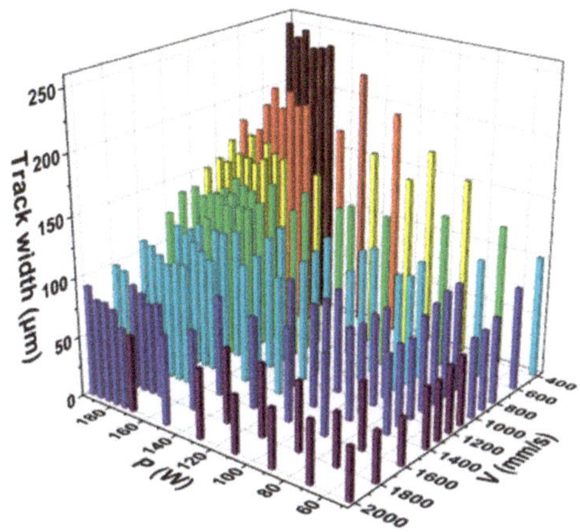

Figure 7. Width of TA15 single-tracks with different laser powers and scanning speeds.

The morphology of the molten pool above the substrate gradually shifts from a semi-elliptical shape to a spherical shape with increasing scanning speed when the laser power is 195 W, as illustrated in Figure 8, and the depth of the molten pool below the baseplate shows a sharp downward trend. The morphology of the molten pool also gradually changes from a "V" shape to a "circular arc" shape. Additionally, at a scanning speed of 200–600 mm/s, large holes appear in the middle and lower parts of the melt pool. At a scanning speed of 200 mm/s, the size of the holes reaches around 120 μm. As the speed continues to increase, the size of the hole gradually decreases. The hole disappears when the speed surpasses 800 mm/s.

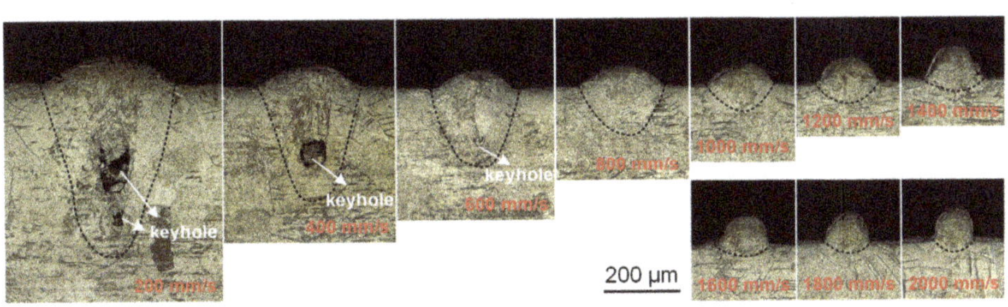

Figure 8. Cross-sectional morphologies of single-track molten pools at different scanning speeds.

When the laser power is 195 W, as Figure 9 shows, the height of the molten pool exhibits a trend of first descending and then growing as the scanning speed increases. The height decreases from 74 μm at 200 mm/s to 46 μm at 800 mm/s and then increases to between 80–90 μm and tends to stabilize. As the scanning speed increases, the molten pool's depth decreases sharply from 455 μm to about 30 μm before stabilizing.

The large holes in the molten pool are similar to keyholes in laser welding [23]. When the aspect ratio of the melt pool exceeds 0.5, it is usually a "keyhole mode" melt pool [22,24,25]; otherwise, it is a "conduction mode" melt pool. From Figure 9, as the depth of the molten pool decreases sharply, the aspect ratio of the molten pool also decreases sharply, from 1.3 to around 0.3, and then it tends to stabilize. The aspect ratio decreases to within 0.5 at a scanning speed of 1000 mm/s. But when the scanning speed increases to 1400 mm/s, the melt pool's height rises to 91 μm, indicating a definite balling phenomenon tendency, and its aspect ratio drops to 0.3. The line of change in the height of the molten pool intersects with the line of change in the depth of the molten pool between 1000 mm/s and 1200 mm/s. Within this range, the molten pool's depth is between 62 μm and 72 μm, its height is from 52 μm to 67 μm, its width is from 142 μm to 155 μm, and its aspect ratio is between 0.43 and 0.46. The single-track has excellent quality and a reasonable characteristic size.

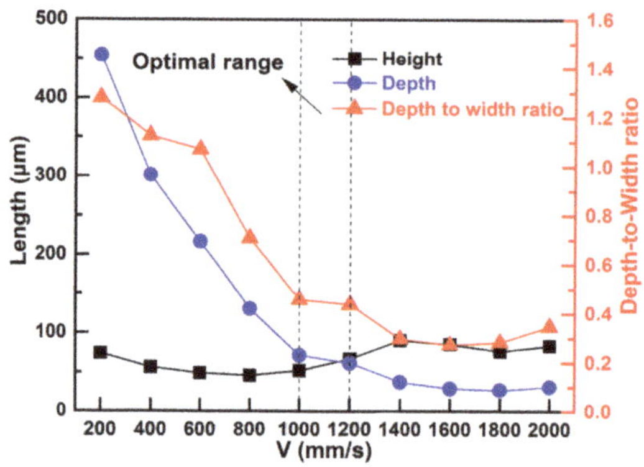

Figure 9. Variation rule of characteristic size of molten pool with different scanning speeds.

In SLM research, it is customary to use the laser linear energy density E_L to describe the combined effect of laser power and scanning speed. The calculation formula is as follows [26]:

$$E_L = \frac{P}{V} \tag{3}$$

Figure 10 shows how the single-track width rises nonlinearly as the laser linear energy density increases, suggesting that changes in laser power and scanning speed also have an impact on the single-track width. Set up a fixed linear energy density of 0.195 J/mm in a single-track experiment.

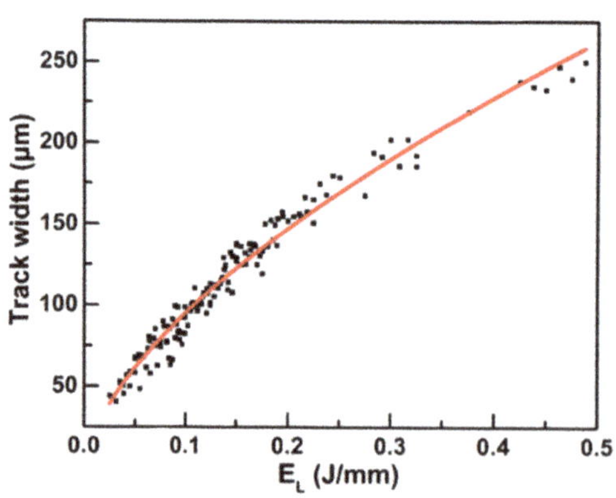

Figure 10. Single-track width with different linear energy densities.

From Table 2 and Figure 11, when the laser linear energy density remains constant and the scanning speed is within the range of 1000–500 mm/s, the morphology and width of the cladding are basically consistent, and the morphology characteristics of the melt pool are also quite similar. The aspect ratio is less than 0.5, and no pores are inside the melt pool. However, the single-track width marginally reduces and the molten pool morphological characteristics remain relatively unchanged when the scanning speed falls under 400 mm/s. One track displayed distortion, with its width decreased to 142 μm, at a scanning speed of 300 mm/s.

Table 2. Single-track width with different scanning speeds when the linear energy density is 0.195 J/mm.

V (mm/s)	1000	900	800	700	600	500	400	300
P (W)	195	176	156	137	117	98	78	59
Single-track width (μm)	155	153	156	156	155	153	147	142

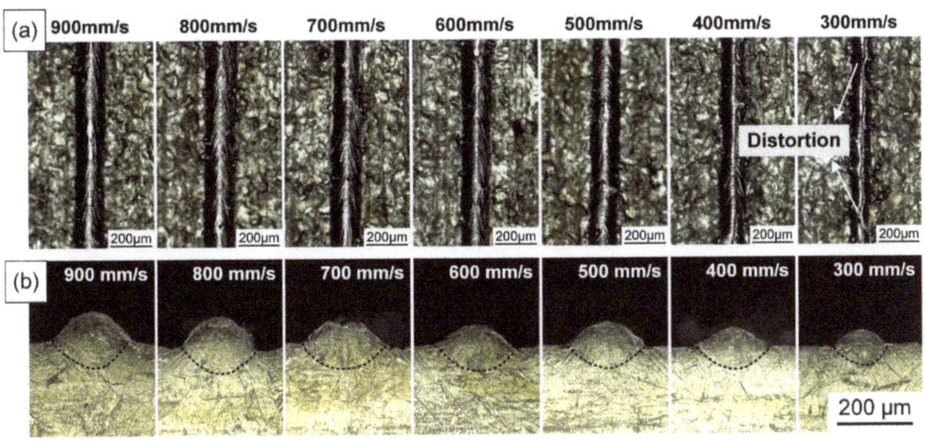

Figure 11. (**a**) Single-track surface morphologies and (**b**) molten pool morphologies with different scanning speeds when the linear energy density is 0.195 J/mm.

The inconsistency of the single-track width formed with the same linear energy density also indicates that the influence of laser power and scanning speed on the single-track width is not consistent. By curve fitting the data in Figure 10, the regular relationship between the single-track width W and the laser linear energy density is obtained:

$$W = 406.85 \times E_L^{0.6311} \tag{4}$$

For a Gaussian laser beam with a fixed spot size heating a metal surface, the formula for calculating the peak temperature T at the center of the beam is as follows [27]:

$$T = \frac{\sqrt{2}AI\sigma}{K\sqrt{\pi}} \tan^{-1}\frac{\sqrt{2Dt}}{\sigma} \tag{5}$$

where A is the absorptivity of the material, I is the laser intensity, σ is the diameter of the laser spot, k is the thermal conductivity of the material, D is the thermal diffusivity of the material, and t is the residence time of the laser.

The formula for calculating the laser intensity I is as follows:

$$I = \frac{P}{2\pi\sigma^2} \tag{6}$$

For a laser beam with a scanning speed of v, the laser dwell time is approximately σ/v, so the formula for calculating the peak temperature T is as follows:

$$T = \frac{AP}{\sigma k\pi\sqrt{2\pi}} \tan^{-1}\sqrt{\frac{2D}{v\sigma}} \tag{7}$$

From the above formula, it can be concluded that a higher laser power or a lower scanning speed results in greater energy input in the localized area. This raises the temperature in the middle of the melt pool, which causes the liquid phase to rise and the cladding track to widen. Furthermore, scanning speed has less impact on the peak temperature than does laser power. While scanning speed mostly influences the energy intake per unit of time, laser power primarily affects the peak temperature. Therefore, even with the same linear energy density, changing the scanning speed will still result in variations in single-track width.

Excessive laser energy density can lead to a "keyhole mode" melt pool, resulting in porosity defects, and the aspect ratio of the melt pool should not exceed 0.5. Conversely, low energy density reduces melting capability, leading to insufficient molten material and causing distortion and a balling phenomenon of the cladding track. Based on the intersection range of the melt pool height and depth variation lines shown in Figure 9, the suitable linear energy density for single-track formation is between 0.165 and 0.195 J/mm. Figure 11 shows that the melt pool width and aspect ratio show similar features at equivalent linear energy densities. Consequently, when the linear energy density is nearly identical, variations in laser power and scanning speed warrant careful consideration due to their significant influence on the formation density and microstructural properties.

3.2. Research on the Formability of TA15 Blocks Formed by SLM

The single-track formation quality is determined by the laser power and scanning speed. The entire surface can be formed through the overlapping of multiple single-tracks. The degree of overlap can be represented by the overlap ratio η, calculated using the following formula:

$$\eta = \frac{W - H}{W} \times 100\% = \frac{D_W}{W} \times 100\% \tag{8}$$

A fixed laser power of 195 W and a moderate scanning speed of 1100 mm/s were used to examine the effect of scanning spacing on the density of the samples while maintaining the quality of single-track formation. Block sizes of 10 × 10 × 5 mm and scanning spacings of 50, 100, 150, and 200 μm were used for fabricating the TA15 samples. The samples' relative density was measured, and their cross-sectional morphology was observed, as shown in Figure 12.

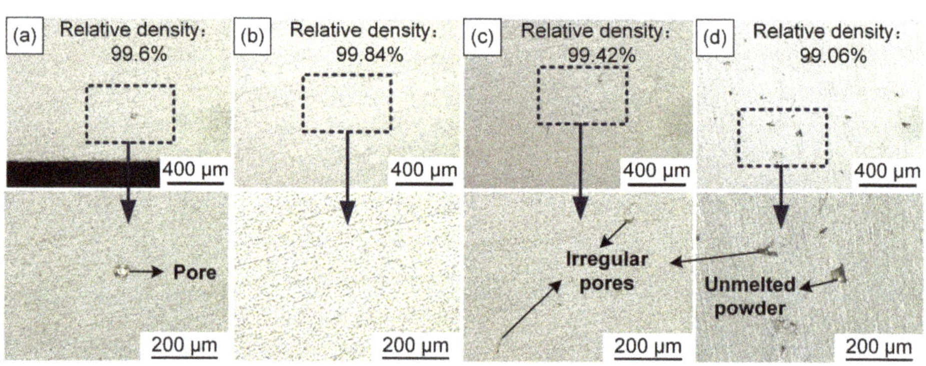

Figure 12. Surface morphology of the sample when the laser power was 195 W, the scanning speed was 1100 mm/s, and the scanning spacing was (**a**) 50 μm, (**b**) 100 μm, (**c**) 150 μm, and (**d**) 200 μm, respectively.

It is obvious from Figure 12a that a small number of circular holes form at the sample's edges when the scanning spacing is 50 μm. With a laser power of 195 W and a scanning speed of 1100 mm/s, the single-track melt width and melt depth are 150 μm and 70 μm, respectively, resulting in an aspect ratio of less than 0.5 and no keyholes. As a result, the single-track formation process is not responsible for the holes. A smaller scanning spacing increases the overlap of the clad tracks, leading to an expanded laser remelting area and heat accumulation [28], especially in the edge regions. The shorter laser dwell time results in more pronounced heat, forming a larger melt pool [29]. The increased Marangoni force and recoil pressure within the melt pool cause unstable flow of the molten material [30], which may ultimately entrap gases, leading to pore formation. The sample surface has no porosity and reaches a relative density of up to 99.84% when the scanning space is 100 μm. Unmelted zones result from the overlap width of the clad tracks being almost zero at a scanning interval of 150 μm, as shown in Figure 13. As a result, the sample surface has a few irregular holes, as shown in Figure 12c, which results in a decreased relative density of 99.42%. The size and quantity of the defects resulting from inadequate fusion increase at a scanning spacing of 200 μm, further reducing the sample's relative density to 99.06%.

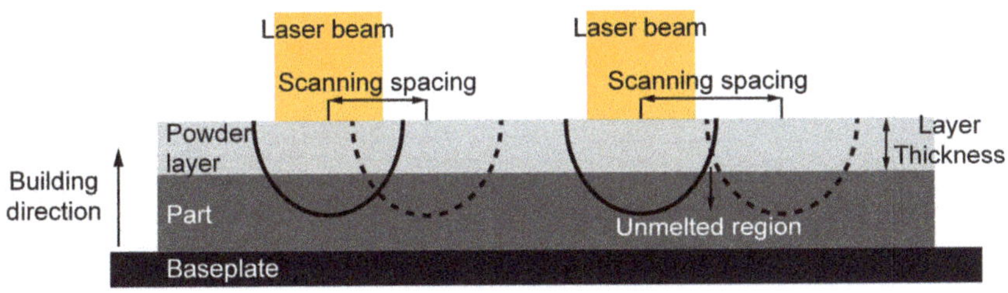

Figure 13. Schematic diagram of fusion deficiency defect generation.

Through experimental validation, selecting appropriate process parameters and overlap ratio ensures continuous and smooth single-track formation, with a suitable melt pool aspect ratio and reasonable scanning spacing, thereby achieving high relative density parts. For further verification, a forming process was selected with similar laser linear energy densities and single-track widths around 150 μm but with different scanning speeds, while fixing the scanning spacing at 100 μm to form the TA15 samples. The surfaces of the samples formed at different scanning speeds under similar linear energy densities were smooth, with good overlap of the clad tracks and no significant defects or powder adhesion, as shown in Figure 14. The surface morphologies of samples 1–4 are similar, and as the scanning speed decreases, the surface solidification ripples transition from a conical shape to an arc shape, which is related to the change in the shape of the melt pool [31]. All six sample groups show excellent relative density, as shown in Table 3. With the quality of single-track formation and reasonable scanning spacing ensured, high relative density TA15 samples can be formed. This also confirms that determining single-track formability through single-track experiments and selecting appropriate forming intervals is an effective method for optimizing the SLMed TA15 alloy.

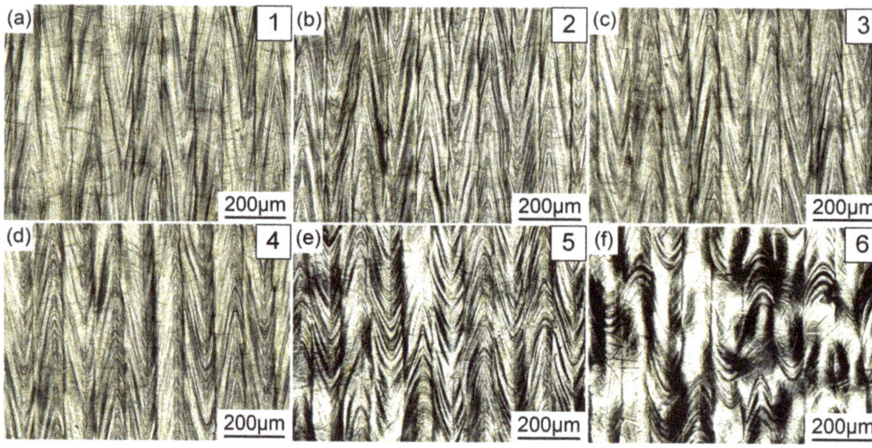

Figure 14. The surface morphologies of the samples with different scanning speeds under a similar linear energy density, sample No. 1 (**a**), sample No. 2 (**b**), sample No. 3 (**c**), sample No. 4 (**d**), sample No. 5 (**e**), and sample No. 6 (**f**).

Table 3. The density and relative density of the samples with different scanning speeds under a similar linear energy density.

Sample	Scanning Speed (mm/s)	Laser Power (W)	Scanning Space (μm)	Linear Energy Density (J/mm)	Single-Track Width (μm)	Density (g/cm^3)	Relative Density (%)
1	1050	190	100	0.18	147	4.450	99.98
2	1000	175	100	0.175	147	4.446	99.89
3	900	160	100	0.178	146	4.444	99.84
4	800	150	100	0.189	149	4.447	99.91
5	600	110	100	0.183	152	4.450	99.98
6	400	90	100	0.225	151	4.449	99.96

3.3. The Influence of Process Parameters on the Mechanical Properties and Microstructure of SLMed TA15

Figure 15 illustrates the tensile curves and mechanical properties of specimen numbers 1 through 6, with the process parameters of the specimens corresponding to those used for density testing. Table 4 presents the mechanical property data for all specimens. As can be observed from Figure 15a, all specimens exhibit a prolonged yield platform. From Figure 15b, it is evident that within the laser scanning speed range of 900–1050 mm/s, the tensile properties of specimens 1 through 3 show minimal variation, all demonstrating high ultimate tensile strength (1289–1299 MPa) and relatively poor ductility (6.4–6.8%). Table 4 indicates that the hardness values of samples 1–3 range between 385.1 and 389.2, with little variation observed. The strength and hardness of samples 3–6 show a decreasing trend with decreasing scanning speed, while the elongation rate exhibits an increasing trend. Compared with sample 3, sample 6's ultimate tensile strength, yield strength, and hardness decreased by 58 MPa, 88 MPa, and 21 MPa, respectively, while the elongation rate increased by 2%. Figure 16 displays the fracture morphologies of all tensile specimens, revealing that the fractures of the six specimens exhibit numerous fine dimples, indicative of a mixed ductile-brittle fracture mode. Distinctively, the fractures of specimens 1 through 4 are characterized by extensive flat, large cleavage platforms; the fracture of specimen 4 shows relatively rough dimpled regions; the fractures of specimens 5 and 6 transition to a terraced tearing ridge feature; and the fracture of specimen 6 presents larger dimples. This suggests that specimens 5 and 6 exhibit a tendency toward a ductile fracture mode.

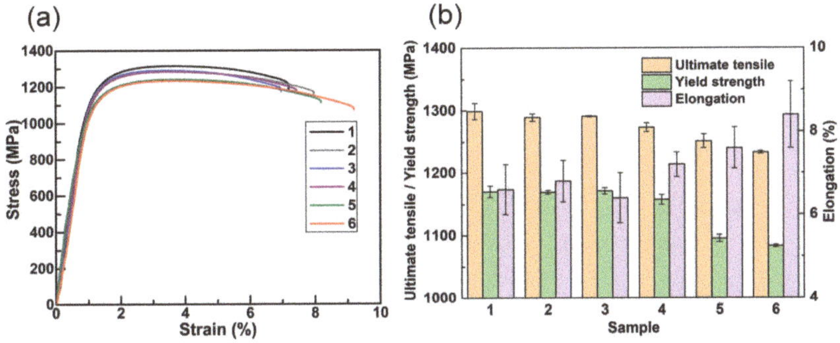

Figure 15. Tensile curves (**a**) and mechanical properties (**b**) of TA15 samples with different SLM forming processes.

Table 4. Mechanical property data of sample Nos. 1–6.

Sample	V (mm/s)	P (W)	H (μm)	σ_U (MPa)	$\sigma_{0.2}$ (MPa)	ε (%)	HV_1
1	1050	190	100	1299 ± 13	1170 ± 9	6.6 ± 0.6	389.2 ± 5.3
2	1000	175	100	1289 ± 6	1169 ± 3	6.8 ± 0.5	387.3 ± 5.3
3	900	160	100	1291 ± 1	1171 ± 5	6.4 ± 0.6	385.1 ± 5.2
4	800	150	100	1273 ± 7	1157 ± 8	7.2 ± 0.3	382.2 ± 4.5
5	600	110	100	1251 ± 11	1095 ± 6	7.6 ± 0.5	373.3 ± 1.9
6	400	90	100	1233 ± 2	1083 ± 2	8.4 ± 0.8	364.4 ± 5.1

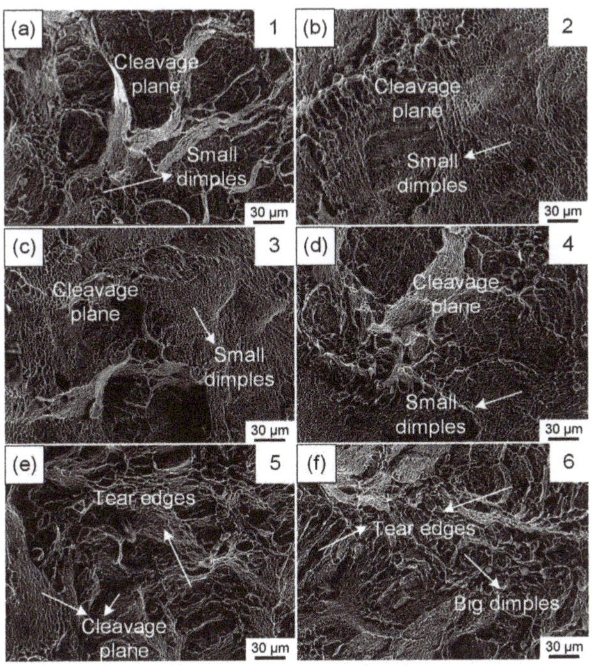

Figure 16. Fracture morphologies of sample No. 1 (**a**), sample No. 2 (**b**), sample No. 3 (**c**), sample No. 4 (**d**), sample No. 5 (**e**), and sample No. 6 (**f**).

Samples 1, 2, 5, and 6, which show notable performance variations, were chosen for microstructural investigation to examine the causes of the variations in mechanical properties brought about by process parameters. As can be observed from Figure 17, the microstructures of sample 1 and sample 2 exhibit a fine acicular morphology with distinct grain boundaries. The fine acicular structure has been previously identified in related studies on SLMed TA15 [6–9,12,32] as α' martensite, with the grain boundaries being those of the primary β phase. Samples 5 and 6 display fine, needle-like α' martensite structures similar to those in samples 1 and 2, as shown in Figure 17. However, compared with samples 1 and 2, the needle-like structures in samples 5 and 6 appear coarser in certain zones. Additionally, the primary β grain boundaries in samples 5 and 6 are not distinct, requiring differentiation of the primary β grains through the orientation of the α' martensite lath.

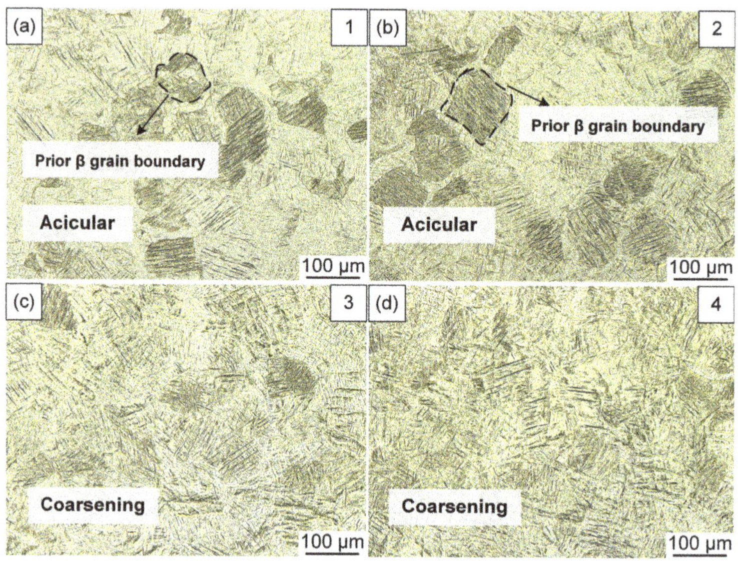

Figure 17. The metallographic diagram of the cross-section of SLM formed TA15: sample No. 1 (**a**), sample No. 2 (**b**), sample No. 5 (**c**), and sample No. 6 (**d**).

The α' martensite in the three samples has an obvious size gradation and different grain orientations, as shown in Figure 18a–c. With values of 1.2% and 1.4%, respectively, compared with 0.4% in sample 1, Figure 18d–f shows that the β phase content in samples 5 and 6 is higher than that in sample 1. The average grain sizes for samples 5 and 6 are 1.32 μm and 1.4 μm, respectively, as shown in Figure 18g–i. This represents a minor increase over the average grain size of sample 1, which is 1.14 μm. Due to the presence of the β isomorphous elements V and Mo in the TA15 alloy, the metastable α' martensite within the SLM-ed TA15 specimens decomposes upon heating in the form of $\alpha' \rightarrow \alpha + \beta$ [5]. During this decomposition process, the β phase nucleates heterogeneously, primarily at the grain boundaries of the α' martensite and at substructures with high densities of dislocations and twins within the martensite. As solute elements diffuse, the body-centered cubic elements V and Mo gradually dissolve into the β phase, and the composition of the α' martensite gradually approaches that of the α phase in equilibrium, leading to the nearly homogeneous nucleation of the α phase. The decomposition of the fine acicular α' martensite results in the formation of fine lamellar α + β phases, and existing research has found that in situ decomposition of α′ martensite occurs during the SLM process [33].

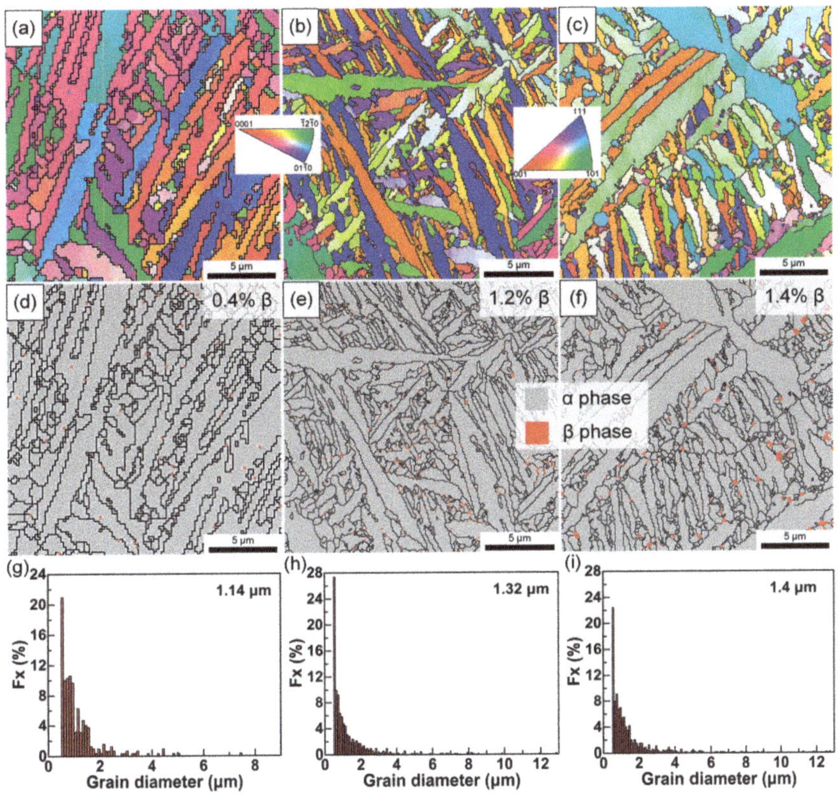

Figure 18. The IPF diagram, phase distribution diagram, and grain size distribution diagram of the cross-section of TA15 formed by SLM, where (**a**,**d**,**g**) are sample No. 1, (**b**,**e**,**h**) are sample No. 5 and (**c**,**f**,**i**) are sample No. 6.

The alterations in the microstructure of the samples are closely related to the thermo-physical processes during SLM formation. Although the single-track widths of samples 3–6 are similar, the fish-scale-like "solidification ripples" left on the surface of the single-track after the molten metal cooling is different, as shown in Figure 19. As scanning speed decreases, the ripple structure progressively shifts from conical to circular. It has been established that the form of the "solidification ripples" corresponds to the melt pool profile [34]. At similar energy densities, the width of the melt pool produced by the laser is approximately the same. However, an increase in scanning speed results in changes in heat flow, leading to a more slender melt pool shape [35], transitioning from a "near-spherical" to a "teardrop" shape, as shown in Figure 19e,f. This transition shows that a decrease in scanning speed can result in a decrease in cooling rate even at similar linear energy densities. When the laser repeatedly heats layers that have already been deposited, a slower rate of cooling is more favorable for accumulating temperatures that are appropriate for the in situ breakdown of α martensite [33].

Consequently, in contrast to processes 1 through 4, the SLMed TA15 alloy under processes 5 and 6 undergoes a transformation wherein the α′ martensite decomposes into α + β phases, and the microstructure evolves from a fine acicular to a fine lamellar morphology. The coarsening of the sample microstructure will reduce the effect of fine grain strengthening. Additionally, the body-centered cubic β phase possesses more slip systems compared with the hexagonal, close-packed α/α′ phase [36]. Thus, the higher β phase content helps to improve the samples' capacity for plastic deformation. The fine, acicular α′ martensite contributes to the specimen's higher strength but lower plasticity [37], whereas the coarser lamellar α + β phases can enhance the specimen's plasticity while concurrently reducing its strength [38,39]. This explains why samples 5 and 6 have superior plasticity but comparatively lower strength and hardness.

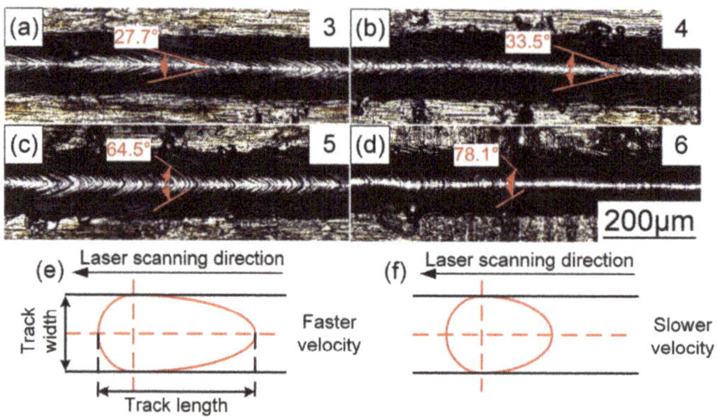

Figure 19. Single-track morphology (**a**–**d**) and schematic diagram of molten pool formation (**e**,**f**) of sample Nos. 3–6.

The experimental results indicate that the SLMed process parameters have a certain influence on the microstructure and properties of the TA15 samples. This also confirms the feasibility of controlling the microstructure of TA15 alloys through process parameter adjustments. The sample's elongation was 8.4% with a laser power of 90 W and a scanning speed of 400 mm/s, which is comparable to that of conventionally forged TA15. Additionally, the tensile strength was measured at 1233 MPa, significantly surpassing that of forged components. Table 5 shows the room temperature tensile properties of SLMed TA15 from related studies. It can be observed from the table that the microstructure of the TA15 specimens in their fabricated state is generally consistent, predominantly consisting of acicular α′ martensite, with high tensile strength and relatively low elongation. In this study, by employing similar laser linear energy densities and varying the laser power and scanning speed, TA15 specimens with higher elongation were obtained. This also indicates that during the SLM forming of TA15 components, different forming processes can be utilized according to the performance requirements of different parts of the component, thereby achieving parts with varying mechanical properties at different locations.

Table 5. The room temperature tensile properties of SLMed TA15 from previous studies.

Processing	Microstructure	σ_U (MPa)	ε (%)	Ref.
SLM	Acicular α′	1350 ± 35	6.0 ± 0.9	[13]
SLM	Acicular α′	1234.2 ± 53.1	7.3 ± 0.7	[9]
SLM	Acicular α′	1214 (XY) 1234 (XY)	9.2 (Z) 34.1 (Z)	[8]
SLM	Acicular α′	1250	3.1	[32]
SLM	Acicular α′	1062.7–1294.9	No tests.	[40]

4. Conclusions

1. The single-track width of SLMed TA15 increases with a higher laser power or lower scanning speed. At the same linear energy density, the single-track width tends to decrease as the laser power decreases, indicating a distinct relationship between linear energy density and single-track width.
2. At the same laser power, the depth of the melt pool in SLMed TA15 increases as the scanning speed decreases, while the depth-to-width ratio also increases sharply. When the aspect ratio exceeds 0.5, the molten pool transitions from "conduction mode" to "keyhole mode". When the linear energy density is the same, a single melt pool has a similar aspect ratio.
3. When the linear energy density and single-track width are close, combined with an appropriate overlapping rate, high-density TA15 alloy samples can be effectively formed.
4. Under similar laser linear energy density conditions, lower scanning speeds lead to heat accumulation during the SLM process, resulting in the in situ decomposition of α' martensite. This results in more β phase and coarser α layer lamellar microstructures in the samples, leading to a decrease in strength and hardness while improving plasticity.

Author Contributions: Conceptualization, J.J.; methodology, J.J. and H.C.; formal analysis, J.J. and Y.C.; investigation, Y.W. and H.C.; validation, F.Z. and P.W.; resources, J.J., J.X. and F.Z.; data curation, C.L., J.X. and P.W.; supervision, D.Z.Z.; writing—original draft preparation, J.J., C.L. and Y.C.; writing—review and editing, J.J., Y.W. and D.Z.Z. funding acquisition, J.J. and Y.W. All authors have read and agreed to the published version of the manuscript.

Funding: This research was funded by the Key Research Projects of Higher Education Institutions in Henan Province of China (24A450003, 24A460027), the Natural Science Foundation in Henan province of China (242300421468), and the Henan Provincial Key Laboratory of Intelligent Manufacturing of Machinery and Equipment Open Subjects (IM202308).

Data Availability Statement: The original contributions presented in this study are included in the article. Further inquiries can be directed to the corresponding authors.

Conflicts of Interest: Author Yuanchao Chen was employed by the company Zhengzhou RuiFei Biotechnology Co., Ltd. The remaining authors declare that the research was conducted in the absence of any commercial or financial relationships that could be construed as a potential conflict of interest.

References

1. Sun, Q.J.; Xie, X. Microstructure and mechanical properties of TA15 alloy after thermo-mechanical processing. *Mater. Sci. Eng. A* **2018**, *724*, 493–501. [CrossRef]
2. Wu, H.; Sun, Z.; Cao, J.; Yin, Z. Formation and evolution of tri-modal microstructure during dual heat treatment for TA15 Ti-alloy. *J. Alloys Compd.* **2019**, *786*, 894–905. [CrossRef]
3. Huang, D.M.; Wang, H.L.; Chen, X.; Chen, Y.; Guo, H. Influence of forging process on microstructure and mechanical properties of large section Ti–6.5Al–1Mo–1V–2Zr alloy bars. *Trans. Nonferrous Met. Soc. China* **2013**, *23*, 2276–2282. [CrossRef]
4. Zhichao, S.; He, Y. Microstructure and mechanical properties of TA15 titanium alloy under multi-step local loading forming. *Mater. Sci. Eng. A* **2009**, *523*, 184–192. [CrossRef]
5. Zhang, X.; Zhao, Y.; Bai, C. *Titanium Alloy and Its Applications*; Chemical Industry Press: Beijing, China, 2005.
6. Wu, X.; Cai, C.; Yang, L.; Liu, W.; Li, W.; Li, M.; Liu, J.; Zhou, K.; Shi, Y. Enhanced mechanical properties of Ti-6Al-2Zr-1Mo-1V with ultrafine crystallites and nano-scale twins fabricated by selective laser melting. *Mater. Sci. Eng. A* **2018**, *738*, 10–14. [CrossRef]
7. Jiang, J.; Ren, Z.; Ma, Z.; Zhang, T.; Zhang, P.; Zhang, D.Z.; Mao, Z. Mechanical properties and microstructural evolution of TA15 Ti alloy processed by selective laser melting before and after annealing. *Mater. Sci. Eng. A Struct. Mater. Prop. Misrostruct. Process.* **2020**, *772*, 138742. [CrossRef]
8. Huang, S.; Sun, B.; Guo, S. Microstructure and property evaluation of TA15 titanium alloy fabricated by selective laser melting after heat treatment. *Opt. Laser Technol.* **2021**, *144*, 107422. [CrossRef]

9. Cai, C.; Wu, X.; Liu, W.; Zhu, W.; Chen, H.; Qiu, J.C.D.; Sun, C.-N.; Liu, J.; Wei, Q.; Shi, Y. Selective laser melting of near-α titanium alloy Ti-6Al-2Zr-1Mo-1V: Parameter optimization, heat treatment and mechanical performance. *J. Mater. Sci. Technol.* **2020**, *57*, 51–64. [CrossRef]
10. Jiang, J.; Chen, J.; Ren, Z.; Mao, Z.; Ma, X.; Zhang, D.Z. The Influence of Process Parameters and Scanning Strategy on Lower Surface Quality of TA15 Parts Fabricated by Selective Laser Melting. *Metals* **2020**, *10*, 1228. [CrossRef]
11. Yin, A.; Yu, W.; Li, W.; Zhu, W.; Ji, V.; Jiang, C.; Wang, C. Microstructural and thermal relaxation of residual stress in dual peened TA15 titanium alloy fabricated by SLM. *Mater. Charact.* **2024**, *218*, 114496. [CrossRef]
12. Xu, W.; Du, Z.; Wang, Z.; Ran, X.; Li, Z.; Chen, R.; Zhu, X.; Zhang, X.; Zhang, J.; Lu, X. A comparative study on the corrosion, wear, and tribocorrosion performances of Ti-6.5Al–2Zr–1Mo–1V (TA15) alloy fabricated by selective laser melting and laser metal deposition. *J. Mater. Res. Technol.* **2024**, *30*, 6245–6253. [CrossRef]
13. Xu, W.; Zhu, X.; Wang, Z.; Ran, X.; Li, Z.; Chen, R.; Du, Z.; Zhang, X.; Singh, H.; Zhang, J. Comparative study of Ti-6.5Al-2Zr-1Mo-1V (TA15) alloy fabricated by selective laser melting and laser metal deposition. *Opt. Laser Technol.* **2024**, *176*, 111052. [CrossRef]
14. Zhou, L.; Yuan, T.; Li, R.; Tang, J.; Wang, G.; Guo, K.; Yuan, J. Densification, microstructure evolution and fatigue behavior of Ti-13Nb-13Zr alloy processed by selective laser melting. *Powder Technol.* **2019**, *342*, 11–23. [CrossRef]
15. Sun, J.; Yang, Y.; Wang, D. Parametric optimization of selective laser melting for forming Ti6Al4V samples by Taguchi method. *Opt. Laser Technol.* **2013**, *49*, 118–124. [CrossRef]
16. Khorasani, A.M.; Gibson, I.; Awan, U.S.; Ghaderi, A. The effect of SLM process parameters on density, hardness, tensile strength and surface quality of Ti-6Al-4V. *Addit. Manuf.* **2019**, *25*, 176–186. [CrossRef]
17. Zhou, Y.; Wen, S.F.; Song, B.; Zhou, X.; Teng, Q.; Wei, Q.S.; Shi, Y.S. A novel titanium alloy manufactured by selective laser melting: Microstructure, high temperature oxidation resistance. *Mater. Des.* **2016**, *89*, 1199–1204. [CrossRef]
18. Song, B.; Dong, S.; Zhang, B.; Liao, H.; Coddet, C. Effects of processing parameters on microstructure and mechanical property of selective laser melted Ti6Al4V. *Mater. Des.* **2012**, *35*, 120–125. [CrossRef]
19. Wei, K.; Wang, Z.; Zeng, X. Preliminary investigation on selective laser melting of Ti-5Al-2.5Sn α-Ti alloy: From single tracks to bulk 3D components. *J. Mater. Process. Technol.* **2017**, *244*, 73–85. [CrossRef]
20. Hang, B.; Li, C.; Shi, L.; Qiu, G. *China Materials Engineering Canon. Volume 4, Nonferrous Metal Materials Engineering. Part One*; Chemical Industry Press: Beijing, China, 2006.
21. *GB/T 228.1-2010*; Metallic Materials—Tensile Testing—Part 1: Method of Test at Room Temperature. Standards Press of China: Beijing, China, 2010.
22. Yang, J.; Han, J.; Yu, H.; Yin, J.; Gao, M.; Wang, Z.; Zeng, X. Role of molten pool mode on formability, microstructure and mechanical properties of selective laser melted Ti-6A1-4V alloy. *Mater. Des.* **2016**, *110*, 558–570. [CrossRef]
23. Rai, R.R.; Elmer, J.; Palmer, T.; Debroy, T. Heat transfer and fluid flow during keyhole mode laser welding of tantalum, Ti–6Al–4V, 304L stainless steel and vanadium. *J. Phys. D Appl. Phys.* **2007**, *40*, 5753. [CrossRef]
24. Wei, K.; Wang, Z.; Zeng, X. Effect of heat treatment on microstructure and mechanical properties of the selective laser melting processed Ti-5Al-2.5Sn α titanium alloy. *Mater. Sci. Eng. A* **2018**, *709*, 301–311. [CrossRef]
25. King, W.E.; Barth, H.D.; Castillo, V.M.; Gallegos, G.F.; Gibbs, J.W.; Hahn, D.E.; Kamath, C.; Rubenchik, A.M. Observation of keyhole-mode laser melting in laser powder-bed fusion additive manufacturing. *J. Mater. Process. Technol.* **2014**, *214*, 2915–2925. [CrossRef]
26. Gu, D.; Shen, Y. Effects of processing parameters on consolidation and microstructure of W–Cu components by DMLS. *J. Alloys Compd.* **2009**, *473*, 107–115. [CrossRef]
27. Bäuerle, D. *Laser Processing and Chemistry*; Springer Science & Business Media: Berlin/Heidelberg, Germany, 2013.
28. Dong, Z.; Liu, Y.; Wen, W.; Ge, J.; Liang, J. Effect of Hatch Spacing on Melt Pool and As-built Quality During Selective Laser Melting of Stainless Steel: Modeling and Experimental Approaches. *Materials* **2018**, *12*, 50. [CrossRef]
29. Xia, M.; Gu, D.; Yu, G.; Dai, D.; Chen, H.; Shi, Q. Influence of hatch spacing on heat and mass transfer, thermodynamics and laser processability during additive manufacturing of Inconel 718 alloy. *Int. J. Mach. Tools Manuf.* **2016**, *109*, 147–157. [CrossRef]
30. Qiu, C.; Panwisawas, C.; Ward, M.; Basoalto, H.C.; Brooks, J.W.; Attallah, M.M. On the role of melt flow into the surface structure and porosity development during selective laser melting. *Acta Mater.* **2015**, *96*, 72–79. [CrossRef]
31. DebRoy, T.; Wei, H.L.; Zuback, J.S.; Mukherjee, T.; Elmer, J.W.; Milewski, J.O.; Beese, A.M.; Wilson-Heid, A.; De, A.; Zhang, W. Additive manufacturing of metallic components–Process, structure and properties. *Prog. Mater. Sci.* **2018**, *92*, 112–224. [CrossRef]
32. Wu, X.; Zhang, D.; Guo, Y.; Zhang, T.; Liu, Z. Microstructure and mechanical evolution behavior of LPBF (laser powder bed fusion)-fabricated TA15 alloy. *J. Alloys Compd.* **2021**, *873*, 159639. [CrossRef]
33. Xu, W.; Brandt, M.; Sun, S.; Elambasseril, J.; Liu, Q.; Latham, K.; Xia, K.; Qian, M. Additive manufacturing of strong and ductile Ti–6Al–4V by selective laser melting via in situ martensite decomposition. *Acta Mater.* **2015**, *85*, 74–84. [CrossRef]

34. Yang, T.; Liu, T.; Liao, W.; MacDonald, E.; Wei, H.; Chen, X.; Jiang, L. The influence of process parameters on vertical surface roughness of the AlSi10Mg parts fabricated by selective laser melting. *J. Mater. Process. Technol.* **2019**, *266*, 26–36. [CrossRef]
35. Chen, S.; Guillemot, G.; Gandin, C.-A. Three-dimensional cellular automaton-finite element modeling of solidification grain structures for arc-welding processes. *Acta Mater.* **2016**, *115*, 448–467. [CrossRef]
36. Liang, Z.; Sun, Z.; Zhang, W.; Wu, S.; Chang, H. The effect of heat treatment on microstructure evolution and tensile properties of selective laser melted Ti6Al4V alloy. *J. Alloys Compd.* **2019**, *782*, 1041–1048. [CrossRef]
37. Kumar, P.; Prakash, O.; Ramamurty, U. Micro-and meso-structures and their influence on mechanical properties of selectively laser melted Ti-6Al-4V. *Acta Mater.* **2018**, *154*, 246–260. [CrossRef]
38. Zhang, X.-Y.; Fang, G.; Leeflang, S.; Böttger, A.J.; Zadpoor, A.A.; Zhou, J. Effect of subtransus heat treatment on the microstructure and mechanical properties of additively manufactured Ti-6Al-4V alloy. *J. Alloys Compd.* **2018**, *735*, 1562–1575. [CrossRef]
39. Vrancken, B.; Thijs, L.; Kruth, J.-P.; Van Humbeeck, J. Heat treatment of Ti6Al4V produced by Selective Laser Melting: Microstructure and mechanical properties. *J. Alloys Compd.* **2012**, *541*, 177–185. [CrossRef]
40. Lu, S.; Li, Y.; Zhen, S.; Yu, D.; Zhang, L.; Chen, R.; Wang, Y.; Li, S. Influence of process parameters on the microstructures, residual stresses and mechanical properties of TA15 titanium alloy fabricated by L-PBF. *Mater. Sci. Eng. A* **2025**, *927*, 147990. [CrossRef]

metals

MDPI

Article

Modeling and Monitoring of the Tool Temperature During Continuous and Interrupted Turning with Cutting Fluid

Hui Liu [1,*], Markus Meurer [1] and Thomas Bergs [1,2]

[1] Manufacturing Technology Institute (MTI), RWTH Aachen University, Campus-Boulevard 30, 52074 Aachen, Germany; m.meurer@mti.rwth-aachen.de (M.M.); t.bergs@mti.rwth-aachen.de (T.B.)
[2] Fraunhofer Institute for Production Technology IPT, Steinbachstr. 17, 52074 Aachen, Germany
* Correspondence: h.liu@mti.rwth-aachen.de; Tel.: +49-241-80-24975

Abstract: In metal cutting, a large amount of mechanical energy converts into heat, leading to a rapid temperature rise. Excessive heat accelerates tool wear, shortens tool life, and hinders chip breakage. Most existing thermal studies have focused on dry machining, with limited research on the effects of cutting fluids. This study addresses that gap by investigating the thermal behavior of cutting tools during continuous and interrupted turning with cutting fluid. Tool temperatures were first measured experimentally by embedding a thermocouple in a defined position within the tool. These experimental results were then combined with simulations to evaluate temperature changes, heat partition, and cooling efficiency under various cutting conditions. This work presents novel analytical and numerical models. Both models accurately predicted the temperature distribution, with the analytical model offering a computationally more efficient solution for industrial use. Experimental results showed that tool temperature increased with cutting speed, feed, and cutting depth, but the heat partition into the tool decreased. In continuous cutting, cooling efficiency was mainly influenced by feed rate and cutting depth, while cutting speed had minimal impact. Interrupted cutting improved cooling efficiency, as the absence of chips and workpieces during non-cutting phases allowed the cutting fluid to flow over the tool surface at higher speeds. The convective cooling coefficient was determined through inverse calibration. A comparative analysis of the analytical and numerical simulations revealed that the analytical model can underestimate the temperature distribution for complex tool structures, particularly non-orthogonal hexahedral geometries. However, the relative error remained consistent across different cutting conditions, with less error observed in interrupted cutting compared to continuous cutting. These findings highlight the potential of analytical models for optimizing thermal management in metal turning processes.

Keywords: tool temperature; turning; cooling; cutting fluid; temperature simulation; analytical model; numerical model

1. Introduction

During metal cutting, considerable heat is generated due to intense friction between the tool and workpiece, as well as the plastic deformation of the workpiece material in the cutting zone. Thermomechanical studies conducted by Taylor and Quinney revealed that only a small fraction of plastic work is stored as internal energy in the metal, with the majority dissipated as heat [1]. Extensive experimental studies have demonstrated that approximately 90% of the mechanical work is converted to heat during metal cutting [2]. Part of this heat transfers to the tool, causing a rapid temperature rise near the tool tip. Elevated temperatures can reduce the hardness and strength of the tool material, promote chemical reactions like oxidation and diffusion between the tool and workpiece materials, and accelerate tool wear [3]. For the workpiece, high cutting temperatures combined with mechanical stresses can lead to surface damage, including oxidation, corrosion, phase transformation, and recrystallization. Liu et al. [4] examined the effects of temperature on the workpiece surface through entropy changes. Further details on temperature effects on workpiece quality are available in review papers [5,6]. In interrupted cutting or when using cutting fluid, tool temperature can fluctuate significantly, leading to thermal fatigue and potentially causing tool fracture or failure [3]. Furthermore, high temperatures in the cutting area also negatively affect the surface quality of the workpiece. Therefore, understanding heat generation and transmission in metal cutting is crucial for improving workpiece quality and tool life.

To investigate the impact of cutting temperature on machining outcomes, various methods have been developed. Early approaches began with the balance (calorimetric) method, introduced in 1949 [7]. In this method, chips formed in the cutting zone are collected in a calorimetric vessel filled with water. By applying the energy conservation law and thermal balance, the temperature difference is measured between the vessel with water alone and the vessel with water and chips. Although applicable to any workpiece and tool materials, this method is now largely of historical interest due to its low accuracy and only approximate estimations of mean chip temperatures [8]. With advancements in sensor technology, modern temperature measurement techniques for machining processes are primarily categorized into infrared (IR) radiometry and thermocouple measurements [9]. Infrared radiometry detects surface-emitted radiation, enabling temperature estimation via the Stefan–Boltzmann law. This method offers fast response times and direct temperature distribution measurement without contact. Heigel et al. [10] used high-speed thermography during orthogonal cutting of titanium alloys, finding that peak temperature increased with cutting speed, while cutting and thrust forces remained unchanged. Similar results were observed by Soler et al. [11] in AISI 4140 cutting tests. Soloer et al. [11] also employed high-speed thermography to measure tool flank temperatures, achieving less than 3% deviation in numerical simulations of tool/chip contact temperatures under varying cutting conditions. Recent findings by Barzegar et al. [12] indicated that increasing the cutting edge radius significantly raises the temperature. While infrared thermography offers significant advantages, it also has limitations. Accurate measurements depend on precise surface emissivity data, which requires temperature calibration of the machined surface. In addition, not all machine tools and machining processes are optically accessible for radiometric measurements, especially those that work with cutting fluids.

A solution for overcoming optical accessibility issues in temperature measurement is to embed a thermocouple or pyrometer fiber directly into the cutting insert, which is typically installed into small holes in the cutting tool made via Electrical Discharge Machining (EDM). Chen et al. [13] investigated the dry turning of AISI 52100 using a CBN tool with an embedded thermocouple. Their measurements, combined with numerical modeling, revealed that the peak temperature on the machined surface in hard turning occurs along the intersection of the cutting edge and the machined surface. Kryzhanivskyy et al. [14] introduced a method to embed multiple thermocouples simultaneously, enabling reverse derivation of the tool temperature distribution and quantifying heat inflow during dry turning. Similarly, Zhang et al. [15] used multiple thermocouples to study the effect of tool coating on temperature during turning. Similar methods can also be applied to pyrometers. Han et al. [16] embedded a ratio pyrometer fiber into the tool to examine the impact of wear on tool temperature, finding that tool wear increased significantly with rising cutting temperatures. However, ratio pyrometers are only effective at temperatures above 150 °C. Current research demonstrates that the embedded multiple thermocouple technique is robust and adaptable to various cutting conditions, including those involving cutting fluids. The main limitation of thermocouples is their response time, with the fastest thermocouples requiring 10 milliseconds or more, making them unsuitable for capturing rapid temperature fluctuations during interrupted cutting or transient temperature changes. Additionally, since thermocouples provide point measurements, estimating the overall temperature field requires inverse derivation using a reliable temperature model.

Point temperature measurements can be used to calibrate temperature models, allowing for the inverse determination of temperature distribution. Transient temperature models are typically categorized into numerical and analytical approaches. Norouzifard et al. [17] embedded K-type thermocouples in the tool holder to measure temperatures at various locations during the machining of AISI 1045 and AISI 304. Their numerical models accurately predicted temperature distributions, showing strong agreement with experimental data. Sharma et al. [18] utilized thermography to measure temperatures during dry machining and thermocouples for turning AISI 304 steel with cutting fluid. By inversely calculating the heat transfer into the tool from dry machining temperatures and applying computational fluid dynamics (CFD) simulations, they achieved an error margin of 5.79% between simulation and experiment. Yi et al. [19] measured tool temperatures with K-type thermocouples and simulated them using finite element (FEM) software, reporting a 4.04% discrepancy between simulated and experimental results. Existing studies demonstrate that numerical simulations can accurately model tool temperature during the turning process, even when accounting for the effects of cutting fluids. However, such simulations often require commercial software, substantial computing resources, and expertise in simulation, which can significantly limit their applicability, particularly in industrial environments.

Analytical temperature models provide a faster and more user-friendly alternative to numerical models, as they typically do not require specialized software. These models simulate temperature variations caused by non-uniform and intermittent heat sources through the superposition of point heat sources. Karaguzel et al. [20] validated the analytical model for tracking rapidly changing tool temperatures during face milling using miniature thermocouples. Oliveira et al. [21] further advanced the analytical model by considering the effects of tool coatings on tool temperature during orthogonal cutting. More recently, the authors of this paper have introduced a transient model that incorporates convective conditions for orthogonal cutting and improves the prediction accuracy of the temperature distribution under the cutting fluid in both continuous and interrupted cutting scenarios [22]. Despite their advantages, analytical models simplify tool geometry, often assuming a rectangular parallelepiped shape, which can lead to inaccuracies for non-rectangular geometries. Additionally, no temperature model currently accounts for convective cooling during turning. This work aims to address this research gap.

In this study, steady-state tool temperature during continuous and interrupted turning of AISI 1045 steel was measured under various cutting and cooling conditions using thermocouples embedded in carbide inserts. These measurements helped determine heat transfer to the tool and the average convective cooling coefficient of the cutting fluid on the tool surface via a numerical model. Additionally, a new transient analytical turning tool temperature model was developed. This research project is intended to answer the following questions: What is the relationship between heat partition to the tool and cutting conditions? How does this relationship vary between continuous and interrupted cutting? How accurate is the analytical model in simulating turning tool temperature, given the assumption of a rectangular parallelepiped tool geometry? The scope of this paper is limited to the temperature of the tool and the heat transferred to it and does not cover chip formation mechanisms or the overall temperature distribution in the cutting area.

The next section outlines the principles of the temperature measurement experiments and describes the tools and workpieces used. Section 3 details the principles and parameters of the numerical and analytical models. The derivation of the analytical model is presented in Appendix A. Section 4 compares the relationship between heat partition and cutting parameters derived from the models, highlighting differences between numerical and analytical results. This paper concludes with a summary of the main findings and suggestions for future research.

2. Experimental Setup For Model Validation

In this study, external longitudinal turning of AISI 1045 steel was performed using carbide tools under various process parameters and cooling conditions. Tool temperature and process force components were measured during the turning process with the application of cutting fluid. The experimental results serve as a dataset for validating numerical and analytical models. This section provides a detailed description of the experimental setup for temperature measurements and the characteristics of the tools and workpieces used.

2.1. Experimental Setup for Temperature Measurement in Longitudinal Turning

The tool temperature is a key parameter in quantifying convective cooling with cutting fluid. The heat distribution along the cutting edge depends on the tool's geometry and engagement conditions. In the turning process, varying chip thickness along the cutting edge results in uneven heat distribution and shifts in the temperature gradient. The measurement method depicted in Figure 1 was employed to capture this temperature distribution.

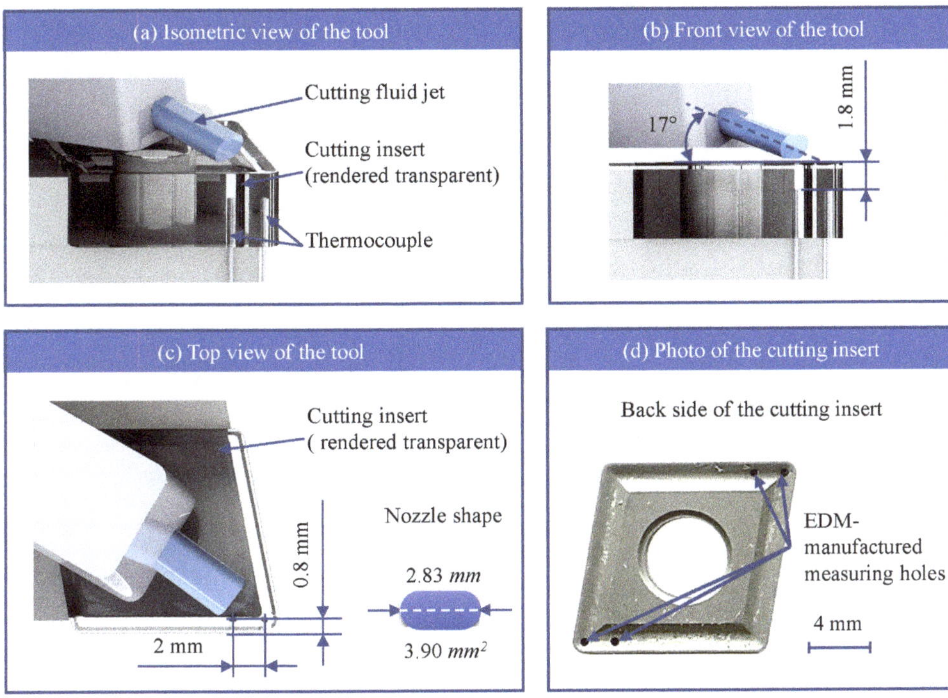

Figure 1. Temperature measuring hole and installation position of the sensors.

The temperature profile was measured at two points on the tool, as shown in Figure 1a. To position the sensors, a 0.5 mm diameter hole was drilled using Electrical Discharge Machining (EDM), located 1.8 mm from the top surface and 0.8 mm from the tool flank face. Two thermocouples were placed 2 mm apart along the main cutting edge. The choice of thermocouples was determined by the required measuring range and rate. Previous measurements in orthogonal cutting [22] showed that the tool temperature near the cutting edge rises by over 200 °C within one second, reaching a steady state after about 5 s. To measure both transient and steady-state temperatures, a Type K thermocouple from TC Mess- und Regeltechnik GmbH (Mönchengladbach, Germany) was selected, offering a measuring range of 0 to 1100 °C and a 15 ms response time. The thermocouple has a diameter of 0.25 mm. Data acquisition was conducted using the NI 9214 acquisition card from National Instruments GmbH (Austin, TX, USA), which features 16 channels for synchronized recording of multiple sensor signals.

Machining tests were performed on a DMG NEF-600 CNC lathe from DMG Mori AG (Bielefeld, Germany), as shown in Figure 2. PCLNR-type ISO tool holders Sandvik Tooling Deutschland GmbH (Düsseldorf, Germany), featuring a cooling channel and a nozzle for rake face cooling, were used and mounted on a Kistler (Sindelfingen, Germany) 9129B piezoelectric dynamometer for cutting force measurements. Cutting fluid was supplied by a ChipBLASTER (Leonberg, Germany) WVHP6-60 high-pressure unit, with a bypass valve for pressure regulation. The cutting fluid used was Vasco TP 519, a water-soluble coolant from Blaser Swisslube AG (Rüegsau, Switzerland), at an 8% concentration, specifically optimized for machining carbon steels and nickel-based alloys.

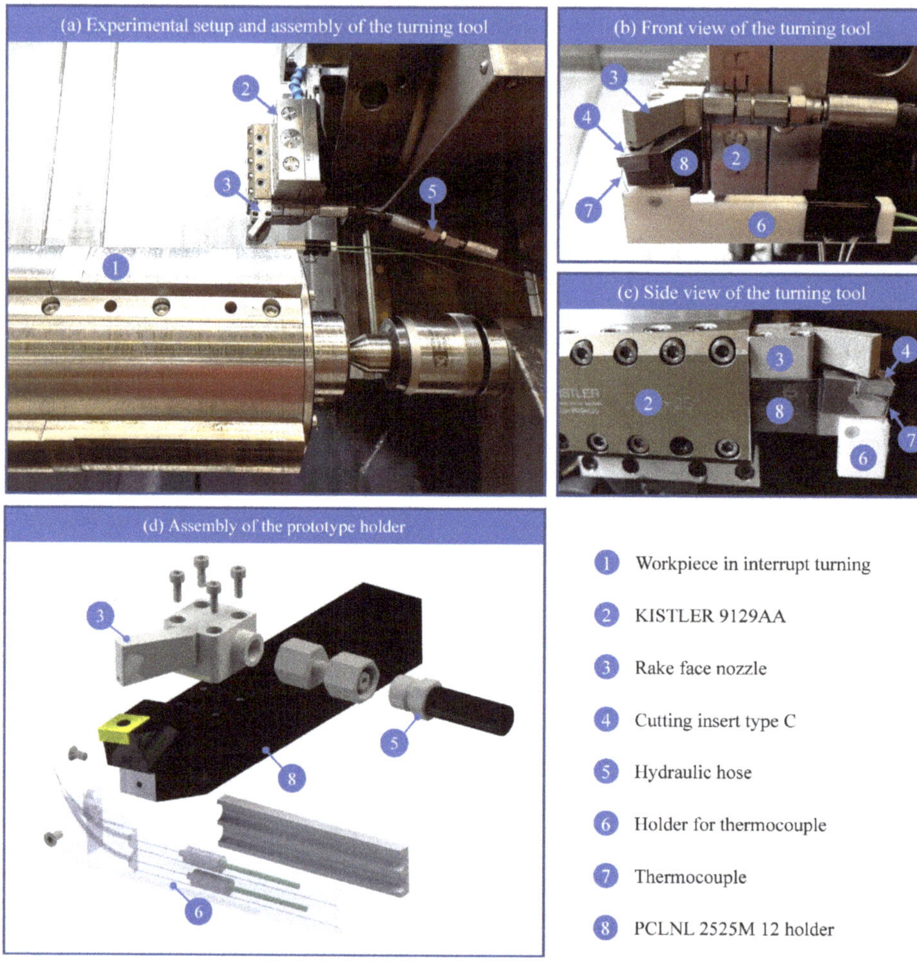

Figure 2. Experimental setup for measuring the tool temperature during turning under cutting fluid conditions.

The thermocouple was initially secured with a sensor holder, as shown in Figure 2d. The holder includes two guide channels to direct the measuring probe to the cutting insert and serves two main functions: protecting the thermocouple from chips and cutting fluid, and preventing the measuring probe from slipping out. The thermocouple was then routed to the cutting insert through the EDM-drilled hole and lubricated with thermally conductive paste to ensure reliable contact with the tool. Subsequent experiments confirmed that this setup securely fixed the sensor in place.

The tests used an indexable insert with C geometry, made of H13A carbide, which is suitable for both finishing and roughing over a wide range of parameters. H13A carbide is specifically designed for machining ISO S (heat-resistant superalloys) and ISO M (stainless steels). This tool material was selected for two main reasons: its thermal properties were previously measured by the author [22], facilitating modeling, and it is widely used in wear studies [23–25], enabling straightforward comparison with other research findings. The large corner angle ($\epsilon_r = 80°$) also facilitated the accurate positioning of measurement sensors. The workpiece material was AISI 1045 carbon steel, with its physical properties summarized in Figure 3. Both microstructure and hardness were measured along the central axis and radially at various points, confirming a homogeneous distribution.

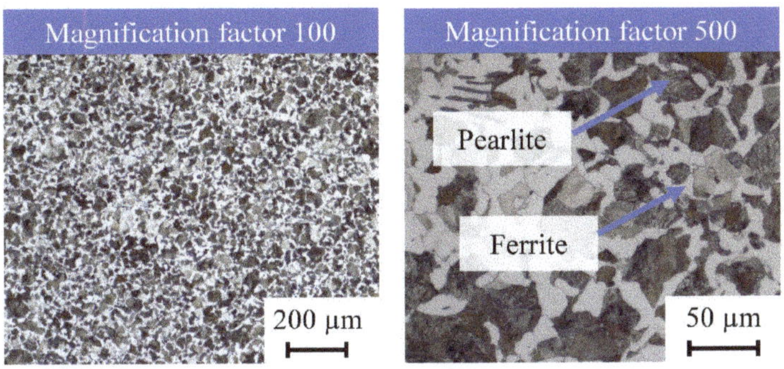

Element	C	Si	Mn	P	S
wt / %	0.469	0.145	0.686	0.0247	0.0261
Element	Cr	Ni	Cu	Mo	Al
wt / %	0.337	0.0866	0.0555	0.0356	0.0288
Physical properties	E *GPa*	R_m *MPa*	$R_{p0.2}$ *MPa*	k *W/m·k*	Hardness *HB*
AISI 1045	210	695	392	51.9	200

Figure 3. Microstructure, mechanical properties, and chemical composition of AISI 1045 workpiece.

A solid shaft with a 250 mm diameter was used for the continuous turning. For the interrupted turning, a VDI 3324-standard fixture was employed, as illustrated in Figure 4. This fixture securely clamps four 55 mm wide AISI 1045 workpiece bars.

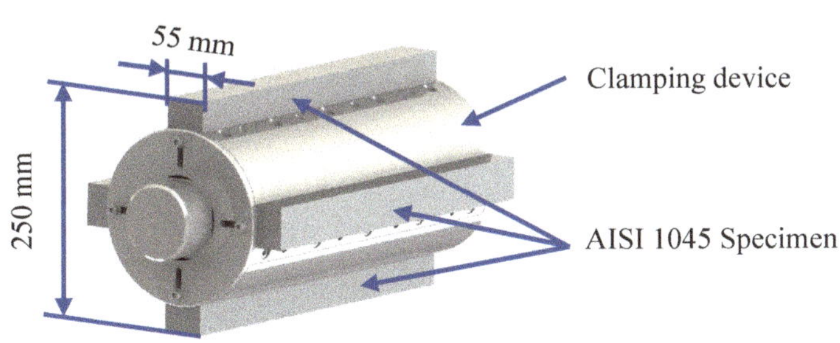

Figure 4. AISI 1045 specimen for interrupted turning according to VDI 3324.

2.2. Experimental Procedure and Test Plan

The experimental procedure was divided into two groups: dry machining and machining with cutting fluid. Each group was further subdivided into continuous and interrupted turning. Initially, machining tests were conducted under dry conditions to measure tool temperature. These results were used both to determine heat partitioning under different cutting conditions and as a reference for assessing the impact of cutting fluid. Subsequently, tests with cutting fluid were performed using the same cutting parameters. The cooling effect was evaluated by comparing tool temperatures between tests with and without cutting fluid

For continuous cutting, a cutting fluid pressure of 80 bar was selected, which corresponds to the typical cutting fluid supply pressure in modern CNC lathes. The experimental results are in good agreement with industrial production conditions. For interrupted cutting, a lower pressure of 10 bar was used. This choice was based on the fact that during the non-cutting phases of interrupted cutting, the cutting fluid cools the tool sufficiently, and the intermittent nature of the process results in short chips that do not require high pressure for effective chip breaking. The investigated parameters are summarized in Table 1.

Table 1. Cutting parameters for temperature analysis.

v_c /m/min	f /mm	a_p /mm	Cutting Fluid	Process Type
70	0.1; 0.2; 0.3	0.8; 2.5	Dry	Continuous
120	0.1; 0.2; 0.3	0.8; 2.5	Dry	Interrupted
70	0.1; 0.2; 0.3	0.8; 2.5	80 bar	Continuous
120	0.1; 0.2; 0.3	0.8; 2.5	10 bar	Interrupted

The test plan includes both continuous and interrupted cutting, with two cutting speeds (v_c), three feed rates (f), and two cutting depths (a_p), each conducted under dry conditions and with cutting fluid. For both continuous and interrupted cutting, there are 24 combinations (2 × 3 × 2 × 2), totaling 48 test conditions. Each condition is tested twice to ensure statistical reliability. This design evaluates the influence of process parameters and cutting fluid on tool temperature and heat partitioning. The empirical results will later support the evaluation and validation of simulation models.

3. Model Setup for Tool Temperature Simulation

This paper presents two models for simulating tool temperature during turning with cutting fluid: an analytical model and a numerical model. As discussed in the introduction, the numerical model accurately accounts for the actual tool geometry and predicts tool temperature distribution with an error margin of less than 10%. Therefore, it is used to determine heat flow into tool and derive the temperature distribution within the tool. The numerical simulation results serve as a benchmark for evaluating the accuracy of the analytical model developed in this paper. The following subsections outline the principles of both models and the steps involved in calculating the temperature field.

3.1. Numerical Model for Temperature Simulation Considering The Cutting Fluid

This work developed a numerical model using the commercial FEM software ABAQUS version 2021. The geometric 3D model of the tool was first measured with an Alicona (Raaba, Austria) Infinite Focus G5 microscope and then imported into the simulation software via a STEP file. The heat source on the tool surface corresponded to the contact area between the tool and the chip, determined through a Keyence (Neu-Isenburg, Germany) VHX optical microscope based on the color change of the tool surface, as shown in Figure 5.

In machining processes, heat is generated mainly in three zones: the primary shear zone, the tool–chip interface, and the tool-workpiece interface. The primary shear zone produces heat due to plastic deformation as the material shears to form a chip. At the tool–chip interface, known as the secondary deformation zone, friction between the tool and chip generates significant heat. Additional heat arises at the tool flank face, or the tool-workpiece interface, due to friction with the newly machined surface [26]. Among these zones, the tool–chip interface is the main point where heat transfers to the tool, as this area has the largest contact surface and highest temperature gradient. However, heat in this zone is not uniformly distributed. The heat flux decreases from the cutting edge toward the boundary of the contact area, influenced by variations in friction, contact pressure, and surface topography. With a contact length of less than 0.5 mm, which is very short relative to the cutting tool width of 127 mm, the model was simplified by treating the heat source as uniformly distributed.

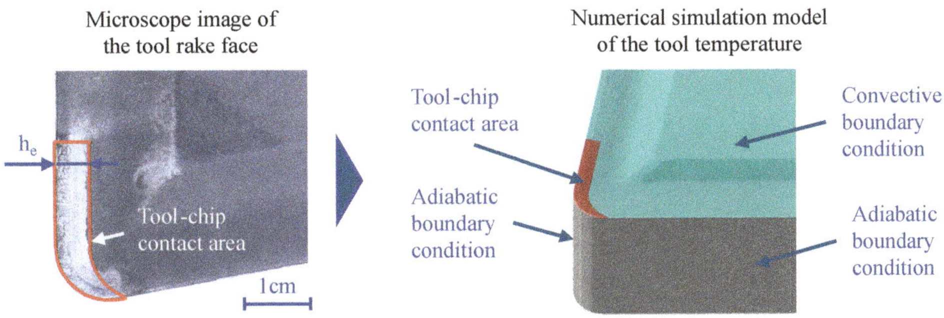

Figure 5. Setup of the heat source and boundary conditions of the numerical temperature model.

On the major flank face of the cutting tool, heat is generated due to friction between the workpiece and the rear surface of the tool. This contact distance is relatively small when the tool is new, but it gradually increases with tool wear, especially as flank-face wear expands. In this study, the tool and experimental conditions were based on an unworn tool, so the heat source distribution on the flank face was minimal compared to that on the rake face. To reduce model complexity and improve calculation efficiency, the heat source distribution on the flank face was therefore ignored.

The cutting fluid primarily cooled the tool rake face, while the rest of the tool remained shielded by the workpiece and did not receive direct cooling. The top surface of the tool (highlighted in Figure 5) was modeled as a convective boundary, and the major and minor flank faces were treated as adiabatic boundaries. Contact heat transfer occurred between the lower and rear surfaces of the cutting tool, which were in contact with the tool holder. The tool holder was maintained at room temperature. Given that the tool holder is made of stainless steel, a heat transfer coefficient of 3.8 $kW/m^2{\cdot}K$ between the cutting insert and holder was assumed [27]. Frekers et al. [28] experimentally observed that, under contact pressures below 10 MPa, the heat transfer coefficient between contacting carbon steels is typically less than 5 $kW/m^2{\cdot}K$, supporting the validity of this assumption. To further improve simulation accuracy in future studies, this assumption should be experimentally verified as proposed in [28]. An implicit solver was used for the temperature simulation, with an initial time increment of 6×10^{-4} s, a maximum increment of 0.6 s, and a minimum increment of 6×10^{-6} s. Time increment control was set to automatic. This step size selection is based on the results of stability tests. To accurately capture the tool's geometric details, a tetrahedral unstructured mesh (DC3D4) with a global mesh size of 0.1 mm was applied, totaling 2,169,379 elements. The mesh configuration follows standard guidelines for cutting simulations to ensure high-resolution temperature distribution [29]. Table 2 summarizes the physical parameters of the tool materials used in the simulation.

Table 2. Properties of H13A carbide (at 27 °C) [22].

Specific Heat Capacity c_p/J/(kg·K)	Density ρ/kg/m^3	Thermal Conductivity κ/W/(m·K)
198	14,800	116

3.2. Analytical Model for Temperature Simulation Considering the Cutting Fluid

The analytical model in this paper assumes the cutting tool to be a rectangular, three-dimensional block heated by a square heat source, as shown in Figure 6. This simplification enables direct integration of the governing equation without requiring a numerical simulation solver. Like the numerical model, the analytical model focuses on cooling of the rake face, applying convective boundary conditions to the upper surface, while the major and minor flank faces are exposed to room temperature air and act as adiabatic boundaries.

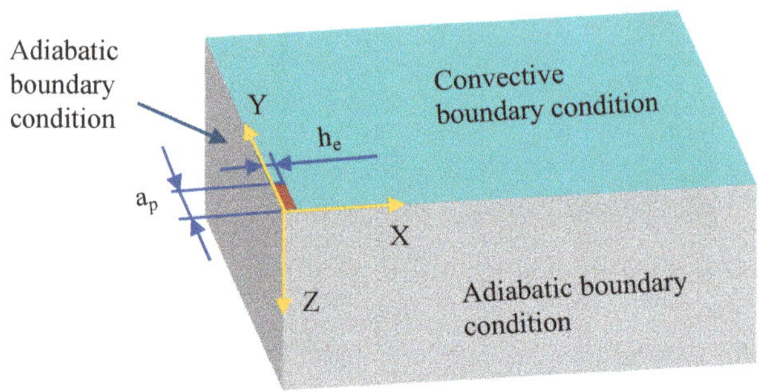

Figure 6. Setup of the heat source and boundary conditions of the analytcial temperature model.

The tool connects to the tool holder in the X, Y, and Z directions, effectively treating it as a semi-infinite body. This boundary condition differs from that in the numerical simulation, which excludes heat transfer between the tool and holder. Since the tool holder remains at room temperature, heat flows continuously through the underside of the cutting tool, approximating an infinite boundary—a common simplification in analytical models [22,30]. However, as ABAQUS 2021 does not support infinite boundary conditions, the numerical model instead applies a contact heat transfer condition. While these boundary differences affect model results, the impact on error is minimal. The evaluation focuses on the steady-state temperature, and the underside of the tool is sufficiently far from the heat source, resulting in a small temperature difference with the surrounding area. The influence of boundary conditions will be explored further in future work.

Equation (1) describes the tool's temperature, where $\dot{q}(\tau)$ represents time-varying heat sources, α is the thermal diffusivity, κ is the thermal conductivity, and h is the convection coefficient. Appendix A provides the derivation of Equation (1).

$$
\begin{aligned}
T(x,y,z,t) = \frac{\alpha}{\kappa} \cdot \int_0^t \frac{1}{4} \cdot \dot{q}(\tau) \cdot \Bigg\{ \frac{2}{\sqrt{4\pi\alpha(t-\tau)}} \exp\left[-\frac{z^2}{4\alpha(t-\tau)}\right] - \frac{h}{\kappa} \exp\left[\frac{h}{\kappa} z + \alpha \frac{h^2(t-\tau)}{\kappa^2}\right] \cdot \\
\operatorname{erfc}\left(\frac{z}{\sqrt{4\alpha(t-\tau)}} + \frac{h}{\kappa}[\alpha(t-\tau)]^{\frac{1}{2}}\right)\Bigg\} \cdot \left\{-\operatorname{erf}\left(\frac{y-a_p}{\sqrt{4\alpha(t-\tau)}}\right) + \operatorname{erf}\left(\frac{y+a_p}{\sqrt{4\alpha(t-\tau)}}\right)\right\} \cdot \\
\left\{-\operatorname{erf}\left(\frac{x-h_e}{\sqrt{4\alpha(t-\tau)}}\right) + \operatorname{erf}\left(\frac{x+h_e}{\sqrt{4\alpha(t-\tau)}}\right)\right\} d\tau + T_r
\end{aligned}
\quad (1)
$$

The temperature distribution was determined by integrating the equation over time, resulting in a first-order integral with a single variable. A key advantage of this simulation model was its simplicity, allowing for temperature distribution calculations at any time without requiring commercial software. In this study, the Python SciPy version 1.14 integrator solved the model, with solution times of approximately 1 min for a 10-s continuous cutting process and around 3 min for interrupted cutting. In contrast, numerical simulations required about 2 h for continuous cutting and 72 h for interrupted cutting. The numerical simulation in this paper uses an industry-standard scheme, though it is not the most efficient from a research perspective. Research by Ulutan et al. indicates that the finite difference method can significantly increase computational efficiency in numerical temperature simulations [31]. However, as this method requires programming or specialized software, it can be challenging for industry to adopt. In the following sections, both numerical and analytical models are used to analyze tool temperature during turning.

4. Experimental and Simulation Results and Discussion

This section discusses the experimental and simulation results in two subsections. First, the process force components and tool temperatures during continuous turning under varying cutting conditions are analyzed. By combining the simulation model with measured temperature data, the heat input to the tool and the heat partition in the cutting zone are determined. The model also accounts for convective cooling, allowing the convective cooling coefficient on the tool rake face from the cutting fluid to be inversely derived from the temperature measurements. The second subsection examines heat transfer and convective cooling during interrupted turning. The constantly changing loads on the tool during interrupted cutting significantly affect the heat input and temperature fluctuations. In addition to analyzing the thermomechanical loads on the tool, the results of both numerical and analytical models are compared, and possible reasons for discrepancies are discussed.

4.1. Results of Continuous Turning

During cutting, the cutting force not only indicates the mechanical load on the tool but also directly influences the temperature in the cutting zone. Approximately 90% of the mechanical energy generated during metal cutting is converted into heat [2]. This mechanical energy can be quantified by the product of the cutting force and cutting speed. Figure 7 illustrates the process force components during continuous turning under various cutting parameters, comparing dry machining with the use of a $p_{cf} = 80$ bar cutting fluid supply.

The force measurement results show that the cutting force increased significantly with higher cutting depth a_p or feed rate f in both dry and fluid-assisted machining. This is due to the larger material removal rate and uncut chip cross-section, both of which require greater energy for chip formation. However, when cutting depth and feed rate were held constant, the influence of cutting speed on the cutting force was minimal. This can be explained by the fact that cutting force is primarily influenced by the uncut chip cross-section, material removal rate, and friction at the tool–chip interface. At higher cutting speeds, thermal softening of the material can occur, facilitating easier material shearing and counteracting any potential force increase, leading to a relatively constant cutting force.

The use of cutting fluid generally reduced cutting force by lowering friction and decreasing the load on the cutting edge. This reduction became more pronounced at lower cutting speeds and greater depths of cut. At lower speeds, the longer contact time allowed the cutting fluid to penetrate the cutting zone more effectively, thereby reducing friction. In contrast, at higher speeds, shorter contact times limited the fluid's effectiveness. Similarly, at greater depths of cut, increased friction and heat generation made the friction-reducing properties of the cutting fluid more critical. In shallow cuts, the lower friction resulted in a less pronounced effect from the cutting fluid.

Figure 8 illustrates the maximum contact thickness (h_e) and contact area (A) between the tool and workpiece under various conditions, measured with an optical microscope (see Figure 5). The maximum tool–chip contact thickness generally increased with feed rate, displaying a consistent trend across conditions. The results show that the contact thickness was approximately 2.5 times the feed. At a feed rate of $f = 0.1$ mm, the contact thickness remained relatively stable with increasing speed. However, at feed rates of $f = 0.2$ mm and 0.3 mm, the contact thickness decreased notably.

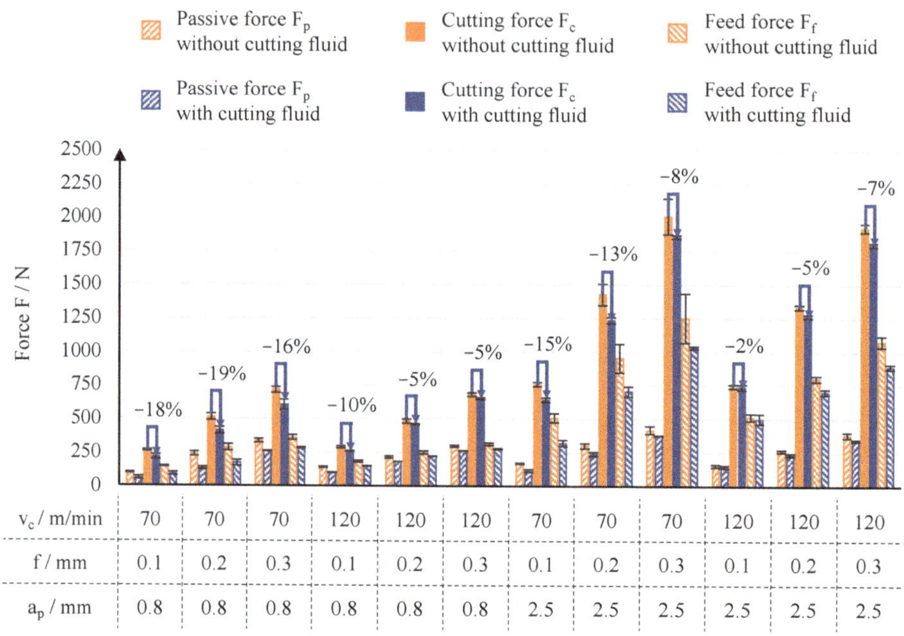

Figure 7. Process force components measured during continuous turning with and without cutting fluid.

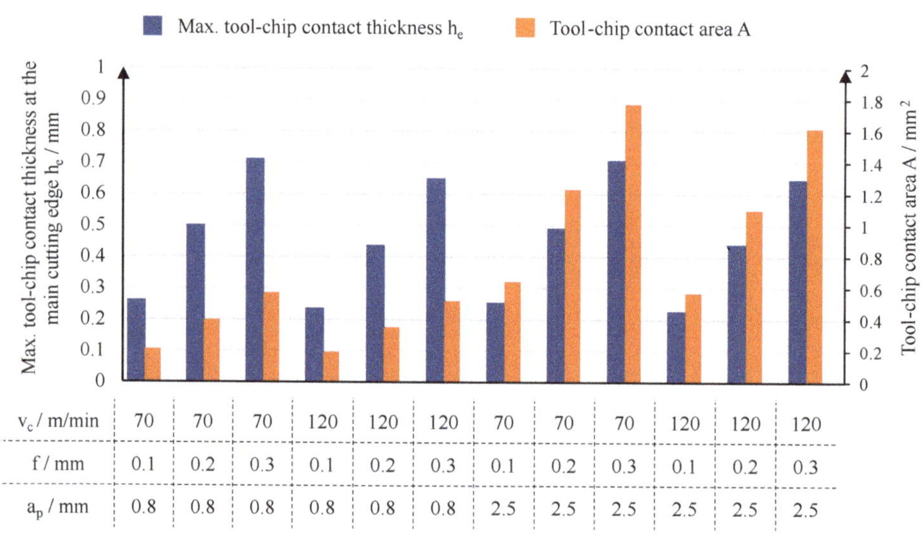

Figure 8. Tool–chip contact thickness and contact area measured using an optical microscope.

The reduction in contact thickness with increasing cutting speed occurred due to several factors. Higher speeds generated more heat at the tool–chip interface, causing thermal softening of the workpiece material. This softening reduced the material's strength, making it easier to shear, which decreased chip compression and shortened the contact length. Additionally, the elevated temperature reduced adhesion between the tool and chip, allowing the chip to slide more easily over the tool surface. Together, thermal softening and reduced adhesion explained the observed decrease in contact thickness at higher speeds, particularly at feed rates of f = 0.2 mm and 0.3 mm.

The orange bars in Figure 8 represent the measured tool–chip contact area, corresponding to the second shear zone. This zone generated significant heat due to friction between the tool and chip. The pressure distribution in this area was uneven, leading to a non-uniform distribution of frictional and contact heat transfer. However, since this contact area was very small (less than 1.8 mm^2) compared to the entire tool surface, the model analysis in this paper assumed uniform heat distribution within this zone. At a cutting depth of a_p = 0.8 mm, the entire contact area remained confined to the tool corner. At a_p = 2.5 mm, the main cutting edge engaged with the workpiece over a larger area, making the feed rate more influential on the contact area due to geometric factors.

Cutting parameters influenced both the contact thickness and the chip shape. Figure 9 illustrates the chip morphology during dry machining under various cutting conditions. At a shallow cutting depth a_p = 0.8 mm and a low feed rate f = 0.1 mm, the chip was thinnest, and due to its low ductility and strength, helical chip segments formed. When the cutting depth remained at a_p = 0.8 mm but the feed rate increased to f = 0.2 mm, long helical chips were produced. At a greater cutting depth a_p = 2.5 mm and a feed rate below f = 0.1 mm, long spiral and snarled chips appeared.

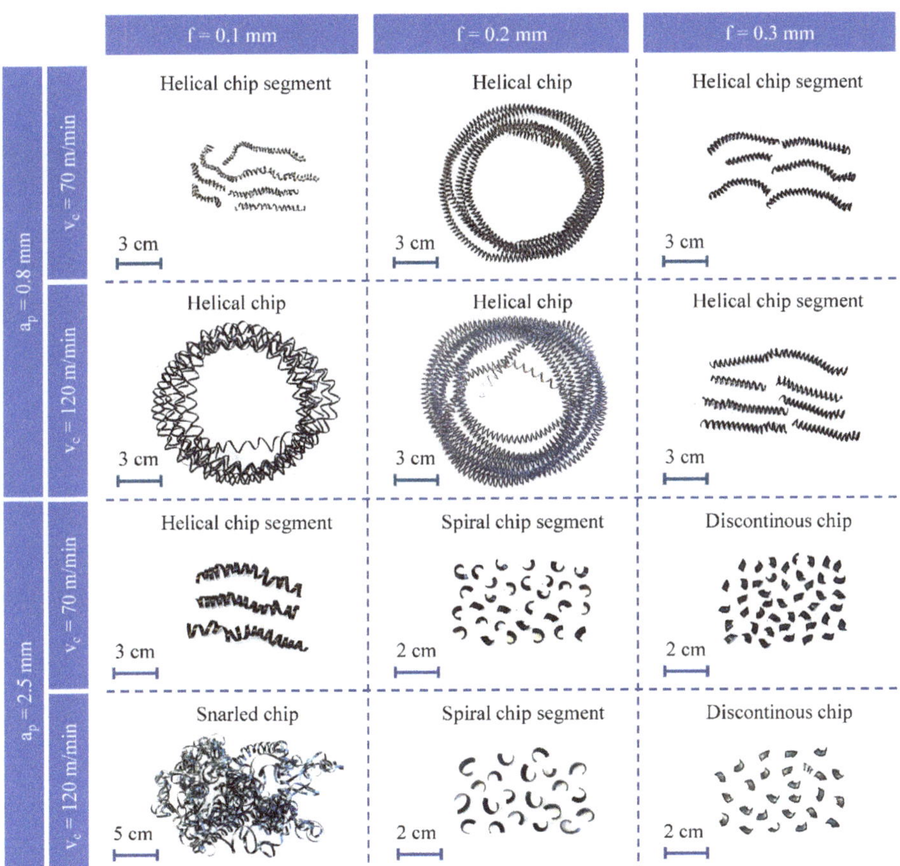

Figure 9. Chip shapes under different cutting conditions during continuous dry turning.

Increasing the feed rate promoted chip breakage as thicker chips formed, leading to higher stress concentrations at the tool–chip interface. The resulting increase in stress and strain rate caused fracture due to enhanced shear forces and ductile sliding fracture. At greater depths of cut, upward-curved spiral chip segments formed because the corner radius had minimal influence, allowing the chip to flow almost orthogonally to the chip groove geometry in the tool corner, causing it to curve upward. Conversely, at lower depths of cut, the corner radius had a stronger effect, resulting in pronounced lateral curvature of the chip and the formation of predominantly helical chips [32].

The cutting force closely related to the temperature in the cutting area, as the product of cutting force and cutting speed represented the total mechanical work of the cutting process. Approximately 90% of this mechanical work converted into heat during cutting [2]. In this study, the heat transferred to the tool was determined indirectly by measuring the tool temperature. Observation points were first defined in the simulation model, corresponding to the thermocouple positions used in the experiment. The heat source was assumed to be uniformly distributed across the tool–chip contact area, as illustrated in Figure 5. The heat source intensity was then iteratively adjusted until the simulated temperature matched the measured temperature. Once this match was achieved, the heat source represented the heat transferred to the tool ($\dot{Q}_{Tool}$). Comparing this value to 90% of the total mechanical work, as described in Equation (2), allowed for the calculation of the heat partition into the tool B_{Tool}.

$$B_{Tool} = \frac{\dot{Q}_{Tool}}{90\% \cdot P_{Mech}} = \frac{\dot{Q}_{Tool}}{90\% \cdot F_c \cdot v_c} \tag{2}$$

Figure 10 presents the measured temperatures alongside the numerical and analytical model results. The temperature T_1, located near the tool nose, was significantly higher than T_2, which was positioned 2 mm away. This occurred because the area near the tool nose was closer to the heat source, where heat accumulated and dissipated less efficiently, leading to higher temperatures. As cutting speed increased, the temperature rose due to increased heat generation. Similarly, increasing the feed rate or cutting depth raised the cutting force, resulting in a higher material removal rate and further elevating the temperature. Guimaraes et al. observed similar trends, noting that tool temperature increased with both cutting speed and cutting depth [33]. However, due to differences in measurement methods, a direct quantitative comparison with the results in this paper was not possible.

A comparison between experimental and simulated temperatures shows that both models accurately captured the temperature differences between the two measurement points, confirming that both the analytical and numerical models effectively represented the temperature distribution near the cutting area. However, the numerical model incorporated detailed tool geometry, while the analytical model simplified it to an orthogonal hexahedron, potentially leading to deviations in estimating heat input to the tool.

Figure 11 illustrates the heat partition to the tool. As the feed rate increased, the heat partition decreased. Similarly, greater cutting depth at the same speed and feed rate further reduced the heat partition to the tool. Overall, the analytical model predicted higher heat input to the tool than the numerical model for the same temperature distribution. This discrepancy arose because the numerical model accounted for the tool rake angle and chip breaker geometry, while the analytical model assumed a flat surface with a 0° rake angle. Additionally, the numerical model used an 80° tool nose, whereas the analytical model assumed 90°. The sharper corners and narrower geometric features in the numerical model led to localized heat buildup, requiring less heat input to match the temperature distribution. The relative difference in heat partition between the two models was around 40%. Specific heat source intensity values are provided in Table 3.

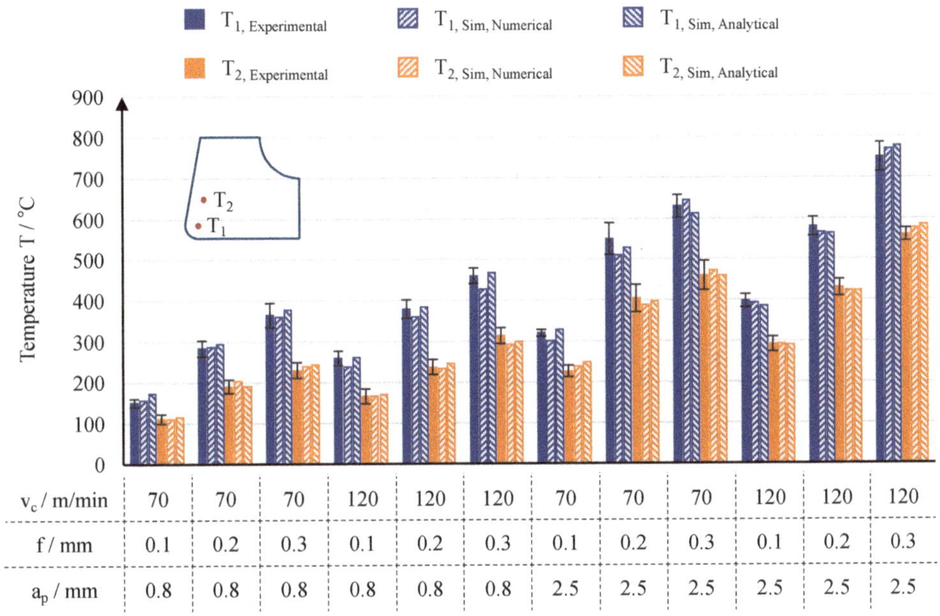

Figure 10. Simulated and measured tool temperature for continuous dry turning.

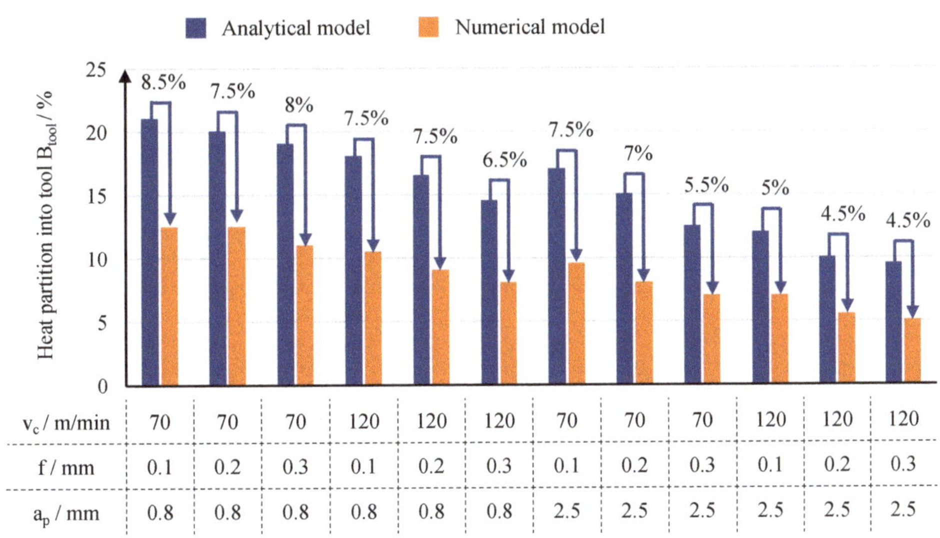

Figure 11. Heat partition into the tool for continuous turning without cutting fluid.

To reduce temperature and promote chip fracture, cutting fluid is commonly used. In this study, a cutting fluid supply pressure of p_{cf} = 80 bar was applied, which is a level typically achievable with the internal pumps of modern CNC machine tools. The cutting fluid influences chip formation through two main mechanisms. First, it reduces friction in the secondary shear zone, increasing chip curvature. Second, it helps dissipate some of the heat generated in the cutting area. The top of the chip is effectively cooled, while the bottom remains mostly unwetted due to close contact with the tool rake face. This creates a temperature gradient that enhances the bending resistance of the chip compared to dry machining, making it more likely to collide with the cutting edge and fracture [3]. Additionally, the force of the cutting fluid jet plays an important role. When directed into the gap between the rake face and the chip, the fluid creates a pressure wedge that lifts the chip, potentially causing it to break off [29].

Table 3. Heat source intensity q as determined by numerical simulation.

v_c	70	70	70	120	120	120	70	70	70	120	120	120
f	0.1	0.2	0.3	0.1	0.2	0.3	0.1	0.2	0.3	0.1	0.2	0.3
a_p	0.8	0.8	0.8	0.8	0.8	0.8	2.5	2.5	2.5	2.5	2.5	2.5
q	164	168	144	287	223	189	118	97	84	164	120	107

Unit: v_c : m/min, f : mm, a_p : mm, q : W/mm^2.

Figure 12 shows the chip shapes under cutting fluid application, using the same cutting parameters as in dry cutting. With cutting fluid supply, all chips broke into small segments. At a cutting depth of a_p = 0.8 mm and a feed rate of f = 0.1 mm, the chip size was approximately 2 mm. As the feed rate increased, the chip size increased slightly. At a cutting depth of a_p = 2.5 mm, chips at feed rates of f = 0.1 mm and 0.2 mm were similar, forming spiral chips with a radius of about 1 cm. However, when the feed rate increased to f = 0.3 mm, the chips became smaller and broke into spiral segments and discontinuous chips.

The frequency and size of chip breakage are directly related to the cooling efficiency of the cutting fluid. When chips broke, the fluid was no longer obstructed by the chip and could flow freely through the cutting area, cooling the cutting area more effectively. Additionally, as the chip curvature increased, the angle between the chip and the tool grew, further facilitating fluid flow in the cutting area and improving cooling performance.

Figure 13 shows the tool temperature at the measurement point when cutting fluid is applied. Compared to the dry cutting temperatures in Figure 10, there is a significant overall decrease in tool temperature. The relationship between tool temperature and cutting parameters follows a similar trend to dry cutting, with temperature increasing as feed rate, cutting depth, and speed rise. In the simulation, the heat partition into the tool was assumed to remain the same as in dry cutting, with only the convective cooling coefficient on the tool surface adjusted. The results indicate that the simulation effectively reflected the temperature distribution under the influence of cutting fluid.

The cooling efficiency of cutting fluid was evaluated using the convective cooling coefficient determined through simulation. Figure 14 presents the convective cooling coefficient obtained via inverse calibration. Both numerical and analytical models showed higher convective cooling at smaller cutting depths, with a slight decrease as the feed rate increased. Smaller cutting depths and feed rates produced smaller chips, which obstructed the cutting fluid less, allowing more effective cooling. As cutting depth or feed rate increased, chip size grew, and chip curvature decreased, covering more of the tool surface and restricting fluid access to the cutting zone, thereby reducing cooling efficiency. The analytical model indicated a greater cooling effect than the numerical model. This discrepancy occurred because the analytical model tended to overestimate the heat transferred to the tool, requiring a stronger cooling effect to match the measured tool temperatures. Nonetheless, the analytical model accurately captured the influence of different cutting parameters on cooling efficiency, and the differences between the two models remained consistent across various conditions.

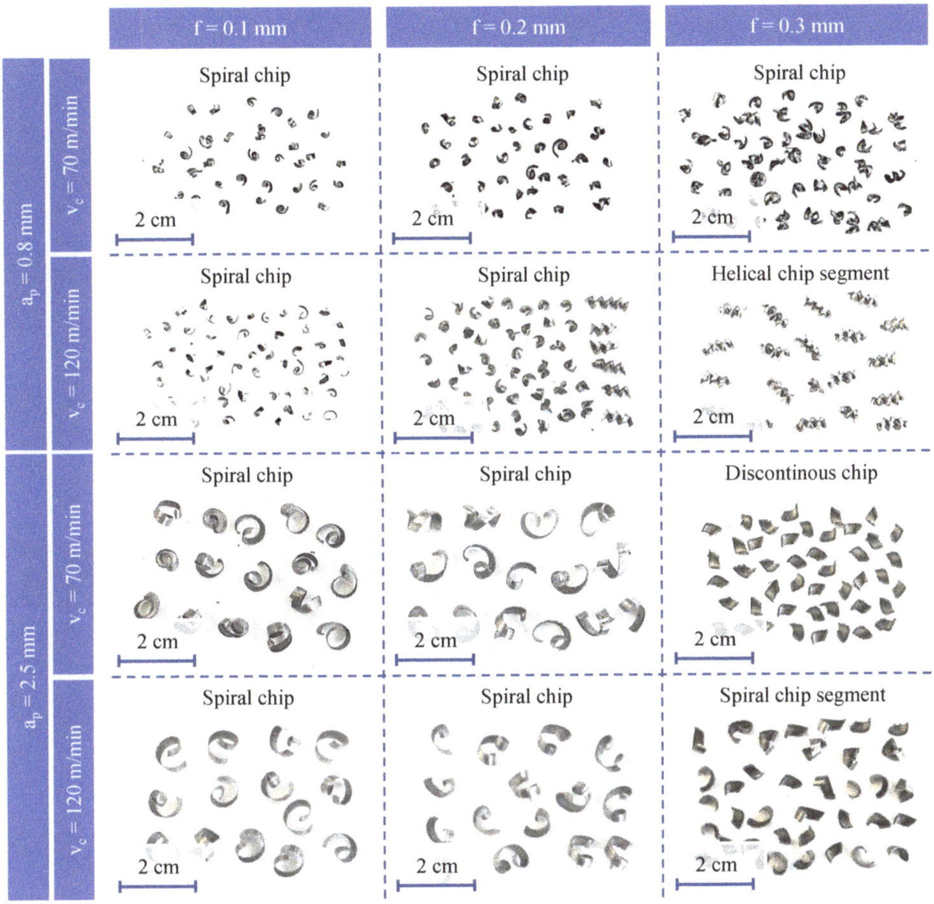

Figure 12. Chip shapes under different cutting conditions during continuous turning with cutting fluid (p_{cf} = 80 bar).

Simulation methods were also be used to analyze the overall temperature distribution of the tool. Figure 15 shows the temperature of the tool rake under the influence of cutting fluid. Both numerical and analytical simulations show similar temperature distributions, with high-temperature areas concentrated in the contact zone between the tool and the chip. Outside this zone, the tool remained close to room temperature due to the cooling effect. Although the measurement point was less than 1.8 mm from the tool surface, the maximum surface temperature was significantly higher than the measured value at this point. These findings suggest that simulations provide a solid foundation for assessing the overall temperature load on the tool, enabling process optimization based on temperature management. Additional temperature distribution results are provided in Appendix B.

While the temperatures between the analytical and numerical models appear to be in good agreement, this agreement is primarily due to the inverse calibration used to fit the measured temperature data rather than any fundamental similarity in the behavior of the models. In particular, each model makes different assumptions about heat partitioning ratios (as shown in Figure 11) and convection coefficients (as shown in Figure 14), resulting in different interpretations of the thermal behavior. The analytical model simplifies the tool–chip interface by applying a uniform heat source distribution and a generalized convective cooling factor, whereas the numerical model incorporates a more complex boundary condition with variable heat transfer parameters that more accurately reflect specific geometric and material interactions. Thus, while the inverse calibration produces similar temperature outputs, this does not imply physical congruence between the two models. Instead, it highlights how each model uniquely balances thermal inputs and boundary conditions to achieve comparable temperature results, and underscores the importance of understanding these differences when interpreting model accuracy and application limitations.

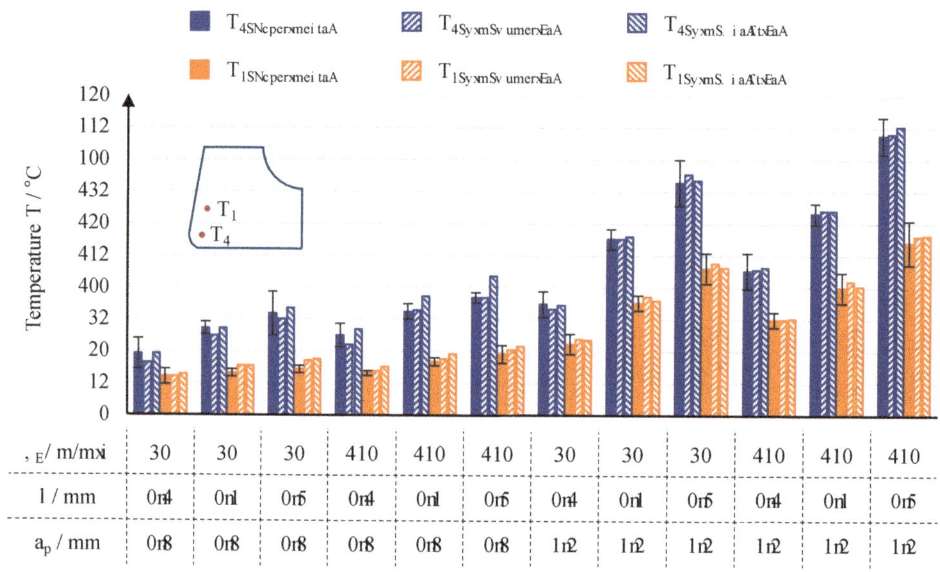

Figure 13. Simulated and measured tool temperature for continuous turning with cutting fluid (p_{cf} = 80 bar).

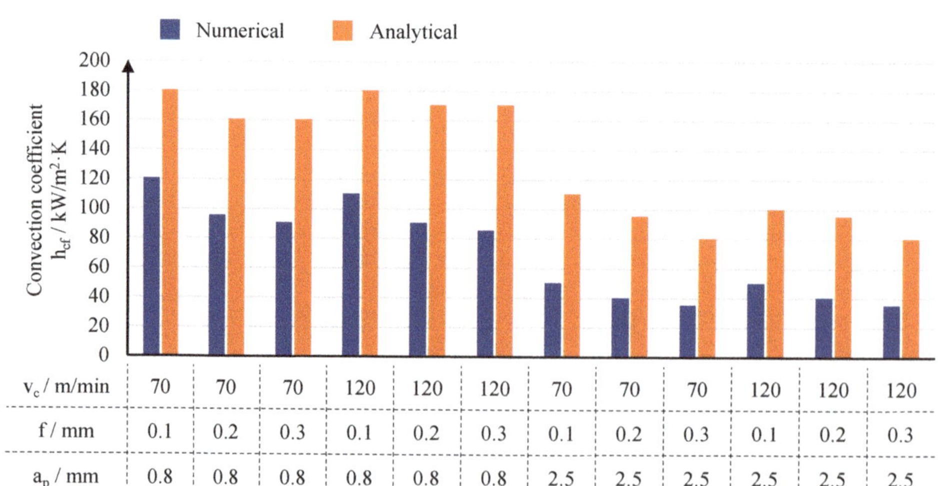

Figure 14. Comparison of the convective cooling coefficient of the tool rake face for continuous turning, determined by the tool temperature using the numerical and analytical models.

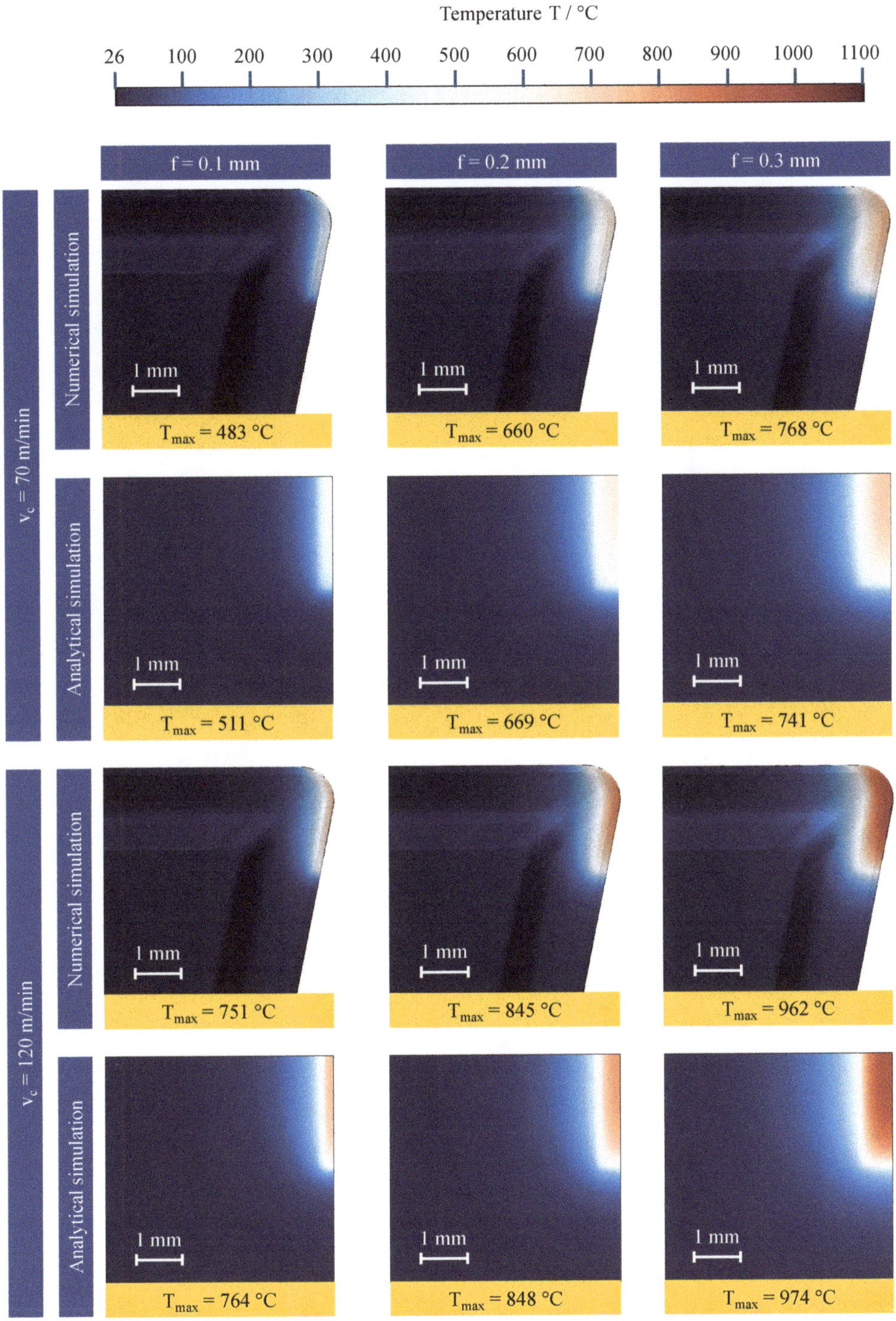

Figure 15. Tool rake face temperature at a_p = 2.5 mm, continuous turning with cutting fluid.

4.2. Results of Interrupted Turning With and Without Cutting Fluid

During interrupted cutting, the tool periodically disengages from the workpiece. This prevents the formation of long continuous chips and allows the tool to be effectively cooled during the non-cutting phase. The aim of investigating interrupted cutting is twofold: to assess its impact on tool forces and temperatures, and to evaluate the accuracy of temperature analysis using simulation and analytical models.

Figure 16 shows the process force components measured during interrupted cutting. Both the magnitude of the forces and their relationship to the cutting parameters are similar to those observed in continuous cutting. This suggests that the cutting force components are primarily determined by the uncut chip geometry, with the interruptions having barely any effect. Although the cutting fluid fully wets the tool during the non-cutting phase, the process forces do not indicate improved lubrication. This is partly because the 8% emulsion Vasco TP 149519 from Blaser Swisslube AG (Rüegsau, Switzerland) is primarily for cooling rather than lubrication. In addition, any lubricating effect is minimal because the cutting process quickly removes the lubricating layer from the tool surface.

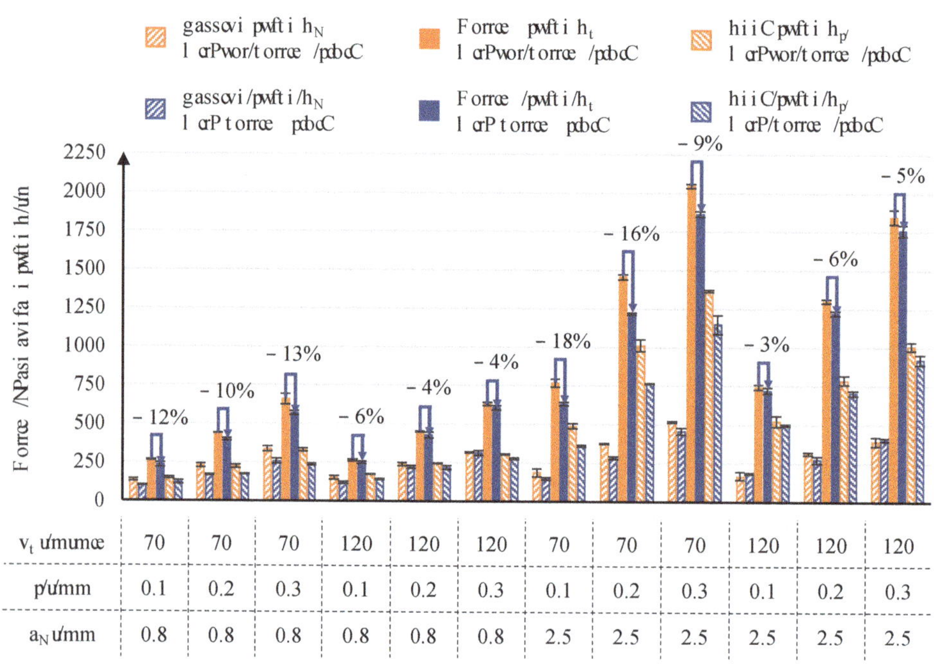

Figure 16. Process force components measured during interrupted turning with and without cutting fluid.

During interrupted cutting, heat generation is intermittent, resulting in fluctuating temperature rises, as shown in Figure 17. Due to the response time of the thermocouple, it captures the temperature fluctuations but cannot fully reflect the actual fluctuation range, making direct temperature comparisons difficult for assessing the heat input into tool. After 8 s of process time, the temperature fluctuation range stabilizes, and the average temperature remains constant. This stabilization time aligns with that measured by Han et al. using a pyrometer [34]. Therefore, the midpoint of the fluctuation range was used for model calibration in this study. Figure 17 compares the calibrated simulated temperatures with the measured data, showing that the simulated temperature fluctuation range is significantly larger than the measured values, which may more accurately reflect the real temperature profile.

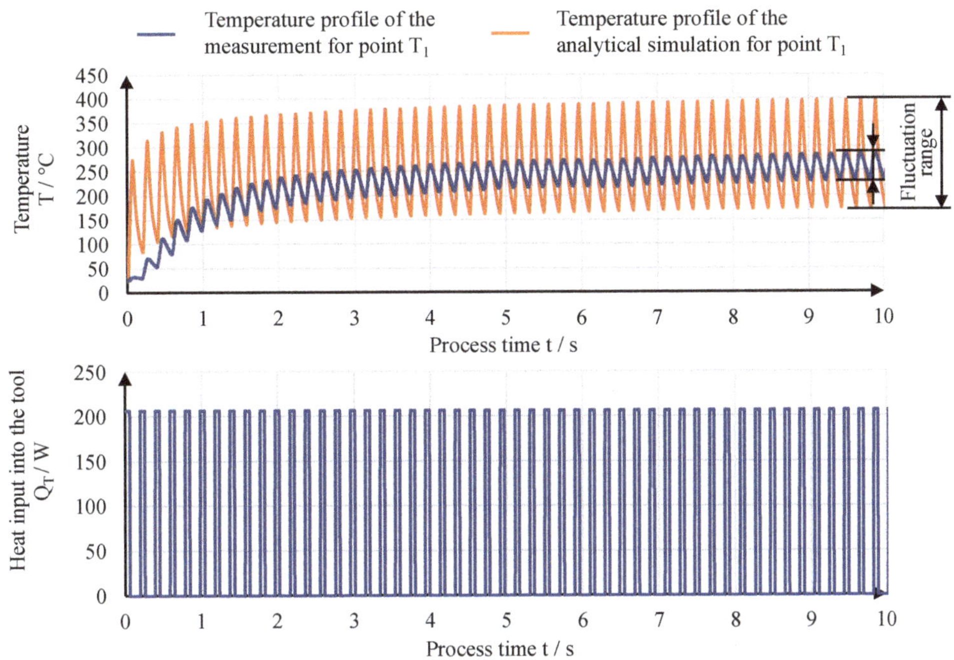

Figure 17. Example of the temperature development from measurement and simulation, and heat input into the tool for the simulation. (v_c = 70 m/min, a_p = 2.5 mm, f = 0.3 mm).

Figure 18 shows the range of temperature fluctuations at the measurement points during dry cutting. The mean temperature at the tool tip increases with cutting depth, speed, and feed rate due to greater heat generation during the cutting process. The mean temperature at point T_1, near the tool corner, is higher than at point T_2, reflecting heat accumulation.

In terms of fluctuation range, the range at T_2 is significantly smaller than at T_1, with both points showing larger fluctuations at higher cutting depth. This indicates that the temperature fluctuation range is greater the closer the point is to the heat source. Additionally, when comparing fluctuation at different cutting speeds, increasing speed slightly reduces the fluctuation range.

A comparison of simulation and experimental results shows that both numerical and analytical simulations accurately capture the median temperature value. However, the simulation results exhibit a significantly larger fluctuation range than the measured values, which is likely due to the previously discussed response time of the temperature sensor. Additionally, the analytical model shows a noticeably larger fluctuation range than the numerical model. This is because the analytical simulation simplifies tool geometry and ignores details of the tool rake face, requiring more heat to reach the same temperature, leading to larger temperature fluctuations.

Differences in the consideration of geometric details lead not only to variations between the analytical and numerical models in terms of temperature fluctuations but also in the amount of heat required to reach the same temperature, as shown in Figure 19. The analytical model requires more heat than the numerical model to achieve the target temperature, which is consistent with the results for continuous cutting. However, for discontinuous cutting, the difference between the heat partition in the analytical and numerical models is within 4%, which is significantly less than for continuous cutting. This suggests that the omission of geometric details in discontinuous cutting has less effect on the heat partition than in continuous cutting.

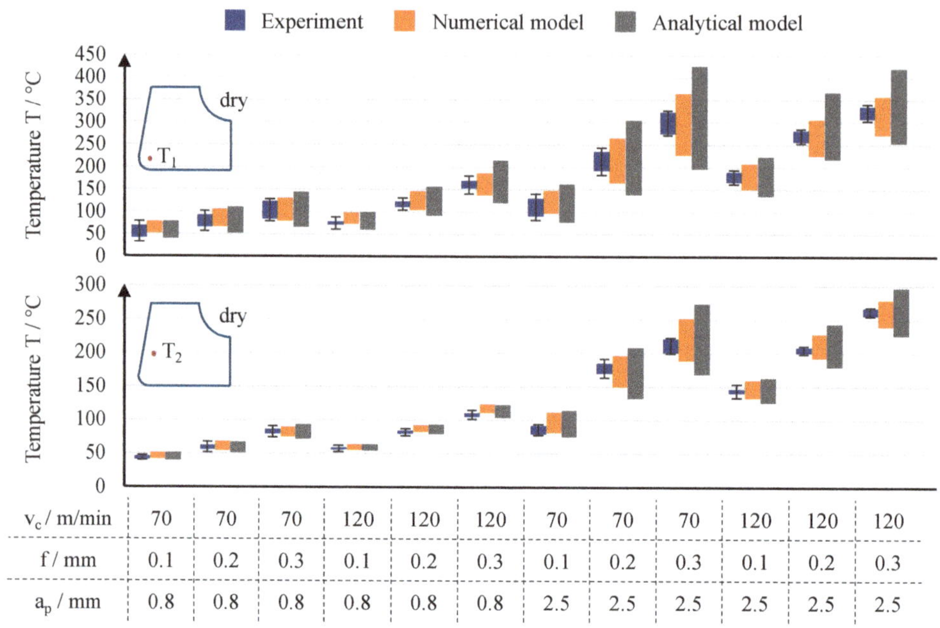

Figure 18. Temperature fluctuations from measurement and simulation for interrupted turning without cutting fluid.

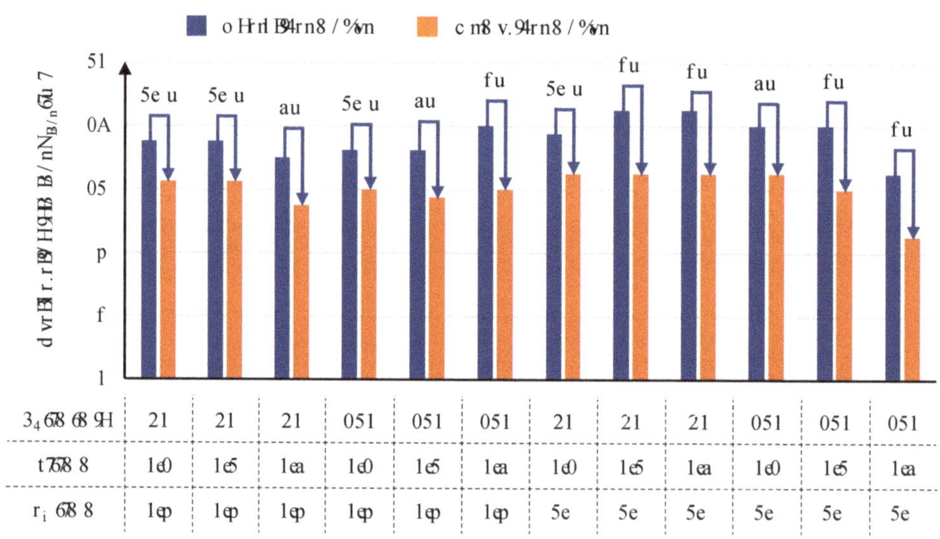

Figure 19. Heat partition into the tool for interrupted turning without cutting fluid.

Geometric details of a tool, such as rake angles, flank angles, and chip breaker, influence how heat accumulates, dissipates, and distributes across the tool's body. In continuous cutting, where heat generation is sustained, these details play a crucial role in temperature distribution. Constant heat input without interruptions leads to a thermal equilibrium where temperature gradients become more pronounced around geometric features, especially those that face higher frictional or cutting forces. As a result, areas like tool edges or corners may experience higher localized temperatures, making the geometric details significantly impactful in controlling temperature behavior.

In contrast, with alternating temperature loads as seen in interrupted cutting, the tool experiences periodic cooling phases between cuts, allowing much of the accumulated heat to dissipate before the next cutting phase. This intermittent cooling diminishes the impact of geometric features on heat retention because each cooling phase moderates the temperature rise across these areas, preventing extreme temperature gradients from forming. The temperature load oscillates rather than accumulating continuously, so the geometric details have less time to impact heat distribution and are less likely to become localized hotspots. For these reasons, geometric details play a lesser role in the thermal dynamics of discontinuous cutting than they do in continuous cutting. This difference is a result of the physical characteristics of heat generation and dissipation unique to each cutting mode, rather than merely a simplification choice in modeling.

Interrupted turning allows the tool to be effectively cooled by the cutting fluid during the non-cutting phase, resulting in a significantly lower overall tool temperature compared to continuous turning. Figure 20 shows the range of tool temperature fluctuation at a cutting fluid pressure of p_{cf} = 10 bar. The experimental results indicate that the mean tool temperature at both measurement points is below 100 °C. At a cutting depth of a_p = 0.8 mm, both points are cooled to near room temperature during the non-cutting phase, with minimal temperature fluctuations. However, at a cutting depth of a_p = 2.5 mm, the temperature fluctuation range increases.

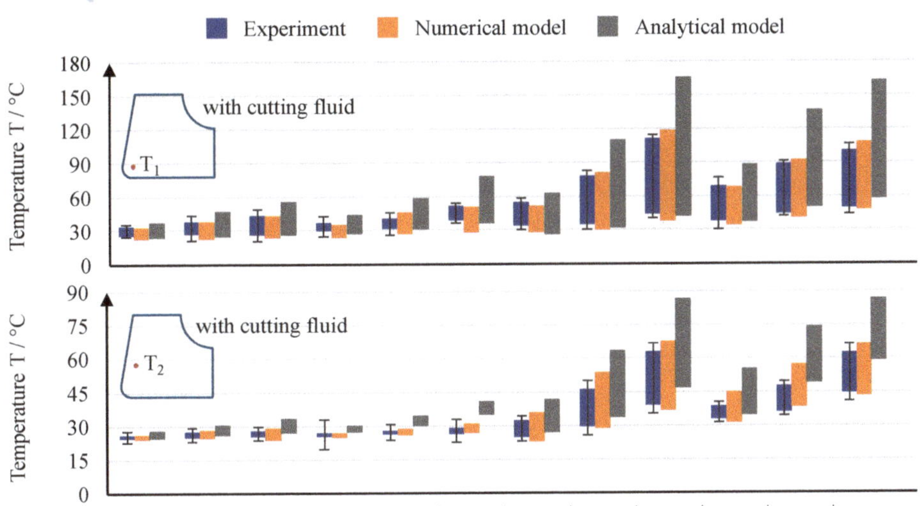

v_c / m/min	70	70	70	120	120	120	70	70	70	120	120	120
f / mm	0.1	0.2	0.3	0.1	0.2	0.3	0.1	0.2	0.3	0.1	0.2	0.3
a_p / mm	0.8	0.8	0.8	0.8	0.8	0.8	2.5	2.5	2.5	2.5	2.5	2.5

Figure 20. Temperature fluctuations from measurement and simulation for interrupted turning with cutting fluid (p_{cf} = 10 bar).

The convective cooling coefficient in the numerical simulation was determined inversely based on temperature data. The simulated mean temperature aligns well with the experimental results. Compared to dry cutting, the difference between the simulated and experimental temperature fluctuation ranges is minimal. To further compare the analytical and numerical models, the analytical model uses the same convective cooling coefficient as the numerical model. In general, the analytical model tended to overestimate the temperature. However, at lower cutting depths the difference was small, at around 10 °C. At a cutting depth of a_p = 2.5 mm, the difference increases with higher speed and feed rate, but the maximum difference in mean values is only about 30 °C. The temperature fluctuation range in the analytical model is also slightly larger than in the numerical model, though the difference is much smaller compared to dry cutting.

Figure 21 shows the convective cooling coefficient determined through numerical simulation. Compared to continuous cutting, the convective cooling coefficient is significantly higher under the same cutting parameters. This is because, during the non-cutting phase, the tool is fully cooled by the cutting fluid. The cutting depth directly affects the convective coefficient: when the depth is large, part of the tool corner is covered by the workpiece, restricting fluid flow into the cutting area and reducing cooling efficiency. Since only small, similarly shaped chips are generated during interrupted cutting, feed rate and cutting speed have little impact on the convective cooling coefficient. A photograph of the chips is provided in Appendix C. Thus, it can be concluded that cooling efficiency in interrupted cutting under flood cooling is primarily influenced by the cutting depth, with minimal effect from cutting speed and feed rate.

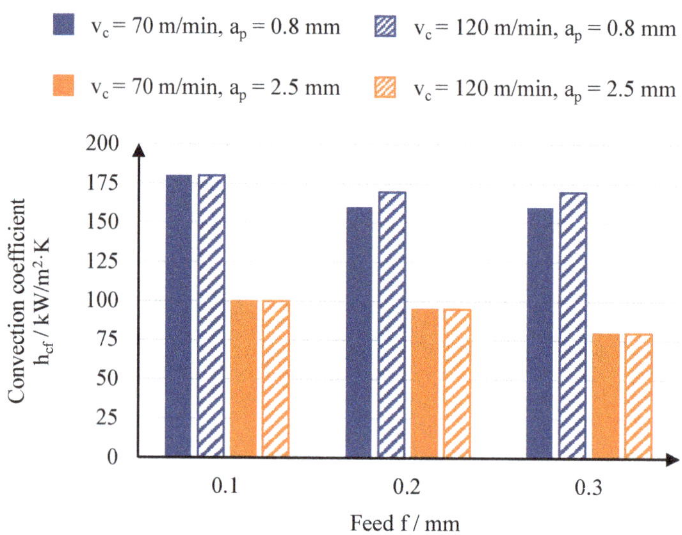

Figure 21. Convective cooling coefficient of the tool rake face for interrupted turning, determined by the tool temperature using the numerical model.

The analysis of interrupted cutting in this subsection reveals that thermocouples cannot fully capture the rapid temperature changes of the tool. To address this issue, the actual temperature fluctuations can be derived inversely using the simulation method outlined in this paper. Additionally, at low cutting depth and feed rates, the difference between the analytical and numerical model results is minimal. As the analytical model is less computationally intensive than the numerical simulation, it shows great potential for use in interrupted cutting analysis.

5. Conclusions and Outlook

This study provides a comprehensive investigation into the tool temperature behavior during both continuous and interrupted turning processes using cutting fluid. The integration of experimental thermocouple measurements with numerical and analytical simulations allowed for a detailed assessment of heat generation, heat partition, and cooling efficiency under varying cutting conditions.

Key findings include the confirmation that thermocouples are effective at capturing steady-state tool temperatures but fall short in detecting rapid fluctuations, particularly in interrupted turning operations. Both numerical and analytical models demonstrated good accuracy in predicting temperature behavior, especially at low cutting depth and feed rates, with the analytical model being computationally more efficient.

The convective cooling coefficient, critical for optimizing tool cooling, was derived from the inverse calibration of experimental data. Interrupted cutting proved advantageous in cooling efficiency, as the tool had opportunities to cool during non-cutting phases. However, the study also highlighted that the analytical model tends to overestimate heat partition, especially when tool geometry complexities are simplified.

In calculating the convective cooling factor, it was assumed that heat partition in the cutting area remains unaffected by the cutting fluid. However, in interrupted cutting with coolant, the wetting effect of the fluid may impact this assumption. The cutting fluid could act as a lubricant between the chip and rake face, altering frictional work and heat generation in the chip, which may affect the accuracy of the cooling coefficient. Additionally, tool temperature under coolant action reflects both heat outflow induced by the cutting fluid and heat inflow from the cutting area. Thus, evaluating the coolant's effect on heat partition based solely on tool temperature remains a complex and unresolved scientific challenge.

The results highlight the potential of analytical models for industrial applications due to their lower computational requirements. However, the numerical model remains more accurate for scenarios requiring detailed geometric considerations. The analytical model proposed in this paper is suitable only for cutting tools with shapes approximating an orthotropic hexahedron; its accuracy diminishes significantly for tools with complex geometries. Additionally, the model cannot calculate the temperature of worn cutting tools, as it does not account for heat sources on the tool flank face. This limitation presents an avenue for future model development. Furthermore, the reliability of the thermocouple-based measurement used in this study requires validation across a broader range of cutting conditions.

Future research should focus on enhancing these models for greater accuracy in complex geometries and transient thermal conditions. First, as discussed in Section 3.1, a systematic comparison and analysis of temperature effects under varying boundary conditions is essential. Additionally, model accuracy could be improved by refining the heat source region to account for changes in tool–chip contact, particularly near the tool edge radius as chip load decreases, which requires a precise mathematical description of heat source distribution. Employing 3D optical scanning, such as with an Alicona microscope, could further enhance accuracy by providing detailed measurements of the contact surface shape.

Accurately defining the heat source on the tool flank face is crucial for analyzing tool temperature under wear conditions. Expanding the model to include complex material properties and wear mechanisms will enhance its relevance in real-world machining. Additionally, this study focuses solely on the tool temperature model, but the method presented can be extended to simulate workpiece and chip temperatures, enabling analysis of overall temperature distribution in the chip region and contact heat transfer between the chip and tool.

Author Contributions: H.L. was responsible for conceptualization, methodology, software, validation, formal analysis, investigation, data curation, visualization, project administration, and original draft preparation. Supervision and funding acquisition were handled by T.B. All authors contributed to writing, review, and editing. All authors have read and agreed to the published version of the manuscript.

Funding: This research was funded by German Research Foundation (DFG) grant number 494849240.

Data Availability Statement: The original contributions presented in this study are included in the article. Further inquiries can be directed to the corresponding author.

Acknowledgments: We appreciate the computing time provided by the NHR Center NHR4CES at RWTH Aachen University (project number p0020236), funded by the Federal Ministry of Education and Research and the participating state governments through GWK resolutions for national high performance computing at universities (www.nhr-verein.de/unsere-partner).

Conflicts of Interest: The authors declare that they have no known competing financial interest or personal relationships that could have appeared to influence the work reported in this paper.

Abbreviations

The following abbreviations are used in this manuscript:

CBN	Cubic Boron Nitride
CFD	Computational Fluid Dynamics
CNC	Computerized Numerical Control
EDM	Electrical Discharge Machining
FEM	Finite Element Method

Appendix A. Derivation of Tool Temperature Model

The model in Figure 6 is governed by the following equation:

$$\frac{\partial^2\theta}{\partial x^2}+\frac{\partial^2\theta}{\partial y^2}+\frac{\partial^2\theta}{\partial z^2}+\frac{1}{\kappa}\cdot g(x,y,z)=\frac{1}{\alpha}\cdot\frac{\partial\theta}{\partial t} \tag{A1}$$

The surface heat source g is defined as:

$$g(x,y,z)=\begin{cases}\dot{q}\cdot\delta(x) & \text{for } 0<x<x_1;\, y_1<y<y_2;\, z=0\\ 0 & \text{for other cases}\end{cases}$$

The boundary conditions are:

$$\left.\frac{\partial\theta}{\partial x}\right|_{x=0}=0;\quad \left.\frac{\partial\theta}{\partial y}\right|_{y=0}=0;\quad -\kappa\left.\frac{\partial\theta}{\partial z}\right|_{z=0}+h\cdot\theta=0$$

The initial condition is:

$$\theta(x,y,z,0)=0$$

Here, θ represents the temperature difference from room temperature. Using the generalized Transient Green's Function Method (TGFM), the solution to Equation (A1) is:

$$\theta(x,y,z,t)=\frac{\alpha}{\kappa}\cdot\int_{\tau=0}^{t}\int_{x'=0}^{x_1}\int_{y'=y_1}^{y_2}\int_{z'=0}^{\infty}G_{X20}\cdot G_{Y20}\cdot G_{Z30}\cdot g(x',y',z',\tau)dz'dy'dx'd\tau \tag{A2}$$

where G denotes the Green's function for specific boundary conditions, and $g(x',y',z',\tau)$ represents the heat source distribution. The variable τ is the effective time of the heat source, and t is the process time. The Green's function for transient heat conduction is well established in the literature for various boundary conditions. According to [35], the Green's function used here is expressed as:

$$G_{X20}(x,t|x',\tau)=\frac{1}{\sqrt{4\pi\alpha(t-\tau)}}\cdot\left\{\exp\left[-\frac{(x-x')^2}{4\alpha(t-\tau)}\right]+\exp\left[-\frac{(x+x')^2}{4\alpha(t-\tau)}\right]\right\} \tag{A3}$$

$$G_{Y20}(y,t|y',\tau) = \frac{1}{\sqrt{4\pi\alpha(t-\tau)}} \cdot \left\{ \exp\left[-\frac{(y-y')^2}{4\alpha(t-\tau)} \right] + \exp\left[-\frac{(y+y')^2}{4\alpha(t-\tau)} \right] \right\} \tag{A4}$$

$$\begin{aligned} G_{Z30}(z,t|z',\tau) &= \frac{1}{\sqrt{4\pi\alpha(t-\tau)}} \cdot \left\{ \exp\left[-\frac{(z-z')^2}{4\alpha(t-\tau)} \right] + \exp\left[-\frac{(z+z')^2}{4\alpha(t-\tau)} \right] \right\} \\ &- \frac{h}{\kappa} \exp\left[\frac{h}{\kappa}(z+z') + \alpha \frac{h^2(t-\tau)}{\kappa^2} \right] \cdot \mathrm{erfc}\left[\frac{z+z'}{\sqrt{4\alpha(t-\tau)}} + \frac{h}{\kappa}\sqrt{\alpha(t-\tau)} \right] \end{aligned} \tag{A5}$$

The coordinates (x,y,z) refer to the observation point, while (x',y',z') indicate the heat source location. The heat source term in the Green's function, modeled as the product of heat source density and a delta function, simplifies to:

$$\int_{-\infty}^{\infty} \dot{q} \cdot \delta(z') G_{Z30}(z,t|z',\tau) dz' = \dot{q} \cdot G_{Z30}(z,t|0,\tau) \tag{A6}$$

Substituting Equation (A6) into Equation (A2) gives:

$$\theta(x,y,z,t) = \frac{\alpha}{\kappa} \int_{\tau=0}^{t} \dot{q} \cdot G_{Z30}(z,t|0,\tau) \cdot \int_{x'=0}^{x_1} \int_{y'=y_1}^{y_2} G_{X20} \cdot G_{Y20} dx' dy' d\tau \tag{A7}$$

where Equation (A5) can be further simplified as:

$$\begin{aligned} G_{Z30}(z,t|0,\tau) = &\frac{2}{\sqrt{4\pi\alpha(t-\tau)}} \exp\left[\frac{z^2}{4\alpha(t-\tau)} \right] - \frac{h}{\kappa} \exp\left[\frac{h}{\kappa} z + \alpha \frac{h^2(t-\tau)}{\kappa^2} \right] \cdot \\ &\mathrm{erfc}\left[\frac{z}{\sqrt{4\alpha(t-\tau)}} + \frac{h}{\kappa}\sqrt{\alpha(t-\tau)} \right] \end{aligned}$$

Since the three-dimensional Green's function problem in Cartesian coordinates can be decomposed into one-dimensional problems, the temperature equation becomes:

$$\theta(x,y,z,t) = \frac{\alpha}{\kappa} \int_{\tau=0}^{t} \dot{q} \cdot G_{Z30}(z,t|0,\tau) \cdot \int_{x'=0}^{h_e} G_{X20} dx' \cdot \int_{y'=0}^{a_p} G_{Y20} dy' d\tau \tag{A8}$$

Integrating Equations (A3) and (A4) separately gives:

$$\int_0^{h_e} G_{X20}(x,t|x',\tau) dx' = \frac{1}{2} \cdot \left[\mathrm{erf}\left(\frac{x+x'}{\sqrt{4a(t-\tau)}} \right) - \mathrm{erf}\left(\frac{x-x'}{\sqrt{4a(t-\tau)}} \right) \right] \Bigg|_0^{h_e} \tag{A9}$$

$$\int_0^{a_p} G_{Y20}(y,t|y',\tau) dy' = \frac{1}{2} \cdot \left[\mathrm{erf}\left(\frac{y+y'}{\sqrt{4a(t-\tau)}} \right) - \mathrm{erf}\left(\frac{y-y'}{\sqrt{4a(t-\tau)}} \right) \right] \Bigg|_0^{a_p} \tag{A10}$$

By substituting Equations (A9) and (A10) into Equation (A8), the temperature Equation (1) is derived.

Appendix B. Simulation Results of Temperature Distribution on the Tool Rake Face

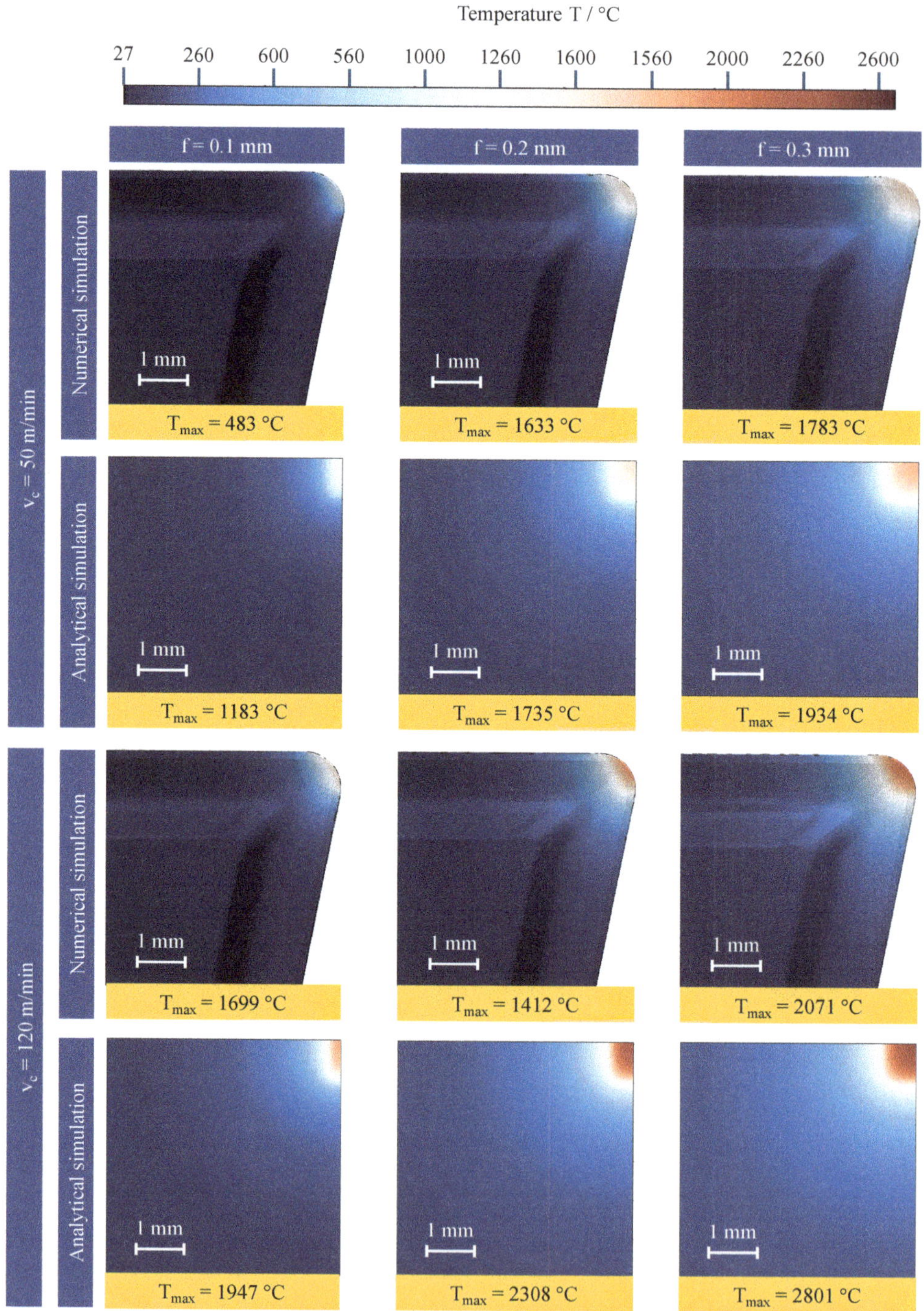

Figure A1. Tool rake face temperature at a_p = 0.8 mm, continuous turning without cutting fluid.

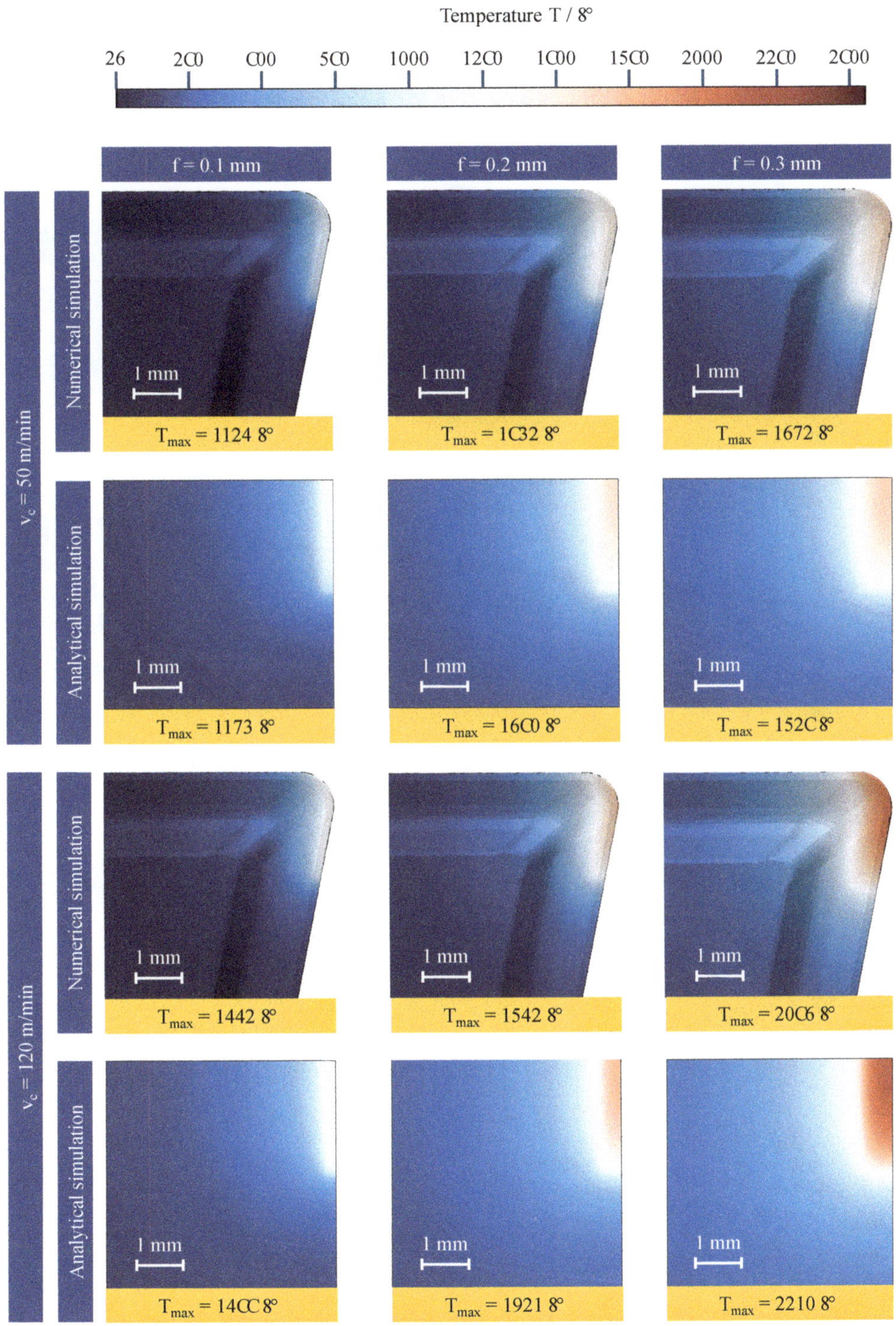

Figure A2. Tool rake face temperature at a_p = 2.5 mm, continuous turning without cutting fluid.

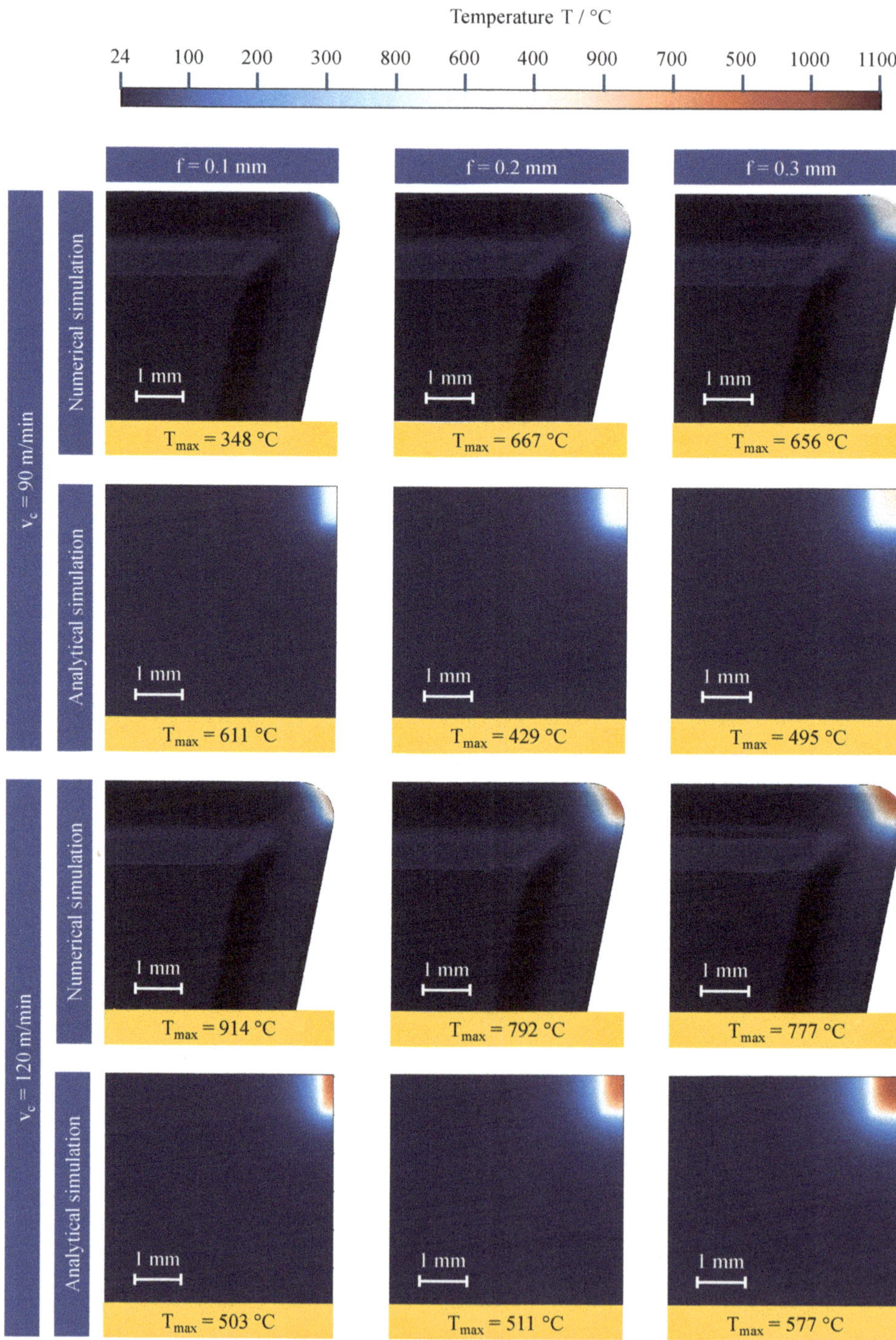

Figure A3. Tool rake face temperature at a_p = 0.8 mm, continuous turning with cutting fluid.

Appendix C. Chip Shape from Interrupted Turning

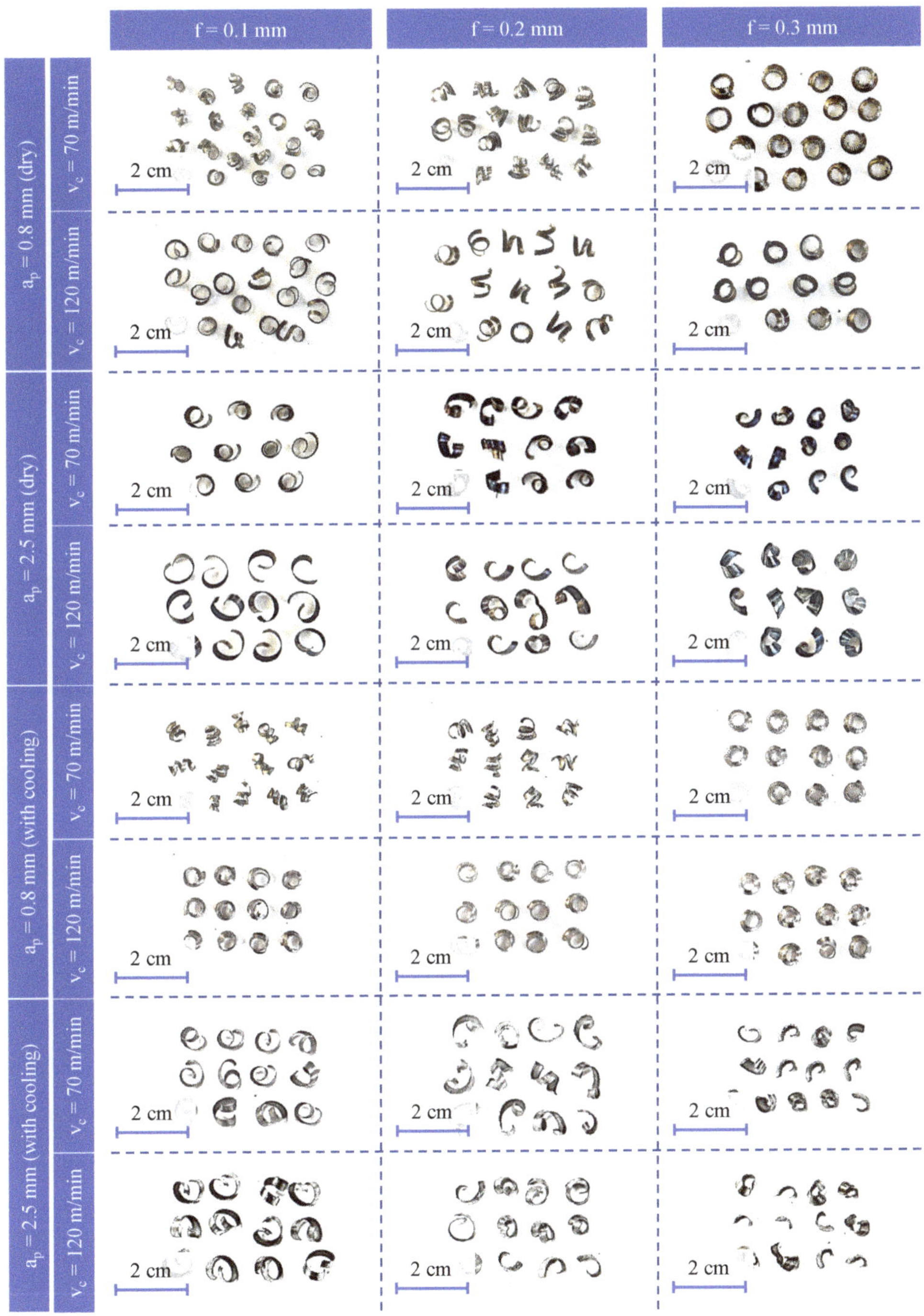

Figure A4. Chip shapes during interrupted turning with and without cutting fluid.

References

1. Taylor, G.I.; Quinney, H. The latent energy remaining in a metal after cold working. *Proc. R. Soc. Lond. A* **1934**, *143*, 307–326.
2. Rittel, D.; Zhang, L.; Osovski, S. The dependence of the Taylor–Quinney coefficient on the dynamic loading mode. *J. Mech. Phys. Solids* **2017**, *107*, 96–114. [CrossRef]
3. Klocke, F. *Manufacturing Processes 1: Cutting*, 1st ed.; Springer: Berlin, Germany, 2013; pp. 78–87.
4. Liu, H.; Meurer, M.; Bergs, T. Surface Integrity Analysis in Orthogonal Milling of Inconel 718. *Procedia CIRP* **2024**, *123*, 191–196. [CrossRef]
5. Jawahir, I.; Brinksmeier, E.; M'Saoubi, R.; Aspinwall, D.; Outeiro, J.; Meyer, D.; Umbrello, D.; Jayal, A. Surface integrity in material removal processes: Recent advances. *CIRP Ann.* **2011**, *60*, 603–626. [CrossRef]
6. Thakur, A.; Gangopadhyay, S. State-of-the-art in surface integrity in machining of nickel-based super alloys. *Int. J. Mach. Tools Manuf.* **2016**, *100*, 25–54. [CrossRef]
7. Schmidt, A., Gilbert, W.; Boston, O. A Thermal-Balance Method and Mechanical Investigation for Evaluating Machinability. *J. Fluids Eng.* **1945**, *67*, 225–232.
8. Cichosz, P.; Karolczak, P.; Waszczuk, K. Review of Cutting Temperature Measurement Methods. *Materials* **2023**, *16*, 6365. [CrossRef]
9. Pereira Guimarães, B.; Da Silva Fernandes, C.; Amaral de Figueiredo, D.; Da Correia Pereira Silva, F.; Macedo Miranda, M. Cutting temperature measurement and prediction in machining processes: Comprehensive review and future perspectives. *Int. J. Adv. Manuf. Technol.* **2022**, *120*, 2849–2878. [CrossRef]
10. Heigel, J.; Whitenton, E.; Lane, B.; Donmez, M.; Madhavan, V.; Moscoso-Kingsley, W. Infrared measurement of the temperature at the tool–chip interface while machining Ti–6Al–4V. *J. Mater. Process. Technol.* **2017**, *243*, 123–130. [CrossRef]
11. Soler, D.; Aristimuño, P.; Buruaga, M.; Garay, A.; Arrazola, P. New calibration method to measure rake face temperature of the tool during dry orthogonal cutting using thermography. *Appl. Therm. Eng.* **2018**, *137*, 74–82 [CrossRef]
12. Barzegar, Z.; Ozlu, E. Analytical prediction of cutting tool temperature distribution in orthogonal cutting including third deformation zone. *J. Manuf. Process.* **2021**, *67*, 325–344. [CrossRef]
13. Chen, L.; Tai, B.; Chaudhari, R.; Song, X.; Shih, A. Machined surface temperature in hard turning. *Int. J. Mach. Tools Manuf.* **2017**, *121*, 10–21. [CrossRef]
14. Kryzhanivskyy, V.; Bushlya, V.; Gutnichenko, O.; M'Saoubi, R.; Stahl, J. Heat flux in metal cutting: Experiment, model, and comparative analysis. *Int. J. Mach. Tools Manuf.* **2018**, *134*, 81–97. [CrossRef]
15. Zhang, J.; Meng, X.; Du, J.; Xiao, G.; Chen, Z.; Yi, M.; Xu, C. Modelling and Prediction of Cutting Temperature in the Machining of H13 Hard Steel of Transient Heat Conduction. *Materials* **2021**, *14*, 3176. [CrossRef]
16. Han, J.; Liu, Z.; Cao, K.; Xu, L.; Shi, T.; Liao, G. Cutting temperature measurement in turning using fiber-optic multi-spectral radiation thermometry and its application in tool wear status recognition. *Measurement* **2022**, *198*, 111413. [CrossRef]
17. Norouzifard, V.; Hamedi, M. A three-dimensional heat conduction inverse procedure to investigate tool–chip thermal interaction in machining process. *Int. J. Adv. Manuf. Technol.* **2014**, *74*, 1637–1648. [CrossRef]
18. Sharma, A.; Tiwari, A.; Dixit, A. Prediction of temperature distribution over cutting tool with alumina-MWCNT hybrid nanofluid using computational fluid dynamics (CFD) analysis. *Int. J. Adv. Manuf. Technol.* **2018**, *97*, 427–439. [CrossRef]
19. Yi, S.; Li, N.; Solanki, S.; Mo, J.; Ding, S. Effects of graphene oxide nanofluids on cutting temperature and force in machining Ti-6Al-4V. *Int. J. Adv. Manuf. Technol.* **2019**, *103*, 1481–1495. [CrossRef]
20. Karaguzel, U.; Bakkal, M.; Budak, E. Modeling and Measurement of Cutting Temperatures in Milling. *Procedia CIRP* **2016**, *46*, 173–176. [CrossRef]
21. Oliveira, G.; Ribeiro, S.; Guimaraes, G. An inverse procedure to estimate the heat flux at coated tool-chip interface: A 3D transient thermal model. *Int. J. Adv. Manuf. Technol.* **2021**, *112*, 3327–3341. [CrossRef]
22. Liu, H.; Rodrigues, L.; Meurer, M.; Bergs, T. A three-dimensional analytical model for transient tool temperature in cutting processes considering convection. *CIRP J. Manuf. Sci. Technol.* **2023**, *43*, 1–14. [CrossRef]
23. Da Silva, L.; Da Silva, O.; Dos Santos, F.; Duarte, F.; Veloso, G. Wear mechanisms of cutting tools in high-speed turning of Ti6Al4V alloy. *Int. J. Adv. Manuf. Technol.* **2019**, *103*, 37–48. [CrossRef]
24. Gariani, S.; El-Sayed, M.; Shyha, I. Optimisation of cutting fluid concentration and operating parameters based on RSM for turning Ti–6Al–4V. *Int. J. Adv. Manuf. Technol.* **2021**, *117*, 539–553. [CrossRef]
25. Hatt, O.; Crawforth, P.; Jackson, M. On the mechanism of tool crater wear during titanium alloy machining. *Wear* **2017**, *374–375*, 15–20. [CrossRef]
26. Astakhov, V. Tribology of Cutting Tools. In *Tribology in Manufacturing Technology*; Springer: Berlin/Heidelberg, Germany, **2013**; pp. 1–66.
27. Madhusudana, C. *Thermal Contact Conductance*, 2nd ed.; Springer: Cham, Switzerland, 2016.
28. Frekers, Y.; Helmig, T.; Burghold, E.; Kneer, R. A numerical approach for investigating thermal contact conductance. *Int. J. Therm. Sci.* **2017**, *121*, 45–54. [CrossRef]
29. Liu, H.; Peng, B.; Meurer, M.; Schraknepper, D.; Bergs, T. Three-dimensional multi-physical modelling of the influence of the cutting fluid on the chip formation process. *Procedia CIRP* **2021**, *102*, 216–221. [CrossRef]
30. Augspurger, T. *Thermal Analysis of the Milling Process*; Apprimus-Verl: Aachen, Germany, 2018.

31. Ulutan, D.; Lazoglu, I.; Dinc, C. Three-dimensional temperature predictions in machining processes using finite difference method. *J. Mater. Process. Technol.* **2009**, *209*, 1111–1121. [CrossRef]
32. Essig, C. *Vorhersage von Spanbruch bei der Zerspanung Mit geometrisch Bestimmter Schneide mit Hilfe Schädigungsmechanischer Ansätze*; Apprimus-Verl: Aachen, Germany, 2010.
33. Guimarães, B.; Rosas, J.; Fernandes, C.M.; Figueiredo, D.; Lopes, H.; Paiva, O.C.; Silva, F.S.; Miranda, G. Real-Time Cutting Temperature Measurement in Turning of AISI 1045 Steel through an Embedded Thermocouple—A Comparative Study with Infrared Thermography. *J. Manuf. Mater. Process.* **2023**, *7*, 50. [CrossRef]
34. Han, J.; Tan, X.; Li, T.; Tang, Z.; Liao, G.; Shi, T. In-situ measurement of cutting edge temperature and its effect on tool wear in turning by a near-infrared fiber-optic two-color pyrometer. *Procedia CIRP* **2021**, *101*, 89–92. [CrossRef]
35. Cole, K. *Heat Conduction Using Greens Functions*, 2nd ed.; Taylor I & Francis Inc.: Milton Park, UK, 2010.

metals

MDPI

Article

Metallic Metamaterials with Auxetic Properties: Re-Entrant Structures

Julian Plewa [1,*], Małgorzata Płońska [1] and Grzegorz Junak [2]

1 Faculty of Science and Technology, Institute of Materials Engineering, University of Silesia in Katowice, 75 Pułku Piechoty Str., 41-500 Chorzów, Poland; malgorzata.plonska@us.edu.pl
2 Faculty of Materials Engineering, Silesian University of Technology, Krasinskiego, 8, 40-019 Katowice, Poland; grzegorz.junak@polsl.pl
* Correspondence: julian.plewa@us.edu.pl

Abstract: The present article is an exploration of metamaterial structures exhibiting auxetic properties. The study shows the effect of three geometric parameters of re-entrant auxetic cells, namely, the internal initial cell angle (θ_0), the strut length ratio h/l, and the degree of opening of the unit cells expressed by the change in the $\Delta\theta$ angle, on the value of the Poisson's ratio. It combines theoretical insights into physical re-entrant auxetic structures with the demonstration of structures that can be subjected to cyclic loading without being damaged. The experimental section features the results of the compression tests of a symmetrical structure made up of four re-entrant cells and tensile tests of a flat mesh structure of size 4×4. In the mesh structure, a modification was applied to the re-entrant cells, creating arched strut connections. It was shown that the value of the maximum load for such structures depends on the bending angle and the length of the inclined strut. The mesh structure was created using torsion springs. Its cyclic tension for different amplitudes yielded Poisson's ratio values in the range of −1.4 to −1.7. These modifications have enabled stable, elastic, and failure-free cyclical changes of the structure's dimensions under load.

Keywords: auxetic structures; re-entrant unit cells; springback

1. Introduction

We are accustomed to the fact that when stretching a solid material, it elongates in the direction of stretching and contracts in the orthogonal direction. This mainly applies to flexible materials and is not easily observed in many other materials, e.g., in most metals, since it is very small. By applying a relatively higher tensile force, a more visible elongation can be achieved just before the tearing of the metal sample. If one applies such tensile force to a so-called metamaterial auxetic structure, it can be observed that it expands in both orthogonal directions when stretched. This effect of mechanical metamaterials is expressed by a negative value of Poisson's ratio ν (NPR), which, in contrast, is positive for most materials. Typically, most materials have a positive Poisson's ratio of about 0.5 for rubber and 0.3 for glass and steel [1].

The concept of a *metamaterial* was propagated in the late 1990s [2–4]. The term itself comes from the Greek words "meta" and "material", where "meta" refers to something that is beyond usual, rearranged, changed, or innovative. Metamaterials are artificially engineered materials that are designed to have properties that are rarely observed in nature. These are, e.g., negative electrical permittivity and negative magnetic permeability, a negative refractive index, negative density, a negative acoustic modulus, a negative elastic modulus, a zero shear modulus, as well as a **negative Poisson's ratio**. Metamaterials are classified into groups such as electromagnetic metamaterials, mechanical metamaterials, and optical metamaterials.

Auxetic structures exhibit an NPR and thus expand orthogonally to the direction of loading instead of contracting. This unique behavior is not caused by the material itself, but by the structure, and thus, it provides entirely new functionality and design possibilities.

In general, metamaterials are marked by the possibility of their material constants taking on negative values. It has been known since a study by Vaselago in 1968 [4] that it is theoretically possible to have a negative refractive index (optical metamaterials). Experimental research on metamaterials covers various fields, including optics, acoustics, and mechanics [5]. Magnetic metamaterials can theoretically exhibit negative values of electric permittivity (ε) and magnetic permeability (μ) [6], whereas thermal metamaterials can exhibit a negative heat transfer coefficient, κ [7]. These examples of artificially designed metamaterials with negative material constants have already borne some experimental fruit, including auxetic structures. In general, these are continuous systems built from repeating structural units [8,9]. From the engineering perspective, mechanical metamaterials constitute a homogenized arrangement of interconnected unit cells.

Nowadays, there are many types of unit cells that are known [10], which form auxetic structures when linked together. Among these, the so-called rotating squares [11], as well as the re-entrant honeycombs [12], are particularly appealing and frequently utilized in research. Structures built from these unit cells have become the most popular types of auxetics. Concave unit cells, as opposed to convex hexagonal honeycomb cells, have a shape resembling a bowtie. Each re-entrant unit cell consists of struts and their joints, with the inclined struts joined at the vertices to form the distinct structural shape of a concave honeycomb structure [12]. At each vertex of the re-entrant cell, the struts (ribs) h and l are joined together to become part of another neighboring cell. Such structures have the form of trusses consisting of struts with movable joints connecting them.

A 2D structure in the form of a multi-element lattice has regularly arranged concave unit cells (Figure 1). A particular property of re-entrant cell structures is that they must be made of elastic material and the arrangement of their unit cells must allow them to shift, i.e., the structure needs to have empty spaces that can be filled as a result of deformation.

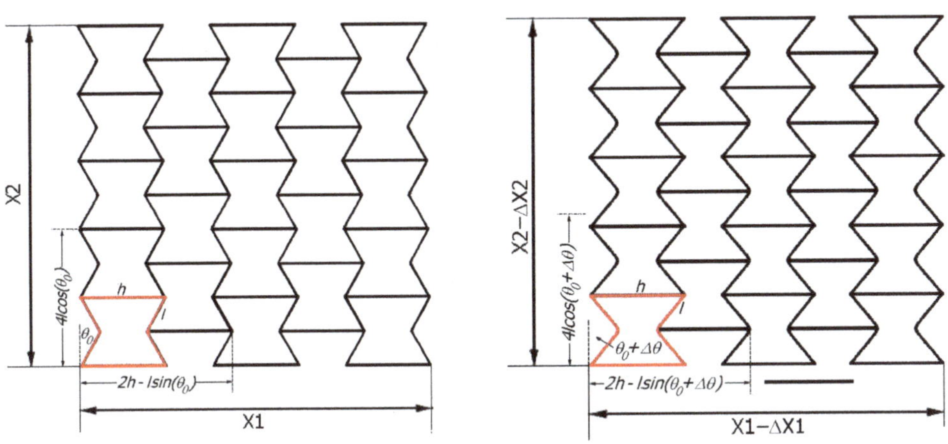

(**a**) (**b**)

Figure 1. Auxetic structure 5 × 5 made of re-entrant cells (bowties) in the initial state (**a**), and after compression (**b**).

Figure 1 compares the initial state of the structure and the deformed state created through compression. This change is especially true for the bending mechanism of the strut joints, although theoretically, a hinged mechanism may be assumed here. Moreover, it should be assumed that the struts do not bend or break during compression. Assuming a rigid connection, the struts can be viewed as bars fixed at both ends. Figure 1 shows the length of the vertical strut (h), the length of the inclined strut (l), the re-entrant angle between the horizontal strut and the inclined strut (θ), and the characteristic linear dimensions of the re-entrant honeycomb structure. The featured diagram of the re-entrant structure (for n = 5, m = 3) shown in Figure 1 considers its initial state with an angle of θ_0, which changes under an external compressive force.

The first theoretical work on structures made of re-entrant unit cells and exhibiting auxetic behavior came from Masters and Evans [13] and Gibson and M. Ashby [14]. They developed a model that describes the Poisson's ratio for small amounts of deformation, also accounting for the linear elastic deformations of the structure. In terms of the nomenclature, the word "auxetic" comes from the Greek word "auxetikos", which means "that which tends to grow", and was proposed by Evans in 1991 [9]. In practice, this applies to materials that increase their transverse dimension under axial tensile stress, resulting in the NPR.

A particular advantage of the structures built from these unit cells is that they can theoretically achieve relatively large linear deformations, especially since, as a result of plastic deformation, under dynamic loading, these materials demonstrate high energy-absorption capabilities [15–17]. These structures owe their unusual properties to both their geometry and the material from which they are made, with each individual unit cell, as well as the cell connections, able to deform by stretching or bending. If there is only reversible deformation under tensile stress, it is due to the elastic properties of the material of the unit cells and their connections. At higher stresses above the yield point, irreversible deformation occurs, which is particularly important for energy absorption. The synthetic parameter for both of these types of deformation is the Poisson's ratio, which is defined as the ratio of the relative change in the linear dimensions of the material in the longitudinal direction and in the direction that is orthogonal to it. This engineering approach indicates that the Poisson's ratio is related to the dimensional change and, therefore, to the shifting of the metamaterial structure. It should be added that the Poisson's ratio is commonly referred to as a material constant related to the elastic properties of the material and corresponds to the range of validity of Hooke's law.

The interest in auxetics stems from their multifaceted potential applications, such as lightweight structures in transportation [18,19], medical devices (knee and hip implants and stents [20,21]), shock absorbers and protective devices, or as sound attenuators [22]. The potential applications of metamaterials also include vibration isolators and thermal insulators [23]. There is also great interest in the possibility of implementing designs of auxetic structures as protective devices to absorb energy and reduce the impact load [23,24].

Mechanical metamaterials can be found in important architectural structures [25], though usually not involving auxetics. It is emphasized, however, that most of the potential applications are only theoretical [26,27]. From this point of view, "auxetic models" are mainly mathematical models with the main purpose of helping to predict auxetic behavior [28].

There is now a huge number of publications available on auxetics that address these structures through theoretical simulation, e.g., [29–34]. This approach involves unreal objects existing in the virtual domain, which only there acquire their unusual properties. Simulation studies enable the analysis of such structures from many different angles. Thus, for example, one of the earlier works [35] has contributed to the understanding of the behavior mechanism of an auxetic structure subjected to loading at different compression velocities. Such studies are expected to contribute to the eventual implementation of auxetic structures that could be utilized in technology.

There is also a very large number of publications devoted to auxetic structures built from re-entrant unit cells. Limiting the discussion to metallic materials, one can mention structures cut from sheet metal [36–41], 3D printed from metal powders and sintered [42–48] or cast [49,50], produced with thin film techniques [23,51], or manufactured using wire [52]. The processes of manufacturing metallic auxetic structures are discussed in detail in a review paper [53], which testifies to the large extent of the research in the field of auxetic metamaterials. It is also emphasized that despite the advanced state of development of these techniques, they have not been applied in manufacturing commercial products to date.

It can be agreed that metallic auxetic metamaterials have a wide application potential due to their superior mechanical properties compared with auxetic materials based on elastomers [50], especially when it comes to high plastic deformation. However, the disadvantage of most auxetics involves difficulties in their manufacturing due to their complex geometric structure. This, together with the lack of structural stability, is the main cited reason for most functional applications remaining in the theoretical stage [28]. The exception is the manufacture and use of auxetic metal foams, which have been pioneered by Lakes [54]. The art of making these structures has been described in an extensive review paper [26], where it was also pointed out that although auxetic foams provide a greater resistance to cracking than conventional foams, most of the potential applications and research work on the former is theoretical.

The present work adopts an engineering approach to auxetic structures of re-entrant cells. It analyzes a model based on geometric transformations, yielding a familiar formula for the Poisson's ratio. By changing the unit cell parameters and finding areas for high negative values of the Poisson's ratio, the design space of these functional materials was expanded.

The physical models of re-entrant auxetic structures assembled from elastic steel provide a complete picture of realistically designed and fabricated auxetic structures. This has demonstrated the possibility of designing and fabricating the required structure through an assembly method that can be readily automated. This can help researchers and designers to develop a variety of applications, including reconfigurable structures that exhibit stable and failure-free operation with cyclic dimensional changes.

Our research aims to describe re-entrant auxetic structures and demonstrate their real-world functioning using fabricated metal model structures as examples. This engineering approach fundamentally differs from the many interesting theoretical works using only simulations. In the belief that only the analysis of real structures can lead to their popularization and real attempts at their application, we present and emphasize the importance of this type of research.

2. Geometric Analysis of an Auxetic Structure of Made of Re-Entrant (Bowtie) Unit Cells

Taking the initial angle θ_0 and the angle change $\Delta\theta$, similar to what has already been performed in the previous work [17], according to the diagram shown in Figure 1, the changes in the linear dimensions are as follows [55]:

$$\frac{\Delta X1}{X1} = \frac{\sin\theta_0 - \sin(\theta_0 \pm \Delta\theta)}{\frac{n}{n-1}\frac{h}{l} - \sin\theta_0}, \tag{1}$$

$$\frac{\Delta X2}{X2} = \frac{\cos(\theta_0 \pm \Delta\theta) - \cos\theta_0}{\cos\theta_0}, \tag{2}$$

where n is the number of unit cells in the horizontal direction. For a large number of unit cells, the following is true:

$$\frac{\Delta X1}{X1} = \frac{\sin\theta_0 - \sin(\theta_0 \pm \Delta\theta)}{\frac{h}{l} - \sin\theta_0}. \tag{3}$$

Formula (3) follows on from the fact that the ratio $n/(n-1)$ tends toward unity, which can be practically applied for $n > 7$. In compression, the angle change is positive ($+\Delta\theta$), while in tension, it is negative ($-\Delta\theta$). This means that stretching leads to a change in the angle according to the following relationship: $\theta_0 + \Delta\theta \rightarrow 0$.

The Poisson's ratio is given by the following relationships:

$$\nu_{12} = -\frac{\frac{\Delta X1}{X1}}{\frac{\Delta X2}{X2}}, \quad \nu_{21} = -\frac{\frac{\Delta X2}{X2}}{\frac{\Delta X1}{X1}}. \tag{4}$$

These are general analytical formulas for describing the Poisson's ratio for auxetic structures made of unit cells and subjected to deformation. The relationships (4) reach high NPR values if their denominator takes very small values. This means that the change in the size in one direction is very small relative to the change in the size in the orthogonal direction.

After substituting (3) into (4), a useful formula for calculating the Poisson's ratio is obtained:

$$\nu_{12} = -\frac{\sin\theta_0 - \sin(\theta_0 \pm \Delta\theta)}{\frac{h}{l} - \sin\theta_0} * \frac{\cos\theta_0}{\cos(\theta_0 \pm \Delta\theta) - \cos\theta_0}. \tag{5}$$

Calculated from Formula (5), the Poisson's ratio is a quantity independent of the number of unit cells in the structure and depends only on its parameters, namely h/l and θ_0, and on the angle change $\Delta\theta$. The lack of a functional relationship between the Poisson's ratio and the number of cells in the structure is particularly useful for many practical application designs, which usually feature a limited number of cells.

Analyzed in Formula (5), the change in angle $\Delta\theta$ is indicative of the movement performed by the structure under loading. In practical terms, the Poisson's ratio requires dynamic dimensional changes. On the other hand, considering the fact that the structure is usually subjected to very small angle changes, within the limit of $\Delta\theta \to 0$, this theoretically means that for $\Delta\theta = 0$ the structure remains locked. In that case, a limit Poisson's ratio value can be calculated using L'Hôpital's rule.

$$\nu_{12}(\text{limit}) = -\frac{\cos^2(\theta_0)}{[\frac{h}{l} - \sin(\theta_0)] * \sin(\theta_0)}, \tag{6}$$

This means that the Poisson's ratio limit ν_{12}(limit) can be approached with very small deformations of the structure. It should be added that Formula (6) is often used, although it gives errors for large deformations.

Formula (6) has the same form as that of a 2D-type structure proposed in an earlier study [14,56] and is often cited by other authors.

Geometric Analysis of a Structure with Dimensions of 5 × 5 (Figure 1)

The study has estimated the effects of the interaction of geometric parameters such as the strut length ratio h/l and angles θ_0 and $\Delta\theta$ with the value of the Poisson's ratio.

Using the formulae given above, it is possible to determine the necessary physical quantities associated with stretching and compressing a given structure. Note that these types of structures can theoretically be stretched to a maximum angle of $\Delta\theta = \theta_0$, and be compressed by an angle of $\Delta\theta = 90^\circ - \theta_{(\text{limit})}$. Whereby $\theta_{(\text{limit})}$ results from the blocking of the inclined struts for the ratio of h/l < 2 and is equal to:

$$\frac{h}{2l} = \sin(\theta_{limit}), \tag{7}$$

For h/l > 2 in compression, the maximum value of the angle change can be as follows: $\Delta\theta = 90^\circ - \theta_0$. Compressive deformation for h/l < 2 can occur up to the critical point at which the inclined struts come into contact.

Using Formula (5), the effect of the geometric parameters of h/l on the value of the Poisson's ratio was analyzed. Two values of the initial angle θ_0 were chosen, and both the processes of compression ($\Delta\theta > 0$) and tension ($\Delta\theta < 0$) were taken into account.

The presented relationships indicate that the Poisson's ratio of the auxetic re-entrant structure strongly depends on the level of strain expressed by $\Delta\theta$. These relationships show a monotonic trend with respect to the changes in angle $\Delta\theta$. It can be seen that at large strains, the value of the Poisson's ratio is not constant and changes markedly throughout the range of deformation. However, its change is monotonic—without any extremes. The negative value of the Poisson's ratio decreases significantly with large compressive deformation and increases strongly in tension.

It can also be seen (Figure 2) that the geometric parameter of the concave cell, i.e., h/l, has a very significant effect on the magnitude of the Poisson's ratio, and for more elongated cells, its negative value increases. From the relationships shown in Figure 2, it can be seen that the analyzed structure achieves a limit value of the Poisson's ratio ν_{limit} according to Formula (6). The highest NPR values are obtained for tension, while for compression, the NPR decreases from the limit value toward zero. In addition, it becomes clear that the more extended the re-entrant unit cells are, the higher the NPR values that can be obtained. Also, there is a strong effect of the θ_0 angle on the NPR. As the angle θ_0 increases, the NPR values decrease. It can be thus concluded that when stretching the structure shown in Figure 1, the value of the NPR increases—i.e., the Poisson's ratio ν12 becomes more negative and, conversely, when compressing the structure, the value of the NPR becomes smaller.

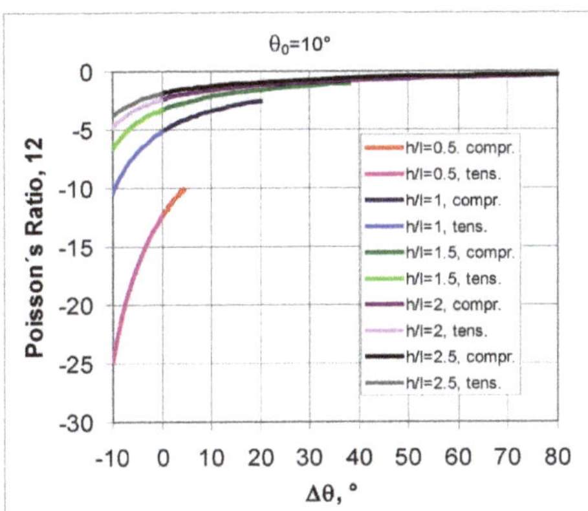

(a)

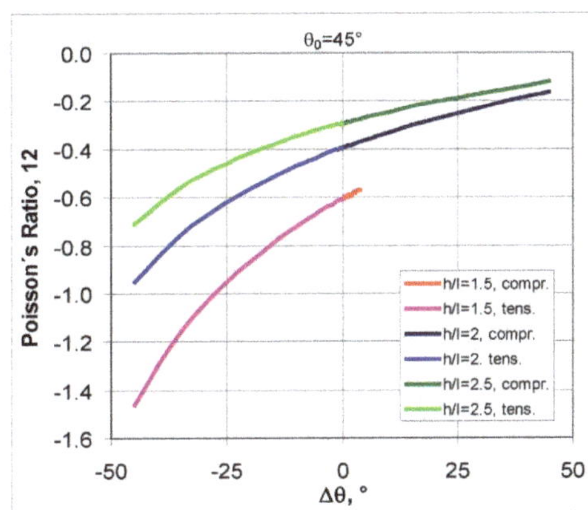

(b)

Figure 2. Changes in the Poisson's ratio as a function of the angle change $\Delta\theta$ for $\theta_0 = 10°$ (**a**) and $\theta_0 = 45°$ (**b**) for a 5 × 5 structure with parameters h/l = 0.5, 1, 1.5, 2, and 2.5.

The relative changes in the dimensions ΔX1/X1 of the structure in the horizontal direction, in the considered range of the angle change $\Delta\theta$ are similar for both compression and tension, and for $\theta_0 = 10°$, they do not theoretically exceed about 40%. However, for $\theta_0 = 45°$, there is a fundamentally big difference between tension and compression. In tension, one can expect a much larger change in the linear dimensions than in compression (Figure 3b).

If one relates the presented theoretical relationships to practical conditions, one should consider small ranges of linear dimensional changes—usually less than 0.1, which are due to the limited elastic properties of the material of the re-entrant unit cells.

Figure 4 compares the relationship between the Poisson's ratio and the relative elongation (+) or relative contraction (−) for this range and in three structures (h/l = 1.5, 2, and 2.5, respectively). This comparison shows that lower angle values θ_0 lead to a higher NPR. On the other hand, flattening the re-entrant unit cells (higher h/l ratio) results in lower NPR values.

The presented relationships indicate a definite trend in the functional relationship between the Poisson's ratio and the h/l parameter, which is also evident for the theoretical value of $\Delta\theta = 0$. For instance, for a structure with an initial angle $\theta_0 = 45°$ and h/l = 1.5 (Figure 4b), one obtains a limit value of the Poisson's ratio (for $\Delta\theta = 0°$ and $\Delta X/X = 0$) equal to $\nu_{12} = -0.605$. In this case, due to the limited possibility of changing the angle in compression, a maximum change in the linear dimensions in the horizontal direction of only $\Delta X1/X1 = 0.063$ can be achieved. Additional practical information that can be derived from the diagrams of Figure 4 is that for the mentioned structure in tension, NPR values higher than -0.605 are obtained, with the reverse being true in compression. In general, one can reiterate that the Poisson's ratio for auxetic re-entrant structures reaches a constant value at very small deformations.

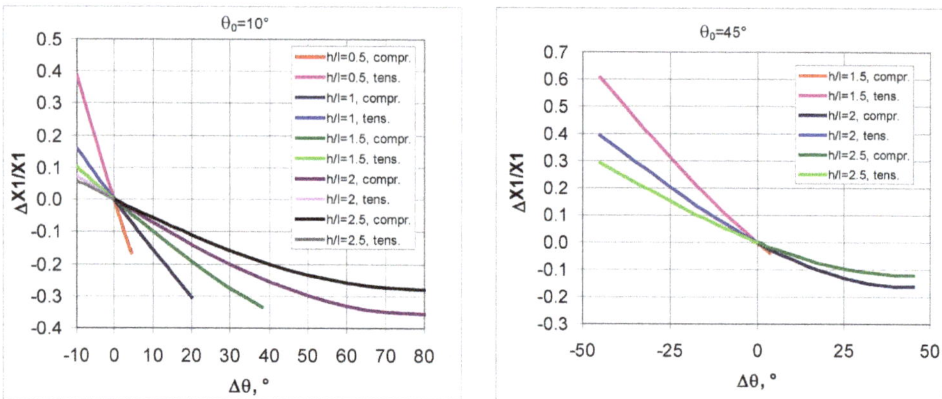

(a) (b)

Figure 3. Relative change in the linear dimensions in the horizontal direction as a function of the angle change $\Delta\theta$ for $\theta_0 = 10°$ (**a**) and for $\theta_0 = 45°$ (**b**).

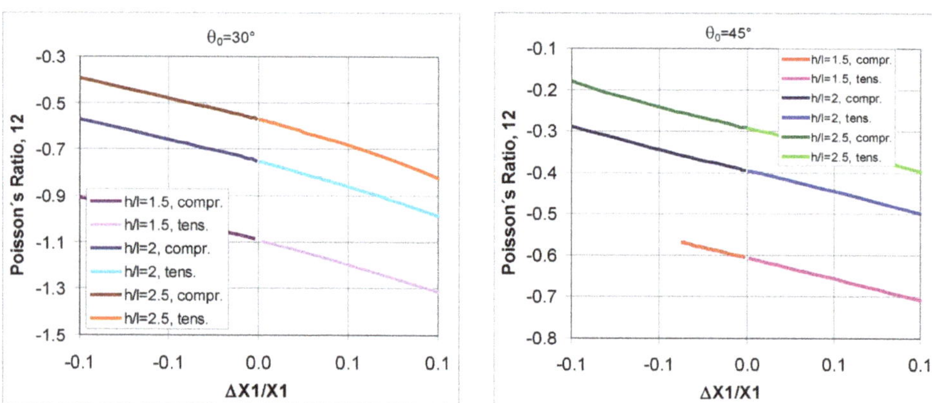

(a) (b)

Figure 4. Changes in the Poisson's ratio as a function of change in the relative linear dimensions in the horizontal direction for two structures, with angle $\theta_0 = 30°$ (**a**) and angle $\theta_0 = 45°$ (**b**).

Another interesting issue is the choice of the shape of the re-entrant unit cells, since their width determines the value of the NPR. Analyzing the structures with the parameter $h/l > 2$ and with the initial angle $\theta_0 = 45°$ one can obtain full ranges ($\Delta\theta = 45°$), which are the same for both compression as well as tension. A detailed interpretation of the relationships considered yields the following conclusions.

Firstly, the NPR values decrease with an increasing parameter of h/l—Figure 5a, whereas compressing the structure gives smaller changes in the linear dimensions in the horizontal direction than in the vertical direction—Figure 5b. Regardless of the h/l value, tension leads to a constant value of dimensional change—Figure 5b, and to smaller NPR values than compression—Figure 5a. On the other hand, large changes in the dimensions in the horizontal direction (ΔX1/X1) accompanying tension correspond to higher NPR values than those accompanying compression—Figure 5d.

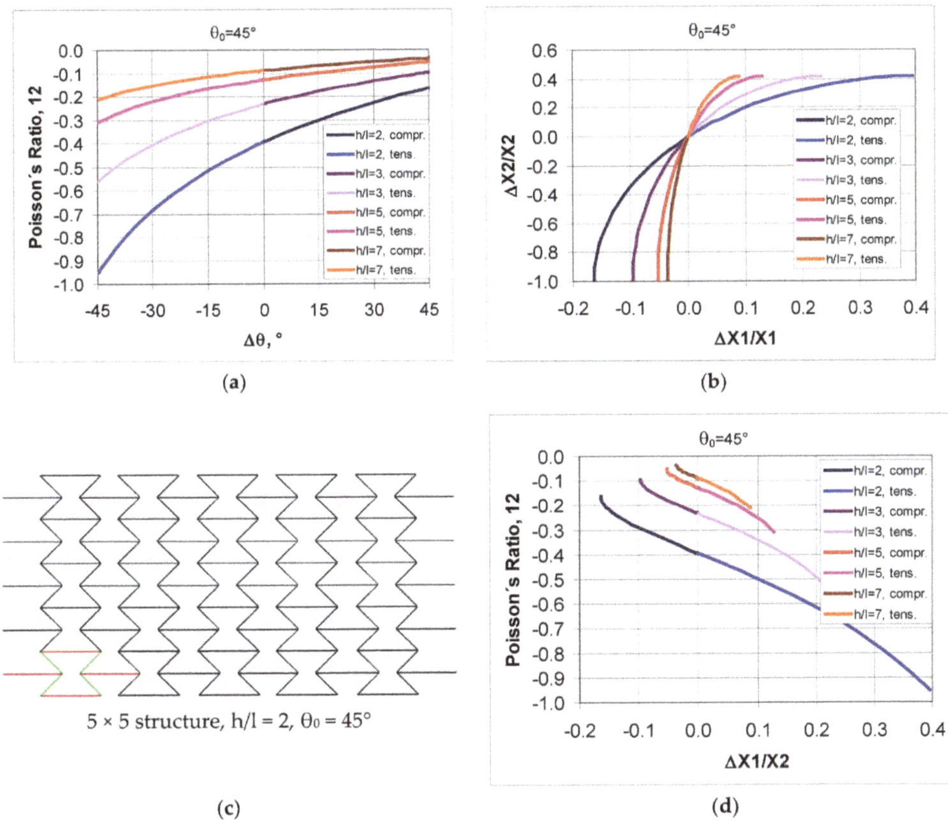

Figure 5. The dependence of the Poisson's ratio on the change in angle Δθ (**a**) and the corresponding relationship between the relative change in the linear dimensions in the horizontal direction (X1) and in the vertical direction (X2)—(**b**), for $h/l \geq 2$, accompanied by the structure's diagram (**c**) and the dependence of the Poisson's ratio on the change in dimensions in the horizontal direction—(**d**).

The diagram of the structure shown in Figure 5c can help one to visualize the stretching process to better attest to the above observations. If the analyzed structures are made of wide re-entrant unit cells ($h/l \geq 2$) and with an inclined angle $\theta_0 = 45°$, relatively small NPR values can be obtained—Figure 5d. However, for tension, NPR values are higher than those for compression. It can be noted that the presented relationships show a monotonic trend of the Poisson's ratio, both as a function of the opening angle Δθ and of the parameter h/l—Figure 5a. Also, there is a monotonic curvilinear relationship between the change in the linear dimensions in the horizontal direction (X1) and in the vertical direction (X2)—Figure 5b. It should be noted that for such structures, the dimensional change in the vertical direction can theoretically reach 100%, while the maximum change in the horizontal direction can reach about 41%.

On the other hand, for structures formed from narrow re-entrant unit cells, i.e., for $h/l < 2$, very high NPR values can be obtained. In this case, however, it is not possible to obtain greater Δθ angle changes than those from Formula (7).

The presented relationships for structures made of narrow re-entrant unit cells (Figure 6a) yield large NPR values—Figure 6b. For example, for a structure made of re-entrant unit cells with the parameters, $h/l = 0.5$ and $\theta_0 = 5°$, theoretically for the angle change $\Delta\theta = -5°$, a Poisson's ratio of −42.4 is obtained in tension, whereas in compression to an angle $\Delta\theta = +4.5°$, the Poisson's ratio is −10.8. It should be noted that the maximum relative elongation in the horizontal direction is much greater than the relative elongation in the vertical direction and, in this case, it is equal to $\Delta X1/X1 = 0.16$ for tension, and $\Delta X1/X1 = -0.3$ for compression.

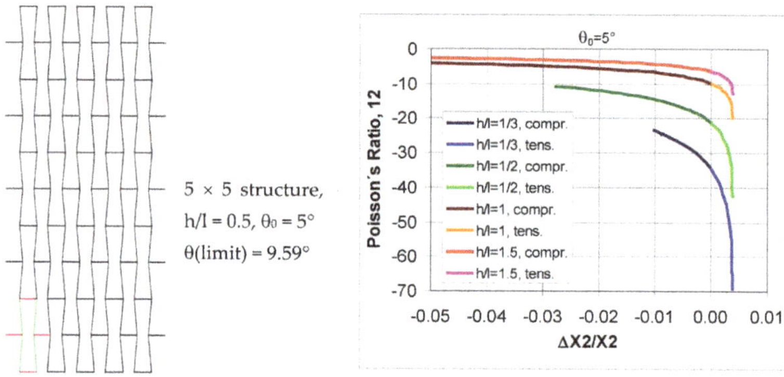

Figure 6. Diagram of the structure (**a**) and the dependence of the Poisson's ratio on the change of linear dimensions in the vertical direction (X2)—(**b**), for $h/l < 2$.

In general, as the h/l parameter decreases, higher NPR values are obtained, with a strong decrease in the elongation values. Taking the Poisson's ratio ν_{12}, it reaches large negative values with very little relative change in the linear dimensions in the vertical direction (Figure 6b)—which stems from θ(limit). While in this case, the NPR values for the parameter $h/l > 1$ do not undergo large changes, they increase strongly for longer unit cells $h/l < 1$—Figure 6b.

The presented geometric analysis of the structure made of re-entrant unit cells shows that such structures exhibit monotonic relationships of the Poisson's ratio, both as a function of the angle change $\Delta\theta$, and as a function of the dimensional change $\Delta X/X$, and that the Poisson's ratio takes negative values. Structures made of re-entrant unit cells thus have auxetic properties, i.e., the NPR effect emerges. To verify these findings, numerous physical models have been assembled. The following section presents selected models of auxetic structures made of metallic materials.

Two types of re-entrant auxetic structures are presented below, with the first type in the form of a spatial model, which is more favorable for the realization of the compression process (free of buckling), while the second type is a flat model chosen for the realization of the tension process.

3. Models of Auxetic Structures

First, the behavior of individual re-entrant unit cells was studied (Figure 7a). Re-entrant-type unit cells made of elastic steel sheet were subjected to compression tests.

The spatial structures were prepared from strips of a Rostfreier Federbandstahl 1.4310 steel sheet, 0.45 mm thick (h + s Präzisions-Folien GmbH, Vohenstrauß, Germany). The strips and the holes in them were produced using a laser, with the sheet strips bent and connected with M1.5 screws to obtain the re-entrant cells. The holes for the connecting screws were 1.6 mm in diameter. The manual fabrication of the structures allowed us to obtain models that are only a practical illustration, and for detailed mechanical studies, a more expert approach is needed for the construction of the structures.

The observed dimensional changes are illustrated in the photographs below. It was demonstrated that a single re-entrant unit cell ($h/l = 1.5$, $\theta_0 = 15°$) subjected to compression can exhibit auxetic behavior as well as bulge.

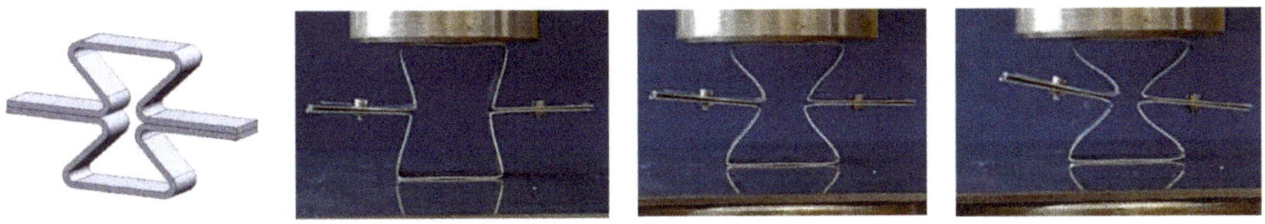

(a) (b) (c) (d)

Figure 7. Schematic diagram of the unit cells (**a**) and a re-entrant unit cell of the steel sheet (parameters $h/l = 30/20$, $\theta_0 = 15°$) in a compression test (**b–d**).

The deformation of the re-entrant steel sheet unit cell shown in Figure 7b consisted of changing its dimensions in the vertical direction by 8 mm (Figure 7c) and by 12 mm (Figure 7d). The observed non-uniform deformation was due to small differences in the unit cell symmetry.

Similar behavior was observed for narrow unit cells ($h/l = 2/3$, $\theta_0 = 15°$) for which the geometric limit is $\theta_{limit} = 19.3°$ (Figure 8).

(a) (b) (c) (d)

Figure 8. Re-entrant cell of the steel sheet ($h/l = 20/30$, $\theta_0 = 15°$) in a compression test (**a–d**).

In this case, the deformation occurred in two ways. Changing the height of the unit cell by 8 mm resulted in either auxetic behavior—Figure 8a,b, or non-auxetic behavior—Figure 8c,d. The difference was caused by the deformation of the short horizontal strut. Its contact surface changed from straight to the shape of an outstretched letter "w" (Figure 8d). It was found that the "w"-shape-forming effect of the horizontal strut can also occur in compression in wider cells, especially when pressure is applied at a particular point or when the bottom of the cell contacting the lower plate of the testing machine leads to this effect.

While a single steel sheet cell does not undergo lateral buckling under compression, planar structures of such cells are easily subjected to it. By combining the four re-entrant unit cells (Figure 9a), a volumetric structure was obtained, which was more appropriate for the squeezing process. A structure of this type is very sensitive to possible inaccuracies in the dimensions and requires very careful manufacturing, especially regarding the horizontal struts being in perfect contact with each other. The structures were made of steel sheets (15 mm wide, 0.25 mm thick) joined to form a volumetric structure, with cell parameters of $h/l = 1.433$, $\theta_0 = 30°$ (Figure 9b) and $h/l = 0.5$, as well as $\theta_0 = 5°$ (Figure 9c).

During the performed measurement, it was found that the compression of the structure from the initial state to state X (locking of the inclined struts, i.e., the 'contact point') corresponding to an angle of 41° occurs within the elastic range, which was connected to the cyclic compression and release.

(a) (b) (c)

Figure 9. Structure 4 × 1 of two different types of re-entrant unit cells, where (**a**) schematic diagram, and of the steel sheet of the unit cells, with cell parameters h/l = 1.433, θ_0 = 30° (**b**), and h/l = 0.5, θ_0 = 5° (**c**).

In this case, one can speak of a springback of the structure, which returns to its original shape when the pressure is released. From the force vs. elongation curves shown below, it is clear that springback changes at different angle values and at different lengths of the inclined cell.

A larger angle θ_0 results in greater springback (Figure 10a), which is expressed by a wider hysteresis loop. Its width is proportional to the elastic energy stored in the structure when compressed.

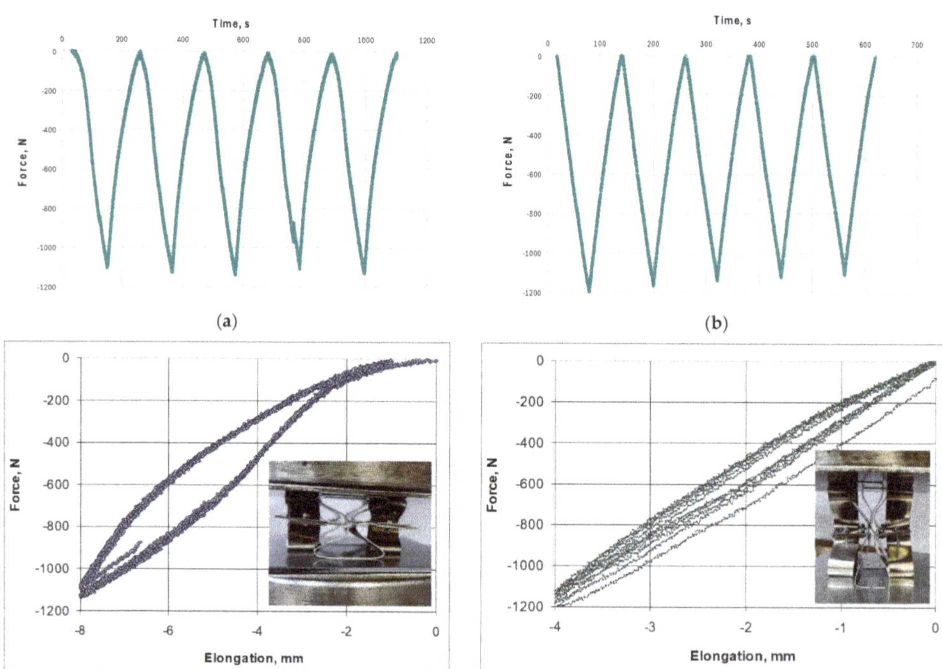

Figure 10. Curves of cyclic compression and release of structures (**a**,**b**), and hysteresis showing changes in the heights of the structures depending on the applied force, for structures with parameters h/l = 1.433, θ_0 = 30° (**c**), and with h/l = 0.5, θ_0 = 5° (**d**), with photographs of the structures after compression—position X.

The cyclic compression process followed a characteristic pattern for both the tested structures. The presented curves for the changes in force over time for the compression process (Figure 10a,b) reflect the differences in the geometry of the structures. One can see a more curvilinear course for a structure with parameters h/l = 1.433, θ_0 = 30° (shorter structure) and smaller values of the force required to squeeze this structure to the X position compared with a longer structure (with parameters h/l = 0.5, θ_0 = 5°). For the longer structure, there was a tendency to continuously decrease the maximum value of the force needed to achieve the given amount of compression (Figure 10b).

A comparison of the experimental relationships in Figure 10c,d shows that the angle θ_0 and the length of the inclined strut l affect the amount of stress and strain. This was also reflected in the width of the obtained hysteresis loop, which was narrower the greater the length of the inclined strut. In the structure with such elongated unit cells (Figure 10d), there was an observed bending of the inclined strut, which meant that a slightly higher force was required to compress this structure. It can also be taken into account that the width of the hysteresis loop was also affected by the fatigue of the material and slight plastic deformation of the structures because after the tests had been completed, a slight shortening of the structures was observed, from 40 mm to 36 mm (Figure 10a) and from 66 mm to 62 mm (Figure 10b).

In position X, for a structure with parameters h/l = 1.433 and θ_0 = 30°, the Poisson's ratio value was estimated to be about −1, and for a structure with parameters h/l = 0.5 and θ_0 = 5°, the Poisson's ratio value was as high as approximately −12. Nevertheless, the force vs. elongation curves obtained from the compression tests testify to the repeatability of the compression and release process, which is characteristic of the elastic range. It has also been found that, in the elastic range, all four connected unit cells were evenly deformed.

In tensile tests, on the other hand, a flexible 4 × 4 structure made of double torsion springs was used (Figure 11). This idea allows for large flexible deformations of metal structures by bending the strut joints more easily. In the initial state, the dimensions of the structure were X1 = 265 mm, X2 = 208 mm, for h = 40 mm, l = 30, θ_0 = 30°.

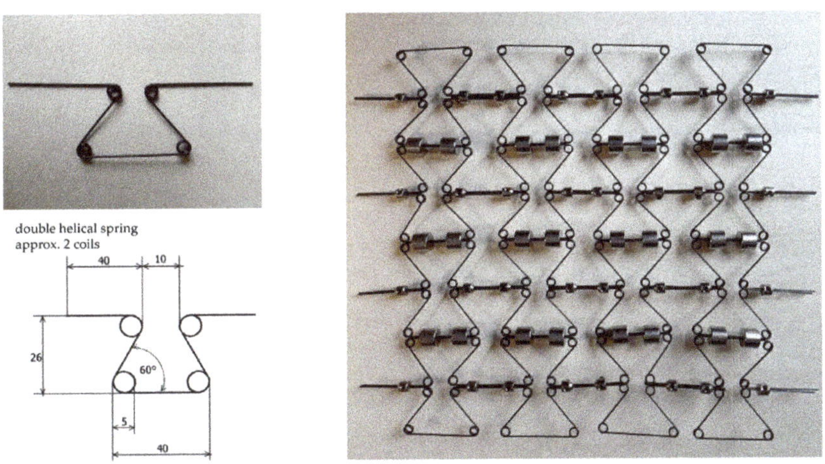

(a) (b)

Figure 11. Torsion spring and its parameters (**a**), and the assembled structure 4 × 4 (**b**).

The components from which the re-entrant cells are made include torsion springs made from 0.8 gauge DH steel wire (Reiner Schmid Produktions GmbH, Solingen, Germany). These concepts arise from the bending mechanism, which is involved in the deformation of the re-entrant structures. Torsion springs exhibit a natural bending mechanism and elastically retain mechanical energy when bent.

Tensile tests were conducted for the given values of maximum elongation in the vertical direction (Figure 12).

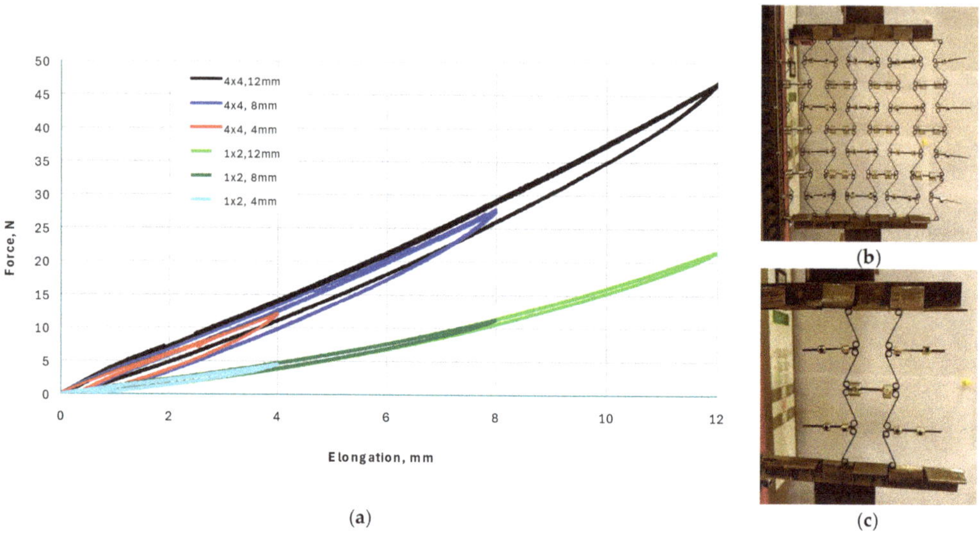

Figure 12. Experimental relationship between tensile force and elongation obtained for cyclic stretching (**a**) and release of structure 4 × 4 (upper curves) and structure 1 × 2 (lower curves), and corresponding photographs from the testing machine (**b**,**c**).

The tensile curves of the structures in the elastic range shown in Figure 12 illustrate successive attempts to increase the amplitude of the tension and release cycle from 4 mm to 12 mm. The tension and release cycle was repeated five times. It can be assumed that the tensile force vs. elongation curves for a given structure coincide. Much less force was needed for stretching a structure made of two unit cells than for stretching a 4 × 4 structure. The slight hysteresis observed may be due to the friction occurring in the torsion springs. It was found that for the range of stretching of the structures, the Poisson's ratio varies slightly from about −1.4 to −1.7.

4. Conclusions

The presented examples of metallic mechanical metamaterials using re-entrant unit cells address a wide range of elastic properties. Through cyclical loading and release, the elastic properties lead to energy accumulation. The occurring changes in the linear dimensions of the structures lead to negative Poisson's ratio values.

The geometric analysis performed showed the role of the h/l parameter in the obtained Poisson's ratio. Large NPR values are possible for low values of h/l, although the initial angle of the structure θ_0 and the degree of opening of the structure expressed by the change in angle $\Delta\theta$ are also relevant here. At the same time, it should be remembered that the re-entrant structures can undergo non-damaging deformation only for certain changes in angle $\Delta\theta$, namely, in tension from the initial angle θ_0 to 0°, and in compression from the initial angle θ_0 to θ_{limit} (for h/l < 2) and up to 90° (for h/l > 2). For structures made of unit cells with h/l < 2, there is a limit angle value θ_{limit}. This angle limit θ_{limit} also cannot be exceeded in the implementation of structures if h/l < 2. The value of this angle is $\theta_{(limit)}$ = arcsin(h/2l) [55]. This entails that the initial angle θ_0 affects the strut length [57].

Both the parameter h/l and the initial angle θ_0 affect the values of the obtained dimensional changes in tension and compression; for example, for θ_0 = 10°, large elongation is obtained for h/l = 0.5 and large compression for h/l = 2.5. On the other hand, high NPR values are possible for structures made of longer unit cells, i.e., for h/l < 1. It is also clear that the re-entrant cell structures exhibit auxetic behavior, i.e., the NPR.

In the produced models, thanks to the solutions applied, there was a reduction in the stress concentration at the strut joints. Both curved connections for structures in compression tests and spring connections for structures in tensile tests have performed well in this regard. A method of assembling structures has been successfully tested, which also has the potential for mass and automated production.

The assembled structures subjected to compression tests, especially those of longer unit cells (Figure 9b), showed little bending of the inclined struts, which, at a larger scale of this phenomenon, can diminish the auxetic effect [58].

Both the structures under compression and those under tension exhibited hysteresis loops of elastic strain energy, with the structures showing little change in linear dimensions after cyclic loading. This may indicate the presence of a small proportion of irreversible deformation. It has been confirmed that the cycle of loading and relaxation can be repeated multiple times, which is known as the low-cycle compression test [59].

The presented experimental models of structures, illustrating the phenomenon of auxetic behavior, should be regarded as an easy-to-implement proposal, with the imperfections occurring causing the noted deviations in the experimental curves (Figures 10 and 12). In further studies, we intend to provide a better instrumentation for making elementary cells, as well as for combining them and assembling model structures.

In summary, the auxetic structures studied could potentially be used as "shock absorbers" for dissipating mechanical energy. In particular, the new concept of unit cells in which the struts are connected with springs can be used to develop new and reconfigurable structures with auxetic properties.

Author Contributions: Conceptualization: J.P.; Methodology: J.P.; Software: G.J. and J.P.; Validation: J.P. and G.J.; Formal Analysis: J.P.; Investigation: J.P. and G.J.; Resources: J.P.; Data Curation: J.P.; Writing—Original Draft Preparation: J.P.; Writing—Review & Editing: J.P. and M.P.; Visualization: M.P.; Supervision: J.P. and M.P.; Project Administration: M.P.; Funding Acquisition: J.P. All authors have read and agreed to the published version of the manuscript.

Funding: The research activities were co-financed by the funds granted under the Research Excellence Initiative of the University of Silesia in Katowice.

Data Availability Statement: The original contributions presented in the study are included in the article, further inquiries can be directed to the corresponding author.

Conflicts of Interest: The authors declare that they have no known conflicting social interests or personal connections that may have arisen and influenced the work reported in this paper. The study is part of a strategic area of interest covering the analysis and assembly of auxetic systems.

References

1. Callister, W.D., Jr. *Materials Science and Engineering. An Introduction*, 7th ed.; John Wiley and Sons Inc.: Hoboken, NJ, USA, 2007.
2. Ziolkowski, R.W. Metamaterials: The early years in the USA. *EPJ Appl. Metamater.* **2014**, *1*, 5. [CrossRef]
3. Pendry, J.; Zhou, J.; Sun, J. Metamaterials: From Engineered Materials to Engineering Materials. *Engineering* **2022**, *17*, 1–2. [CrossRef]
4. Boardman, A. Pioneers in metamaterials: John Pendry and Victor Veselago. *J. Opt.* **2011**, *13*, 020401. [CrossRef]
5. Yu, X.; Zhou, J.; Liang, H.; Jiang, Z.; Wu, L. Mechanical metamaterials associated with stiffness, rigidity and compressibility: A brief review. *Prog. Mater. Sci.* **2018**, *94*, 114–173. [CrossRef]
6. Sun, J.; Liu, X.; Zhou, J.; Kudyshev, Z.; Litchinitser, N.M. All-Dielectric Transition Magnetic Metamaterials. *Sci. Rep.* **2015**, *5*, 16154. [CrossRef]
7. Sklan, S.R.; Li, B. Thermal metamaterials: Functions and prospects. *Natl. Sci. Rev.* **2018**, *5*, 138–141. [CrossRef]
8. Available online: http://people.ee.duke.edu/~drsmith/metamaterials/metamaterials_homogenization.htm (accessed on 20 August 2024).
9. Smith, D.R.; Pendry, J.B. Homogenization of metamaterials by field averaging. *J. Opt. Soc. Am. B* **2006**, *23*, 321. [CrossRef]
10. Elipe, A.C.; Lantada, A.D. Comparative study of auxetic geometries by means of computer-aided design and Engineering. *Smart Mater. Struct.* **2012**, *21*, 105004. [CrossRef]
11. Grima, J.N.; Evans, K.E. Auxetic behavior from rotating squares. *J. Mater. Sci. Lett.* **2000**, *19*, 1563–1565. [CrossRef]

12. Almgren, R.F. An isotropic three-dimensional structure with Poisson's ratio = −1. *J. Elast.* **1985**, *15*, 427–430. [CrossRef]
13. Masters, I.G.; Evans, K.E. Models for the elastic deformation of honeycombs. *Compos. Struct.* **1996**, *35*, 403–422. [CrossRef]
14. Gibson, L.; Ashby, M. *Cellular Solids: Structure and Properties*, 2nd ed.; Cambridge University Press: Cambridge, UK, 1999.
15. Tarlochan, F. Sandwich Structures for Energy Absorption Applications: A Review. *Materials* **2021**, *14*, 4731. [CrossRef] [PubMed]
16. Francisco, M.B.; Pereira, J.L.J.; Oliver, G.A.; da Silva, L.R.R.; Cunha, S.S., Jr.; Gomes, G.F. A review on the energy absorption response and structural applications of auxetic structures. *Mech. Adv. Mater. Struct.* **2022**, *29*, 5823–5842. [CrossRef]
17. Hu, L.L.; Zhou, M.Z.; Deng, H. Dynamic crushing response of auxetic honeycombs under large deformation: Theoretical analysis and numerical simulation. *Thin-Walled Struct.* **2018**, *131*, 373–384. [CrossRef]
18. Han, D.; Ren, X.; Zhang, Y.; Zhang, X.Y.; Zhang, X.G.; Luo, C.; Xie, Y.M. Lightweight auxetic metamaterials: Design and characteristic study. *Compos. Struct.* **2022**, *293*, 115706. [CrossRef]
19. Han, D.; Zhang, Y.; Zhang, X.Y.; Xie, Y.M.; Ren, X. Lightweight auxetic tubular metamaterials: Design and mechanical characteristics. *Comp. Struct.* **2023**, *311*, 116849. [CrossRef]
20. Shirzad, M.; Zolfagharian, A.; Bodaghi, M.; Nam, S.Y. Auxetic metamaterials for bone-implanted medical devices: Recent advances and new perspectives. *Eur. J. Mech.-A Solids* **2023**, *98*, 104905. [CrossRef]
21. Fardan, M.F.; Lenggana, B.W.; Ubaidillah, U.; Choi, S.-B.; Susilo, D.D.; Khan, S.Z. Revolutionizing Prosthetic Design with Auxetic Metamaterials and Structures: A Review of Mechanical Properties and Limitations. *Micromachines* **2023**, *14*, 165. [CrossRef]
22. Yang, C.; Vora, H.D.; Chang, Y. Behavior of auxetic structures under compression and impact forces. *Smart Mater. Struct.* **2018**, *27*, 025012. [CrossRef]
23. Curtis, S.M.; Dengiz, D.; Bumke, L.; Quandt, E. Shape Memory Alloy Thin Film Auxetic Structures. In Proceedings of the International Conference on Shape Memory and Superelastic Technologies, SMST 2022, Carlsbad, CA, USA, 16–20 May 2022; ASM International: San Diego, CA, USA, 2022; pp. 11–13.
24. Isaac, C.W.; Duddeck, F. Recent progress in 4D printed energy-absorbing metamaterials and structures. *Virtual Phys. Prototyp.* **2023**, *18*, e2197436. [CrossRef]
25. Naboni, R.; Mirante, L. Metamaterial computation and fabrication of auxetic patterns for Architecture. *Digit. Manuf. Rapid Prototyp.* **2015**, *2*, 129–136. [CrossRef]
26. Jiang, W.; Ren, X.; Wang, S.L.; Zhang, X.G.; Zhang, X.Y.; Luo, C.; Xie, Y.M.; Scarpa, F.; Alderson, A.; Evans, K.E. Manufacturing, characteristics and applications of auxetic foams: A state-of-the-art review. *Compos. Part B* **2022**, *235*, 109733. [CrossRef]
27. Li, X.; Peng, W.; Wu, W.; Xiong, J.; Lu, Y. Auxetic mechanical metamaterials: From soft to stiff. *Int. J. Extrem. Manuf.* **2023**, *5*, 042003. [CrossRef]
28. Balan, M.; Mertens, J.; Bahubalendruni, M.R. Auxetic mechanical metamaterials and their futuristic developments: A state-of-art review. *Mater. Today* **2023**, *34*, 105285. [CrossRef]
29. Yin, H.; Xiao, Y.; Wen, G.; Qing, Q.; Deng, Y. Multiobjective optimization for foam-filled multi-cell thin-walled structures under lateral impact. *Thin-Walled Struct.* **2015**, *94*, 1–12. [CrossRef]
30. Necemer, B.; Klemenc, J.; Glodež, S. The computational LCF-analyses of chiral and Re-entrant auxetic structure using the direct cyclic algorithm. *Mater. Sci. Eng. A* **2020**, *789*, 139618. [CrossRef]
31. Ghiasvand, A.; Khanigi, A.F.; Guerrero, J.W.G.; Derazkola, H.A.; Tomków, J.; Janeczek, A.; Wolski, A. Investigating the Effects of Geometrical Parameters of Re-Entrant Cells of Aluminum 7075-T651 Auxetic Structures on Fatigue Life. *Coatings* **2023**, *13*, 405. [CrossRef]
32. Dong, Z.; Li, Y.; Zhao, T.; Wu, W.; Xiao, D.; Liang, J. Experimental and numerical studies on the compressive mechanical properties of the metallic auxetic reentrant honeycomb. *Mater. Des.* **2019**, *182*, 108036. [CrossRef]
33. Zhang, J.; Lu, G.; Wang, Z.; Ruan Alomarah, D.A.; Durandet, Y. Large Deformation of an Auxetic Structure in Tension: Experiments and Finite Element Analysis. *Compos. Struct.* **2017**, *184*, 92–101. [CrossRef]
34. Sangsefidi, A.R.; Dibajian, S.H.; Kadkhodapour, J.; Anaraki, A.P.; Schmauder, S.; Schneider, Y. An Abaqus plugin for evaluation of the Auxetic structure performance. *Eng. Comput.* **2022**, *38* (Suppl. S2), S1681–S1704. [CrossRef]
35. Xiao, D. Insight into the negative Poisson's ratio effect of metallic auxetic reentrant honeycomb under dynamic compression. *Mater. Sci. Eng. A* **2019**, *763*, 138151. [CrossRef]
36. Francesconi, L.; Baldi, A.; Dominguez, G.; Taylor, M.L. An Investigation of the Enhanced Fatigue Performance of Low-porosity Auxetic Metamaterials. *Exp. Mech.* **2020**, *60*, 6093–6107. [CrossRef]
37. Amin, F.; Najabat Ali, M.; Ansari, U.; Mir, M.; Minhas, M.A.; Shahid, W. Auxetic Coronary Stent Endoprosthesis: Fabrication and Structural Analysis. *J Appl. Biomater. Funct. Mater.* **2015**, *13*, e127–e135. [CrossRef]
38. Zhou, Y.; Zhou, C.; Shu, Z.; Jia, L.-J. Effect of Two-Dimensional Re-Entrant Honeycomb Configuration on Elastoplastic Performance of Perforated Steel Plate. *Appl. Sci.* **2020**, *10*, 3067. [CrossRef]
39. Nedoushan, R.J. Improvement of energy absorption of expanded metal tubular structures under compressive loads. *Thin-Walled Struct.* **2020**, *157*, 107058. [CrossRef]
40. Arifurrahman, F.; Critchley, R.; Horsfallm, I. Experimental numerical study of auxetic sandwich panels on 160 g of PE4 blast loading. *J. Sandw. Struct. Mater.* **2021**, *23*, 3902–3931. [CrossRef]
41. Box, F.; Johnson, C.G.; Pihler-Puzović, D. Hard auxetic metamaterials. *Extrem. Mech. Lett.* **2020**, *40*, 100980. [CrossRef]
42. Dallago, M.; Raghavendra, S.; Luchin, V.; Zappini, G.; Pasini, D.; Benedetti, M. Geometric assessment of lattice materials built via selective Laser Melting. *Mater. Today Proc.* **2019**, *7*, 7353–7361. [CrossRef]

43. Kranz, J.; Herzog, D.; Emmelmann, C. Design guidelines for laser additive manufacturing of lightweight structures in $TiAl_6V_4$. *J. Laser Appl.* **2015**, *27*, S14001. [CrossRef]
44. Bourell, D.; Kruth, J.P.; Leu, M.; Levy, G.; Rosen, D.; Beese, A.M.; Clare, A. Materials for additive manufacturing. *CIRP Ann. Manuf. Technol.* **2017**, *66*, 659–681. [CrossRef]
45. Yánez, A.; Cuadrado, A.; Martel, O.; Afonso, H.; Monopoli, D. Gyroid porous titanium structures: A versatile solution to be used as scaffolds in bone defect reconstruction. *Mater. Des.* **2018**, *140*, 21–29. [CrossRef]
46. Yang, L.; Harrysson, O.; West, H.; Cormier, D. Compressive properties of Ti–6Al–4V auxetic mesh structures made by electron beam melting. *Acta Mater.* **2012**, *60*, 3370–3379. [CrossRef]
47. Brown, T.D.; Dalton, P.D.; Hutmacher, D.W. Melt electrospinning today: An opportune time for an emerging polymer process. *Prog. Polym. Sci.* **2016**, *56*, 116–166. [CrossRef]
48. Hassanin, H.; Abena, A.; Elsayed, M.A.; Essa, K. 4D printing of NiTi auxetic structure with improved ballistic performance. *Micromachines* **2020**, *11*, 745. [CrossRef] [PubMed]
49. Shivdeo, A.; Hardikar, A.; Bhikule, R.; Wagholikar, S.; Borhade, O.; Walame, M.V. Analysis of 2D Auxetic Metamaterial as a Variable Macro and Micro Structural Filter. *Int. Res. J. Eng. Technol.* **2019**, *6*, 695–700.
50. Ghaedizadeh, A.; Shen, J.; Ren, X.; Xie, Y.M.M. Tuning the Performance of Metallic Auxetic Metamaterials by Using Buckling and Plasticity. *Materials* **2016**, *9*, 54. [CrossRef]
51. Dengiz, D.; Goldbeck, H.; Curtis, S.M.; Bumke, L.; Jetter, J.; Quandt, E. Shape Memory Alloy Thin Film Auxetic Structures. *Adv. Mater. Technol.* **2023**, *8*, 2201991. [CrossRef]
52. He, G.; Tan, Q.; Jiang, G.; Li, Q. A novel mechanism for auxetic behavior in entangled materials with a spiral wire structure. *Smart Mater. Struct.* **2014**, *23*, 095011. [CrossRef]
53. Joseph, A.; Mahesh, V.; Harursampath, D. On the application of additive manufacturing methods for auxetic structures: A review. *Adv. Manuf.* **2021**, *9*, 342–368. [CrossRef]
54. Friis, E.A.; Lakes, R.S.; Park, J.B. Negative Poisson's ratio polymeric and metallic foams. *J. Mater. Sci.* **1988**, *23*, 4406–4414. [CrossRef]
55. Plewa, J.; Płońska, M.; Junak, G. Experimental study of auxetic structures made of reentrant cells ("bow" type). *Materials* **2024**, *17*, 3061. [CrossRef] [PubMed]
56. Smith, C.W.; Grima, J.N.; Evans, K.E. A novel mechanism for generating auxetic behaviour in reticulated foams: Missing rib foam model. *Acta Mater.* **2000**, *48*, 4349–4356. [CrossRef]
57. Levy, O.; Goldfarb, I. Design considerations for negative Poisson ratio structures under large deflection for MEMS applications. *Smart Mater. Struct.* **2006**, *15*, 1459–1466. [CrossRef]
58. Lekesiz, H.; Bhullar, S.K.; Karaca, A.A.; Jun, M.B.G. Mechanical characterization of auxetic stainless steel thin sheets with reentrant Structure. *Smart Mater. Struct.* **2017**, *26*, 085022. [CrossRef]
59. Lvov, V.A.; Senatov, F.S.; Korsunsky, A.M.; Salimon, A.I. Design and Mechanical Properties of 3D-Printed Auxetic Honeycomb Structure. *Mater. Today Comm.* **2020**, *24*, 1011173. [CrossRef]

metals

MDPI

Article

Simulation of Localized Stress Impact on Solidification Pattern during Plasma Cladding of WC Particles in Nickel-Based Alloys by Phase-Field Method

Dongsheng Wei [1], Ming Chen [1,*], Chunlin Zhang [2], Xingang Ai [2] and Zhiwen Xie [1,*]

[1] School of Mechanical Engineering and Automation, University of Science and Technology Liaoning, Anshan 114051, China; weidongsheng1999@163.com

[2] School of Materials and Metallurgy, University of Science and Technology Liaoning, Anshan 114051, China; zhangchunlin@email.cn (C.Z.); aixingang@126.com (X.A.)

* Correspondence: chenming@ustl.edu.cn (M.C.); xzwustl@126.com (Z.X.)

Abstract: As materials science continues to advance, the correlation between microstructure and macroscopic properties has garnered growing interest for optimizing and predicting material performance under various operating conditions. The phase-field method has emerged as a crucial tool for investigating the interplay between microstructural characteristics and internal material properties. In this study, we propose a phase-field approach to couple two-phase growth with stress–strain elastic energy at the mesoscale, enabling the simulation of local stress effects on the solidified structure during the plasma cladding of WC particles and nickel-based alloys. This model offers a more precise prediction of microstructural evolution influenced by stress. Initially, the phase field of WC-Ni binary alloys was modeled, followed by simulations of actual local stress conditions and their impacts on WC particles and nickel-based alloys with ProCAST and finite element analysis software. The results indicate that increased stress reduces grain boundary migration, decelerates WC particle dissolution and diffusion, and diminishes the formation of reaction layers and Ostwald ripening. Furthermore, experimental validation corroborated that the model's predictions were consistent with the observed microstructural evolution of WC particles and nickel-based alloy composites.

Keywords: phase field; stress–strain elastic energy; WC particles; ProCAST; local stress

1. Introduction

Nickel-based alloys are indispensable in industries requiring high mechanical properties and corrosion-resistant materials, including the aerospace, automotive, and marine sectors [1]. Although these alloys are robust, the performance of critical components can be compromised under the extreme operating conditions of steel production, including high temperatures, heavy loads, and intense friction. Therefore, enhancing these alloys through surface modification technologies is crucial to improving their performance in harsh environments. Ceramic particles with high hardness, excellent thermal stability, and strong wear resistance (such as Tic, WC, and ZrO_2) are utilized as superior composite coating additives [2–4]. Due to the excellent wettability between WC and nickel-based alloys, WC is considered an ideal reinforcing material. By incorporating WC particles into nickel-based alloys, the wear resistance, hardness, and overall performance of the developed coatings are significantly enhanced, thereby boosting their stability and reliability under such demanding conditions [5–7].

Plasma cladding technology employs high-energy plasma beams to melt metal powder, which is then metallurgically combined with the substrate material to form high-performance composite coatings. Within a high-temperature plasma melt pool, WC particles and nickel-based alloys exhibit cellular alloy reaction layers, eutectic structures, and carbides, significantly impacting the mechanical properties of the coating [8,9].

Due to the temperature influence during the experiment, systematically studying the state of WC particles and nickel-based alloys throughout the entire solidification process is challenging. Additionally, limitations in experimental methods make it difficult to reveal the changes and formation of their microstructures. However, numerical simulation can effectively compensate for these shortcomings. Currently, microstructure simulation methods include cellular automata [10–12], phase-field modeling [13–15], and the Monte Carlo method [16–18], among others. Cellular automata (CA) demonstrate notable simplicity in modeling microstructural evolution, attributed to their high computational efficiency and independence from mesh structure dependencies. Nevertheless, CA exhibit limitations in addressing complex interface morphology and multi-physics field interactions and are particularly deficient in capturing the effects of grain boundary morphology. The Monte Carlo (MC) method enhances the predictive accuracy of microstructural evolution by leveraging statistical sampling techniques, making it particularly effective for systems characterized by pronounced stochastic behavior. Nevertheless, this method is computationally intensive and demands a substantial number of samples. It also exhibits inherent limitations in addressing coupled fields, such as stress and temperature. Moreover, due to its sharp interface model, the MC method struggles to accurately simulate interfacial properties of grain boundaries, resulting in less precise simulations of solid-state diffusive phase transitions compared with the phase-field method. Consequently, its application in modeling multiphysics interactions during solidification is constrained. The phase-field method is distinguished by its ability to incorporate continuously varying field variables across interfaces, enabling a more precise representation of interfacial morphological migration and microstructural changes without the need for the explicit tracking of interfaces [19]. Additionally, it can self-couple temperature, flow fields, and stress, enabling the simulation of the evolution of many complex microstructures. This is beyond the capability of cellular automata and Monte Carlo methods. Therefore, this study employs the phase-field method to investigate the influence of the local stress field during the solidification process of surface overlay welding.

The phase-field model, as an effective mathematical tool, was initially introduced to describe the microstructure transformation of materials without explicitly tracking complex interface boundaries. This method not only conserves significant computing resources but also greatly simplifies model complexity. The phase-field model proposed by Karma and Rappel [20] successfully simulated the anisotropy and tip-branching phenomenon during dendritic growth, providing an important theoretical basis for understanding microstructural evolution during solidification. Chen [21] and Moelans et al. [22] discussed in detail the foundations and applications of phase-field models, emphasizing their effectiveness in addressing interface dynamics issues in materials science. Echebaria et al. [23] study further improved the phase-field model to more accurately describe grain growth under varying temperature and composition conditions. Tegze et al. [24] coupled the phase-field method with multiple physical fields, such as fluid dynamics and electromagnetic fields, enabling a more comprehensive description of the various physical phenomena in the solidification process. In recent years, phase-field modeling has seen substantial advancements in coupling multiple physical fields and in simulating a diverse range of complex physical phenomena. Wang et al. [25] effectively demonstrated the integration of phase-field methods with Monte Carlo simulations and fluid dynamics to achieve a thorough understanding of microstructural evolution under coupled thermal and mechanical conditions, which is vital for materials operating in complex environments. Munoz et al. [26] advanced phase-field models to provide more accurate characterizations of grain growth across varying temperatures and compositions, thereby more effectively simulating metal solidification, crystal growth, and phase transformations, which in turn enhances the predictive capabilities of these models for industrial applications.

Previous studies have not specifically employed the phase-field model to investigate the localized stress–strain effects occurring during the solidification of WC particles in nickel-based alloy composite coatings. In this study, we introduce a numerical model that couples the phase-field method with elastic strain energy to simulate the effects of localized stresses on microstructural evolution, enabling a more detailed analysis of stress-induced microstructural changes. Unlike earlier models that oversimplify stress effects, our approach facilitates a more precise observation of the interplay between the local stress field and microstructural development during solidification. Integrating the macroscopic stress data derived from ProCAST 2019 simulations with a phase-field model yields a more accurate representation of the effects of localized stresses during solidification. This approach, which links macroscopic stress distribution to microstructural evolution, offers deeper insights into the stress mechanisms influencing microstructural changes. The feasibility and accuracy of the phase-field model are validated through a rigorous comparison of simulation results with SEM experimental data. This study serves as a crucial reference for comprehensively understanding the stress mechanisms between WC particles and nickel-based alloys during solidification, thereby offering theoretical support for the optimization and design of related materials.

2. Phase-Field Models

2.1. Phase-Field Model

The development of the phase-field model integrates the simulation of grain growth via the phase-field method with the consideration of elastic strain energy. This approach disregards the chemical reactions and phase transitions among other metals (Cr, Fe, etc.) in nickel-based alloys and WC particles. Firstly, the nickel-based alloy is simplified to pure Ni, constructing a binary system consisting of WC particles and Ni. By introducing field variables and establishing a two-phase growth system, elastic strain energy under varying stress conditions is introduced into the phase-field model. Solving partial differential equations allows for the simulation of the influence of local stress and the grain evolution process of WC particles and nickel-based alloys during the plasma cladding solidification process.

2.2. Phase Field

By utilizing the phase-field method to establish a grain growth model, we initially consider WC particles and nickel-based alloys as two distinct phases during plasma cladding, namely, phase α (WC particles) and phase β (Ni). By employing the multi-order parameter two-phase-field model proposed by Fan and Chen [27], a series of orientation field variables $\psi_1^\alpha(r)$, $\psi_2^\alpha(r)$, …, $\psi_p^\alpha(r)$, $\psi_1^\beta(r)$, $\psi_2^\beta(r)$, …, $\psi_q^\beta(r)$, and the component field $C(r)$ are defined. Here, p and q denote the number of orientation field variables for phases α and β, respectively, with ψ representing the characteristics of the corresponding phase grains. The orientation field variables are non-conservative, describing the orientation of different grains in polycrystalline materials. Simultaneously, these orientation field variables continuously vary in space, with values ranging from 0 to 1. Within a grain, only one orientation field variable ψ attains a value of 1, while all other orientation field variables ψ are 0. At the grain boundary, the orientation field variable ψ transitions continuously from 0 to 1. The component field variable $C(r)$ is conservative, with values of C_α and C_β in the α and β phase grains, respectively, and intermediate values between C_α and C_β at the grain boundaries.

In this model, the total chemical free energy comprises local free energy, interfacial gradient energy, and elastic energy. According to diffusion interface theory and continuum field theory, the total free energy F in a two-phase system can be expressed as

$$F = \int [f_0(C(r);\psi_i^\alpha(r);\psi_i^\beta(r)) + \frac{k_c}{2}(\nabla C(r))^2 + \frac{k_i^\alpha}{2}(\nabla \psi_i^\alpha(r))^2 + \frac{k_i^\beta}{2}(\nabla \psi_i^\beta(r))^2 + f^{elast}]dr^3 \quad (1)$$

Here, f_0 denotes the local free energy; $\nabla C(r)$ signifies the component field variable; $\nabla \psi_i^\alpha(r)$ and $\nabla \psi_i^\beta(r)$ represent the gradient terms of the orientation field variables for the

α and β phases; k_c, k_i^{α}, and k_i^{β} denote the corresponding gradient energy coefficients; and f^{elast} signifies the elastic energy density.

The interfacial energy σ_{gb} between phase α and phase β in the system can be expressed as

$$\sigma_{gb} = \int_{-\infty}^{+\infty} [\Delta f(\psi_i^{\alpha}, \psi_i^{\beta}, C) + \frac{k_c}{2}\left(\frac{dC}{dx}\right)^2 + \frac{k_i^{\alpha}}{2}\left(\frac{d\psi_i^{\alpha}}{dx}\right)^2 + \frac{k_i^{\beta}}{2}\left(\frac{d\psi_i^{\beta}}{dx}\right)^2]dx \quad (2)$$

The driving force of microstructural evolution is derived from the reduction in system free energy. The evolution dynamics of these field variables are described by the Allen–Cahn [28] and Cahn–Hilliard [29] equations.

$$\frac{d\psi_i^{\alpha}(r,t)}{\partial t} = -L_i^{\alpha}(\sigma)\frac{\delta F}{\delta\psi_i^{\alpha}(r,t)} + \varsigma_{\alpha}(r,t) \quad (3)$$

$$\frac{d\psi_i^{\beta}(r,t)}{\partial t} = -L_i^{\beta}(\sigma)\frac{\delta F}{\delta\psi_i^{\beta}(r,t)} + \varsigma_{\beta}(r,t) \quad (4)$$

$$\frac{dC(r,t)}{\partial t} = \nabla \cdot M\nabla\left[\frac{\delta F}{\delta C(r,t)}\right] + \varsigma_C(r,t) \quad (5)$$

$L_i^{\alpha}(\sigma)$ and $L_i^{\beta}(\sigma)$ are kinetic coefficients associated with grain boundary mobility, while M is the kinetic coefficient associated with the atomic diffusion coefficient. $\varsigma_{\alpha}(r,t)$ and $\varsigma_{\beta}(r,t)$ represent the noise terms considering thermal fluctuations. We set the thermal noise term to $\varsigma = 0.001$ (where $i = \alpha$, β, and C). In practical applications, local stress can alter the total free energy of the system, indirectly impacting the interfacial energy and consequently affecting the migration rate L. Increasing stress decreases the migration rate between phase interfaces. Here, we tensorize L, which can be obtained as

$$L(\sigma) = L_0 e^{(-\mu\frac{\sigma}{k})} \quad (6)$$

Here, L_0 represents the reference mobility, μ signifies the stress sensitivity factor, k denotes a material-specific physical constant, and σ refers to the local stress. To facilitate numerical solutions, we construct the free energy density function f_0, defined as

$$f_0 = f(C) + \sum_{i=1}^{p} f(C, \psi_i^{\alpha}) + \sum_{i=1}^{q} f(C, \psi_i^{\beta}) + \sum_{k=\alpha}^{\beta}\sum_{i=1}^{p}\sum_{i=1}^{q} f(\psi_i^k, \psi_i^k) \quad (7)$$

$f(C)$ is a function of the concentration C, and $f(C, \psi_i^{\alpha})$ and $f(C, \psi_i^{\beta})$ describe the combined relations between the orientation field ψ and the concentration field C(*r*). Moreover, $f(\psi_i^k, \psi_i^k)$ is coupled with the orientation field ψ.

$$f(C) = -\frac{A}{2}(C - C_m)^2 + \frac{D_{\alpha}}{4}(C - C_{\alpha})^4 + \frac{D_{\beta}}{4}(C - C_{\beta})^4 + \frac{B}{4}(C - C_{\beta})^4 \quad (8)$$

$$f(C, \psi_i^{\alpha}) = -\frac{\gamma_{\alpha}}{4}(C - C_{\beta})^2 + \frac{\delta_{\alpha}}{4}(\psi_i^{\alpha})^4 \quad (9)$$

$$f(C, \psi_i^{\beta}) = -\frac{\gamma_{\beta}}{4}(C - C_{\beta})^2 + \frac{\delta_{\alpha}}{4}(\psi_i^{\beta})^4 \quad (10)$$

$$f\left(\psi_i^k, \psi_i^k\right) = \frac{\varepsilon}{2}(\psi_i^{\alpha})^2\left(\psi_i^{\beta}\right)^2 \quad (11)$$

In the aforementioned equation, C_{α} and C_{β} represent the concentrations of phase α and phase β at equilibrium, respectively, while D_{α}, D_{β}, δ, $C_m = \frac{C_{\beta}+C_{\alpha}}{2}$, A, B, γ_{α}, γ_{β}, and ε denote the relevant parameters.

2.3. Elastic Strain Energy

Solid-state phase transitions occur during the solidification of binary alloys. As the interface between the two phases advances and solute atoms continue to diffuse, significant changes in phase-field variables and solute concentration are observed near the interface. Simultaneously, differences in lattice parameters between the new phase and the parent phase generate an elastic field between them. Both phenomena manifest themselves as strains that result in significant changes near the phase interface. The elastic strain energy changes the total free energy of the system and influences the evolution of phase and solute fields in a way that is long range. Therefore, it is essential to account for the elastic strain energy in the elastic field near the two-phase interface and integrate it into the governing equations during the two-phase solidification process. The elastic strain energy can be expressed as follows:

$$f^{elast} = \varepsilon_{kl}^{el} C_{ijkl} \varepsilon_{ij}^{el} \tag{12}$$

where ε_{kl}^{el} and ε_{ij}^{el} represent the elastic strain components and C_{ijkl} is the position-dependent elastic tensor. The elastic strain can be expressed as follows:

$$\varepsilon_{ij}^{el} = \varepsilon_{ij} - \varepsilon_{ij}^{0} \tag{13}$$

where denotes the total strain energy and ε_{ij}^{0} represents the position- and component-dependent eigenstrain. We consider the total strain energy ε_{ij} as follows: the intrinsic strain ε_{ij}^{el} linearly related to stress, the intrinsic strain ε_{ij}^{C} due to component inhomogeneity, and the intrinsic strain ε_{ij}^{t} in the presence of a temperature gradient. Thus, the total strain can be expressed as follows:

$$\varepsilon_{ij} = \varepsilon_{ij}^{t} + \varepsilon_{ij}^{C} + \varepsilon_{ij}^{el} \tag{14}$$

By introducing the Kronecker delta, denoted by δ_{ij} $(i = j,\ \delta_{ij} = 1\ ,\ \delta_{ij} \neq 1\ ,$ and $\delta_{ij} = 0\ )$, and the coefficient of thermal expansion ζ, the temperature-induced eigenstrain ε_{ij}^{t} is defined as follows:

$$\varepsilon_{ij}^{t} = \zeta T \delta_{ij} \tag{15}$$

According to Vegard's law, which states that the lattice constant is linearly related to the solid solution concentration, the intrinsic strain ε_{ij}^{C} due to component inhomogeneity can be expressed as follows:

$$\varepsilon_{ij}^{C} = \varepsilon_0 (C - C_m) \delta_{ij} \tag{16}$$

Here, ε_0 represents the lattice expansion coefficient of the element with respect to the matrix element [30]. The total strain in the system can be expressed as follows:

$$\delta\varepsilon_{ij} = \frac{1}{2}\left[\frac{\partial u_i}{\partial r_i} + \frac{\partial u_j}{\partial r_j}\right] \tag{17}$$

u represents the displacement field, and r denotes the position vector.

$$\sigma_{ij}^{el} = C_{ijkl} \varepsilon_{ij}^{el} = C_{ijkl} (\varepsilon_{ij} - \varepsilon_{ij}^{t} - \varepsilon_{ij}^{C}) \tag{18}$$

Assuming that phases α and β are linearly elastic and obey Hooke's law, the stress can be expressed as follows:

$$f^{elast} = (\varepsilon_{kl} - \varepsilon_{kl}^{t} - \varepsilon_{kl}^{c}) C_{ijkl} (\varepsilon_{ij} - \varepsilon_{ij}^{t} - \varepsilon_{ij}^{c}) \tag{19}$$

Assuming that the rate of reaching mechanical equilibrium is much faster than the chemical reaction rate, we solve the mechanical equilibrium equation to obtain the elastic energy, expressed as [31]

$$\frac{\partial \sigma_{ij}^{el}}{\partial r_j} = 0 in V \tag{20}$$

2.4. Solving Parameters

The phase-field method serves as a powerful tool for modeling the evolution of complex microstructures. By solving the variational problem of the generalized free energy function, it is feasible to simulate the motion and phase transition phenomena at phase interfaces across continuous time and space scales. This approach is particularly well suited for addressing the dynamic behavior of phase interfaces in multiphase materials.

Among the various methods for solving variational problems, common techniques include finite difference methods [32–34], Fourier spectral methods [35–37], and finite element methods [38,39]. The finite difference method is widely used due to its concise theoretical foundation, simple meshing, and ease of constructing and programming discrete equations. The finite difference method is characterized by significant localization, providing high adaptability and accuracy in addressing variational problems with substantial local variations. In contrast, the Fourier spectral method excels at handling problems with periodic boundary conditions, while the finite element method demonstrates its unique and powerful capability in addressing problems with complex geometries and non-uniform material properties. Thus, the finite difference method is chosen as the primary research tool in this study.

We apply Equation (1) to Equations (3)–(5) to solve the equations by using the Euler–Lagrange equations:

$$\frac{\delta F}{\delta \psi} = -\frac{d}{d_x}\left(\frac{\partial F_v}{\partial(\nabla \psi_x)}\right) + \frac{\partial F_v}{\partial \psi} \tag{21}$$

From the above equation, with $F_v = f_0 + \frac{k_c}{2}(\nabla C(r))^2 + \frac{k_i^\alpha}{2}(\nabla \psi_i^\alpha(r))^2 + \frac{k_i^\beta}{2}\left(\nabla \psi_i^\beta(r)\right)^2 + f^{elast}$, the following expression can be derived:

$$\frac{d\psi_i^\alpha(r,t)}{\partial t} = -L_i^\alpha(\sigma)\left[\frac{\partial f_0}{\partial \psi_i^\alpha} - k_i^\alpha \nabla^2 \psi_i^\alpha + \frac{\partial f^{elast}}{\partial \psi_i^\alpha}\right] + \varsigma_\alpha(r,t) \tag{22}$$

$$\frac{d\psi_i^\beta(r,t)}{\partial t} = -L_i^\beta(\sigma)\left[\frac{\partial f_0}{\partial \psi_i^\beta} - k_i^\beta \nabla^2 \psi_i^\alpha + \frac{\partial f^{elast}}{\partial \psi_i^\beta}\right] + \varsigma_\beta(r,t) \tag{23}$$

$$\frac{dC(r,t)}{\partial t} = \nabla \cdot M\nabla\left[\frac{\partial f_0}{\partial C} - k_c \nabla^2 C + \frac{\partial f^{elast}}{\partial C}\right] + \varsigma_C(r,t) \tag{24}$$

The numerical solution of the system of equations is achieved by discretizing the Ginzburg–Landau and Cahn–Hilliard equations in both space and time. Spatially, we employ the finite difference method (FDM) to partition the solution region into a discrete grid, replacing the continuous solution region with a finite number of grid nodes. We utilize the nine-point lattice discrete Laplace operator [40], which can be expressed as follows:

$$\nabla^2 \psi_i = \frac{1}{\Delta x^2}\left[\frac{1}{2}\sum_j^{mm}(\psi_j - \psi_i) + \frac{1}{4}\sum_k^{nm}(\psi_k - \psi_i)\right] \tag{25}$$

Here, ψ_i represents the field variable at lattice point i, Δx is the spatial step size, and j and k denote the nearest and next-nearest neighbor lattice points of lattice point i. In this study, we employ a square lattice to describe the micro-evolutionary process, resulting in the values of mm and nm both being 4.

For temporal discretization, we employ the explicit Euler method, which can be expressed as follows:

$$\psi_i(t+\Delta t) = \psi_i(t) + \frac{d\psi_i}{dt} \times \Delta t \quad (26)$$

In the above equation, Δt represents the time step.

This study aims to investigate the impact of localized stress on the microstructural evolution of WC grains during the solidification of nickel-based alloys by using the phase-field method and to compare the results with experimental observations. The influence of varying levels of localized stress on grain growth and the solidification characteristics of grains during Ostwald ripening are examined. Therefore, we assume that the kinetic processes of the microstructural evolution of grains are entirely governed by the kinetic coefficients. Parameters such as gradient energy coefficients and grain boundary mobility coefficients are referenced from the $Al_2O_3-ZrO_2$ two-phase system [27]. The specific parameters for the phase-field simulation are detailed in Table 1.

Table 1. Relevant parameters employed in the simulation process.

Symbol	Description	Value	Ref.
Δx	Space step	2.0	[27]
Δt	Time step	0.1	
k_i^α	Gradient energy coefficient of phase α	2.5	
k_i^β	Gradient energy coefficient of phase β	2.0	
k_c	Gradient energy coefficient of concentration	0.5	
L_0	Benchmark migration rate	3.0	
A	-	2.0	
B	-	9.88	
D_α	-	1.52	
γ_α	-	1.23	
δ	-	1.0	
μ	Stress sensitivity coefficient	0.06	-
k	Physical constant	1.025	
ε_0	Lattice expansion coefficient	2.6×10^{-6}	[30]
ζ	Thermal expansion coefficient	9.6×10^{-6}	[41]

To visualize the microstructure generated by the computer simulation, the following functions are defined:

$$\varphi(r) = 0.75\left(\sum_{i=1}^{p}[\psi_i^\alpha(r)]^2\right) + \sum_{i=1}^{q}[\psi_i^\beta(r)]^2 \quad (27)$$

3. Macro-Simulation

In this section, ProCAST finite element software is utilized to simulate the local stress during the solidification process of plasma cladding. By analyzing the temperature field and equivalent stress field of the casting, the equivalent stress at different positions during solidification is introduced into the phase field. Consequently, the effect of local stress on WC particles at various positions is determined. The flowchart of the simulation using ProCAST is illustrated in Figure 1.

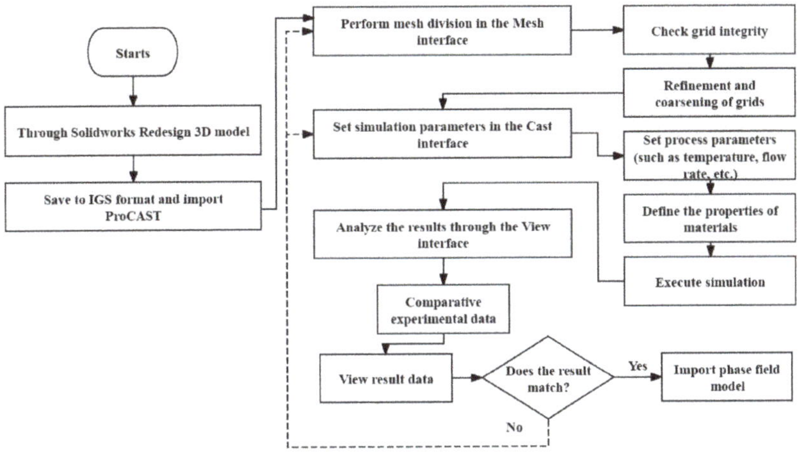

Figure 1. ProCAST simulation flowchart.

3.1. Model Establishment and Grid Partitioning

To mitigate the difference in material properties between the cladding layer and the substrate during the experimental process, a nickel-based alloy transition layer is initially prepared on the substrate, followed by a WC-enhanced nickel-based alloy hard layer on the surface of the transition layer, as shown in Figure 2. To realistically simulate the experimental process and facilitate the simulation, we use a nickel-based alloy as the substrate for Ni40A-45 wt.% WC casting. Firstly, 3D modeling is performed by using SolidWorks 2020, as shown in Figure 3. Here, Figure 3a denotes the upper mold size, 60 mm × 40 mm × 8 mm; Figure 3b represents the lower mold size, 60 mm × 40 mm × 4 mm; the middle oval long casting bar is for the casting with a size of 56 mm × 12 mm × 5 mm; and the small rectangle indicates the casting mouth size, 2 mm × 12 mm × 2 mm.

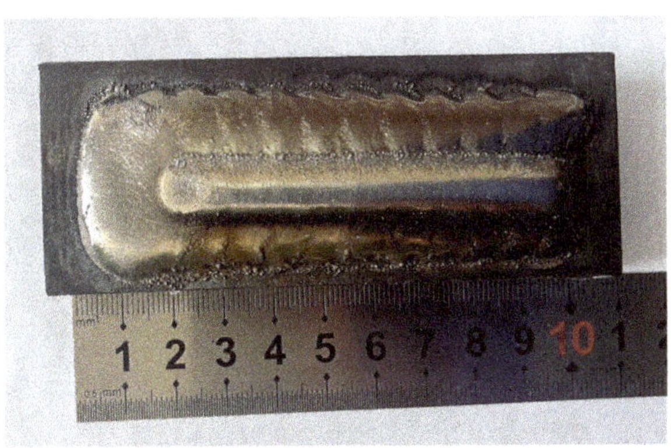

Figure 2. Sample picture of plasma cladding.

The 3D solid model is then imported into ProCAST for mesh division. To ensure the accuracy of the simulation results, the casting mesh is finely detailed. The total number of 2D surface mesh elements is 6956, and the total number of 3D mesh cells is 59,447, as shown in Figure 4.

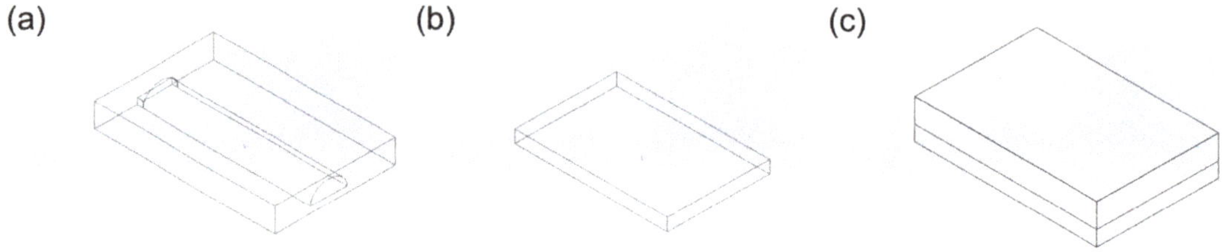

Figure 3. Mold structure of 3D model: (**a**) upper mold structure diagram, (**b**) lower mold structure diagram, and (**c**) overall mold structure diagram.

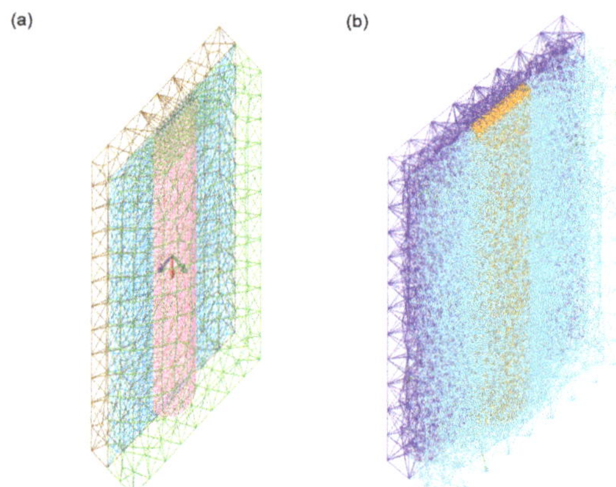

Figure 4. Mesh partitioning: (**a**) 2D mesh partitioning and (**b**) 3D mesh partitioning.

3.2. Boundary Conditions

ProCAST contains only a limited selection of commonly used materials and does not include Ni40A-45 wt.% WC for plasma cladding or Ni40A metal materials for the cladding layer. However, we can utilize the thermodynamic database to input the basic elemental ratios of the casting and the required metal molds. This allows us to generate data to view the phase diagrams of the materials, coefficients of thermal expansion, Young's modulus, Poisson's ratio, and other relevant material properties, as shown in Table 2. The upper mold material is set as Resin Bonded Sand, while the lower mold and casting materials are selected as Ni40A and Ni40A-45 wt.% WC. Additionally, to better simulate the experiment, the properties of the casting material are set to be elastic–plastic.

Table 2. Chemical composition of Ni40A and Ni40A-45% WC alloys (wt.%).

Element	C	Cr	B	Si	Fe	Cu	W	Ni
Ni40A	0.07	11.09	1.23	3.46	3.01	3.66	-	Bal
Ni40A-45 wt.% WC	1.83	0.1	1.24	1.76	1.44	-	43.2	Bal

At the beginning of casting, the bottom plate is preheated to 200 °C for 30 min, while the casting temperature is automatically calculated based on the thermodynamic database data, yielding a temperature of 1362 °C, which we round to 1400 °C. The casting time is set to 6 s, and the heat transfer coefficients vary for different contact materials. For the heat transfer coefficient between the molds, we use h = 500; for the coefficient between the casting and the upper mold, we use h = 1000; and for the coefficient between the casting and the lower mold, we use $h = 2000$. Additionally, we set the boundary conditions, with the periphery of the mold being naturally cooled by air.

3.3. Analysis of the Process

By configuring the material properties, flow rate and boundary conditions, we can obtain the calculation results of the solidification process of the casting. To facilitate the analysis of subsequent stress values, we randomly select a cross-section during the solidification process of Ni40A-45 wt.% WC and choose three points in the upper, middle, and lower layers near the cross-section, as depicted in Figure 5.

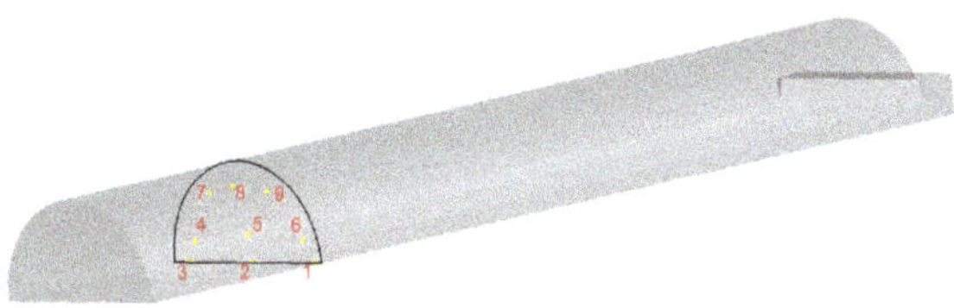

Figure 5. Schematic diagram of selected points of solidification process.

The equivalent stress-versus-temperature curves for each point can be plotted by using the Viewer interface of ProCAST. To ensure more accurate stress values, we take the average temperature and stress for each layer, as shown in Figure 6. From Figure 6a, it is evident that the temperature gradient exists across the cross-section during the same time period, with the upper layer having the highest temperature and the lower layer having the lowest temperature. Simultaneously, the effective stress during solidification increases as the temperature decreases.

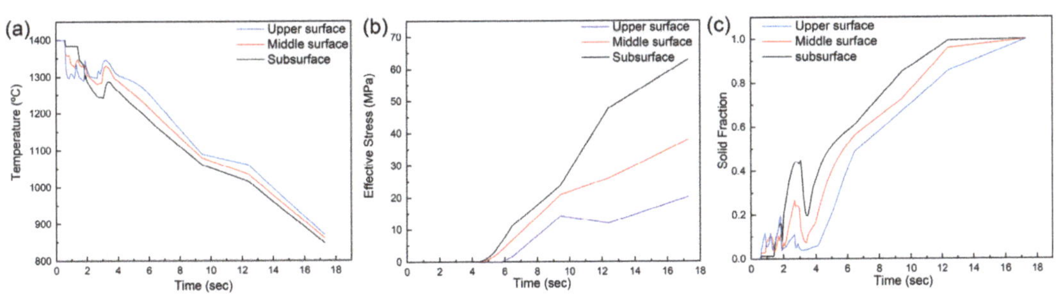

Figure 6. Temperature, equivalent force, and solid-phase rate graphs: (**a**) temperature change graph, (**b**) equivalent force change graph, and (**c**) solid-phase rate change graph.

From the curves of the upper layers in Figure 6a,b between 10 s and 12 s, it can be concluded that when the temperature changes slowly, the effective stress during solidification not only does not increase but also decreases. Figure 6c shows that the solidification rate of the lower layer is higher compared with the middle and upper layers during the same time period. Additionally, Figure 6b indicates that the effective stress is relatively higher at locations with a higher solidification rate. The effective stress changes drastically at the beginning of solidification and then tends to level off over time.

4. Simulation Results and Discussion

The phase-field simulation of the WC-Ni binary system is conducted in a two-dimensional region by using a square grid with periodic boundaries to simulate grain growth under varying stresses. The spatial step is set to $\Delta x = 20$, and the time step size of $\Delta t = 0.1$ is selected to ensure numerical stability. A square grid of 300 × 300 is used for the simulation. The number of orientation field variables is set to $p = q = 36$ [42], and the initial values of the orientation field variables are normally distributed random numbers between −0.001 and 0.001.

The selection of stress parameters was determined through a comprehensive comparison of experimental data and simulation results. As depicted in Figure 7a, during the cladding process at a preheating temperature of 200 °C, the WC particles in the Ni40A-45 wt.% WC composite predominantly concentrate in the middle layer. Figure 7b reveals that under low-magnification backscattered imaging from a scanning electron microscope, the upper layer of the cladding exhibits only sporadic distributions of WC particles, with the lower layer showing an almost complete absence of these particles. Moreover, as observed in Figure 7a,b, most WC particles exhibit round shapes; however, a small number of particles possess irregular shapes. When these particles deviate from a spherical form, significant singular charges and fields may develop at their edges and tips. According to Olyslager [43], in bi-isotropic materials, the formation of such singular fields is closely linked to the material's electromagnetic properties. Similarly, Valagiannopoulos [44] examined the formation of singular fields near the edges of metals and suggested that these fields can be effectively mitigated through material and geometric optimization.

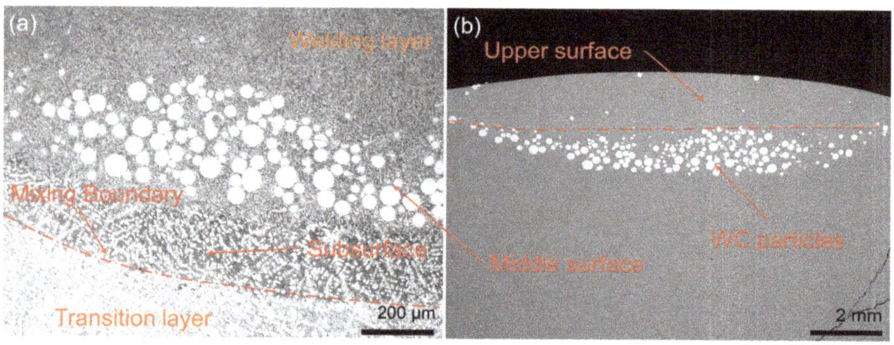

Figure 7. Cross-sectional images of the cladding layer: (**a**) optical micrograph showing the cladding layer, transition layer, and subsurface regions and (**b**) SEM micrograph highlighting the distribution of WC particles across the upper and middle surfaces.

For our study, this suggests that non-spherical WC particles may cause localized stress concentrations, which could impact the microstructural evolution during solidification and lead to localized anomalies in material properties. To mitigate these effects, we optimize the particle shape during simulation to make it more rounded, thereby preventing the formation of singular fields at the tips and edges. For the singular shapes generated in the experiments, in future studies, we can consider optimizing the formation of these singular fields by changing the particle shapes or choosing suitable materials to avoid local anomalies in the materials.

In practical applications, process and environmental influences cause stress variations even in adjacent areas. Figure 6 illustrates that the stress levels vary at different depths, even when subjected to the same temperature. Our primary focus is on investigating the impact of localized stress on WC particles and nickel-based alloys during the plasma cladding process. Therefore, we refer to the effective stress curves of the layers shown in Figure 6b and select the stress values corresponding to a solid-phase ratio of 0.7–0.9, as illustrated in Figure 6c. We then use stress values ranging from 10 to 30 MPa to simulate the effects of different localized stresses on WC particles.

By incorporating these stress conditions into the phase-field model, the results are depicted in Figure 8. Figure 8a–e illustrate the microstructural evolution around the WC particles under various stress conditions. In these images, the dark-gray regions represent WC particles, the white grains signify the nickel-based alloy, and the light-gray areas around the circles indicate the reaction layer between the WC particles and the nickel-based alloy.

This suggests that under high-temperature conditions, WC particles partially dissolve and precipitate, integrating with the nickel-based alloy to form a new phase. To determine the composition of this phase, the physical phase of the sample was analyzed by using X-ray diffraction, as depicted in Figure 9. The analysis revealed that the primary components are $\gamma - Ni/Fe$, M_7C_3, $M_{23}C_6$, Fe_3C, and various carbides, as indicated by the intensities of the diffraction peaks. However, this study primarily focuses on the quantity of phase formation under varying stress conditions rather than the phase transformation between the two phases.

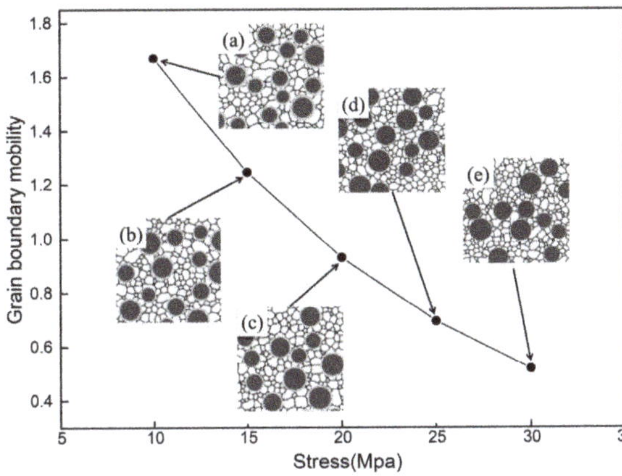

Figure 8. Evolution of WC-Ni phase field under 2D conditions: (**a**–**e**) extent of dissolution and diffusion of WC particles under varying stresses and evolution of surrounding grains.

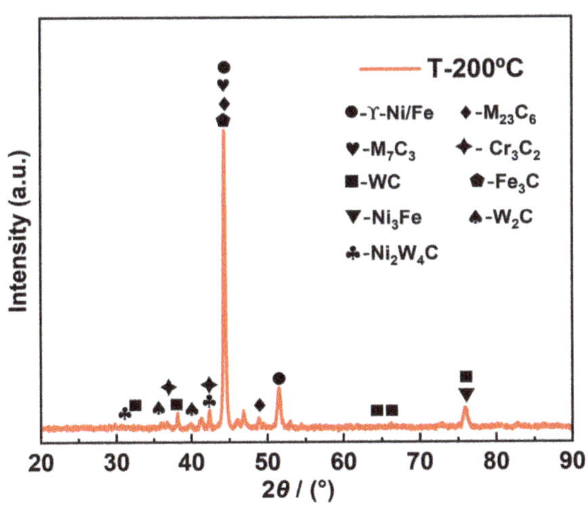

Figure 9. X-ray diffraction (XRD) pattern of cladding layer.

As illustrated in Figure 8a, at a local stress of 10 MPa, a distinct reaction layer forms between the WC particles and the nickel-based alloy, and the surrounding grain size is large. According to Equation (6), under low-stress conditions, grain boundary mobility is higher, and the dissolution and diffusion processes of WC particles are less influenced. Consequently, during solidification, WC particles form a distinct reaction layer with the nickel-based alloy. Additionally, due to the minimal change in the system's free energy, the formation of large grains through the merging of neighboring grains becomes more pronounced, a phenomenon known as Ostwald ripening.

As depicted in Figure 8e, when the local stress increases to 30 MPa, grain boundary mobility decreases significantly, the total free energy of the system reduces, WC particles dissolve only slightly during solidification, the reaction layer diminishes, and the formation of large grains by the merging of neighboring grains is weakened.

In summary, when the local stress is low, the system's elastic energy decreases marginally, leading to a minor change in the total free energy. Since the movement of microstructures occurs alongside a reduction in free energy, the extent of this reduction determines the energy available for grain boundary migration. Under higher stress, the magnitude of the elastic energy change increases, resulting in a greater decrease in the total free energy of the system, which in turn reduces grain boundary migration. Consequently, the reaction layer between the WC particles and the nickel-based alloy diminishes during solidification, and the Ostwald ripening effect among the grains is weakened. Conversely, under low stress, the intrinsic strain within the lattice is less affected, leading to minor changes in the system's total free energy. This results in an increased reaction layer, more pronounced changes in the shape of WC particles, and a more evident Ostwald ripening phenomenon.

To more effectively observe the changes in the two phases during solidification, we employ 125 × 125 grid points while maintaining constant time steps, space steps, and the number of orientation field variables. This setup allows us to simulate the influence of and changes in individual WC particles under localized stresses during solidification. We select stress levels of 10 MPa, 20 MPa, and 30 MPa, representing upper surface, middle surface, and subsurface, respectively, to simulate the changes in the grains surrounding the WC particles and the reaction layer between the WC particles and the nickel-based alloy, as illustrated in Figure 10.

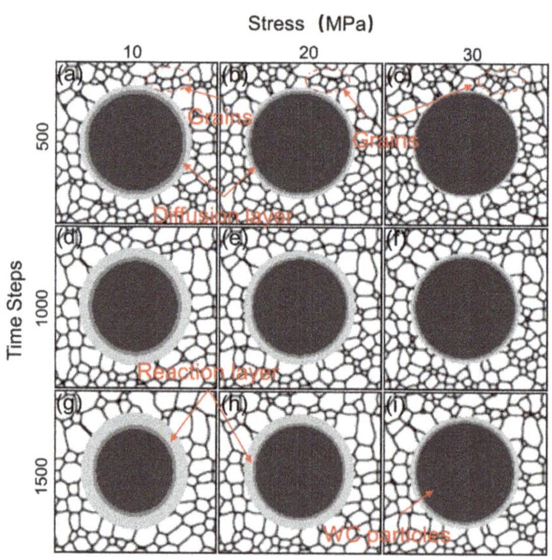

Figure 10. The evolution of individual WC particles under different stress conditions: (**a**,**d**,**g**) WC particles and nickel-based alloy grain morphology at 10 MPa, (**b**,**e**,**h**) WC particles and nickel-based alloy grain morphology at 20 MPa, and (**c**,**f**,**i**) WC particles and nickel-based alloy grain morphology at 30 MPa.

By comparing Figure 10a,c, it is evident that for the same number of steps, higher local stress results in slower grain evolution around the WC particles, with relatively smaller grain sizes. Additionally, excessive local stress significantly impacts the total free energy of the system, leading to a decrease in free energy, reduced grain boundary mobility, an increase in the number of grains around the WC particles, and a reduction in their average size. As illustrated in Figure 10g, under lower stress, the decomposition and diffusion of WC particles are pronounced, and the formation of the reaction layer is prominent. Additionally, the lower local stress introduces a certain degree of randomness in the direction of the dissolution and diffusion of the WC particles, altering their shape from circular to elliptical. As depicted in Figure 10i, at high local stress, the dissolution and diffusion of WC particles are significantly influenced, resulting in a reduced thickness of the reaction layer with the nickel-based alloy. Additionally, the small grains near the large grains are not engulfed but retain their shapes, while the geometry of the WC particles remains largely unchanged.

As illustrated in Figure 11a, during the cladding process, the WC particles near the upper surface exhibit significant dissolution and diffusion due to slower temperature changes and lower stress. This leads to the formation of a thicker reaction layer with the surrounding nickel-based alloy. The lower stress and higher grain boundary mobility in the upper layer introduce more randomness into the dissolution and diffusion process of WC particles, resulting in a distinct ellipsoidal shape and relatively larger grain size around the WC particles. As depicted in Figure 11b, the localized stress in the middle layer is relatively moderate, slowing down the dissolution and diffusion processes of WC particles, which results in a thinner reaction layer compared with the upper layer. Additionally, the increased stress affects realm migration, reducing both grain boundary mobility and the Ostwald ripening effect. As illustrated in Figure 11c, at this stage, contact between the cladding layer and the substrate leads to more drastic temperature changes and higher local stress. This significantly reduces the dissolution and diffusion of WC particles, resulting in the thinnest reaction layer with the surrounding nickel-based alloy. The higher stress significantly restricts grain boundary migration, thereby preventing the engulfment of smaller grains by larger ones and preserving the shape of the WC grains.

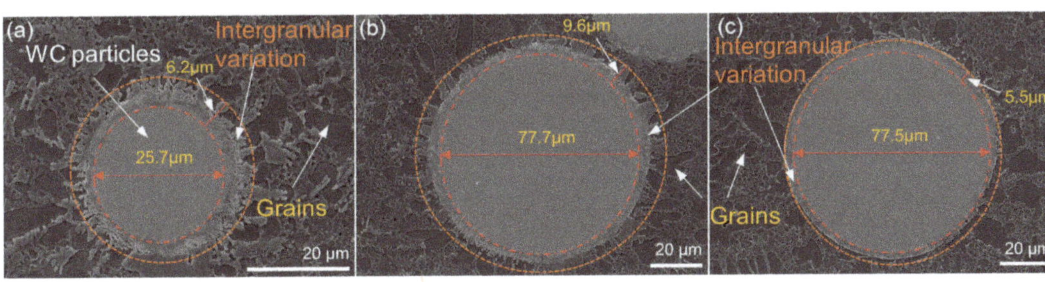

Figure 11. Morphological organization of WC particles: (**a**) morphology of WC particles on the upper surface, (**b**) morphology of WC particles on the middle surface, and (**c**) morphology of WC particles on the subsurface.

By comparing Figure 10g with Figure 11a, the simulation results indicate that localized stress in the upper level is lower, resulting in pronounced dissolution and diffusion of WC particles and a prominent reaction layer. Additionally, the relative size of the grains surrounding the WC particles is larger, and the internal WC particles form distinct ellipsoids, consistent with experimentally observed changes. Comparing Figures 10h and 11b, the simulation results demonstrate that localized stresses in the intermediate layer are relatively moderate, reducing the reaction layer formed between the WC particles and the nickel-based alloy and resulting in smaller grain sizes. Additionally, the shape of the WC particles changes slightly, aligning with experimentally observed results. Comparing Figures 10i and 11c, the simulation results reveal that the localized stress at the lower level is higher, significantly reducing the reaction layer around the WC particles. The relative size of the surrounding grains is smaller, and the shape of the WC particles remains unchanged, consistent with experimentally observed changes. The comparison between experimental and simulation results indicates a high degree of consistency.

5. Conclusions

This study investigates the effect of local stress on the microstructural evolution of WC particles and nickel-based alloys during the solidification process of plasma cladding by using the phase-field method combined with ProCAST software. Throughout the solidification of WC particles within nickel-based alloys, localized stresses have a pronounced impact on grain boundary mobility, particle dissolution and diffusion, and the Ostwald ripening of grains. The following conclusions are drawn:

1. When the localized stress is high, the mobility of grain boundaries within the system decreases significantly. Consequently, the dissolution and diffusion of WC particles are markedly hindered, leading to smaller surrounding grain sizes and preserving the shape of WC particles. This stability minimizes the formation of WC singularities and reduces stress concentration, thereby contributing to a more stable microstructural configuration of the material.
2. At lower levels of localized stress, WC particles undergo pronounced dissolution and diffusion, resulting in an expanded reaction layer with the surrounding nickel-based alloys. Additionally, the Ostwald ripening of grains around WC particles becomes more evident, potentially forming new phases and structures that enhance the overall performance of the material.
3. This study demonstrates that optimizing the microstructure during the solidification process can enhance the wear resistance, hardness, and overall performance of coatings. These findings offer a theoretical foundation for advancing material properties through surface modification techniques, which are particularly beneficial for materials exposed to extreme conditions in metallurgical, aerospace, and automotive applications.

Future research should investigate the stresses induced by phase transitions during the solidification of tungsten carbide particles in nickel-based alloys, the influence of temperature, and the impact of the singularity of tungsten carbide particles. Efforts should be directed towards optimizing computational models, broadening the scope of experimental validation, and employing a multiscale simulation approach.

Author Contributions: Conceptualization, M.C. and X.A.; Methodology, M.C. and Z.X.; Software, D.W. and X.A.; Validation, D.W. and C.Z.; Formal analysis, M.C. and C.Z.; Investigation, D.W. and M.C.; Resources, M.C. and Z.X.; Data curation, D.W. and C.Z.; Writing—original draft, D.W.; Writing—review and editing, M.C. and C.Z.; Visualization, D.W. and M.C.; Supervision, X.A. and Z.X.; Project administration, Z.X.; Funding acquisition, Z.X. All authors have read and agreed to the published version of the manuscript.

Funding: This research was supported by National Key Research and Development Program (No. 2021YFB3702003).

Institutional Review Board Statement: Not applicable.

Informed Consent Statement: Not applicable.

Data Availability Statement: The data presented in this study are available on request from the corresponding author.

Conflicts of Interest: The authors declare no conflicts of interest.

References

1. Fayed, E.M.; Brailovski, V.; Jahazi, M.; Medraj, M. Stability of the Microstructure and Elevated-Temperature Mechanical Properties of Additively Manufactured Inconel 718 Superalloy Subjected to Long-Term in-Service Thermal Cycling. *Mater. Sci. Eng. A-Struct. Mater. Prop. Microstruct. Process.* **2022**, *838*, 142790. [CrossRef]
2. Xu, S.; Cai, Q.; Li, G.; Lu, X.; Zhu, X. Effect of Scanning Speed on Microstructure and Properties of Tic/Ni60 Composite Coatings on Ti6al4v Alloy by Laser Cladding. *Opt. Laser Technol.* **2022**, *154*, 108309. [CrossRef]

3. Kim, G.; Kim, Y.C.; Cho, J.E.; Yim, C.H.; Yun, D.S.; Lee, T.G.; Park, N.K.; Chung, R.H.; Hong, D.G. Microstructural Evolution, Hardness and Wear Resistance of Wc-Co-Ni Composite Coatings Fabricated by Laser Cladding. *Materials* **2024**, *17*, 2116. [CrossRef] [PubMed]
4. Obadele, B.A.; Andrews, A.; Olubambi, P.A.; Mathew, M.T.; Pityana, S. Effect of Zro 2 Addition on the Dry Sliding Wear Behavior of Laser Clad Ti6al4v Alloy. *Wear* **2015**, *328–329*, 295–300. [CrossRef]
5. Shen, X.; Peng, H.; Xue, Y.; Wang, B.; Su, G.; Zhu, J.; Li, A. Microstructure and Properties of Wc/Ni-Based Laser-Clad Coatings with Different Wc Content Values. *Materials* **2022**, *15*, 6309. [CrossRef]
6. Chen, B.J.; Xu, M.K. Preparation and Performance Analysis of Micro-Nano Composite Coatings Reinforced with Wc Particles. *Mater. Res. Express* **2024**, *11*, 026508. [CrossRef]
7. Nomoto, S.; Kusano, M.; Kitano, H.; Watanabe, M. Multi-Phase Field Method for Solidification Microstructure Evolution for a Ni-Based Alloy in Wire Arc Additive Manufacturing. *Metals* **2022**, *12*, 1720. [CrossRef]
8. Li, M.; Yang, J.; Han, B.; Song, L.; Li, P.; Dong, W.; Xue, X. Comparative Investigation on Microstructures and Properties of Wc/Cr3c2 Reinforced Laser Cladding Ni-Based Composite Coatings Subjected to Ultrasonic Impact Treatment. *Mater. Today Commun.* **2023**, *34*, 105219. [CrossRef]
9. Wang, Y.; Zhou, J.; Zhang, T.; Li, P.; Zhu, H.; Meng, X. Effects of Wc Particles on the Microstructure of In718/Wc Composite Coatings Fabricated by Laser Cladding: A Two-Dimensional Phase-Field Study. *Coatings* **2023**, *13*, 432. [CrossRef]
10. Teferra, K.; Rowenhorst, D.J. Optimizing the Cellular Automata Finite Element Model for Additive Manufacturing to Simulate Large Microstructures. *Acta Mater.* **2021**, *213*, 116930. [CrossRef]
11. Zinovieva, O.; Zinoviev, A.; Patel, M.N.; Molotnikov, A.; Easton, M.A. Modelling Grain Refinement under Additive Manufacturing Solidification Conditions Using High Performance Cellular Automata. *Mater. Des.* **2024**, *245*, 113248. [CrossRef]
12. Ushmaev, D.; Liao, Z.; Notron, A.; Axinte, D. On the Importance of Interface Stability in Cellular Automata Models: Planar and Dendritic Solidification in Laser Melted Ysz. *Mater. Des.* **2022**, *219*, 110823. [CrossRef]
13. Tourret, D.; Liu, H.; Llorca, J. Phase-Field Modeling of Microstructure Evolution: Recent Applications, Perspectives and Challenges. *Prog. Mater. Sci.* **2022**, *123*, 100810. [CrossRef]
14. Lv, S.J.; Wu, H.H.; Wang, K.Y.; Zhang, C.L.; Zhu, J.M.; Wang, S.Z.; Wu, G.L.; Gao, J.H.; Yang, X.S.; Mao, X.P. Phase Field Simulation of Eutectoid Microstructure During Austenite-Pearlite Phase Transformation. *J. Mater. Res. Technol.* **2023**, *26*, 8922–8933. [CrossRef]
15. Wu, Y.; Xiong, J.L.; Luo, Q.; Chen, J.B.; Zeng, R.T.; Wang, S. Estimation of the Critical Value of the Second-Phase Particles in the Microstructure of Az31 Mg Alloy by Phase-Field Methods. *Crystals* **2022**, *12*, 1504. [CrossRef]
16. Schwen, D.; Schunert, S.; Jokisaari, A. Evolution of Microstructures in Radiation Fields Using a Coupled Binary-Collision Monte Carlo Phase Field Approach. *Comput. Mater. Sci.* **2021**, *192*, 110321. [CrossRef]
17. Zhang, Z.; Ge, P.; Li, J.Y.; Ren, D.X.; Wu, T. Monte Carlo Simulations of Solidification and Solid-State Phase Transformation during Directed Energy Deposition Additive Manufacturing. *Prog. Addit. Manuf.* **2022**, *7*, 671–682. [CrossRef]
18. Zheng, Y.; Liu, J.P.; Liang, Y.F.; Wu, P.P. Monte-Carlo-Assisted Phase Field Simulations of Grain Structure Evolution During the Welding Process. *Metals* **2023**, *13*, 623. [CrossRef]
19. Browne, D.J.; Hunt, J.D. A Fixed Grid Front-Tracking Model of the Growth of a Columnar Front and an Equiaxed Grain during Solidification of an Alloy. *Numer. Heat Transf. Part B Fundam.* **2004**, *45*, 395–419. [CrossRef]
20. Karma, A.; Rappel, W.-J. Quantitative Phase-Field Modeling of Dendritic Growth in Two and Three Dimensions. *Phys. Rev. E* **1998**, *57*, 4323–4349. [CrossRef]
21. Chen, L.-Q. Phase-Field Models for Microstructure Evolution. *Annu. Rev. Mater. Res.* **2002**, *32*, 113–140. [CrossRef]
22. Moelans, N.; Blanpain, B.; Wollants, P. An Introduction to Phase-Field Modeling of Microstructure Evolution. *Calphad* **2008**, *32*, 268–294. [CrossRef]
23. Echebarria, B.; Folch, R.; Karma, A.; Plapp, M. Quantitative Phase-Field Model of Alloy Solidification. *Phys. Rev. E Stat. Nonlin. Soft Matter Phys.* **2004**, *70 Pt 1*, 061604. [CrossRef] [PubMed]
24. Tegze, G.; Bansel, G.; Tóth, G.I.; Pusztai, T.; Fan, Z.; Gránásy, L. Advanced Operator Splitting-Based Semi-Implicit Spectral Method to Solve the Binary Phase-Field Crystal Equations with Variable Coefficients. *J. Comput. Phys.* **2009**, *228*, 1612–1623. [CrossRef]
25. Wang, Y.B.; Zhang, Y.; Liu, X.T.; Wang, J.X.; Xie, X.Y.; Jiang, J.J.; Liu, J.X.; Liu, H.; Wu, Y.J.; Dong, S.; et al. Simulation of Microstructure Evolution in Mg Alloys by Phase-Field Methods: A Review. *Crystals* **2022**, *12*, 1305. [CrossRef]
26. Munoz, E.J.; Attari, V.; Martinez, M.C.; Dickerson, M.B.; Radovic, M.; Arroyave, R. Phase-Field Model of Silicon Carbide Growth During Isothermal Condition. *Comput. Mater. Sci.* **2024**, *242*, 113058. [CrossRef]
27. Chen, L.-Q.; Fan, D. Computer Simulation Model for Coupled Grain Growth and Ostwald Ripening—Application to Al_2O_3-ZrO_2 Two-Phase Systems. *J. Am. Ceram. Soc.* **2005**, *79*, 1163–1168. [CrossRef]
28. Allen, S.M.; Cahn, J.W. A Microscopic Theory for Antiphase Boundary Motion and Its Application to Antiphase Domain Coarsening. *Acta Metall.* **1979**, *27*, 1085–1095. [CrossRef]
29. Cahn, J.W.; Hilliard, J.E. Free Energy of a Nonuniform System. I. Interfacial Free Energy. *J. Chem. Phys.* **1958**, *28*, 258–267. [CrossRef]
30. Zheng, S.; Sokoluk, M.; Yao, G.; de Rosa, I.; Li, X. Fe–Ni Invar Alloy Reinforced by Wc Nanoparticles with High Strength and Low Thermal Expansion. *SN Appl. Sci.* **2019**, *1*, 172. [CrossRef]

31. Chen, L.Q.; Shen, J. Applications of Semi-Implicit Fourier-Spectral Method to Phase Field Equations. *Comput. Phys. Commun.* **1998**, *108*, 147–158. [CrossRef]
32. Gránásy, L.; Börzsönyi, T.; Pusztai, T. Crystal Nucleation and Growth in Binary Phase-Field Theory. *J. Cryst. Growth* **2002**, *237–239*, 1813–1817. [CrossRef]
33. Hu, Z.; Wise, S.M.; Wang, C.; Lowengrub, J.S. Stable and Efficient Finite-Difference Nonlinear-Multigrid Schemes for the Phase Field Crystal Equation. *J. Comput. Phys.* **2009**, *228*, 5323–5339. [CrossRef]
34. Eiken, J.; Bottger, B.; Steinbach, I. Multiphase-Field Approach for Multicomponent Alloys with Extrapolation Scheme for Numerical Application. *Phys. Rev. E Stat. Nonlin Soft Matter Phys.* **2006**, *73 Pt 2*, 066122. [CrossRef] [PubMed]
35. Yoon, S.; Jeong, D.; Lee, C.; Kim, H.; Kim, S.; Lee, H.G.; Kim, J. Fourier-Spectral Method for the Phase-Field Equations. *Mathematics* **2020**, *8*, 1385. [CrossRef]
36. Jaatinen, A.; Achim, C.V.; Elder, K.R.; Ala-Nissila, T. Thermodynamics of Bcc Metals in Phase-Field-Crystal Models. *Phys. Rev. E Stat. Nonlin Soft Matter Phys.* **2009**, *80 Pt 1*, 031602. [CrossRef]
37. Tegze, G.; Granasy, L.; Toth, G.I.; Podmaniczky, F.; Jaatinen, A.; Ala-Nissila, T.; Pusztai, T. Diffusion-Controlled Anisotropic Growth of Stable and Metastable Crystal Polymorphs in the Phase-Field Crystal Model. *Phys. Rev. Lett.* **2009**, *103*, 035702. [CrossRef]
38. Wang, Y.U.; Jin, Y.M.; Cuitiño, A.M.; Khachaturyan, A.G. Nanoscale Phase Field Microelasticity Theory of Dislocations: Model and 3d Simulations. *Acta Mater.* **2001**, *49*, 1847–1857. [CrossRef]
39. Choudhury, A.; Nestler, B. Grand-Potential Formulation for Multicomponent Phase Transformations Combined with Thin-Interface Asymptotics of the Double-Obstacle Potential. *Phys. Rev. E Stat. Nonlin Soft Matter Phys.* **2012**, *85 Pt 1*, 021602. [CrossRef]
40. Oono, Y.; Puri, S. Computationally Efficient Modeling of Ordering of Quenched Phases. *Phys. Rev. Lett.* **1987**, *58*, 836–839. [CrossRef]
41. Xia, M.J.; Gu, D.D.; Ma, C.L.; Zhang, H.M.; Dai, D.H.; Chen, H.Y.; Li, C.P.; Zhou, Z.J.; Chen, G.F.; Kelbassa, I. Fragmentation and Refinement Behavior and Underlying Thermodynamic Mechanism of Wc Reinforcement During Selective Laser Melting of Ni-Based Composites. *J. Alloys Compd.* **2019**, *777*, 693–702. [CrossRef]
42. Fan, D.; Chen, L.Q. Computer Simulation of Grain Growth Using a Continuum Field Model. *Acta Mater.* **1997**, *45*, 611–622. [CrossRef]
43. Olyslager, F. Overview of the Singular Behavior of Electromagnetic Fields at Edges and Tips in Bi-Isotropic and Special Bianisotropic Media. *Radio Sci.* **1995**, *30*, 1349–1354. [CrossRef]
44. Valagiannopoulos, C.A. On Smoothening the Singular Field Developed in the Vicinity of Metallic Edges. *Int. J. Appl. Electromagn. Mech.* **2009**, *31*, 67–77. [CrossRef]

metals MDPI

Article

Study of Size Effect on $Ni_{60}Nb_{40}$ Amorphous Particles and Thin Films by Molecular Dynamic Simulations

Wenbiao Zhang [1,2,*] **and Yungui Ma** [1]

[1] College of Optical Science and Engineering, Zhejiang University, Hangzhou 310027, China; yungui@zju.edu.cn
[2] Key Laboratory of Spintronics Materials, Devices and Systems of Zhejiang Province, Hangzhou 311300, China
* Correspondence: 0622546@zju.edu.cn

Abstract: $Ni_{60}Nb_{40}$ amorphous particles (APs) and amorphous thin films (ATFs) with various sizes were investigated by molecular dynamic simulations. It is revealed that sample size has effects on both $Ni_{60}Nb_{40}$ APs and ATFs composed of shell or surface and core components. $Ni_{60}Nb_{40}$ APs have an average bond length of 2.57 Å with major fivefold-symmetry atomic packing and low bond-orientation orders of Q_6 and Q_4 in both core and shell components. Ni atoms in $Ni_{60}Nb_{40}$ APs and ATFs prefer to segregate to the shell and surface regions, respectively. Atomic packing structure differences between various-sized $Ni_{60}Nb_{40}$ APs and ATFs affect their glass transition temperatures T_g, i.e., T_g decreases as the particle size or the film thickness decreases in $Ni_{60}Nb_{40}$ APs and ATFs, respectively. Our obtained results for $Ni_{60}Nb_{40}$ APs and ATFs clearly reveal a size effect on atomic packing and glass transition temperature in low-dimensional metallic glass systems.

Keywords: amorphous particles; amorphous thin films; simulations; size effect

1. Introduction

Due to its unique atomic packing structure with no long-range translation or orientation order, metallic glass (MG), a new class of materials, exhibits excellent properties, e.g., high strength, large elastic limit, superior wear and corrosion resistance [1,2]. Concerning the structure in MG, although considerable studies have been reported, due to the complicated interactions between elements and topological complexity in multi-components MG with more than three elements [3,4], a complete understanding of this issue is still lacking. Fortunately, the Ni-Nb binary system, which has good glass-forming ability (GFA) and can be prepared in mm-sized MG by copper mold casting, provides prototypes to investigate the correlation of the structure with the properties of MG, and relatively reliable semi-empirical embedded atom method (EAM) potentials of this binary system were recently reported [5]. Extensive works via experiments and simulations [6] have been carried out to describe the structure or to reveal the effects of various factors, e.g., the correlation between cooling rate and defect concentration [7] and composition [8]. The atomic-level structure of a representative multi-component metallic glass (MG) has been resolved. It was also found that the sample size is an important factor affecting properties of MG. For example, Ref. [9] showed that the yield strength and deformation mechanism in the metallic glass were size-dependent, and Ref. [10] suggest that the transition from brittle to ductile behavior is driven by sample size, while the extent of ductility is driven by the surface state elastic limit [11]. It was experimentally revealed by in situ transmission electron microscopy observations that small-sized $Ni_{60}Nb_{40}$ magnetron-sputtered glassy films confined by a frame exhibited a significant size effect on the elastic strain limits [12]. The mechanical behavior of nanoglass is also affected by the grain boundary thickness and the fraction of atoms at interfaces for a given average grain size [13,14]. However, few studies have been reported of the size effect on the atomic structure of MG. One possible reason for this could be that free-standing amorphous particles (APs) and amorphous thin films (ATFs) with various sizes, having the same thermal history, are very difficult, if not impossible, to experimentally fabricate. Due to the small size of the key unit in systems, nanostructured amorphous alloys have been considerably investigated. The sample size is an important factor affecting the structure and properties of amorphous alloys. However, few studies of the size effect on atomic structure and properties of amorphous alloys have been reported, especially in low-dimensional amorphous particles (APs) and amorphous thin films (ATFs), due to the fact that the experimental fabrication of APs and ATFs with various sizes and the same thermal history is very difficult, if not impossible. Concerning the complicated interactions between elements and topological complexity in multi-component amorphous alloys with more than three elements, a complete understanding of this issue is still lacking. Fortunately, the Ni-Nb binary system, which has good glass-forming ability (GFA) and can be prepared in mm-sized MG by copper mold casting, provides prototypes to investigate the correlation of the structure with the properties of amorphous alloys because its relatively reliable semi-empirical embedded atom potential method was recently developed. So far, the size effects on the atomic structure and glass transition temperature of $Ni_{60}Nb_{40}$ APs and ATFs have not been reported yet. Therefore, in this work, we studied the size effects in Ni60Nb40 APs and ATFs with various sizes and revealed the atomic packing and glass transition temperature differences in Ni60Nb40 APs and ATFs with various particle sizes or thicknesses. In this work, the size effects on the structure and property of $Ni_{60}Nb_{40}$ APs and ATFs were investigated by molecular dynamics simulations. The size effect on structure is mainly characterized by pair distribution functions, coordination number, HA index, and bond-orientation order parameters, while the size effects on glass transition temperature and segregation were also analyzed. Our obtained results for Ni60Nb40 APs and ATFs clearly reveal a size effect on atomic packing and glass transition temperature in low-dimensional amorphous systems and will trigger more experimental and theoretical studies on amorphous alloys.

2. Methods

Molecular dynamics simulations of $Ni_{60}Nb_{40}$ amorphous particles (APs) and amorphous thin films (ATFs) were performed based on the embedded atom method (EAM) potential in the canonical Ensemble (NVT) using the large-scale atomic molecular massively parallel simulator (LAMMPS) code. The initial structure of APs was constructed as follows: (1) creating a cubic zone containing 50, 100, 200, 400, 700, 1000, 4000, and 5000 atoms randomly distributed with a density of about 8.5 g/cm^3. (2) Locating this cubic zone at the center of a big simulation box with the periodic boundary condition, i.e., the simulation box is 62 Å for 50–1000 atoms, and 80 Å for 4000 and 5000 atoms, to achieve a sufficient vacuum layer and negligible interaction between MGPs to mimic isolated particles. (3) The simulation box was melted and equilibrated at 2200 K for 1 ns, then quenched to 300 K with a cooling rate of 1×10^{12} K/s. The centered cubic zone turned into a spherical-like particle which almost remains during the melting and the subsequent cooling process. The time step was 1 fs, i.e., 1×10^6 configurations were generated for 1 ns and we saved 10,000 configurations at intervals of 100 for data analyses. $Ni_{60}Nb_{40}$ APs had various atom numbers of 50, 100, 200, 400, 700, 1000, 4000, and 5000 and their corresponding radii of about 5.5 Å, 7.2 Å, 9.5 Å, 11.7 Å, 13.8 Å, 15.8 Å, 25.1 Å, and 27.0 Å. These particles were then put at the center of simulation box which was under the periodic boundary condition with sizes of 62 Å for 50–1000 atoms, and 80 Å for 4000 and 5000 atoms, to achieve a sufficient vacuum layer and negligible interaction between APs. The $Ni_{60}Nb_{40}$ APs were melted and equilibrated at 2200 K for 1 ns to remove the influence of initial configurations, then quenched to 300 K with a cooling rate of 1×10^{12} K/s. $Ni_{60}Nb_{40}$ APs were equilibrated at 300 K for 1 ns and exhibited an amorphous structure as shown in Figure 1. We further analyzed atomic structures of all studied APs with various sizes, in which two components, i.e., core and shell, were divided. In order to obtain reliable structural information, all structure analyses with 2000 configurations were carried out to obtain average values. Secondly, $Ni_{60}Nb_{40}$ amorphous thin films (ATFs) with different thicknesses were obtained by cutting the bulk box with a size of 56 Å to the given thicknesses from 8.6 Å to 40 Å, and setting two vacuum layers along the thick direction to reach the free-standing condition. After relaxing the thin films for 1 ns at 2200 K, all the systems were quenched down to 300 K with a cooling rate of 1×10^{12} K/s and relaxed at 300 K for 1 ns. The initial bulk box (used for thin films) was generated in a different way for the particles. We started with a Ni single crystal containing 10,000 atoms with a simulation cell approximately 56 Å × 56 Å × 56 Å in X-, Y-, and Z-directions, respectively, and then randomly substituted the atomic percentage (40%) of Ni atoms by Nb atoms. This mixture was then held at T = 2200 K (a temperature well above the melting point of the $Ni_{60}Nb_{40}$ alloy) for 1 ns to allow for relaxation. The simulation time step was 1 fs. The purpose of this relaxation was to ensure the initial system has become a structurally homogeneous liquid. The model was then cooled to T = 300 K at a cooling rate of 10^{12} K/s and then relaxed for 1 ns at 300 K. Periodic boundary conditions were applied to all three dimensions. The films with different thicknesses were cut from the center part of the bulk box to avoid the composition segregation. Compositions of all thin films were found to be close to the average value of 60 at.% for Ni atoms and 40 at.% for Nb atoms in the bulk with a fluctuation less than 0.5 at.%. The thin films with thicknesses from 8.6 Å to 40 Å were relaxed for 1 ns at 300 K. For each size MGP (or MGTF) investigated, we ran MD trajectories. In order to obtain reliable structural information, all structure analyses were carried out with 2000 configurations to obtain average values.

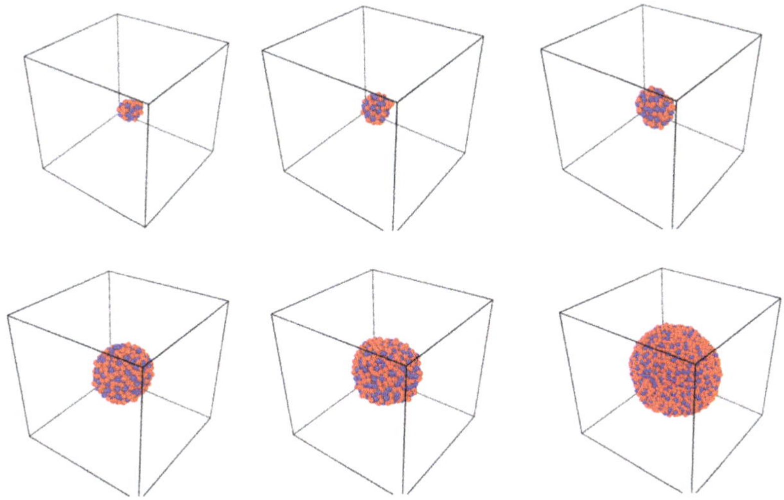

Figure 1. Structure configurations of 50, 100, 200, 1000, 4000, and 5000 amorphous particles, respectively.

3. Results and Discussion

3.1. Amorphous Particles

The pair distribution function $g(r)$ is a powerful method for describing the structure of amorphous materials. For finite-sized materials, such as particles, $g(r)$ can be obtained via:

$$g(r) = \frac{dn(r)}{dV(r) * \frac{N}{V_0}} = \frac{1}{S}\frac{V_0}{N}\frac{dn(r)}{dr} \tag{1}$$

where N represents the total number of atoms, V_0 is the model volume, S is the surface area of the particle shell. Figure 2 shows $g(r)$ for $Ni_{60}Nb_{40}$ amorphous particles (APs), i.e., metallic glassy particles (MGPs) with various sizes from 50 to 5000 atoms which are divided into two parts, core and shell, as shown in the inset picture of Figure 2. $g(r)$ of $Ni_{60}Nb_{40}$ APs exhibits obvious the characteristic of being amorphous, e.g., the relative intensive first peak and a split of the second peak. The intensity of the first peak increases with the size in $Ni_{60}Nb_{40}$ APs. It is found that $g(r)$ in $Ni_{60}Nb_{40}$ APs has a narrow and sharp first peak, while for the second and third peaks of $g(r)$ in the range of 4–7 Å, $Ni_{60}Nb_{40}$ APs have a similar structure for sizes above 400 atoms, i.e., $Ni_{60}Nb_{40}$ APs with 50–400 atoms exhibit more irregular peaks as compared with APs having 700–5000 atoms. Using the core-and-shell model, the thickness of shell in $Ni_{60}Nb_{40}$ APs, determined by the minimum after the first peak in total $g(r)$, is about 3.66 Å. It is found that the intensity of the core component increases while the shell $g(r)$ component decreases as the AP's size becomes larger. The first and second peak positions about 2–5.6 Å of the shell $g(r)$ component in $Ni_{60}Nb_{40}$ APs rarely change with increasing the size while their relative intensities change significantly. When the size becomes larger than 400 atoms, the shell $g(r)$ components of in $Ni_{60}Nb_{40}$ APs do not change much with MGP's size, while two maximums for the second peak are located at about 4.3 Å and 5.0 Å and the third peak is located at about 6.7 Å in $Ni_{60}Nb_{40}$ APs. To search for more detailed structure information contained in $g(r)$, we plotted the partial $g(r)$ of Ni-Ni, Ni-Nb, and Nb-Nb pairs, shown in Figure 3. Again, in $Ni_{60}Nb_{40}$ APs, all three partial $g(r)$ values are similar for sizes larger than 400 atoms. The first peak in three partial $g(r)$ is similar in $Ni_{60}Nb_{40}$ APs, while the peak position of Nb-Nb bonds almost remains unchanged in the range from 50 atoms to 400 atoms. The intensities of the first peak of Ni-Ni and Ni-Nb partial $g(r)$ slightly increase with size, while for Nb-Nb partial $g(r)$ they remain nearly unchanged. Concerning the second neighboring atomic packing structures, several features are detected: (1) for Ni-Ni partial $g(r)$, peaks are located at about 4.05 and 5 Å; (2) for Ni-Nb partial $g(r)$, peaks are located at about 4.3 and 5 Å; for the third neighbor, the peak is located at about 6.4 Å for Ni-Nb partial $g(r)$; (3) For Nb-Nb partial $g(r)$, the peak is mainly located at about 4.5 Å. Table 1 lists all first peak positions in total and partial $g(r)$ for $Ni_{60}Nb_{40}$ APs. It is found that the first peak position of total $g(r)$ becomes slightly larger, from 2.56 Å to 2.59 Å, as particle size increases from 50 atoms to 5000 atoms for $Ni_{60}Nb_{40}$ APs. For partial $g(r)$, Ni-Ni, Ni-Nb, and Nb-Nb also slightly increase as size becomes larger, i.e., Ni-Ni increases from 2.46 Å to 2.49 Å, Ni-Nb increases from 2.59 Å to 2.63 Å, and Nb-Nb increases from 2.92 Å to 2.98 Å.

Coordination number depends on the cutoff length in $Ni_{60}Nb_{40}$ APs. The first minimum in $g(r)$ was used here as the cutoff length for the first nearest shell. The total and the partial coordination number (CN) distributions for the studied $Ni_{60}Nb_{40}$ APs are plotted in Figure 4. Three peaks located at about CN = 6, 9, 12 are found in the total CN distribution. The CN distribution depends on the size of $Ni_{60}Nb_{40}$ APs. For smaller $Ni_{60}Nb_{40}$ APs, e.g., 50, 100, and 200 atoms, the peak at CN = 6 decreases with increasing size and almost disappears for sizes larger than 200 atoms. Meanwhile, the peak of CN = 12 largely increases when the size is larger than 200 atoms. To further investigate local atomic environments in the studied $Ni_{60}Nb_{40}$ APs, CN distributions in both core and shell components were also calculated. In the core component, CN is only distributed in about 11–16, i.e., almost no atomic packing with CN less than 11 was found in the core component. CN distributions having very sharp and irregular peaks for 50- and 100-atom $Ni_{60}Nb_{40}$ APs should be linked with small atomic numbers in the core component. However, after 200 atoms, they become have a steady distribution from CN = 11–16. In the shell component for $Ni_{60}Nb_{40}$ APs, CN has three peaks located at CN = 6, 9, and 12. The intensity of the CN = 6 rapidly decreases with increasing size from 50 to 400 atoms, while the peaks of CN = 9 and 12 do not change much with size. This lower CN in the shell component for $Ni_{60}Nb_{40}$ APs can be explained by the surface effect, i.e., surface atoms do not have enough neighbor atoms as compared to core atoms in general. Considering the possibilities of different atomic packing for various atoms in $Ni_{60}Nb_{40}$ APs, Ni- and Nb-centered CN distributions are plotted in Figure 4. In the core component, CNs of Ni atoms are centered at about 12, while CNs are centered at about 15 for Nb atoms. In the shell component, CN distributions of Ni atoms are similar to the CN distributions of the total, indicating that the shell components in $Ni_{60}Nb_{40}$ APs should have a Ni-rich composition. This is indeed confirmed later. The CN distribution of Nb atoms in the 50-atom $Ni_{60}Nb_{40}$ MGP peaks at CN = 9, and shifts to a slightly high value of CN = 10–11 for larger APs.

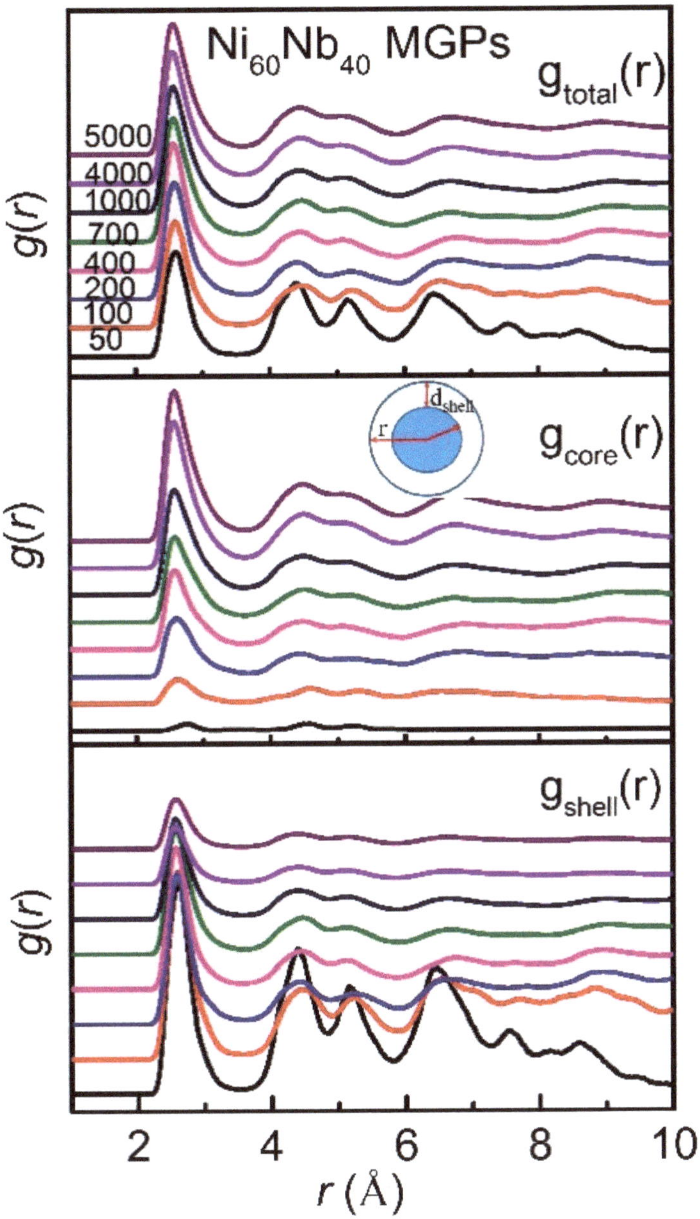

Figure 2. Pair distribution functions $g(r)$ for $Ni_{60}Nb_{40}$ APs with various sizes of core and shell components together; illustration of the core-and-shell model in the inset.

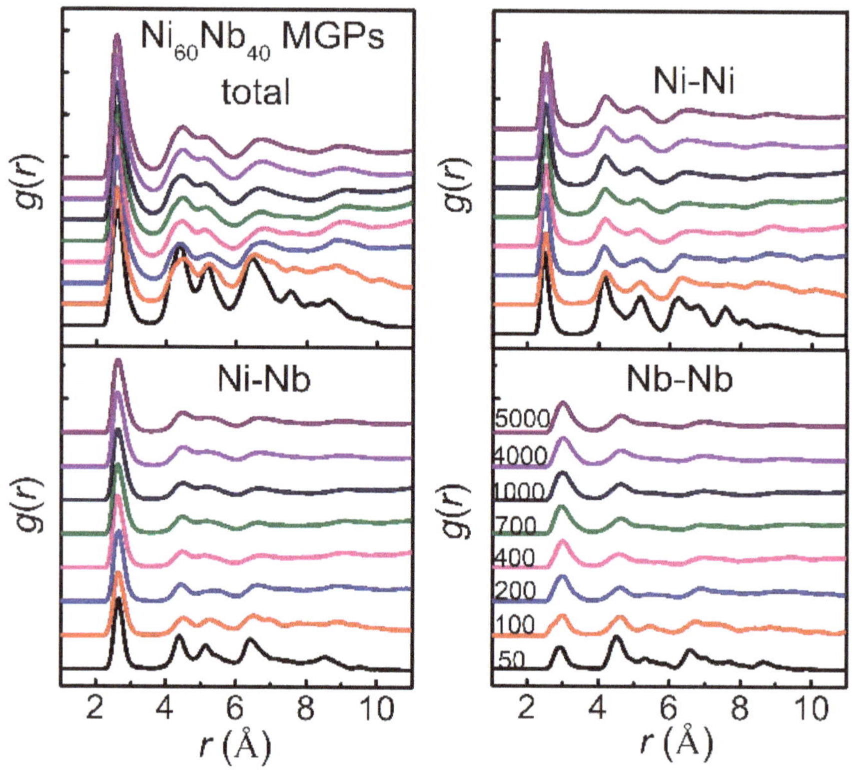

Figure 3. Partial pair distribution functions $g(r)$ for $Ni_{60}Nb_{40}$ APs with various sizes.

Table 1. The first peak positions of total $g(r)$ for $Ni_{60}Nb_{40}$ amorphous particles (APs) and thin films (ATFs).

$Ni_{60}Ni_{36}$ APs		$Ni_{60}Ni_{36}$ ATFs	
Number of Atoms	**Average Bond (Å) (±0.02)**	**Thickness**	**Average Bond (Å) (±0.02)**
50	2.56	8.6 Å	2.58
100	2.57	10.1 Å	2.59
200	2.57	13.7 Å	2.58
400	2.57	15.4 Å	2.58
700	2.58	17.8 Å	2.58
1000	2.57	20 Å	2.57
4000	2.59	25 Å	2.57
5000	2.59	30 Å	2.58

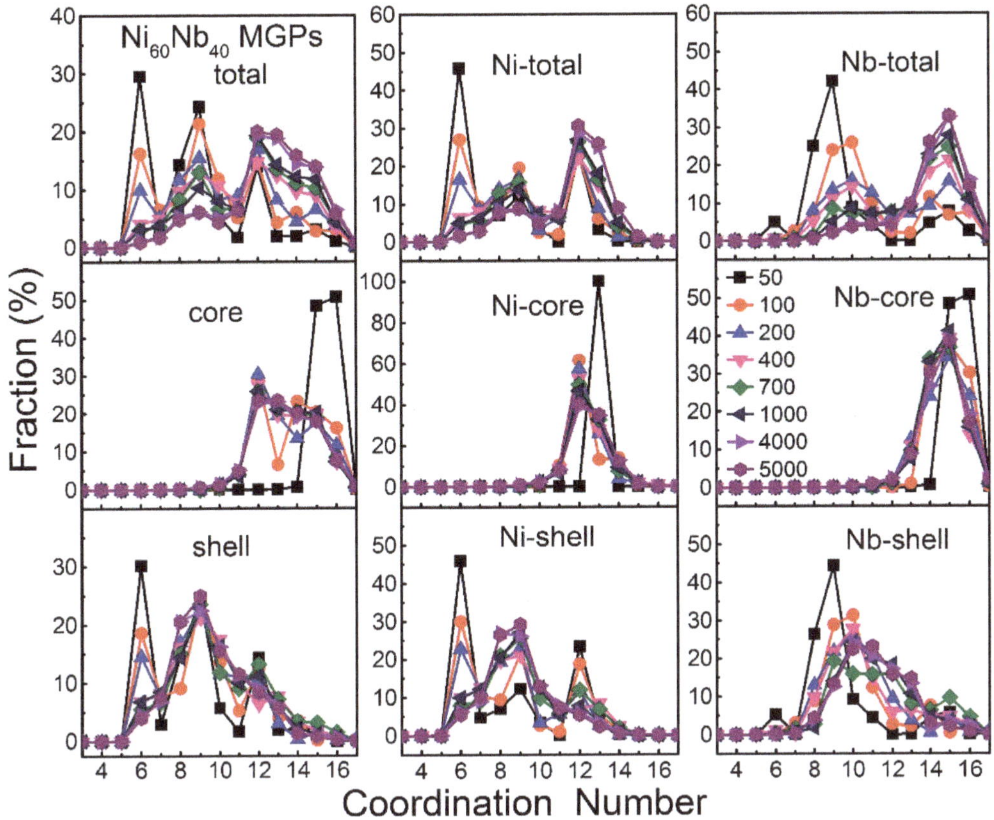

Figure 4. The coordination number distributions of the total sample, core, and shell components, and partial coordination number distributions for $Ni_{60}Nb_{40}$ APs with various sizes.

The Honey–Anderson (HA) index [15] is introduced to investigate local structure of $Ni_{60}Nb_{40}$ APs, as shown in Figure 5. HA has four indices i, j, l, m, and is a method used to characterize the common neighbors of an atom pair. $i = 1$ means atom A and atom B form a bond, and otherwise $i = 2$. j denotes the number of common neighbors which form bonds with atom A and atom B. l represents the number of bonds formed among the neighboring atoms. m is a special index to classify the bonds arrangements. It is found that the 1551 HA indexes account for large proportions that equal more than 40% in total, which is characteristic of the fivefold-symmetry structure, implying that icosahedral-like structures are dominant in $Ni_{60}Nb_{40}$ APs. Meanwhile, the fractions of 1551, 1541, 1661, and 1441 indexes slightly increase with size, while the 1431 and 1321 indexes have the opposite trend of decreasing as size increases from 50 to 5000 atoms. In the core component of $Ni_{60}Nb_{40}$ APs, HA indexes do not depend on size. The 1551 index has a fraction of more than 50%, and the second, third, fourth and fifth highest indexes are 1661, about 20%; 1541, about 10%; 1441, about 8%; and 1431, about 6%, respectively. In the shell component of $Ni_{60}Nb_{40}$ APs, 1321 and 1431 indexes have obvious high fractions, while they are tiny in the core component, although the 1551 index is still the dominant local atomic packing. Bond-orientation order (BOO) parameters, e.g., Q_6 and Q_4, are often used to measure atomic cluster symmetries in disordered systems [16–19]. The averaged parameters (BOO) are sensitive to different bond-orientational symmetries, so they can be used to differentiate the local atomic environments. The complex vector (q_l) can be defined as $q_{lm}(i) = \frac{1}{N_b(i)}\sum_{j=1}^{N_b(i)} Y_{lm}(\hat{r_{ij}})$, where l represents the free integer parameter, m is an integer that runs from $-l$ to l, Y_{lm} are the spherical harmonics, $\hat{r}_{ij}$ is the vector from atom i to atom j, and the sum goes over all neighboring atoms $N_b(i)$, the number of atom i. One can average the spatially local bond order parameters, Q_4 and Q_6, according to the following equation:

$$Q_l(i) = \sqrt{4\pi/(2l+1)}|\hat{q}_{lm} \tag{2}$$

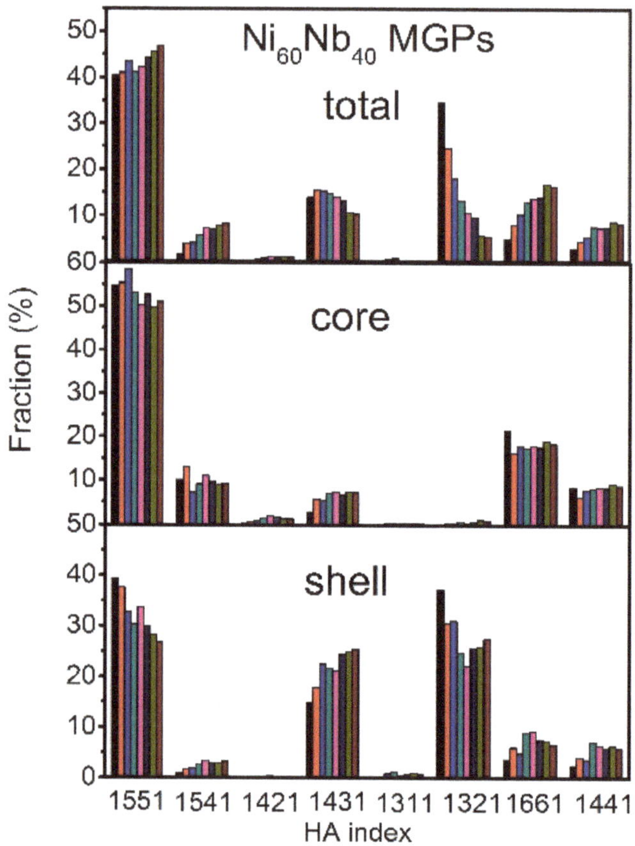

Figure 5. The HA index for total sample, core, and shell components for $Ni_{60}Nb_{40}$ APs with various sizes.

Figure 6 depicts Q_6 and Q_4 values for $Ni_{60}Nb_{40}$ APs with various sizes. It is found that Q_6 and Q_4 values in the shell component are larger than those in the core component in this system. In total particles, both parameters have values of the averages in both shell and core components. These results indicate more ordering in the shell component than in the core component. This might be linked with composition segregation in the shell component. Figure 7 depicts cluster size-dependent Ni contents in the shell component in $Ni_{60}Nb_{40}$ APs, in which the thickness of the shell determined by the minimum after the first peak in total $g(r)$ is about 3.66 Å in $Ni_{60}Nb_{40}$ APs. The Ni composition in the shell is larger than 60 at.% and increases from 64 at.% for 50-atom APs to 73 at.% in 5000-atom APs, suggesting a segregation of Ni atoms in shell components in larger clusters, which are roughly correlated with Q_6 and Q_4 values. Q_6 and Q_4 values in the shell component of

$Cu_{64}Zr_{36}$ APs illustrated in Ref [20] are even larger than those in $Ni_{60}Nb_{40}$ APs, i.e., atomic packing in the shell component of $Cu_{64}Zr_{36}$ APs, roughly speaking, has a slightly higher order than that in $Ni_{60}Nb_{40}$ APs. These results are closely linked with a larger degree of composition segregation in the shell component of $Cu_{64}Zr_{36}$ APs [20].

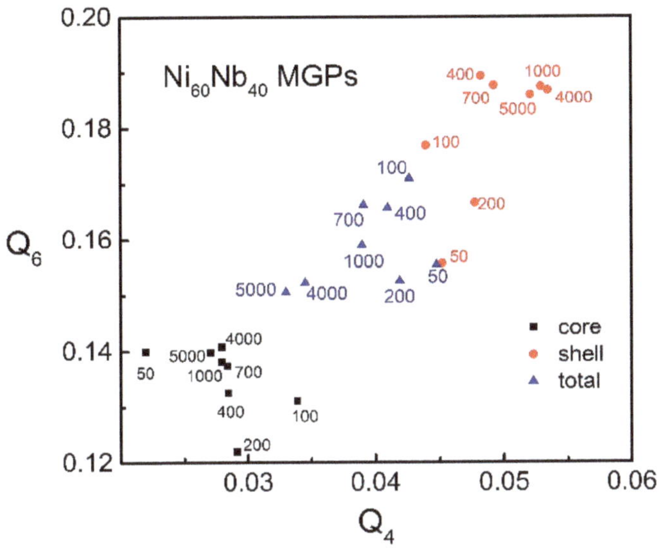

Figure 6. The Q_6-Q_4 distribution of total, core, and shell components for $Ni_{60}Nb_{40}$ APs with various sizes. Numbers of atoms are marked.

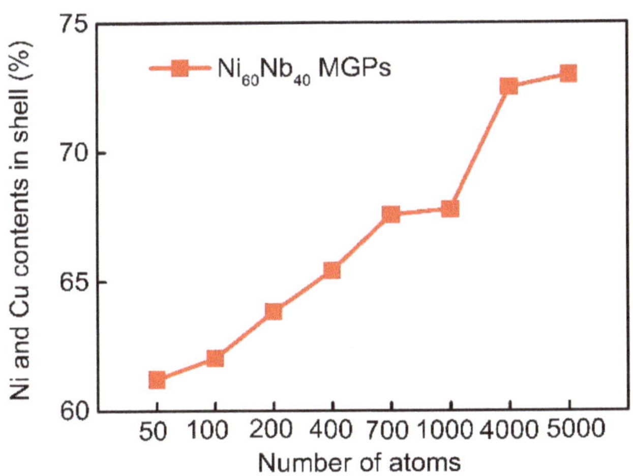

Figure 7. Ni contents in the shell of $Ni_{60}Nb_{40}$ APs with various sizes. The dash line represents the initial composition of APs.

All the above-mentioned results clearly reveal that particle size indeed affects the atomic structure of $Ni_{60}Nb_{40}$ APs. The properties of amorphous materials are always closely related with their structures, which motivated us to compare the size effect on properties of such APs. Hence, we further studied the size dependence of the glass transition temperature, T_g, which is closely related to the atomic structure in disordered systems. The glass transition temperature of $Ni_{60}Nb_{40}$ APs with various sizes was estimated by the intersection of potential energy of low and high temperatures, as plotted in Figure 8a. It is found that T_g increases as the size of the particle increases in $Ni_{60}Nb_{40}$ APs, i.e., the sample size indeed influences the glass transition temperature of $Ni_{60}Nb_{40}$ APs. T_g of all the studied $Ni_{60}Nb_{40}$ APs can be well fitted with a simple equation of $T_g = f_{shell}T_g^{shell} + f_cT_g^c$, where f_{shell} and f_c, and T_g^{shell} and T_g^c, represent the fractions and glass transition temperatures of the shell and core components, respectively. The shell thicknesses were determined by the minimum after the first peak in total $g(r)$ curves of $Ni_{60}Nb_{40}$ APs. Figure 8b shows the fitting curves with a fitting parameter of T_g^{shell} = 630 K and T_g^c = 1090 K for $Ni_{60}Nb_{40}$ APs. The glass transition temperature, T_g, for the shell component is much lower than that for the core component in $Ni_{60}Nb_{40}$ APs, which could be caused by the fast dynamic motion of the shell component.

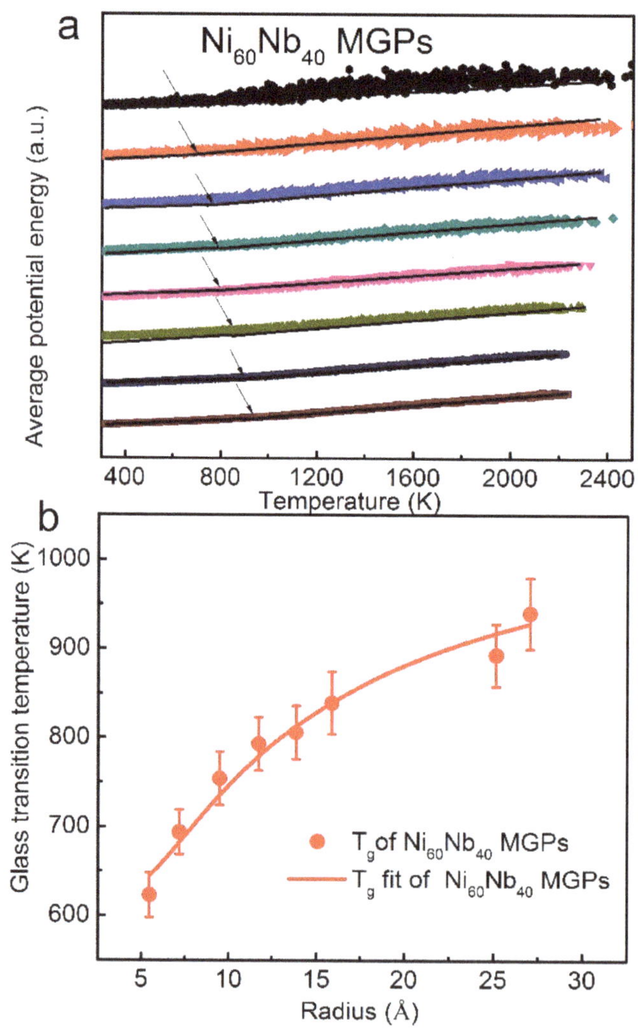

Figure 8. (**a**) Determination of glass transition temperature T_g via an intersection of low- and high-temperature dependence of the potential energy for $Ni_{60}Nb_{40}$ APs with various sizes. (**b**) Fitting of glass transition temperature data for $Ni_{60}Nb_{40}$ APs with various sizes.

3.2. Amorphous Thin Films

Besides the APs discussed above, we also explored size effects on atomic structure and glass transition temperature for $Ni_{60}Nb_{40}$ amorphous thin films (ATFs) i.e., metallic glassy thin films (MGTFs). The structure configuration images from the simulations for the thin films of different thicknesses are shown in Figure 9. Figure 10 shows pair distribution function $g(r)$ curves of $Ni_{60}Nb_{40}$ ATFs at 300 K with different thicknesses from 8.6 Å to 40 Å. The sharp first peak and the split second peak in $g(r)$ curves of films demonstrate that all studied thin films are fully amorphous. $Ni_{60}Nb_{40}$ ATFs have sharp and narrow first peaks in $g(r)$. For the second peak within 4 to 6 Å, the relative intensity of the shoulder peak for $Ni_{60}Nb_{40}$ ATFs becomes weak, indicating that the medium range order might be low in $Ni_{60}Nb_{40}$ ATFs. Similar to $Ni_{60}Nb_{40}$ APs, $Ni_{60}Nb_{40}$ ATFs can also be composed of two regions, i.e., the core region and surface layer region. Figure 11 demonstrates the composition variations along the film thicknesses in $Ni_{60}Nb_{40}$ ATFs with a thickness of 40 Å. The element segregation again appears in the surface layer of the $Ni_{60}Nb_{40}$ MGTFs, which is also observed in $Ni_{60}Nb_{40}$ ATFs with other thicknesses. The fraction of Ni atoms is found to be increased in the surface region, where Ni content is about 75~90%, higher than its nominal content of 60%. In contrast, the compositions in the core regions are close to their nominal values in $Ni_{60}Nb_{40}$ ATFs.

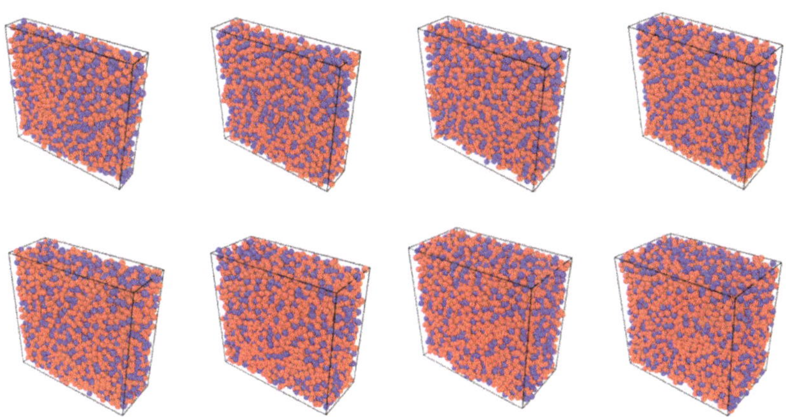

Figure 9. Structure configurations of 8.6 Å, 10.7 Å, 13.1 Å, 15.4 Å, 17.8 Å, 20 Å, 25 Å, and 30 Å thick amorphous thin films, respectively.

Furthermore, the coordination number, HA index, and BOO parameters for $Ni_{60}Nb_{40}$ ATFs were also studied. As plotted in Figure 12, CN distribution depends on the thickness of $Ni_{60}Nb_{40}$ ATFs and two peaks located at about CN = 9, 12 were found in the total CN distribution. For thinner ATFs, e.g., 8.6 Å, 10.7 Å, and 13.1 Å, the peak at CN = 9 decreases with increasing thickness and the peak of CN = 12 increases when the thickness is larger than 15.4 Å. CN distributions in both core and surface components were also calculated. In the core component, CN is distributed within 11–16 and has a peak at CN = 12. In the shell component for $Ni_{60}Nb_{40}$ ATFs, CN has a peak located at CN = 9. This lower CN in the surface component can be explained by the surface effect as explained by the AP findings. Ni- and Nb-centered CN distributions are plotted in Figure 12. In the core component, CNs of Ni atoms are centered at about 12, while CNs are centered at about 14 for Nb atoms. In the surface component, CN distributions of Ni atoms are 9 and those for Nb atoms are 11. As shown in Figure 13, 1551 HA indexes account for large proportions, more than 40% in total, implying that icosahedral-like structures are dominant in $Ni_{60}Nb_{40}$ ATFs, as in $Ni_{60}Nb_{40}$ APs in Figure 5. Meanwhile, the fractions of 1551, 1541, 1661, and 1321 indexes slightly increase with thickness, while the 1421 and 1311 indexes have the opposite trend of decreasing as thickness increases from 8.6 to 40 Å. In the core component of $Ni_{60}Nb_{40}$ ATFs, the 1551 index has a fraction of more than 45%. In the shell component of $Ni_{60}Nb_{40}$ ATFs, 1421 and 1311 indexes also have high fractions above 20%, similar to 1431 and 1321 indexes for the shell component in $Ni_{60}Nb_{40}$ APs in Figure 5. Q_6 and Q_4 values for $Ni_{60}Nb_{40}$ ATFs with various thickness are plotted in Figure 14. It is seen that Q_6 and Q_4 values in the surface component are larger than those in the core component in $Ni_{60}Nb_{40}$ ATFs, which is similar to $Ni_{60}Nb_{40}$ APs in Figure 5, indicating higher ordering in the surface component that in the core one. This phenomenon could be linked to composition segregation in the surface component, as illustrated in Figure 11.

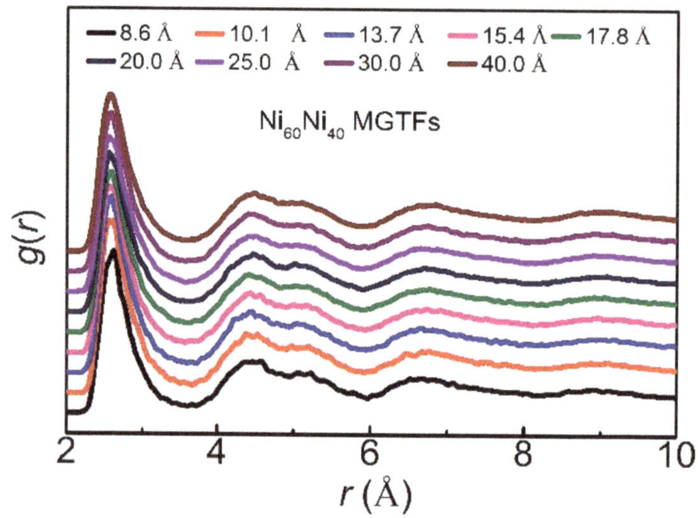

Figure 10. Pair distribution functions *g(r)* for $Ni_{60}Nb_{40}$ ATFs with various thicknesses.

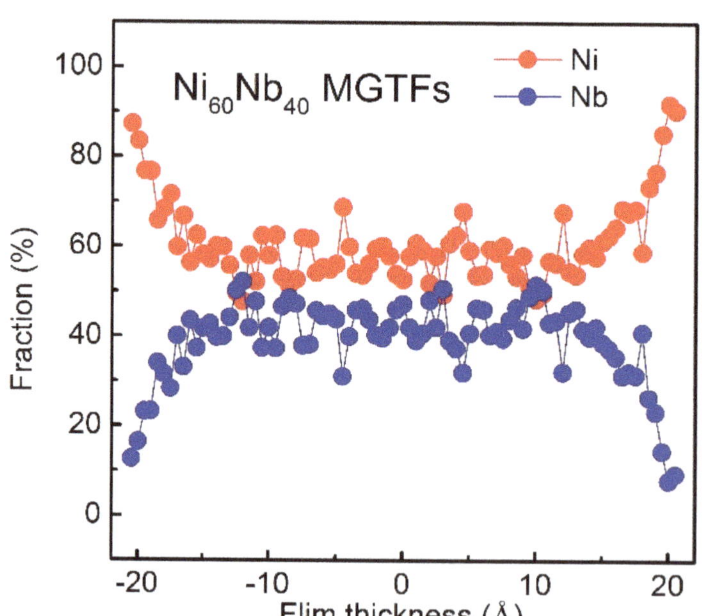

Figure 11. Composition variations along the film thickness for the 40 Å thick $Ni_{60}Nb_{40}$ MGTF.

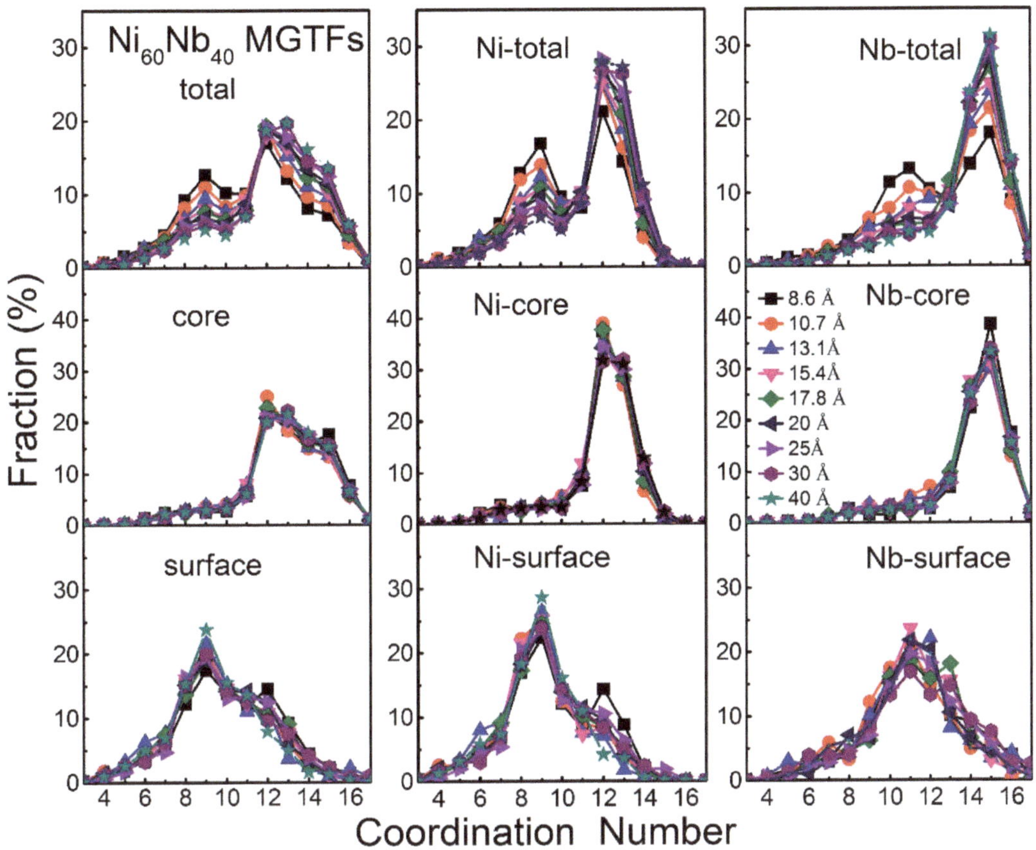

Figure 12. The coordination number distributions of the total sample, core, and surface components, and partial coordination number distributions for $Ni_{60}Nb_{40}$ ATFs with various thickness.

The average atomic internal energy of $Ni_{60}Nb_{40}$ ATFs with various thicknesses as a function of temperature during cooling was studied, from which the glass transition temperature T_g for ATFs was estimated by an intersection of potential energy of low and high temperatures, shown in Figure 15a. It was found again that T_g decreases substantially as the film thickness decreases in $Ni_{60}Nb_{40}$ ATFs, similar to the results observed in $Ni_{60}Nb_{40}$ APs in Figure 8a. The glass transition temperature as a function of film thickness can be well fitted with a simple equation (model II), i.e., $T_g(d) = T_g(\infty)(1 - C/d)$, as shown in Figure 15b, where T_g (∞) and C represent two fitting parameters, and d is the film thickness. The obtained $T_g(\infty)$ and C values are 975 K and 1.6 for $Ni_{60}Nb_{40}$ ATFs, respectively. In fact, T_g of all the studied $Ni_{60}Nb_{40}$ ATFs can also be fitted with the model I of $T_g = f_{surface}T_g{}^{surface} + f_cT_g{}^c$, where $f_{surface}$ and f_c, and $T_g{}^{surface}$ and $T_g{}^c$, represent the fractions and glass transition temperatures of the surface layer and core components, respectively. The surface thicknesses were determined by the minimum after the first peak in the total $g(r)$ curves of $Ni_{60}Nb_{40}$ ATFs. $T_g{}^c$ and $T_g{}^{surface}$ of the studied $Ni_{60}Nb_{40}$ ATFs were found to be about 978 K and 906 K, respectively, while $T_g{}^c$ and $T_g{}^{shell}$ of the studied $Ni_{60}Nb_{40}$ APs in Figure 8b were found to be about 1090 K and 910 K, respectively. These results reveal that the $Ni_{60}Nb_{40}$ APs and ATFs are both composed of the surface (shell) and core components, while their corresponding T_g values are slightly different, most likely resulting from different atomic packing in both components for $Ni_{60}Nb_{40}$ APs and ATFs. The different shapes of the samples, spherical particles and flat thin films, and different element segregation, could cause the different size effects on the atomic packing and T_g of $Ni_{60}Nb_{40}$ APs and ATFs.

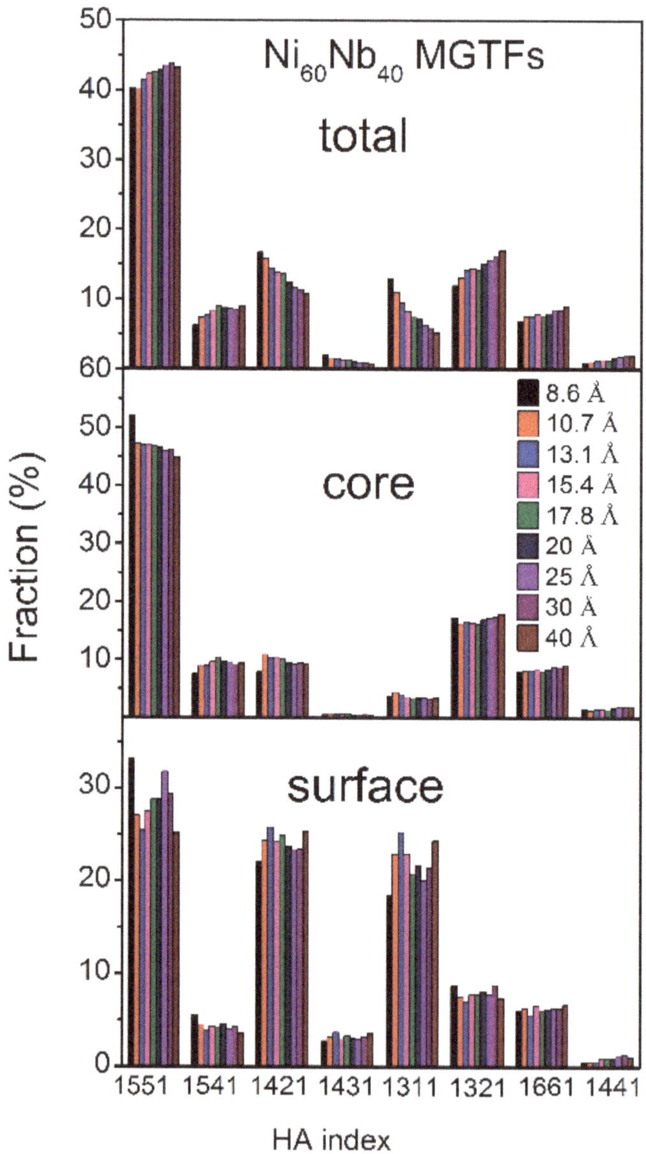

Figure 13. The HA index for total sample, core, and surface components for $Ni_{60}Nb_{40}$ ATFs with various thickness.

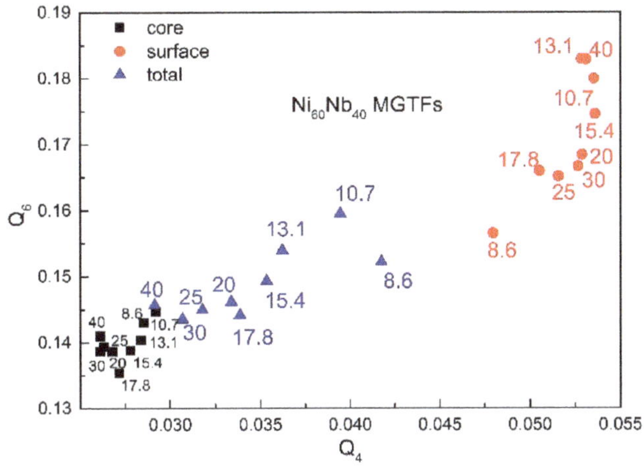

Figure 14. The Q_6-Q_4 distribution of total, core, and surface components for $Ni_{60}Nb_{40}$ ATFs with various thickness. Thickness of films are marked.

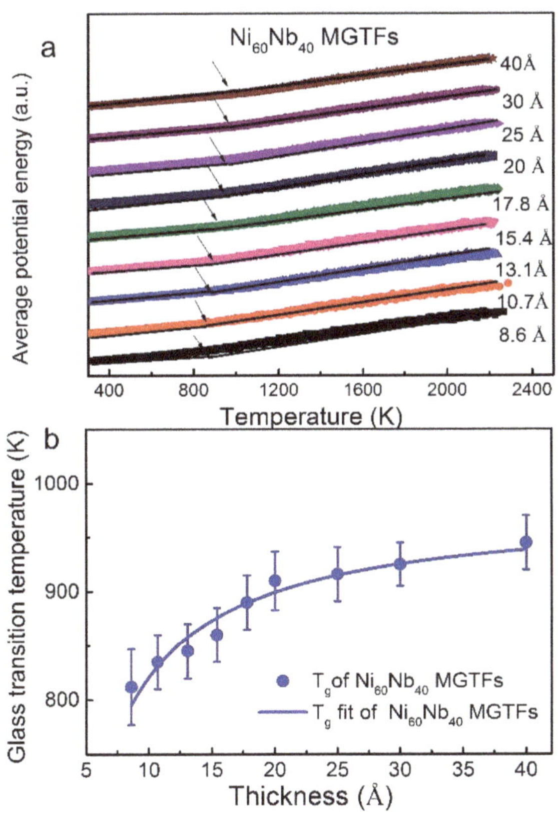

Figure 15. (**a**) Determination of glass transition temperature T_g via an intersection of low- and high-temperature dependence of the potential energy for $Ni_{60}Nb_{40}$ ATFs with different thicknesses. (**b**) Fitting of glass transition temperature T_g for $Ni_{60}Nb_{40}$ ATFs with various thickness. The two models used give almost the same quality of fittings.

4. Conclusions

In conclusion, $Ni_{60}Nb_{40}$ APs with 50 to 5000 atoms and $Ni_{60}Nb_{40}$ ATFs with a thickness range of 8.6–40 Å were systematically investigated by molecular dynamics simulations. The sample size effects on the atomic structure were clearly revealed in both $Ni_{60}Nb_{40}$ APs and ATFs. $Ni_{60}Nb_{40}$ APs have an average bond length of 2.57 Å and atomic packing with fivefold symmetry, i.e., HA indices of 1551 and 1541 are dominant in $Ni_{60}Nb_{40}$ APs. $Ni_{60}Nb_{40}$ ATFs have average bond length of 2.58 Å, and also have a high fraction of the 1551 index following by the 1421 index. Q_6 and Q_4 values in the shell (or surface) component are larger than those in the core component in both $Ni_{60}Nb_{40}$ AP and ATF systems. Ni atoms in $Ni_{60}Nb_{40}$ APs and ATFs prefer to segregate to the shell and surface regions, respectively, causing a higher concentration in the shell of APs or in the surface of ATFs. These atomic structure differences in $Ni_{60}Nb_{40}$ APs and ATFs with various sizes affect their glass transition temperatures, showing that T_g decreases as the MGP gets smaller or MGTF gets thinner. Our obtained results for $Ni_{60}Nb_{40}$ APs and ATFs clearly reveal a size effect on atomic packing and glass transition temperature in low-dimensional metallic glass systems, promoting more studies, especially experimental works, of the subject in other amorphous materials on the nanometer scale, when they will be used in application.

Author Contributions: Data curation, formal analysis, investigation and writing-original draft preparation, W.Z.; Supervision and validation, Y.M. All authors have read and agreed to the published version of the manuscript.

Funding: This research received no external funding.

Data Availability Statement: The raw data supporting the conclusions of this article will be made available by the authors on request.

Conflicts of Interest: The authors declare no conflict of interest.

References

1. Schuh, C.A.; Hufnagel, T.C.; Ramamurty, U. Mechanical behavior of amorphous alloys. *Acta Mater.* **2007**, *55*, 4067–4109. [CrossRef]
2. Zhou, Z.-Y.; Yang, Q.; Yu, H.-B. Toward atomic-scale understanding of structure-dynamics-properties relations for metallic glasses. *Prog. Mater. Sci.* **2024**, *145*, 101311. [CrossRef]
3. Fu, J.; Ma, J. Nanoengineering of metallic glasses. *Adv. Eng. Mater.* **2023**, *25*, 2200659. [CrossRef]
4. Sohrabi, S.; Fu, J.; Li, L.; Zhang, Y.; Li, X.; Sun, F.; Ma, J.; Wang, W.H. Manufacturing of metallic glass components: Processes, structures and properties. *Prog. Mater. Sci.* **2024**, *144*, 101283. [CrossRef]
5. Zhang, Y.; Ashcraft, R.; Mendelev, M.; Wang, C.; Kelton, K.J. Experimental and molecular dynamics simulation study of structure of liquid and amorphous Ni62Nb38 alloy. *J. Chem. Phys.* **2016**, *145*, 204505. [CrossRef] [PubMed]
6. Zhu, Y.; Wang, W.; Song, Y.; Zhang, S.; Li, H.; Wang, A.; Zhang, H.; Zhu, Z. Atomic-scale Nb heterogeneity induced icosahedral short-range ordering in metallic glasses. *J. Mater. Sci. Technol.* **2022**, *108*, 73–81. [CrossRef]
7. Liu, Y.; Bei, H.; Liu, C.T.; George, E.P. Cooling-rate induced softening in a Zr50Cu50 bulk metallic glass. *Appl. Phys. Lett.* **2007**, *90*, 071909. [CrossRef]
8. Cheng, Y.; Ma, E.; Sheng, H. Atomic level structure in multicomponent bulk metallic glass. *Phys. Rev. Lett.* **2009**, *102*, 245501. [CrossRef] [PubMed]
9. Wang, X.; Jiang, F.; Hahn, H.; Li, J.; Gleiter, H.; Sun, J.; Fang, J. Sample size effects on strength and deformation mechanism of Sc75Fe25 nanoglass and metallic glass. *Scr. Mater.* **2016**, *116*, 95–99. [CrossRef]
10. Chen, D.Z.; Jang, D.; Guan, K.M.; An, Q.; Goddard, W.A., III; Greer, J.R. Nanometallic glasses: Size reduction brings ductility, surface state drives its extent. *Nano Lett.* **2013**, *13*, 4462–4468. [CrossRef] [PubMed]
11. Tian, L.; Cheng, Y.-Q.; Shan, Z.-W.; Li, J.; Wang, C.-C.; Han, X.-D.; Sun, J.; Ma, E. Approaching the ideal elastic limit of metallic glasses. *Nat. Commun.* **2012**, *3*, 609. [CrossRef] [PubMed]
12. Jiang, Q.; Liu, P.; Cao, Q.; Wang, C.; Li, X.; Gao, X.; Wang, X.; Zhang, D.; Han, X.; Zhang, Z. The effect of size on the elastic strain limit in Ni60Nb40 glassy films. *Acta Mater.* **2013**, *61*, 4689–4695. [CrossRef]
13. Adibi, S.; Branicio, P.S.; Zhang, Y.-W.; Joshi, S.P. Composition and grain size effects on the structural and mechanical properties of CuZr nanoglasses. *J. Appl. Phys.* **2014**, *116*, 043522. [CrossRef]
14. Lee, S.-W.; Jafary-Zadeh, M.; Chen, D.Z.; Zhang, Y.-W.; Greer, J.R. Size effect suppresses brittle failure in hollow Cu60Zr40 metallic glass nanolattices deformed at cryogenic temperatures. *Nano Lett.* **2015**, *15*, 5673–5681. [CrossRef] [PubMed]
15. Honeycutt, J.D.; Andersen, H.C. Molecular dynamics study of melting and freezing of small Lennard-Jones clusters. *J. Phys. Chem.* **1987**, *91*, 4950–4963. [CrossRef]
16. Steinhardt, P.J.; Nelson, D.R.; Ronchetti, M. Bond-orientational order in liquids and glasses. *Phys. Rev. B* **1983**, *28*, 784. [CrossRef]

17. Rein ten Wolde, P.; Ruiz-Montero, M.J.; Frenkel, D. Numerical calculation of the rate of crystal nucleation in a Lennard-Jones system at moderate undercooling. *J. Chem. Phys.* **1996**, *104*, 9932–9947. [CrossRef]
18. Kawasaki, T.; Tanaka, H. Formation of a crystal nucleus from liquid. *Proc. Natl. Acad. Sci. USA* **2010**, *107*, 14036–14041. [CrossRef] [PubMed]
19. Russo, J.; Tanaka, H. The microscopic pathway to crystallization in supercooled liquids. *Sci. Rep.* **2012**, *2*, 505. [CrossRef]
20. Zhang, W.; Liu, J.; Lu, S.; Zhang, H.; Wang, H.; Wang, X.; Cao, Q.; Zhang, D.; Jiang, J. Size effect on atomic structure in low-dimensional Cu-Zr amorphous systems. *Sci. Rep.* **2017**, *7*, 7291. [CrossRef] [PubMed]

metals

MDPI

Article

Effect of Tool Speed on Microstructure Evolution and Mechanical Properties of Friction Stir Welded Joints of Al-Mg-Si Alloy with High Cu Content

Wangzhen Li [1,2,3], Zhang Luo [1], Youping Sun [1,2,3,*] and Xinyu Liu [1]

1 School of Mechanical and Automotive Engineering, Guangxi University of Science and Technology, Liuzhou 545006, China; liwangzhen@gxust.edu.cn (W.L.)
2 Guangxi Key Laboratory of Automobile Components and Vehicle Technology, Liuzhou 545006, China
3 Guangxi Earthmoving Machinery Collaborative Innovation Center, Liuzhou 545006, China
* Correspondence: syptaiji@126.com

Abstract: OM, SEM, EBSD, and other analytical techniques were utilized to investigate the effects of the rotating speed of a mixing head on the microstructures and mechanical properties of a joint. The results indicate that, compared with the base material, the grain size in the nugget zone is significantly refined. Furthermore, as the rotational speed of the mixing head increases, the grain size in the nugget zone increases noticeably, and the proportion of high-angle grain boundary length initially decreases and then increases. The texture types in different areas of the joint are markedly distinct: the base material primarily consists of recrystallization texture and rolling texture, while the core zone mainly comprises C-shear texture. Among the joints tested at various rotation speeds, the lowest hardness values are observed in the advancing side heat-affected zone, and the tensile properties of the joints are notably reduced due to the dissolution and coarsening of the second phase. The joint exhibits optimal performance at 1000 r/min, with a tensile strength and elongation of 196.3 MPa and 13.5%, respectively.

Keywords: aluminum alloy; EBSD; texture; mechanical property

1. Introduction

As heat-treatable aluminum alloys, 6-series aluminum alloys possess characteristics such as low density, high specific strength, good corrosion resistance, and weldability. These attributes make them widely used across various fields, including aviation, automobiles, and shipbuilding [1]. Friction Stir Welding (FSW) is a solid-state welding process that successfully eliminates the defects associated with traditional methods. It boasts advantages such as energy efficiency, environmental protection, and high-quality connections, thereby being considered a revolutionary welding technology [2,3]. During the FSW process, friction between the tool and the workpiece generates a significant amount of heat, softening the material. Strong plastic deformation occurs under the agitation of the stirring needle, resulting in the formation of fine recrystallized grains [4,5]. The different thermal coupling effects in various areas of the joint lead to an uneven microstructural distribution, and the formation of different grains and textures directly impacts the overall performance of the welded joint. Therefore, it is crucial to study the microstructures, textural evolution characteristics, and mechanical properties of welded joints under different welding process parameters.

For FSW, tool speed, welding speed, tool shape, and other process variables greatly impact the flow and thermal cycling characteristics of materials, leading to various changes in microstructure, texture, and mechanical properties [6]. In the past, some scholars have investigated the effects of welding conditions on the microstructures and mechanical properties of aluminum alloy joints. Dong et al. [7] examined the effect of welding speed on the microstructure and hardness of 6005A-T6 aluminum alloy, finding that the original β'' phase in the nugget zone (NZ) is completely integrated into the α-Al matrix. Moreover, a lower welding speed promotes the formation of a GP zone, thereby increasing the hardness of the welding core zone. The welding speed has little effect on the hardness of the thermos-mechanically affected zone (TMAZ), with the evolution of the precipitated phase being mainly related to the peak temperature. The heat-affected zone (HAZ) has the lowest hardness due to the coarsening of the β'' and Q' phases, but its hardness gradually increases with increasing welding speed. Krasnowski et al. [8] studied the effects of tool shape on the microstructure and mechanical properties of 6082 aluminum alloy, discovering that the joint can achieve good tensile properties with conventional tools; however, when a non-threaded plain needle mixing head is used, the joint performance is poor. Sato et al. [9] investigated the relationship between the microstructures and microhardness of 6063 aluminum alloy joints in relation to tool speed. The researchers concluded that the change in hardness is independent of grain size, and the softening of the joint is due to the coarsening and dissolution of the β'' phase. Additionally, increasing tool rotation speed widens the softening zone of the joint. Ahmad et al. [10] explored the effects of stirring head rotation speed on the mechanical and flexural properties of 6061 aluminum alloy joints. Their research showed that the HAZ range increases with higher stirring head rotation speeds, and both hardness and flexural properties are improved.

The Cu content of 6-series aluminum alloys is generally lower than 0.4%, but it has been shown that an increase in Cu content can reduce the precipitation activation energy of the alloy [11], alter the precipitation sequence of the alloy, and improve its strength [12]. However, no one has studied the effect of higher Cu content on the FSW joints of Al-Mg-Si alloys. Therefore, in this work, the effect of rotational speed on the microstructure and mechanical properties of FSW joints of Al-Mg-Si alloys containing 0.8% Cu was investigated, providing a theoretical basis for practical applications.

2. Experimental Materials and Methods

The material used in the experiment was an Al-0.75Mg-0.75Si-0.8Cu alloy ingot, which was obtained by melting pure aluminum, pure magnesium, 10% aluminum–silicon, and 10% aluminum–copper intermediate alloys in specific proportions. The specimen size was 150 mm × 120 mm × 20 mm. Subsequently, the ingot underwent homogenization treatment at 560 °C for 24 h, after which it was processed into a 100 mm × 80 mm × 14 mm plate by milling. The plates were first rolled to 2 mm through two hot rolling steps (the first pass was 4 mm and the second pass was 8 mm) with a rolling temperature of 400 °C and a holding time of 20 min for each pass. The rolling rate was 10 s^{-1} for both passes. Finally, the plates were subjected to a solution treatment at 510 °C for 80 min and an aging treatment at 195 °C for 13 h [13].

Before welding, all plates were sanded with sandpaper and alcohol to remove the oxidized layer and grease. The diameter of the tool shoulder used for welding was 12 mm, the length of the tool pin was 1.85 mm, and its shape was conical. The large diameter of the tool pin was 3.6 mm, while the small diameter was 2.5 mm. The welding direction (WD) was perpendicular to the rolling direction (RD). An ETM105D electronic universal testing joint (Wance, made in China) was used for mechanical property testing at room temperature (tensile rate of 2 mm/min), and each group of tensile specimens was tested three times to determine the average value. A SIGMA scanning electron microscope was used to observe the fracture morphologies of the welded joints. In the welded joints, cross-section (ND) samples were used for grinding and polishing. A mixture of 70% nitric acid and 30% methanol was used for electrochemical polishing. The polishing voltage was 15 V, the polishing solution was cooled by liquid nitrogen at −35 °C, and the polishing time was 40 s. EBSD experiments were carried out using a field emission scanning electron microscope (SIGMA, Carl Zeiss, Oberkochen, Germany) with an EBSD probe (Aztec-Max80). The microhardness profile was measured in the cross-section perpendicular to the welding direction along the mid-thickness direction of the plate using an HVS-1000A hardness tester under a 200 g load for 10 s.

3. Experimental Results and Analysis

3.1. Microstructure Analysis of Joint

Figures 1–4 illustrate the microstructural morphologies of different regions of the weld cross-section under various rotational speeds. According to Figures 1b and 2b, the NZ grains undergo dynamic recrystallization at high temperatures with strong plastic deformation; thus, the grains are very fine. However, in Figures 3b and 4b, the NZ grains obviously grow due to the increase in heat input with increasing rotational speed. The grains in the TMAZ are subjected to shear extrusion of the grains in the NZ, resulting in the presence of a clearly oriented distribution of the AS-TMAZ. This distribution leads to the appearance of a clear boundary between the HAZ of the joint and the NZ on the forward side, as shown in Figures 1a and 2a. As the rotational speed increases, the effects of temperature and shear on the TMAZ of the joint increase, and the degree of dynamic recrystallization of the grains increases. Conversely, the recrystallized grains in the NZ grow with increasing rotational speed, resulting in blurring of the boundary between the NZ and the TMAZ of the joint, as shown in Figures 3a and 4a. Similarly, the changes in the HAZ of the engine on the backward side are basically consistent with the previous analysis. However, it is interesting to note that when the rotational speed is 750 r/min, the zone of influence of the heat engine expands to the inside of the NZ on the backward side (shown by the arrows in Figure 1c). After the rotational speed is increased, the phenomenon of expanding the zone of influence of the heat engine to the inside of the NZ disappears. A reasonable explanation for this phenomenon is that the material in the upper part moves downward during the FSW process, and part of the material in the forward side accumulates in the backward side under the shearing effect of the stirring head. At low rotational speeds, the heat input is low, the material is less mobile, and less recrystallization occurs. As a result, the material accumulated on the backward side is subjected to a very strong squeezing action, leading to grain expansion in the TMAZ toward the NZ [14].

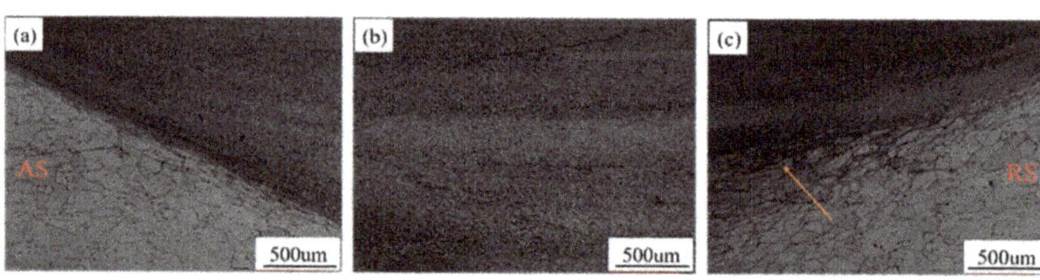

Figure 1. Microstructure of alloy FSW joint at rotation speed of 750 r/min (**a**) AS, (**b**) NZ, (**c**) RS.

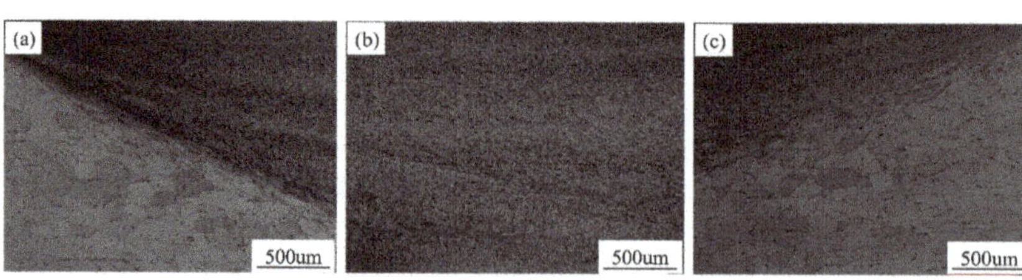

Figure 2. Microstructure of alloy FSW joint at rotation speed of 1000 r/min (**a**) AS, (**b**) NZ, (**c**) RS.

Figure 3. Microstructure of alloy FSW joint at rotation speed of 1250 r/min (**a**) AS, (**b**) NZ, (**c**) RS.

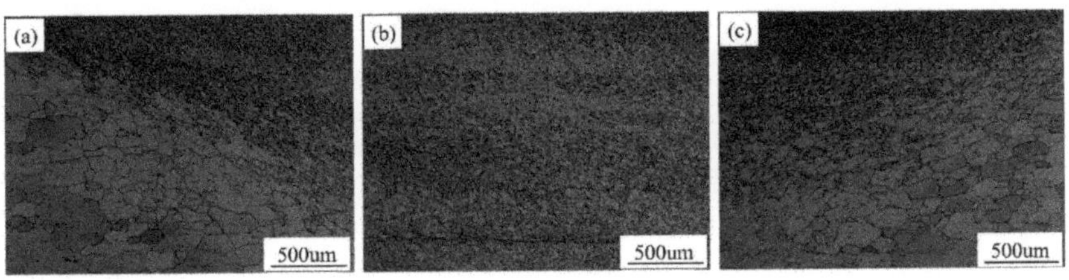

Figure 4. Microstructure of alloy FSW joint at rotation speed of 1500 r/min (**a**) AS, (**b**) NZ, (**c**) RS.

Figures 5–7 show the EBSD characterization and analysis of the different regions of the joint and the BM at a rotational speed of 1000 r/min. The EBSD diagram of HAZ is 5 mm from the exact center of NZ. TMAZ is 2 mm from the center of NZ. The aluminum alloy is rolled into a sheet after T6 heat treatment, and recrystallization occurs. In addition, some grains are elongated along the rolling direction, as shown in Figure 5a. After FSW of the BM, the grains in the HAZ grow significantly under the influence of the welding thermal cycle, with grain growth in the AS-HAZ being more pronounced. Statistically, the average grain growth in the AS-HAZ increases from 85.58 μm to 101 μm in the BM. The grains in the TMAZ are deflected along the direction of shear force under the action of shear extrusion, as shown in Figure 5d,e. The grain organization in the TMAZ is relatively chaotic due to varying temperature and strain gradients [15]. The places far from the stirring pins in the TMAZ are subjected to very small temperatures and strains; thus, very coarse shear bands are generated. Inside these coarse shear bands, there are many grains caused by dislocation slip and strain. There are many low-angle grain boundaries created by dislocation slip and stacking, and most of these low-angle grain boundaries are oriented in the same direction as the shear direction. Closer to the stirring pin, the strain and temperature increase, the proportion of low-angle grain boundaries increases, and the orientation angle increases. However, due to the limited degree of plastic deformation, the grains form inhomogeneous shear bands. Finally, close to the stirring pin, the grains form fine shear bands with a certain aspect ratio under the action of stronger temperature and shear as some dynamic recrystallization occurs. Some of the grains in the HAZ are subjected to shear compression from the grains in the TMAZ. Dynamic restitution occurs under the influence of the weld thermal cycle, producing a number of low-angle grain boundaries (as shown by the black arrows in Figure 5b,c). Conversely, the RS-HAZ is subjected to a relatively high squeezing force, producing additional low-angle grain boundaries; thus, the percentage of high-angle grain boundary length in the base material is higher than that in the HAZ, as shown in Figure 6. As mentioned earlier, the TMAZ produces many low-angle grain boundaries under the effects of temperature and strain, and the percentage of high-angle grain boundary length in the TMAZ is the lowest. Moreover, the high-angle grain boundaries in the AS-TMAZ are greater than those in the RS-TMAZ due to the higher plastic deformation and temperature. These factors lead to the occurrence of dynamic recrystallization to a higher extent in the AS-TMAZ than in the RS-TMAZ.

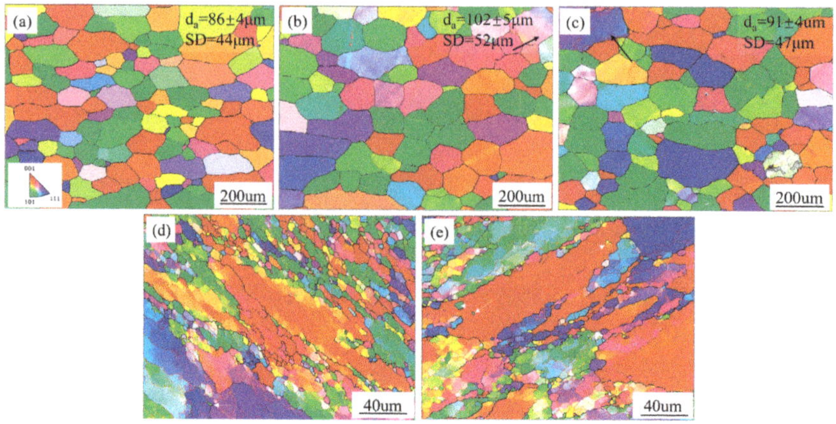

Figure 5. Microstructure of alloy FSW joint at BM and speed of 1000 r/min (**a**) BM, (**b**) AS-HAZ, (**c**) RS-HAZ, (**d**) AS-TMAZ, (**e**) RS-TMAZ.

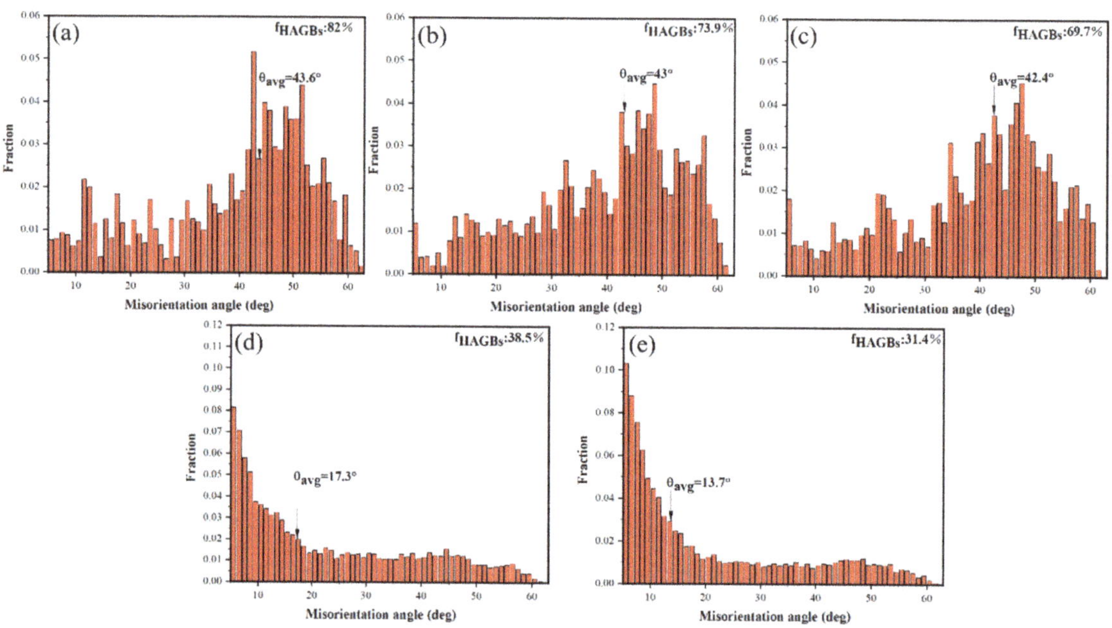

Figure 6. Misorientation distribute maps of the FSW joint at BM and speed of 1000 r/min (**a**) BM, (**b**) AS-HAZ, (**c**) RS-HAZ, (**d**) AS-TMAZ, (**e**) RS-TMAZ.

Grain orientation spread (GOS) is the best technique for distinguishing the degree of grain deformation [16]. It is generally accepted that grains with GOSs of less than 2° are recrystallized grains [17]. TMAZ grains are subjected to large shear extrusion pressure and thermal cycling, which leads to rapid dislocation proliferation and the formation of deformed grains. In the process of dynamic recovery, the deformed grains will transform into subgrains. In addition, the grain near the NZ is subjected to relatively large shear force and thermal cycling, which leads to dynamic recrystallization of a few grains. Therefore, the proportion of recrystallized grains in TMAZ is very low. The plastic deformation on the forward side is higher than that on the backward side, so the capacity for dynamic recrystallization on the forward side is greater than that on the backward side, resulting in a higher percentage of recrystallized grains in the AS-TMAZ than in the RS-TMAZ.

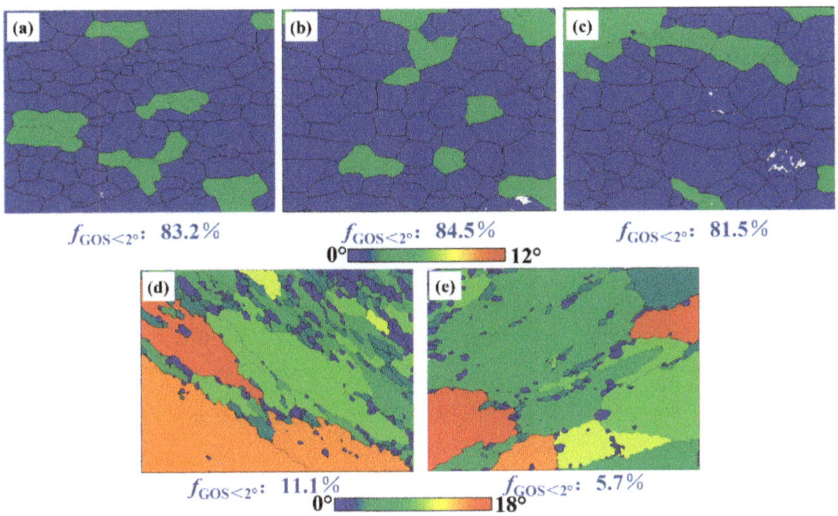

Figure 7. GOS maps of alloy at BM and speed of 1000 r/min (**a**) BM, (**b**) AS-HAZ, (**c**) RS-HAZ, (**d**) AS-TMAZ, (**e**) RS-TMAZ.

The texture is an indication of the orientation of the grains in the alloy material, and the polar diagram is a stereographic projection of the grain orientation and density [18]. Figure 8 shows the polar diagrams of the different areas of the joint and the (100), (110), and (111) directions of the base material at a rotational speed of 1000 r/min. After T6 heat treatment of rolled aluminium alloy plates, the texture components become more complex, retaining the G (110)<100>, B (110)<112>, and S (123)<634> textures generated during rolling but also generating the R-cubeRD (012)<100>, P (110)<211>, and cube (001)<100> recrystallization textures and the C (001)<110> shear texture. The formation of R-cubeRD and P textures may be related to particle-stimulated nucleation (PSN). The texture density in the HAZ is lower than that of the base material, probably because the heat input induces the transition of the rolled texture into a recrystallized texture. Since the grains in the HAZ close to the joint are subjected to less plastic deformation, the HAZ has only one more A(11-1)<1–10> shear texture than the BM. The TMAZ of the joint has approximately the same texture density as the parent material but is subjected to greater shear, making the texture component predominantly shear. The AS-TMAZ of the joint is subjected to a relatively great shear force, mainly forming A and C shear textures. Conversely, the RS-TMAZ of the joint is subjected to less shear force, and A_1^* (11-1)<112> and C shear textures are formed.

The EBSD characterization and analysis of the NZ of the alloy at different stirring head rotation speeds are shown in Figure 9. The presence of serrated grains is shown in Figure 9a,d,g,j. These grains are elongated and divided into segments by low-angle grain boundaries, and a certain amount of Geometric Dynamic Recrystallisation (GDRX) of the NZ grains occurs [15,19]. At 750 r/min, NZ grains produce many serrated crystals under the action of shear force. With the increase in rotational speed, not only do the serrated grains undergo growth, but the number of segments cut by low-angle grain boundaries continues to decrease due to the increased rotational speed improving the heat input and promoting the occurrence of Dynamic Recrystallisation (DRX). Therefore, the serrated crystals produce new DRX grains. Continuous dynamic recrystallization of grains (CDRX) occurs based on a certain degree of GDRX. At the same time, by observing Figure 9i,l, it can be found that the grain size of NZ does not change significantly with the increase in rotational speed, which indicates that CDRX is conducive to grain refinement. The high-angle grain boundaries and the average orientation deviation angle fractions of the grains in the NZ show a pattern of first decreasing and then increasing with increasing stirrer head rotational speed, as shown in Figure 9b,e,h,k. At a stirrer head rotational speed of 1000 r/min, the average orientation deviation angle of the grains in the NZ is the lowest at 24.2°, while the lowest percentage of NZ high-angle grain boundaries length is 57.6% at a rotation speed of 1250 r/min. The NZ high-angle grain boundaries and the average orientation deviation angle are lower at a rotational speed of 1500 r/min than at 750 r/min. It is clear that the increase in stirring head speed increases the plastic deformation and heat input, leading to more dislocations. Additionally, more low-angle grain boundaries are formed during the dynamic recovery process. Therefore, from 750 r/min to 1250 r/min, the number of low-angle grain boundaries gradually increases. At a rotational speed of 1500 r/min, the average grain size of the NZ is 8.22 μm, as shown in Figure 9l. Compared to 1250 r/min, the grain size of the NZ does not change significantly, which is obviously caused by the enhancement in the ability to generate DRX. This phenomenon results in the production of more recrystallized grains. As a result, the number of high-angle grain boundaries increases. According to the statistics of the recrystallized grains in the weld core region at different rotational speeds, the volume fractions are 41.6%, 41.3%, 37.1%, and 43.6% (As shown in Figure 10) and the highest volume fraction of recrystallization is found at 1500 r/min, which is consistent with the previous analysis.

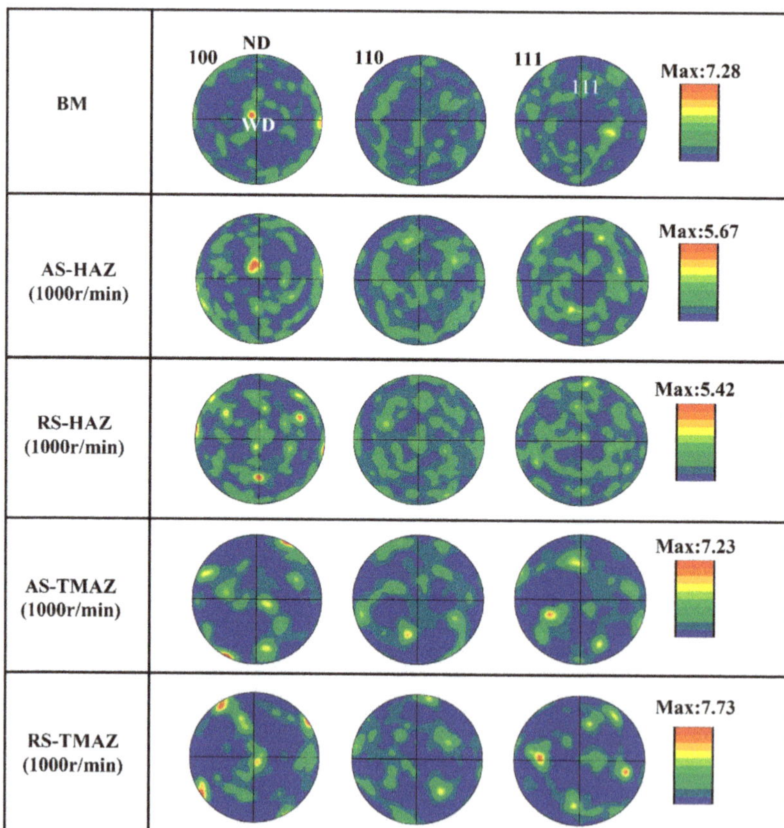

Figure 8. Pole figures of alloy at BM and speed of 1000 r/min.

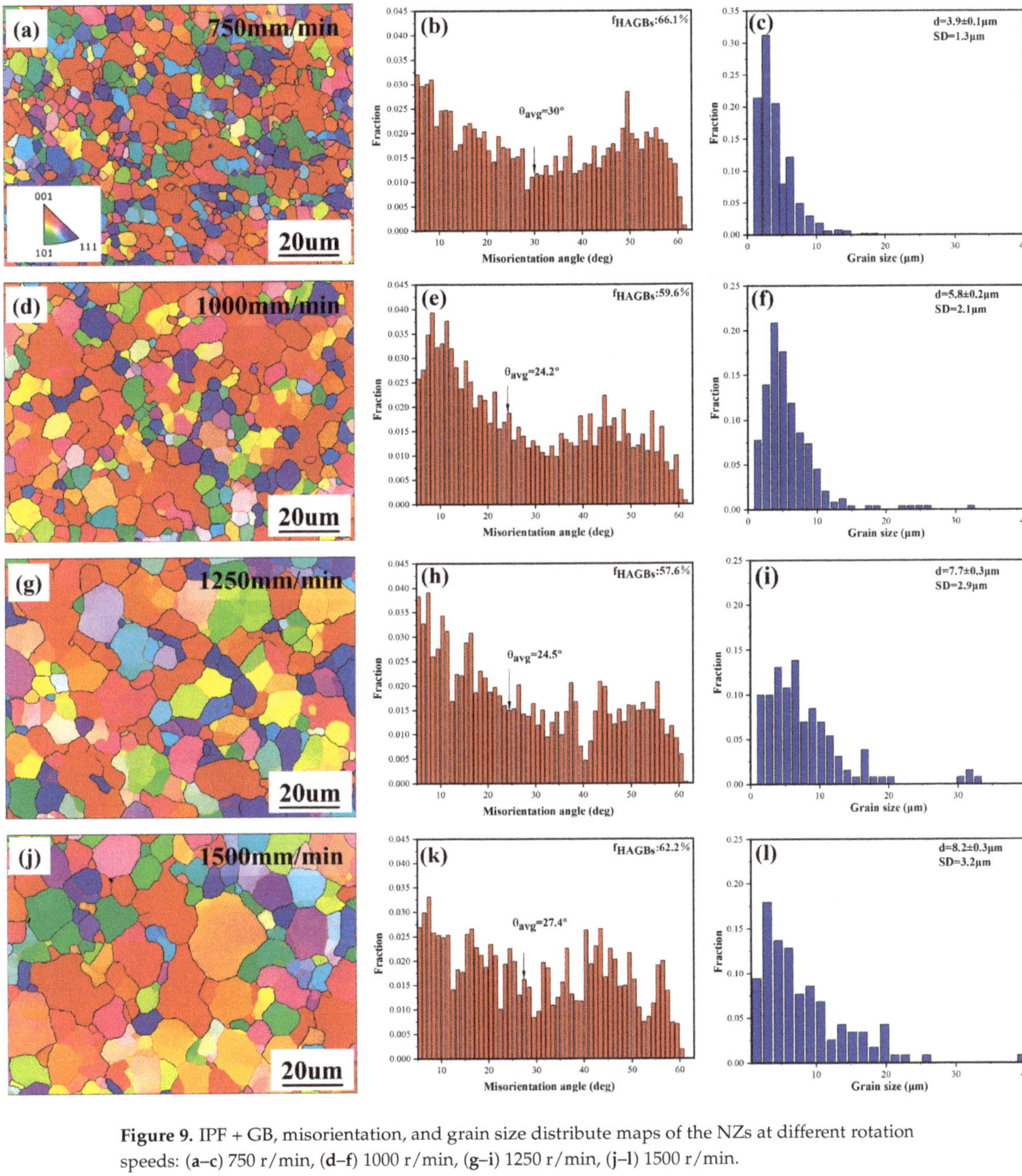

Figure 9. IPF + GB, misorientation, and grain size distribute maps of the NZs at different rotation speeds: (**a–c**) 750 r/min, (**d–f**) 1000 r/min, (**g–i**) 1250 r/min, (**j–l**) 1500 r/min.

The polar patterns of the NZ for different rotation speeds of the alloy are shown in Figure 11. The texture orientation of the NZ obtained in Figure 11 deviates from the ideal state due to the complex constraints of the FSW process. At a rotational speed of 750 r/min, a large number of C{001}<1–10> textures and a small number of A{11-1}<1–10> textures mainly exist in the weld core region. With the increase in rotational speed, at 1000 r/min, the texture density of the C texture increases while the A texture disappears, and a new B{11-2}<1–10> texture is produced with a very low texture density. At 1250 r/min and

1500 r/min, the density of the C texture begins to decrease but remains the dominant texture component in the weld core region. Conversely, the density of the B texture increases slightly. The reasonable explanation for this phenomenon is that in the process of plastic deformation, the aluminum alloy {111} slip plane is in line with the shear face, and the <110> slip direction is parallel to the shear direction. Therefore, it is easy to produce $A/\overline{A}$ texture under shearing force [20]. With the increase in shear force, the {112} slip plane will activate, thus forming the $B/\overline{B}$ texture [21,22]. The understanding of shear textural evolution is provided in some experimental and theoretical studies of FCC metals [23–26]. However, in these experiments, the C texture is dominant, and the texture density is low despite the appearance of the B texture. Since dynamic recrystallization occurs in NZ grains, the type of NZ texture identified in some studies [27,28] is a cube recrystallization texture. The C texture may be formed by the rotational squeezing force of the cube texture by the axial shoulder [29,30].

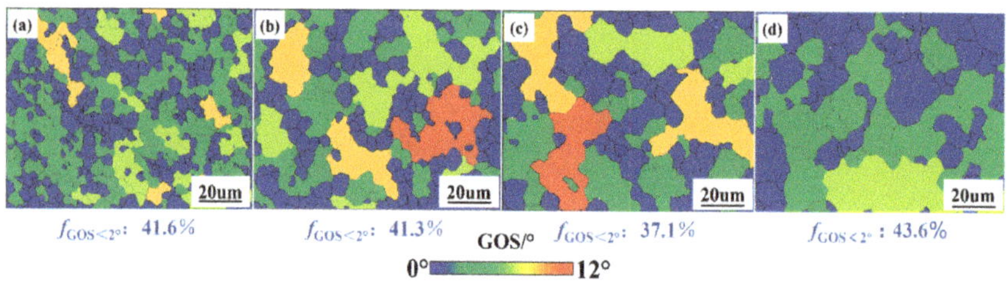

Figure 10. IPF + GB, misorientation, and grain size distribute maps of the NZs at different rotation speeds: (**a**) 750 r/min, (**b**) 1000 r/min, (**c**) 1250 r/min, (**d**) 1500 r/min.

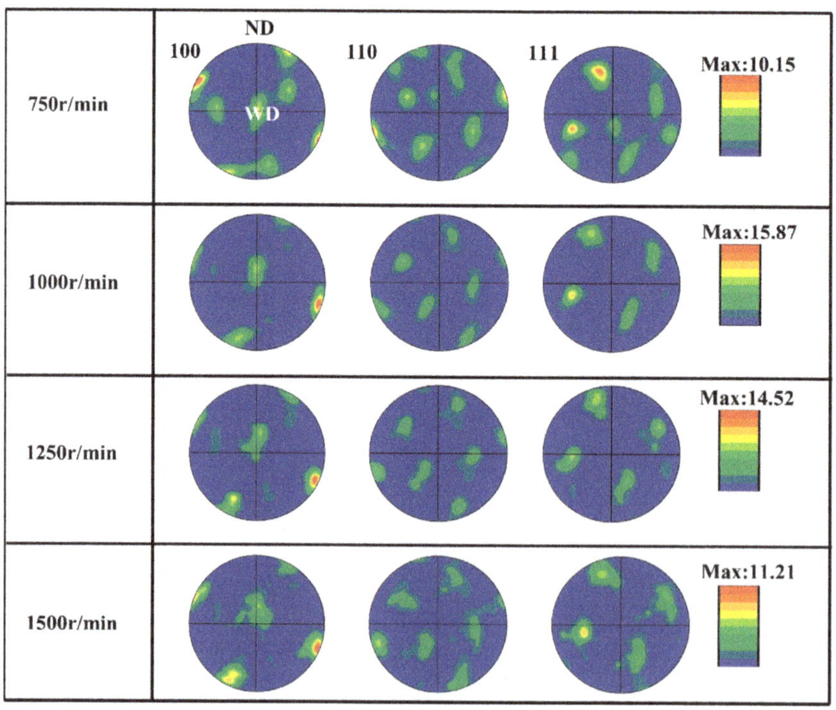

Figure 11. Pole figures of the NZs at different rotation speeds.

3.2. Mechanical Properties

The hardness distribution curve at the center of the cross-section of the FSW joint is shown in Figure 12. The hardness curve of the joint shows a W-shape. During the FSW process, the β″ phase in the NZ is dissolved, and overaged precipitated phases, such as β′ and Q′, are generated. The increase in the tool rotational speed induces these overaged phases to break up and dissolve, while the residual heat after stirring produces a transient aging effect, which causes clusters of solute atoms to precipitate and grow [9,31]. By comparing Figure 12a,b, it can be found that the hardness of NZ increases significantly at 1000 r/min. With the continuous increase of rotational speed, the hardness of NZ does not change much when the error is considered, which may be caused by the joint action of grain size and short aging after welding. At low rotational speeds, the TMAZ and HAZ hardnesses are approximately the same. At high speeds, the hardness of the TMAZ is almost the same as that of the NZ, which may be due to the increase in dislocation density caused by the increase in rotational speed. The HAZ has a low dislocation density, overaged phase coarsening, and grain growth; thus, the hardness of the HAZ is the lowest in the joint. Moreover, the hardness of the AS-HAZ is lower than that of the RS-HAZ due to the pronounced grain size growth.

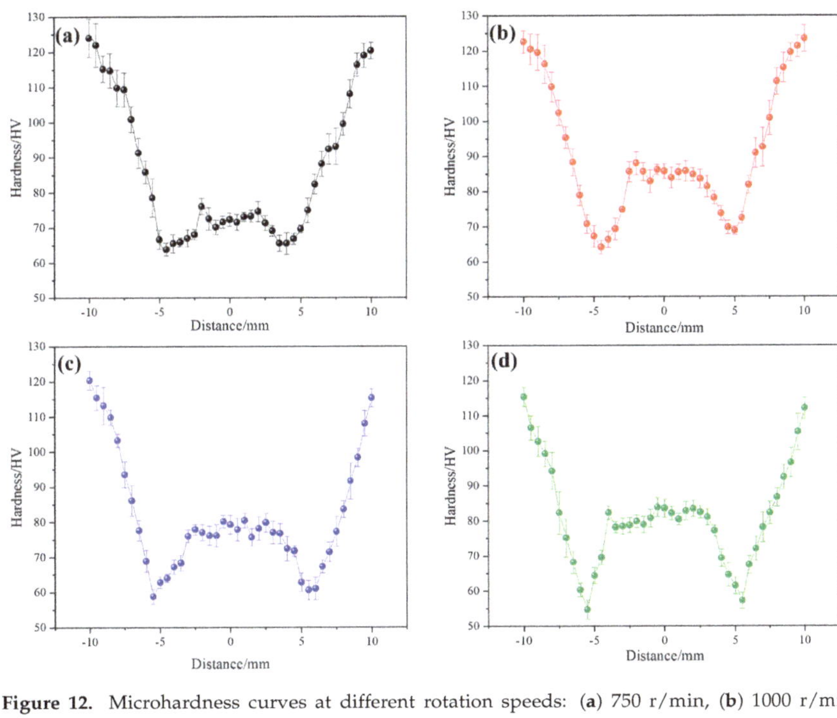

Figure 12. Microhardness curves at different rotation speeds: (**a**) 750 r/min, (**b**) 1000 r/min, (**c**) 1250 r/min, (**d**) 1500 r/min.

The tensile properties of the BM and different welded specimens are shown in Figure 13, and the specific values are given in Table 1. First, the tensile properties of the specimens after FSW decrease drastically due to the dissolution and coarsening of precipitated phases during the thermal cycling of the joint. Second, the UTS first increases and then decreases with increasing rotational speed, while the elongation decreases with increasing rotational speed. The performance of the joints is generally highly related to the hardness. In Figure 12, the highest value of NZ hardness is found at a rotational speed of 1000 r/min, and the lowest value of hardness is approximately the same at 750 r/min and

1000 r/min. At high rotational speeds, a significant decrease in the hardness of the HAZ occurs. In addition, the dissolution of the joint precipitation phase leads to the softening of the joint, improving the plasticity of the alloy. However, with increasing rotational speed, the overaged phase in the HAZ gathers and coarsens, leading to a significant decrease in the properties of the alloy. Thus, at low rotational speeds, the elongation is higher than that of the base material, and with increasing rotational speed, the elongation gradually decreases. At a rotational speed of 750 r/min, the elongation of the joint is higher than that of the base material. At a rotational speed of 1000 r/min, the joint has the highest tensile strength and higher elongation than the parent material, but the joint efficiency is only 61.25% of that of the parent material. In combination with Figure 12, it can be seen that the whole joint undergoes different degrees of softening under different temperature conditions, which indicates that the dissolution and coarsening of the precipitated phase leads to the loss of joint strength. Additionally, all the welded specimens show significant jagged curves during tensile testing, as shown in Figure 13b. Baghdadi [32] believes that this phenomenon occurs due to the uneven distribution of hard particles (Mg_2Si) in the alloy. Due to the obvious temperature gradient in each area of the joint, the degree of dissolution and coarsening of precipitates in different areas is different, resulting in an uneven distribution of precipitates in each area, resulting in the PLC effect.

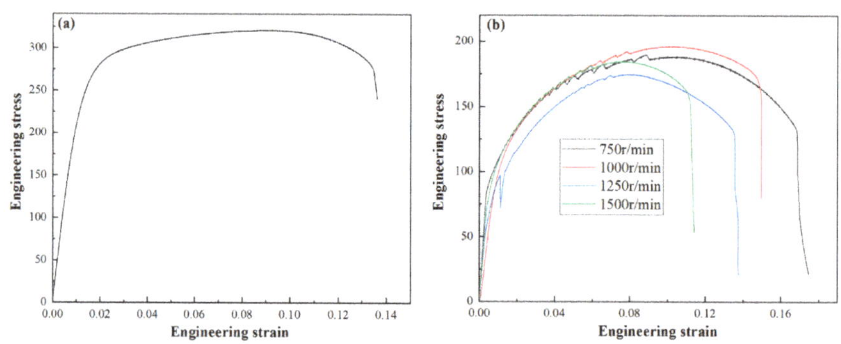

Figure 13. Tensile properties of joints at different welding speeds: (**a**) BM, (**b**) joint.

Table 1. Transverse tensile properties of BM and FSW joints.

Part	Rotational Speeds/(r/Min)	UTS/MPa	Elongation/%
BM	-	320 ± 4	12.6 ± 1.2
FSW joint	750	185 ± 1	15.9 ± 0.7
	1000	196 ± 3	13.5 ± 0.9
	1250	175 ± 2	9.9 ± 1.4
	1500	186 ± 1	8.2 ± 0.8

Figure 14 shows the fracture location of the joint at different rotational speeds. At 750 r/min and 1000 r/min, the fracture location is on the backward side of the joint. However, at 750 r/min, the crack extends along the direction 45° from the tensile axis, and the final fracture location is at the top of the NZ towards the edge of the receding side. This phenomenon occurs due to the complex transformation of the microstructure near the interface between the TMAZ and NZ, resulting in the appearance of low Taylor index grain bands in this region and forming a local crystallographic weak zone, leading to the fracture of the crack after its formation through this weak zone [33,34]. At 1000 r/min, the fracture direction of the crack is opposite to that at 750 r/min, and the fracture region tends to be between the RS-HAZ and RS-TMAZ. This region may be generated due to the low dislocation density and serious microstructural inhomogeneity, as shown in Figure 5e. At 1250 and 1500 r/min, the joints fracture in the position of the AS-HAZ, which is attributed to the overaging phase of overaged phase aggregation and coarsening, leading to easy cracking in this region.

Figure 14. Tensile fracture morphology of samples at different welding speeds: (**a**) 750 r/min, (**b**) 1000 r/min, (**c**) 1250 r/min, (**d**) 1500 r/min.

Figure 15 shows SEM photographs of the fracture of the joint at different rotational speeds. From the figure, it can be seen that both the parent material and the joint undergo ductile fractures. In this case, the tensile fracture of the parent material has both large and small dimples, and there are second-phase particles inside the dimples, making the parent material moderately tough. At 750 r/min, the morphology of the tensile fracture shows the presence of small and dense dimples, and its elongation is highest in this process. At 1000 r/min, there are some large and small dimples, but they are shallow. At higher speeds, the tensile fracture shows a convoluted surface, indicating that the toughness of the joint is poor.

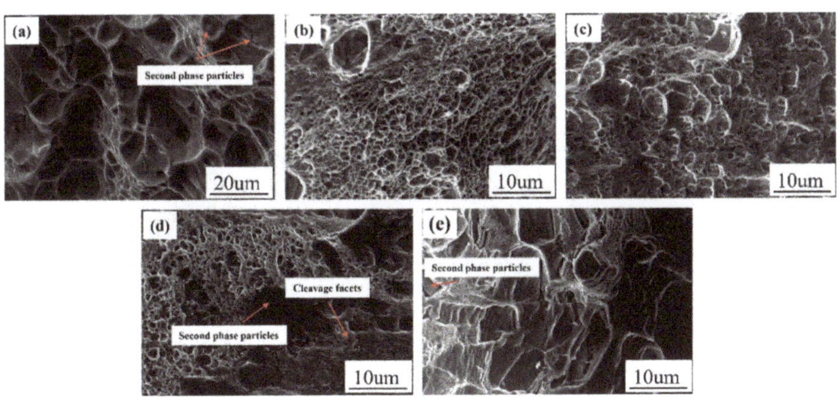

Figure 15. SEM tensile fracture images of BM and FSW joints: (**a**) BM, (**b**) 750 r/min, (**c**) 1000 r/min, (**d**) 1250 r/min, (**e**) 1500 r/min.

4. Conclusions

(1) During the friction stir welding process, the grains of the heat-affected zone are slightly larger than those of the base metal. The thermomechanically affected zone undergoes partial dynamic recrystallization, resulting in an extremely heterogeneous

structure due to the temperature and strain gradients. The base metal mainly consists of rolled and recrystallized textures. The texture density of the heat-affected zone is weaker than that of the base material, and the heat-affected zone has more A-texture. The thermal mechanically affected zone has a shear texture because it is subjected to shear forces. Among them, the advancing side of the thermomechanically affected zone has A and C shear textures, and the receding side of the thermomechanically affected zone has A_1^* and C shear textures.

(2) During the friction stir welding process, the nugget zone structure experiences thermo-mechanical coupling, resulting in geometric dynamic recrystallization and continuous dynamic recrystallization. With increasing rotational speed, the nugget zone grain size gradually increases. Additionally, the percentage of low-angle grain boundaries first increases and then decreases. Finally, the type of texture is gradually converted from A and C to B and C, with the C texture being the primary component of the nugget zone.

(3) With increasing rotational speed, the tensile strength of the AlMgSiCu alloy welded joints first increases and then decreases while the elongation decreases. The optimum performance of the joints is obtained at a rotational speed of 1000 r/min, with tensile strength and elongation of 196 MPa and 13.5%, respectively. Due to the precipitation dissolution and coarsening in different areas of the joint, the strength efficiency of the joint is only 61.25% of that of the base material under the optimal process. The fracture form of the AlMgSiCu alloy joints is ductile fracture.

Author Contributions: Conceptualization, W.L. and Y.S.; Methodology, W.L., Y.S. and Z.L.; Validation, W.L. and Z.L.; Formal analysis, W.L. and X.L.; Data curation, W.L., Z.L. and X.L.; Writing—original draft preparation, W.L.; Writing—review and editing, W.L. and Y.S.; Funding acquisition, Y.S. All authors have read and agreed to the published version of the manuscript.

Funding: This research is funded by Project for Project for Enhancing Young and Middle-aged Teacher's Research Basis Ability in Colleges of Guangxi: Study on the microstructure and failure mechanism of laser welded joints of dissimilar aluminum alloys (2022KY0351); The Guangxi Key R&D Programme (GK AB23026104); The Guangxi Science and Technology Major Special Project (GK AA23062089); The Liuzhou Science and Technology Programme (2023PRJ0102); The Construction of Guangxi Construction Machinery Low Carbon Digital Intelligence Technology Innovation Centre (2023PRJ0101).

Data Availability Statement: The data that support the findings of this study are available from the corresponding author upon reasonable request.

Conflicts of Interest: The authors declare no conflicts of interest.

Abbreviations

FSW	Friction stir welding
TMAZ	Thermal-mechanically affected zone
BM	Base metal
GDRX	Geometric Dynamic Recrystallization
HAZ	Heat affected zone
CDRX	Continuous Dynamic Recrystallization
NZ	Nugget zone
EBSD	Electron back-scattering Patterns
AS	Advancing side
RS	Receding side

References

1. Meng, X.; Huang, Y.; Cao, J.; Shen, J.; Santos, J.F.D. Recent progress on control strategies for inherent issues in friction stir welding. *Prog. Mater. Sci.* **2021**, *115*, 100706. [CrossRef]
2. Mishra, R.S.; Ma, Z.Y. Friction stir welding and processing. *Mater. Sci. Eng. R. Rep.* **2005**, *50*, 1–78. [CrossRef]

Wang, G.; Zhao, Y.; Hao, Y. Friction stir welding of high-strength aerospace aluminum alloy and application in rocket tank manufacturing. *J. Mater. Sci. Technol.* **2017**, *34*, 73–91. [CrossRef]

Liu, H.J.; Hou, J.C.; Guo, H. Effect of WS on Microstructure and mechanical properties of self-reacting friction stir welded 6061–T6 aluminum alloy. *Mater. Des.* **2013**, *50*, 872–878. [CrossRef]

Mosleh, A.O.; Yakovtseva, O.A.; Kishchik, A.A.; Kotov, A.D.; Moustafa, E.B.; Mikhaylovskaya, A.V. Effect of Coarse Eutectic-Originated Particles on the Microstructure and Properties of the Friction Stir-Processed Al-Mg-Zr-Sc-Based Alloys. *J. Miner. Met. Mater. Soc.* **2023**, *75*, 2989–3000. [CrossRef]

Buffa, G.; Hua, J.; Shivpuri, R.; Fratini, L. Design of the friction stir welding tool using the continuum based FEM model. *Mater. Sci. Eng. A* **2006**, *419*, 381–388. [CrossRef]

Dong, P.; Li, H.; Sun, D.; Gong, W.; Liu, J. Effects of welding speed on the microstructure and hardness in friction stir welding joints of 6005A-T6 aluminum alloy. *Mater. Des.* **2013**, *45*, 524–531. [CrossRef]

Krasnowski, K.; Hamilton, C.; Dymek, S. Influence of the tool shape and weld configuration on microstructure and mechanical properties of the Al 6082 alloy FSW joints. *Arch. Civ. Mech. Eng.* **2014**, *15*, 133–141. [CrossRef]

Sato, Y.S.; Urata, M.; Kokawa, H. Parameters controlling microstructure and hardness during friction-stir welding of precipitation-hardenable aluminum alloy 6063. *Metall. Mater. Trans. A* **2002**, *33*, 625–635. [CrossRef]

Fauzi, M.A.; Uday, M.B.; Zuhailawati, H.; Ismail, A.B. Microstructure and mechanicalproperties of alumina-6061 aluminum alloy joined by friction welding. *Mater. Des.* **2010**, *31*, 670–676. [CrossRef]

Li, H.; Wang, X.L.; Shi, Z.X.; Wang, Z.X.; Zheng, Z.Q. Precipitation behaviors of Al-Mg-Si-(Cu) aluminum alloys during continuous heating. *Chin. J. Nonferrous Metals.* **2011**, *21*, 2028–2034.

Wang, Z.X.; Li, H.; Gu, J.H.; Song, R.G.; Zheng, Z.J. Effect of Cu content on microstructures and properties of Al-Mg-Si-Cu alloy. *Chin. J. Nonferrous Met.* **2012**, *22*, 3348–3355. [CrossRef]

Chen, Y.; Wei, W.; Zhao, Y.; Shi, W.; Zhou, X.; Rong, L.; Wen, S.; Wu, X.; Gao, K.; Huang, H.; et al. Effect of solid solution and aging treatment on microstructure and mechanical properties of Al-Mg-Si-Cu alloy. *Trans. Mater. Heat Treat.* **2017**, *38*, 30–37. [CrossRef]

Liu, H.; Hu, Y.; Dou, C.; Sekulic, D.P. An effect of the rotation speed on microstructure and mechanical properties of the friction stir welded 2060-T8 Al-Li alloy. *Mater. Charact.* **2017**, *123*, 9–19. [CrossRef]

Prangnell, P.B.; Heason, C.P. Grain structure formation during friction stir welding observed by the 'stop action technique'. *Acta Mater.* **2005**, *53*, 3179–3192. [CrossRef]

Peng, Y.; Huang, B.; Zhong, Y.; Su, C.; Tao, Z.; Rong, X.; Li, Z.; Tang, H. Electrochemical corrosion behavior of 6061 Al alloy under high rotating speed submerged friction stir processing. *Corros. Sci.* **2023**, *215*, 111029. [CrossRef]

Field, D.P.; Bradford, L.T.; Nowell, M.M.; Lillo, T.M. The role of annealing twins during recrystallization of Cu. *Acta Mater* **2007**, *55*, 4233–4241. [CrossRef]

Shi, L.; Dai, X.; Tian, C.; Wu, C. Effect of splat cooling on microstructures and mechanical properties of friction stir welded 2195 Al–Li alloy. *Mater. Sci. Eng. A* **2022**, *858*, 144169. [CrossRef]

Wang, T.; Zou, Y.; Matsuda, K. Micro-structure and micro-textural studies of friction stir welded AA6061-T6 subjected to different rotation speeds. *Mater. Des.* **2016**, *90*, 13–21. [CrossRef]

Fonda, R.W.; Bingert, J.F. Texture variations in an aluminum friction stir weld. *Scr. Mater.* **2007**, *57*, 1052–1055. [CrossRef]

Sidor, J.J.; Petrov, R.H.; Kestens, L.A.I. Modeling the crystallographic texturechanges in aluminum alloys during recrystallization. *Acta Mater.* **2011**, *59*, 5735–5748. [CrossRef]

Bo, W.; Dan-Qing, Y.; Yu-Qiang, C.; Bin, W.; Zeying, L. Evolution of texture and microstructure of 2E12 aluminum alloy during cold rolling and annealing. *Chin. J. Nonferrous Met.* **2013**, *23*, 3064–3074. [CrossRef]

Canova, G.R.; Kocks, U.F.; Jonas, J.J. Theory of torsion texture development. *Acta Metall.* **1984**, *32*, 211–226. [CrossRef]

Barnett, M.R.; Montheillet, F. The generation of new high-angle boundaries in aluminium during hot torsion. *Acta Mater.* **2002**, *50*, 2285–2296. [CrossRef]

Toth, L.S.; Gilormini, P.; Jonas, J.J. Effect of rate sensitivity on the stability of torsion textures. *Acta Metall.* **1988**, *36*, 3077–3091. [CrossRef]

Yuqi, L.; Tianliang, L.; Zhong, L.; Chuanmei, W.; Zhimin, Z.; Long, J.; Bing, L. Numerical and experimental investigations on dynamic behaviors of a bolted joint rotor system with pedestal looseness. *J. Sound Vib.* **2024**, *571*, 118036.

Suhuddin, U.F.H.R.; Mironov, S.; Sato, Y.S.; Kokawa, H.J.M.S. Grain structure and texture evolution during friction stir welding of thin 6016 aluminum alloy sheets. *Mater. Sci. Eng. A* **2010**, *527*, 1962–1969. [CrossRef]

Chen, S.; Jiang, X. Texture evolution and deformation mechanism in friction stir welding of 2219Al. *Mater. Sci. Eng. A* **2014**, *612*, 267–277. [CrossRef]

Zhang, L.; Wang, X.; Wei, X. Evolution of Grain Structure and Texture for 6082-T6 Aluminum Alloy during Friction Stir Welding. *J. Wuhan Univ. Technol. Mater. Sci. Ed.* **2019**, *34*, 397–403. [CrossRef]

Luo, Z.; Sun, Y.; Li, W.; He, J.; Luo, G.; Liu, H. Evolution of Microstuctures, texture and Mechanical Properties of Al-Mg-SiCu Alloy under Diffierent Welding Speedsduring Friction Stir Welding. *Metals* **2023**, *13*, 1120. [CrossRef]

Tao, W.; Yong, Z.; Xuemei, L.; Matsuda, K. Special grain boundaries in the nugget zone of friction stir welded AA6061-T6 under various welding parameters. *Mater. Sci. Eng. A* **2016**, *671*, 7–16. [CrossRef]

Baghdadi, A.H.; Rajabi, A.; Selamat, N.F.M.; Sajuri, Z.; Omar, M.Z. Effect of post-weld heat treatment on the mechanical behavior and dislocation density of friction stir welded Al6061. *Mater. Sci. Eng. A* **2019**, *754*, 728–734. [CrossRef]

33. Tao, Y.; Ni, D.R.; Xiao, B.L.; Ma, Z.Y.; Wu, W.; Zhang, R.X.; Zeng, Y.S. Origin of unusual fracture in stirred zone for frictio welded 2198-T8 Al-Li alloy joints. *Mater. Sci. Eng. A* **2017**, *693*, 1–13. [CrossRef]
34. Zhang, Z.; Wei, S.; Wang, P.; Qiu, W.; Zhang, G. Progress in applications of laser induced cavitation on surface processing. *Laser Technol.* **2024**, *170*, 110212. [CrossRef]

metals

MDPI

cle

teraction of Mechanical Characteristics in Workpiece ıbsurface Layers with Drilling Process Energy Characteristics

hael Storchak [1,2,*], Larysa Hlembotska [2], Oleksandr Melnyk [2] and Nataliia Baranivska [2]

[1] Institute for Machine Tools, University of Stuttgart, Holzgartenstraße 17, 70174 Stuttgart, Germany
[2] Department of Mechanical Engineering, Zhytomyr Polytechnic State University, Chudnivska str. 103, 10005 Zhytomyr, Ukraine; zid_gle@ztu.edu.ua (L.H.); o.l.melnyk@ztu.edu.ua (O.M.); ktpl_bnv@ztu.edu.ua (N.B.)
* Correspondence: michael.storchak@ifw.uni-stuttgart.de; Tel.: +49-711-685-83831

Abstract: The performance properties of various types of parts are predominantly determined by the subsurface layer forming methods of these parts. In this regard, cutting processes, which are the final stage in the manufacturing process of these parts and, of course, their subsurface layers, play a critical role in the formation of the performance properties of these parts. Such cutting processes undoubtedly include the drilling process, the effect of which on the mechanical characteristics of the drill holes subsurface layers is evaluated in this study. This effect was evaluated by analyzing the coincidence of the energy characteristics of the short hole drilling process with the mechanical characteristics of the drilled holes' subsurface layers. The energy characteristics of the short-hole drilling process were the total drilling power and the cutting work in the tertiary cutting zone, which is predominantly responsible for the generation of mechanical characteristics in the subsurface layers. As mechanical characteristics of the drill holes' subsurface layers were used, the microhardness of machined surfaces and total indenter penetration work determined by the instrumented nanoindentation method, as well as maximal indenter penetration depth, were determined by the sclerometry method. Through an analysis of the coincidence between the energy characteristics of the drilling process and the mechanical characteristics of the subsurface layers, patterns of the effect of drilling process modes, drill feed, and cutting speed, which essentially determine these energy characteristics, on the studied mechanical characteristics have been established. At the same time, the increase in the energy characteristics of the short-hole drilling process leads to a decrease in the total indenter penetration work and the maximum indenter penetration depth simultaneously with an increase in the microhardness of the drilled holes' subsurface layers.

Keywords: cutting; short hole drilling; subsurface layers; nanoindentation; sclerometry; indenter penetration work; indenter penetration depth

1. Introduction

The aspiration to ensure the specified performance properties of critical parts necessitates the search for ways to realize this assurance. One of the possible ways to solve this problem is the application of technological methods, in particular, cutting processes that generate specific subsurface layer mechanical characteristics of parts for various devices and mechanisms in the machining process [1]. These are the characteristics generated by the various cutting processes that are responsible for the sought-performance properties [2]. The relationship between mechanical characteristics and performance properties is particularly significant for critical complex-profile parts [3] functioning in alternating and high contact loads [4]. The mechanical characteristics of the machined subsurface layers of parts that correlate quite closely with various performance properties of these parts include the characteristics of the machined material structure, microhardness of these layers [5], distribution of residual stresses in these layers [6], and the surface integrity of the machined parts [5,7]. Although each of these mechanical characteristics correlates with the performance properties of machined parts separately, this relationship is unilateral. This transforms into a significant problem when it is necessary to consider the interrelation of multiple mechanical characteristics with performance properties simultaneously. It is especially aggravated when analyzing spatial cutting processes with a significant gradient of tool contact conditions with machined material and gradient of cutting modes, for example, when analyzing one of the oldest machining processes—the drilling process [8]. The use of generalized (integral) mechanical characteristics of subsurface layers is of considerable help in these cases. The characteristics determined by instrumented nanoindentation [9] and sclerometry [10] are used as such characteristics.

The presented study is a continuation of earlier research [11] and its development in terms of investigating the coincidence of mechanical characteristics of holes subsurface layers formed as a result of thermomechanical contact between the tool and the machined material under the conditions of the short hole drilling process.

2. Mechanical Characteristics Determination of Drilled Subsurface Layers

One of the most common and complex cutting processes for various materials is the drilling process [12], occupying 30% to 50% of all cutting processes [13]. In this regard, the drilling process has attracted the attention of many researchers [14]. Along with numerous experimental and theoretical studies of various drilling processes using different types of drills, considerable attention has also been paid to the study of mechanical characteristics of the drill holes subsurface layers [15,16]. These investigations include surface integrity studies [5], which involve studies of the microstructure and microhardness of subsurface layers [7], as well as residual stresses in these layers [6] formed as a result of the drilling process [8].

The influence of contact conditions and peculiarities of the drilling process with the insertion of additional energy in the form of vibrations on the surface integrity, in particular on the plastic degree deformation of the machined material, its microhardness distribution, the magnitude of residual stresses and others is the subject of a study by Sun and colleagues [17]. Strodick et al. studied the effect of cutting modes on surface integrity during deep-hole drilling with BTA drills [18]. The focus of Ren and Liu's study was the effect of cutting modes on the work hardening of machined material [19]. The study results of the residual hole surface roughness dependence on cutting modes during the drilling of AISI H11 tool steel are presented in the work of Garg and Goyal [20]. Studies by Lacombe and colleagues proved the significant influence of microhardness, as one of the characteristics of drilled hole surface integrity, on the fatigue strengths of aluminum alloy parts [21]. The authors of the study also noted significant dependence of such characteristics of drill hole surface integrity as machined surface roughness, microhardness, and microstructure of machined subsurface layers on cutting modes [22].

Recently, researchers have also been attracted to the study of surface integrity characteristics of the surface machined by drilling various heterogeneous materials, mainly composite materials. In addition to the accepted surface integrity characteristics, such studies also consider the delamination processes of composite materials [23], dislocation development [24], and dislocation degree [25] of the mentioned composite materials. A significant number of publications are devoted to the study of surface integrity characteristics of holes during the drilling of difficult-to-machine materials, such as titanium alloys and nickel-containing alloys. In the study of titanium alloy drilling, the alloy Ti-6Al-4V was generally used as the machined material. While analyzing the microstructure of the hole subsurface layers during the drilling of titanium alloy Ti-6Al-4V, Varote and Joshi found a relationship between such characteristics of the microstructure as grain size and disorientation angle with drilling modes and the presence of lubricating and cooling fluids [26]. Similar studies were carried out by Haddag et al. [27], using the analysis of the microhardness distribution of subsurface layers to estimate the microstructure of these layers. Wang and colleagues related the microstructure of subsurface layers to their microhardness and residual stress distribution in these layers [28]. A significant part of studies on the drilling process of titanium alloys is devoted to the dependence evaluation of the roughness geometric parameters of drilled holes on the cutting process conditions, particularly the drill feed and cutting speed [29,30]. Considerable attention is also paid to the analysis of microhardness of cutting machined subsurface layers of titanium alloys during drilling [31,32], the influence of various lubricating and cooling fluids and methods of their supply to the cutting zone on the formation of surface integrity characteristics [33], the analysis of the influence of special geometrical parameters of drills on surface integrity [34], and others.

In the study of the surface integrity of drilled holes in nickel-containing alloys, Inconel 718 alloy served as a prime example of the machined material. Lotfi et al. evaluated the relationship of drilling conditions of Inconel 718 alloy, in particular, drill feed and the drill revolutions number, with the microhardness of holes subsurface layers, microstructure, and roughness height parameters [35]. They analyzed the influence of the cutting modes noted above on the microstructure characteristics using a numerical drilling model. A study of the drilling process conditions effect of nickel alloy UDIMET 720 on the surface topography of drill holes' subsurface layers and microhardness is presented in research by Dutilh and colleagues [36]. The influence of tool geometric parameters on the structure of Inconel 718 alloy was evaluated in a study by Xu and colleagues [37]. The study by Rahim and Sasahara [38] concentrated on analyzing the effect of the coolant delivery method and its composition on surface integrity characteristics such as microhardness of subsurface layers, surface roughness of drill holes subsurface layers, and plastic deformation of subsurface layers in the machined Inconel 718 alloy. Karabulut and Kaynak showed the possibilities of improving the machined surface quality compared to the drilled surface in the same deformed alloy by studying the drilling process of additively manufactured Inconel 718 alloy [39]. Its height roughness parameters were used to evaluate the surface quality.

The residual stress distribution in subsurface layers due to the cutting process and the significant influence of this surface integrity characteristic on the fatigue resistance of machined parts is of particular interest to researchers. The methodological peculiarities of determining residual stresses in the holes' subsurface layers formed during drilling were studied by Girinon and colleagues [40]. Nobre and Outeiro developed an experimental and numerical method to estimate the residual stresses in the subsurface layers of the workpiece formed by the drilling process [41]. An analytical model for calculating residual stresses in subsurface layers induced by the drilling process was developed by Rasti and coworkers [42]. The developed model was based on the analytical model of orthogonal cutting, equivalently transformed to the conditions of the main cutting edge of the drill. Experimental values of residual stresses determined by instrumented nanoindentation were used to validate the proposed model. A numerical spatial model to simulate the residual stresses generated by the reaming process is proposed by Leveille et al. [43]. Experimentally measured cutting forces were used as boundary values of the kinetic characteristics of the reaming process. Wu and colleagues investigated the distribution of residual stresses formed in the subsurface layers of aluminum alloy sheets due to the drilling process [44]. To reduce the residual stress magnitudes, the fastening parameters of the sheet material are optimized.

The characteristics of surface integrity that are considered as a result of the drilling process on various structural materials are ambiguous and often contradictorily related to the cutting process conditions. To clarify this relationship and attempt to ensure its unambiguity, generalized (integral) mechanical characteristics of subsurface layers [11] determined by instrumented nanoindentation [45,46] and sclerometry [47,48] have recently been used. The state-of-the-art instrumented nanoindentation method [49] was developed based on the research of Atkins and Tabor [50]. Later, this method was significantly modified by creating new algorithms for analyzing test results [51] and devices for its implementation [52]. The indenter surface scratching test method, called sclerometry, was created mainly for the analysis of various coatings [53], particularly for the evaluation of the retention strength of the coating on the substrate surface [54] and the adhesion ability of coatings [55]. Subsequently, a multi-pass sclerometry method [56,57] was proposed to reduce the uncertainty in estimating the pavement failure load [58]. Significant researchers' attention has also been directed towards the development of new devices to realize instrumented nanoindentation [59,60] and sclerometry [61], as well as improvements to existing devices [62,63]. Besides the modernization of the mechanical and electrical parts of the devices for the realization of the specified test methods, the main direction of development and research of the devices was the development of testing methodology. Research on improving instrument calibration methods [64], as well as methods for determining and evaluating measurable mechanical properties [65,66], fully relate to this.

The instrumented nanoindentation method was mainly used to determine the same surface integrity characteristics discussed above: microhardness of the subsurface layers, strain degree of the machined material, material microstructure parameters, and residual stresses in the subsurface layers. Using the instrumented nanoindentation method of five different materials in terms of deformability, Tsybenko et al. evaluated the microhardness of pre-applied scratches and the effect of test conditions on this microhardness [67]. England et al. investigated the relationship between microhardness and corrosion resistance of AM 316L steel [68]. The influence of microhardness determined using instrumented nanoindentation on precipitation strengthening of steels with different alloying is studied by Moon et al. [69].

Li and colleagues used the instrumented nanoindentation method to obtain flow curves of AISI 1045 steel [70]. Through instrumented nanoindentation, Paul et al. [71] investigated the possibility of evaluating the deformation capacity and fracture behavior of 304L steel by nanoindentation. The mechanical properties of 18CrNiMo7-6 steel were determined via instrumented nanoindentation by Zhou and his colleagues [72]. The instrumented nanoindentation method has also been successfully used to determine residual stresses in various materials, such as copper foil [73], films and coatings [74], austenitic steels [75], and other difficult-to-machine steels and alloys. The study of microstructure and mechanical properties of various materials has also been successfully performed using the sclerometry method. Fan et al. investigated the nanomechanical properties of newly synthesized high entropy alloys [76] using the sclerometry method. The effect of sclerometric modes on the mechanical properties of subsurface layers of Ti-6Al-4V titanium alloy was the focus of a study by Pratap and colleagues [77].

In addition to determining the above-mentioned individual surface integrity characteristics, the instrumental nanoindentation method and the sclerometry method are also used to determine the energetic (integral) mechanical characteristics of the subsurface layers [11,78]. Bezyazychnyy et al. investigated the influence of cutting modes, in particular cutting speed and tool feed, during the turning of chrome-nickel alloy on the accumulated strain energy of the workpiece subsurface layers [79]. The relationship between the microhardness of subsurface layers of various steels and alloys and the strain energy of the machined material was studied by Yamamoto et al. [80]. Ren and Liu [19] have established a relationship between the work hardening of Inconel 718 alloy and the turning conditions of this machined material.

Thus, the analysis of known methods leads to the conclusion that the instrumented nanoindentation method and the sclerometry method can also be used to determine the mechanical characteristics of the hole subsurface layers formed due to the drilling process.

3. Materials and Methods

The formation patterns of physical and mechanical characteristics of the workpiece subsurface layers during real spatial cutting processes, such as, for example, turning, drilling, milling, threading, and others, are of the greatest interest due to the widespread use of these cutting processes in production. This study aims to establish the relationship between the energy characteristics of the short hole drilling process and the mechanical characteristics of the hole subsurface layers formed during the drilling process. This will ensure that the desired mechanical characteristics are subsequently generated by setting the appropriate energy characteristics of the drilling process. The paper studies the formation patterns of mechanical characteristics of subsurface layers in one of the most common cutting processes—the drilling process. The mechanical characteristics study of the hole subsurface layers formed during short hole drilling is a continuation and methodological extension of the previously published study [11] devoted to the orthogonal cutting process. The flowchart of information data flow, which is a methodology for studying the coincidence of mechanical characteristics of subsurface layers formed in the drilling process with the integral characteristics of the cutting process, is presented in Figure 1.

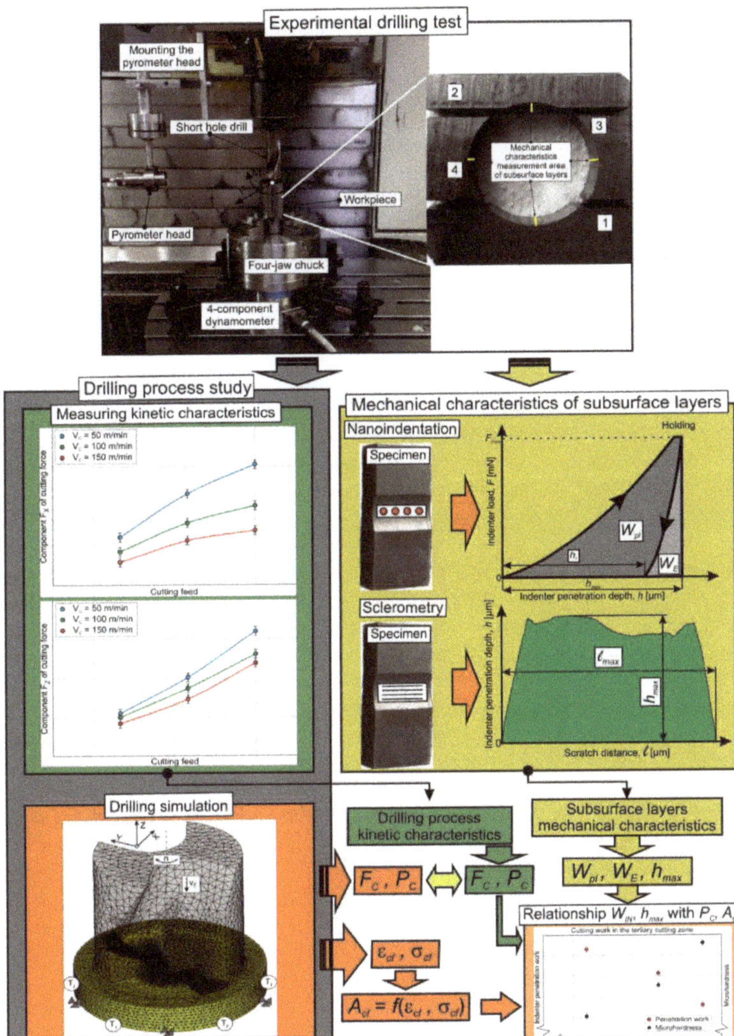

Figure 1. Methodology scheme for determination of physical and mechanical characteristics of drilled subsurface layers.

The experimental studies are divided into two parts: measurement of cutting forces during short hole drilling and determination of mechanical characteristics of drill hole subsurface layers using instrumented indentation and sclerometry. This step is performed as the initial step. The measured components of the cutting forces are used to determine the cutting power P_C. In parallel, the microhardness of the subsurface layers and the total work of indenter penetration W_{IN}, as well as the maximum depth of indenter penetration h_{max} during sclerometry, are determined from the diagrams of indenter penetration into the specimens' machined surface. These studies are performed in the second step. The third step of the study is devoted to determining the cutting work when drilling short holes in the tertiary cutting zone A_{dcf}. Since the determination of cutting work in the specified zone experimentally causes significant difficulties for real spatial cutting processes, in particular drilling, this characteristic is determined using a numerical cutting model. In the last fourth step of the research methodology, the coincidence of the mechanical characteristics of the subsurface layers formed by the short hole drilling process with the integral characteristics of the cutting process (P_C and A_{dcf}) is identified. This is the focus of the entire study.

3.1. Materials

The machining of the test material by short hole drilling was carried out at the machining center UWF 1202 H (Hermle AG, Gosheim, Germany) (see Figure 1, top, left). The well-known AISI 1045 structural steel was used as the machined material. Table 1 summarizes the mechanical and thermal properties of the machined material. The characterization of the experimental setup, the measuring circuit for determining the components of the cutting force, the used tool (a short hole drill equipped with two carbide inserts: outer and inner) and its geometrical parameters, as well as the initial geometry of the machined workpiece, are given in a previously published study [81].

Table 1. Mechanical and thermal properties of the steel AISI 1045 and carbide inserts adapted from Refs. [82,83].

Material	Strength (MPa)		Elastic Modulus (GPa)	Elongation (%)	Hard-ness	Poisson's Ratio	Specific Heat (J/kg·K)	Thermal Expansion (μm/m·°C)	Thermal Conductivity (W/m·K)
	Tensile	Yield							
AISI 1045	690	620	206	12	HB 180	0.29	486	14	49.8

To determine the resulting cutting force F_C when drilling short holes, the three components F_x, F_y, and F_z were measured (see Figure 1, top left). The feed motion during machining was carried out by displacement of the table with the workpiece fixed on a 4-component dynamometer to the drill fixed in the spindle of the machining center. The feed was varied at three levels: 0.05 mm/rev, 0.1 mm/rev, and 0.15 mm/rev. The nominal cutting speed was also varied at three levels: 50 m/min, 100 m/min, and 150 m/min. To achieve reliability of measurements, the drilling process was repeated at least five times for each set of cutting modes. The maximum error in measuring the cutting force components did not exceed 11%.

The mechanical characteristics of the subsurface layers generated during the short hole drilling process were determined using instrumental nanoindentation and sclerometry. The specimens for measuring mechanical characteristics were prepared from drilled workpieces cut into separate parts (see Figure 1, top, right). Schemes for measuring mechanical performance are illustrated in Figure 2. The Fischer Picodentor HM500 (Helmut Fischer Ltd. Institute for Electronics and Measurement Technology, Sindelfingen, Germany) measuring system was used to perform experimental studies using the instrumental indentation method [11,57]. The instrumented indentation was carried out on the specimen hole surfaces drilled at different cutting speeds and tool feeds. Berkovich indenter was used for measurements. The maximum force on the indenter was 450 mN. The measurement accuracy of the indenter applied load was 0.02 mN. The change of indenter load rate was 20 mN/s. After reaching the maximum load on the indenter during the test, a hold at the specified load for 5 s was performed (see Figure 2a). The measurement accuracy of the indenter penetration depth into the test specimen was 5 nm. Tests of the drilled specimen with each set of cutting speeds and drill feeds were repeated at least 10 times. The processing results of each obtained indenter penetration diagram were the Vickers microhardness of the drilled hole subsurface layers and the total work of indenter penetration W_{IN} (see the diagram in Figure 2a). The processing results of loading diagrams were averaged for each set of cutting speeds and drill feeds. The largest error in measurement in the diagrams' results processing did not exceed 10%.

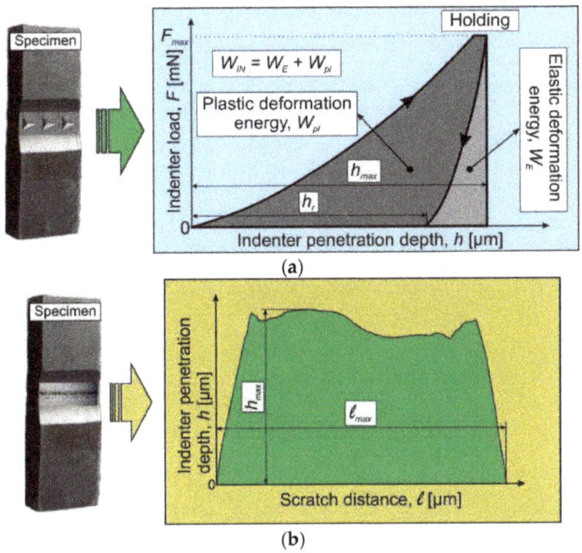

Figure 2. Schematics for mechanical characteristics measurement: (**a**) measurement scheme for instrumented nanoindentation; (**b**) measurement scheme for sclerometry.

Studies of subsurface layers' mechanical characteristics of drilled holes by the sclerometry method were carried out using the device "Micron-gamma" [11,57]. To carry out the studies noted above, a Berkovich indenter was installed so that the projection of one of its edges was parallel to the vector of the indenter movement velocity. The load on the indenter during sclerometry was 100 mN, its movement speed was 20 μm/s, and the length of the produced scratch was 630 μm. The test diagram (sclerogram) determined the maximum depth of indenter penetration along the length of the produced scratch (see the diagram in Figure 2b). Sclerometric tests of the drilled specimen with each set of cutting speeds and drill feeds were repeated at least 10 times. The error in determining the maximum indenter penetration depth resulting from sclerometry did not exceed 11%. The produced scratches were analyzed visually using an optical profilometer, "Micron-alpha" (Micron Ltd., Kyiv, Ukraine) [11,57]. The image vertical resolution provided by the device "Micron-alpha" was 2 nm.

3.2. Methods

The short hole drilling process is a prime example of a real cutting process. In the cutting zones of this machining process, a three-dimensional stress-strain state of the machined material is realized [84]. Under such loading conditions, either adiabatic hardening of the machined material or isothermal softening of this material is realized in the cutting zones [83,85]. Under such loading conditions of the machined material during drilling, the corresponding mechanical characteristics of the subsurface layers are generated. Thus, the patterns of these mechanical characteristic's generation are essentially formed by the loading conditions of the machined material, which, in turn, are determined by the cutting process characteristics. Following the analogy with the first part of the study of the mechanical properties' formation of the machined subsurface layers [11], the total drilling power P_C and the drilling work in the tertiary cutting zone A_{dcf} was chosen as the characteristics determining the loading conditions of the machined material in the drilling process. Following the study methodology, the coincidence of the mentioned integral characteristics of the drilling process with the generated mechanical characteristics of the holes' subsurface layers formed during the short hole drilling process will be identified.

The following relationship is used to determine the total drilling power P_C [84,86]:

$$P_C = F_C \cdot V_C, \tag{1}$$

where F_C—is the resultant cutting force, V_C—is the cutting speed.

The deformation work of the machined material in the tertiary cutting zone A_{dcf} was determined in the dependence of equivalent stresses σ_{dcf} acting in the tertiary cutting zone of the machined material and strains ε_{dcf} of the machined material in this zone:

$$A_{dcf} = V_m \cdot \int_{t_s}^{t_e} \sigma_{dcf} d\varepsilon_{dcf}, \tag{2}$$

where V_m is the material removal volume, t_s and t_e are the simulation start and end times, respectively.

Determination of the deformation work of the machined material in the tertiary cutting zone during short hole drilling was performed using the previously developed numerical cutting model of the drilling process [81]. According to the developed finite element model of the short hole drilling process, the workpiece material was modeled as an isotropic material and the drill as an ideally rigid body [87]. The behavior of the machined material under thermomechanical loading during cutting was described by the Johnson-Cook constitutive equation [88,89].

The Columbian friction model was used to describe the contact interaction conditions between the drill and the machined material [90]. The friction coefficient values in different cutting zones and their areas were determined according to the previously developed methodology [91]. An algorithm embedded in the software [92] was used to separate machined material into the volume of the chip and machined workpiece body. In this regard, it was not necessary to apply a special fracture model of the machined material [93]. Determination of equivalent stresses in the tertiary cutting zone of machined material and material deformations in the process of drilling short holes were performed using tracking points. The layout of the tracking points is shown in Figure 3.

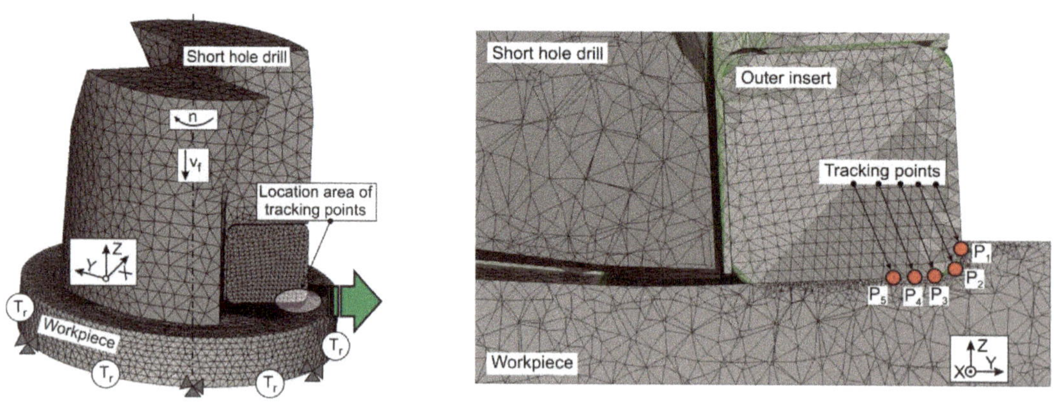

(a) (b)

Figure 3. Layout scheme of tracking points: (**a**) initial geometric model of milling with a mesh and boundary conditions; (**b**) location of tracking points.

Five tracking points (P_1, P_2, P_3, P_4, and P_5) were used to determine the drill work, as shown in Figure 3b. The equivalent stresses and strains of the machined material were averaged over the specified five tracking points.

4. Results

As noted in Section 3.2, the integral characteristics of the cutting process are used to evaluate the effect of the short hole drilling process on the mechanical characteristics of the drill hole subsurface layers. The total drilling power P_C and the cutting work in the tertiary cutting zone A_{dcf} are chosen as such characteristics. The total drilling power P_C was determined using the resultant cutting force F_C (see Equation (1)). Figure 4 shows the effect of drill feed on the resultant cutting force and cutting power at different cutting speeds. The resulting cutting force F_C increases monotonically with increasing drill feed f. At the same time, the increase in cutting speed has a somewhat opposite effect: the increase in cutting speed V_C causes a decrease in the resultant cutting force (see Figure 4a). Increasing the drill feed leads to a significantly greater increase in the total cutting power than in the resultant cutting force (see Figure 4b). An increase in cutting speed causes a proportional increase in cutting power.

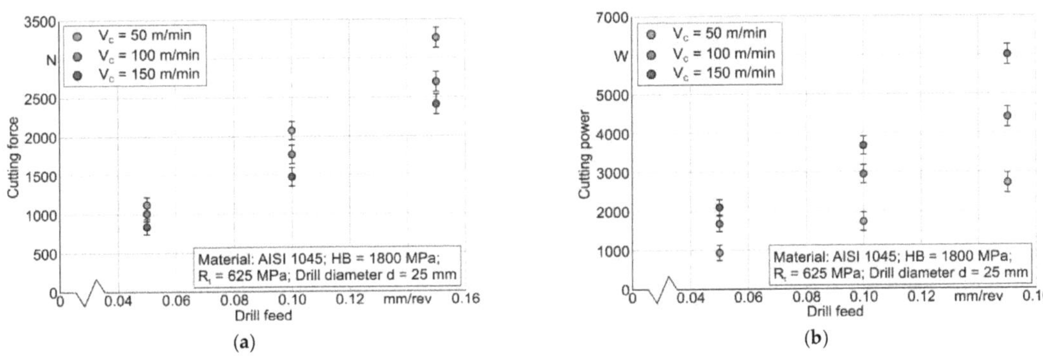

Figure 4. Dependence of the resultant cutting force and cutting power on drill feed and cutting speed: (**a**) cutting modes influence on the resultant cutting force; (**b**) cutting modes influence on cutting power.

To determine the cutting work in the tertiary cutting zone A_{dcf}, the simulated values of equivalent stresses σ_{dcf} and corresponding strains ε_{dcf} of the machined material in the specified cutting zone are used. The stresses and strains required for this calculation are simulated using a numerical drilling model (see Section 3.2). To prove the valid simulation of equivalent stresses and strains, the adequacy of the previously developed finite element drilling model [81] was verified. The adequacy of the drilling model was tested by comparing experimentally determined and simulated values of the resultant cutting force and total drilling power. The results of this comparison are presented in Figure 5. The deviation of the experimentally determined resultant cutting force from its simulated value at cutting speed V_C = 50 m/min ranges from 10.6% to 18.2% for the corresponding drill feed variation from 0.05 mm/rev to 0.15 mm/rev (see Figure 5a). The deviation between the experimentally determined and simulated values of the resulting cutting force at cutting speed V_C = 100 m/min ranges from 12.1% to 22.7% for the corresponding drill feed variation from 0.05 mm/rev to 0.15 mm/rev (see Figure 5b). The specified deviations of the resulting cutting force at a cutting speed of V_C = 150 m/min range from 18.6% to 26% for the corresponding drill feed variation from 0.05 mm/rev to 0.15 mm/rev (see Figure 5c). The same deviations were determined between the simulated and experimentally determined values of the total drilling power (see Figure 5d–f). Since the deviations between the experimental and simulated values of the specified drilling process characteristics range from 10.6% to 26%, it can be stated that the finite element drilling model can be used to simulate equivalent stresses and corresponding strains of machined material in the tertiary cutting zone.

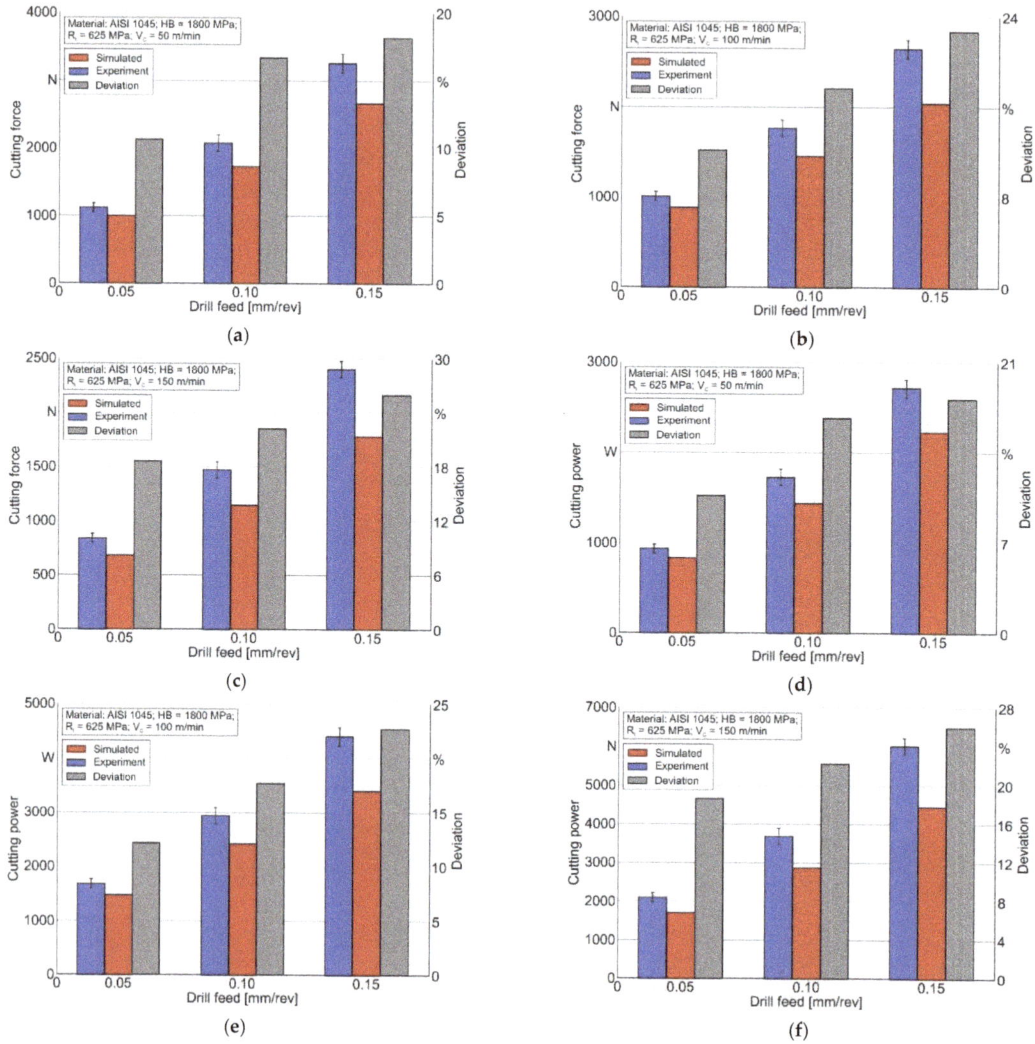

Figure 5. Experimental and simulated values comparison of the resultant cutting force and cutting power at changing of the cutting modes: (**a**) cutting force dependence on drill feed at cutting speed V_C = 50 m/min; (**b**) cutting force dependence on drill feed at cutting speed V_C = 100 m/min; (**c**) cutting force dependence on drill feed at cutting speed V_C = 150 m/min; (**d**) cutting power dependence on drill feed at cutting speed V_C = 50 m/min; (**e**) cutting power dependence on drill feed at cutting speed V_C = 100 m/min; (**f**) cutting power dependence on drill feed at cutting speed V_C = 150 m/min.

The relationship between the simulated values of equivalent stresses and strains of machined material in the tertiary cutting zone determined for five tracking points, for a cutting speed of V_C = 100 m/min and a feed f = of 0.1 mm/rev, is shown in Figure 6a. Figure 6b illustrates the simulated value of the dependence of cutting work on the drill feed for different cutting speeds. Cutting work A_{dcf} increases in proportion to the increase in drill feed and also grows with rising cutting speed.

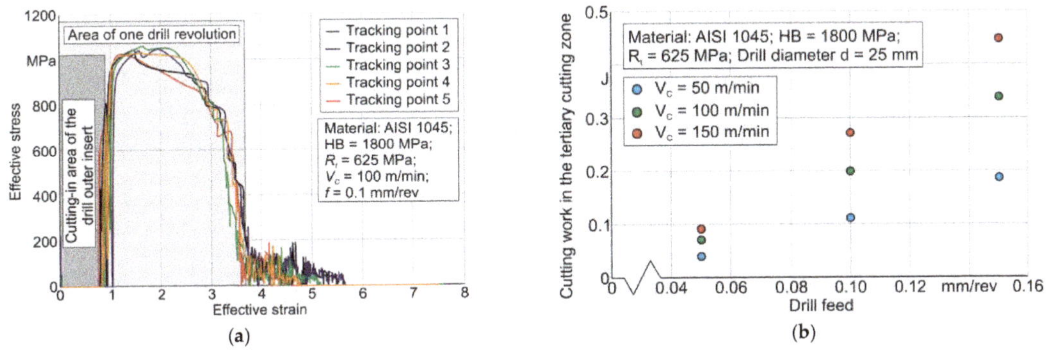

Figure 6. Cutting work A_{cf} in the region of the tertiary cutting zone: (**a**) relationship between effective stress and effective strain for five tracking points; (**b**) effect of drill feed and cutting speed on cutting work A_{cf}.

Following the study methodology (see Section 3, Figure 1), the microhardness and total indentation work, as well as the maximum indentation depth, were considered as the mechanical characteristics of drilled hole subsurface layers. The microhardness and the total indenter penetration work were determined using the instrumented nanoindentation method. The maximal indenter penetration depth was studied using sclerometry (see Section 3.1). The study results of the mechanical characteristics noted above using the instrumented nanoindentation method and a drill feed for different values of cutting speed are presented in Figure 7. The microhardness of the drill hole subsurface layers increases in proportion to the drill feed, as shown in Figure 7a. At the same time, as the cutting speed increases, the microhardness of machined material subsurface layers decreases. Increasing the drill feed leads to a monotonic lowering of the total indenter penetration work W_{IN}, as shown in Figure 7b. Conversely, an increase in cutting speed leads to an increase in work W_{IN}.

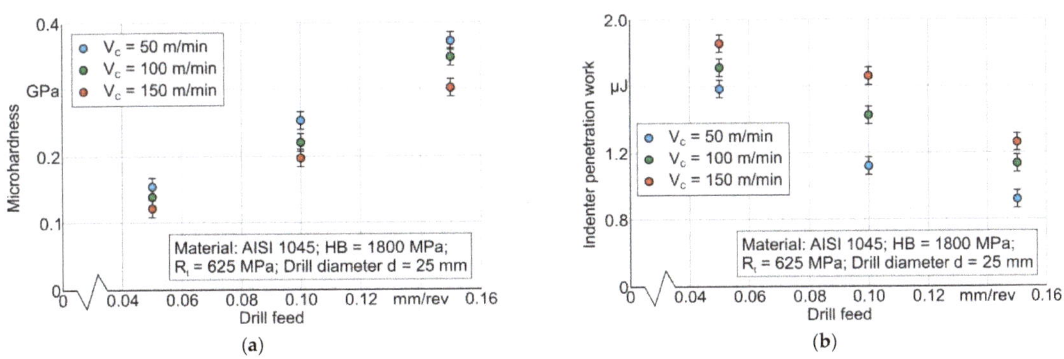

Figure 7. Effect of drill feed and cutting speed on microhardness of drill hole subsurface layers and total indenter penetration work: (**a**) microhardness dependence on cutting modes; (**b**) indenter penetration work dependence on cutting modes.

Cutting modes have a similar effect on the maximum indenter penetration depth h_{max} during sclerometry: with increasing drill feed, the depth h_{max} monotonically decreases, and with increasing cutting speed, the maximum indenter penetration depth increases—Figure 8.

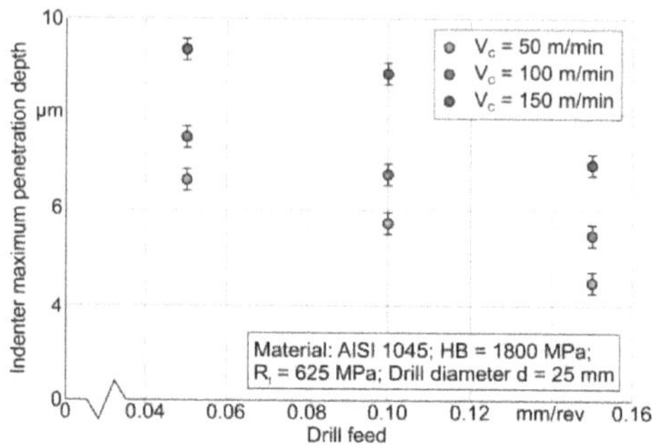

Figure 8. The indenter penetration depth depends on the cutter feed.

5. Discussion

The process of short hole drilling is evident to the external environment, including an outside observer, by various phenomena and characteristics similar to other types of spatial cutting processes. The most sensitive and informative characteristics of the drilling process, which have a significant impact on the mechanical characteristics of cutting machined subsurface layers, are the integral, mainly energy characteristics of this process: total cutting power and cutting work in the tertiary cutting zone. These characteristics of the drilling process are significantly dependent on the cutting modes, in this case, the drill feed and cutting speed. This is shown by the experimental studies presented in Figure 4. The rise in cutting power with increasing drill feed is due to the proportional increase in the volume of machined material removed during the cutting process (see Figure 4a) [84,86]. The decrease in total cutting power with increasing cutting speed is due to the softening of the machined material resulting from a significant increase in cutting temperature [94] due to the increase in cutting speed [95,96]. The same influence mechanisms underlie the variation patterns of cutting work in the tertiary cutting zone as a function of changes in drill feed and cutting speed (see Figure 4b).

Cutting modes in the short hole drilling process logically also have a significant effect on the mechanical characteristics of the subsurface layers since changes in cutting modes, in particular, drill feed and cutting speed, entail significant changes in the thermomechanical effect of the drill on the machined material [95,96]. In turn, the growth of plastic strain causes hardening of the subsurface layers of machined material and, as a consequence, an increase in the microhardness of the machined subsurface layers (see Figure 7a). An increase in cutting speed has the opposite effect on the microhardness of the subsurface layers due to the softening of the machined material resulting from an increase in cutting temperature [84,86] caused by an increase in cutting speed [94]. The considered effect mechanisms of drill feed and cutting speed changes on the state of machined subsurface layers have the opposite effect on the total indenter penetration work W_{IN} (see Figure 7b). This can be explained by the fact that the work W_{IN} decreases with the hardening of the test material and increases with its softening. The effect of cutting modes on the maximum indenter penetration depth h_{max} during sclerometry can also be interpreted similarly (see Figure 8).

The influence of cutting modes on the mechanical characteristics of drill holes' subsurface layers is simplistic and mechanistic since they do not identify the physical sense of the cutting process's impact on these mechanical characteristics. The main factors determining the process of generating mechanical characteristics of subsurface layers as a result of the thermomechanical impact of the drill on the machined material are the hardening and softening processes of subsurface layers [95,96]. Cutting modes, in particular drill feed and cutting speed, influence the hardening or softening of the machined material. However, there is only a correlation between these cutting modes and the mechanical characteristics of the machined subsurface layers. At the same time, there is a lack of possibility to predict which processes—hardening or softening of the machined material—will prevail at which values of cutting modes.

Consequently, there is an absence of the possibility of changing the mechanical characteristics of the machined subsurface layers using a targeted approach. This possibility is available from the integral (energy) characteristics of the cutting process, particularly the characteristics discussed above: the total drilling power and the cutting work in the tertiary cutting zone. This is explained by the fact that the energy characteristics of the cutting process fully numerically reflect the effects of the thermomechanical impact of the tool on the machined material, determining the prevalence of either kinetic impact [85,86] or thermal impact [94,97].

Following the above-mentioned, the coincidence between the mechanical characteristics of the drill holes' subsurface layers and the energy characteristics of the drilling process was analyzed. Figure 9 shows the coincidence results of indenter penetration work with microhardness and total cutting power, as well as cutting work in the tertiary cutting zone at different drill feeds. The specified coincidence of mechanical characteristics and energy characteristics of the drilling process at a drill feed of f = 0.05 mm/rev is shown in Figure 9a, at a drill feed of f = 0.1 mm/rev is shown in Figure 9b, and at a drill feed of f = 0.15 mm/rev is shown in Figure 9c.

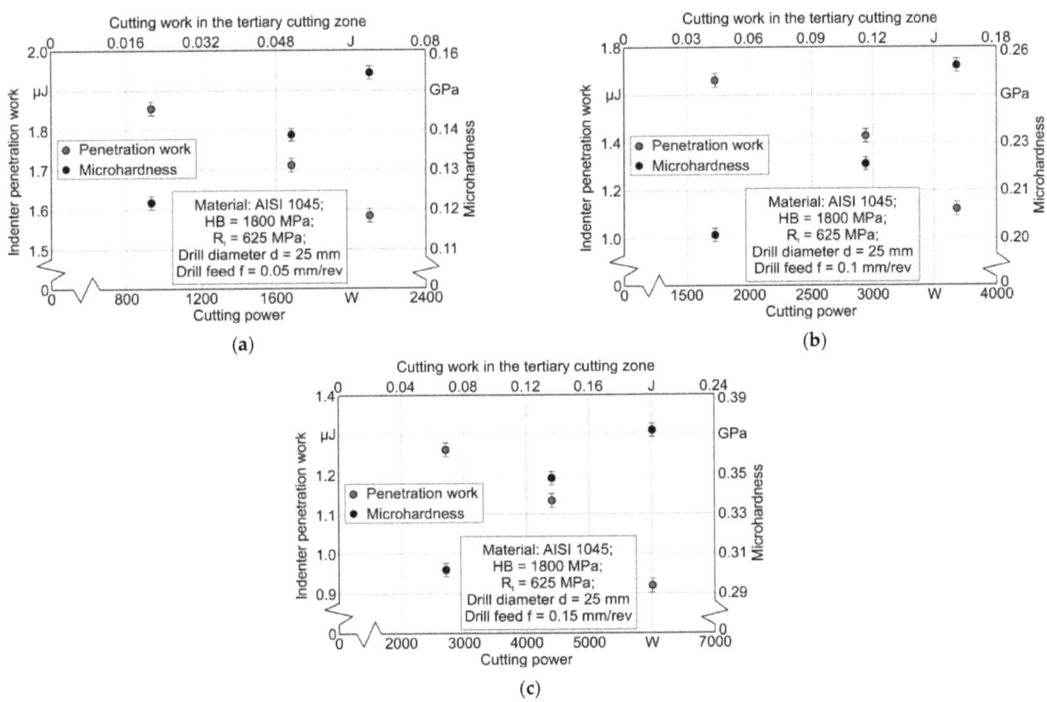

Figure 9. The coincidence of the indenter penetration work and microhardness with the cutting power and cutting work in the tertiary cutting zone: (**a**) at a drill feed of f = 0.05 mm/rev; (**b**) at a drill feed of f = 0.1 mm/rev; (**c**) at a drill feed of f = 0.15 mm/rev.

As increasing power P_C and work A_{dcf} are caused due to increasing cutting speed, the indenter penetration work W_{IN} increases monotonically for all studied values of drill feeds. At the same time, the microhardness of the machined subsurface layers decreases with increasing P_C and work A_{dcf} for all studied values of drill feed (see Figure 9).

Figure 10 shows the coincidence of the maximum indenter penetration depth with the total drilling power P_C and cutting work in the tertiary cutting zone A_{dcf}) at the different drill feeds indicated above. As these energy characteristics increase, the indenter penetration depth decreases.

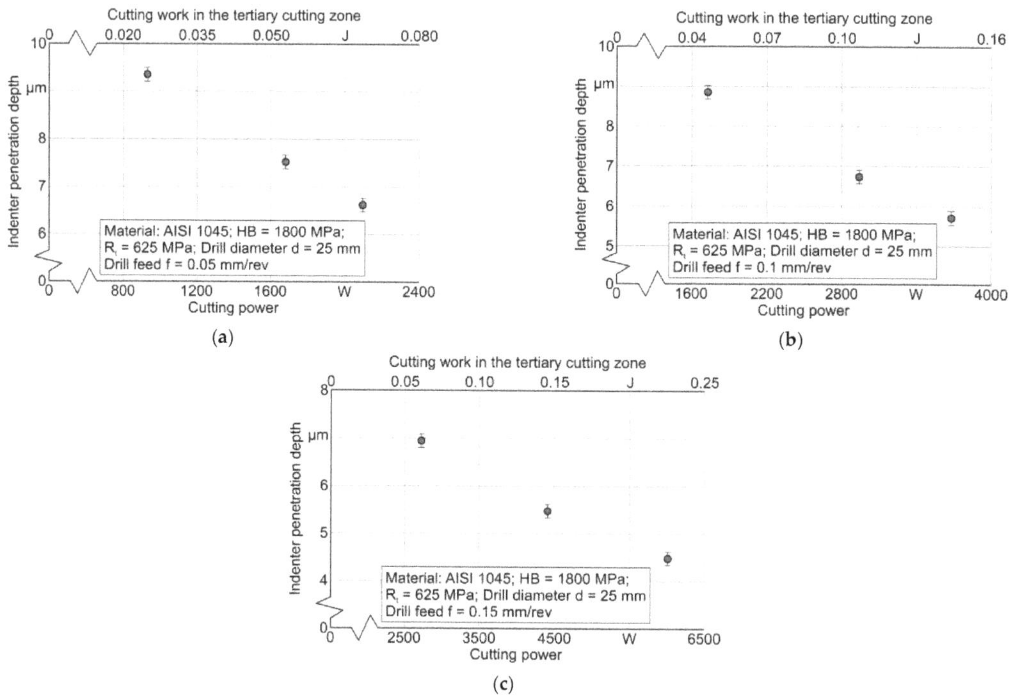

Figure 10. The coincidence of indenter penetration depth during sclerometry with the cutting power and cutting work in the tertiary cutting zone: (**a**) at the drill feed of f = 0.05 mm/rev; (**b**) at the drill feed of f = 0.1 mm/rev; (**c**) at the drill feed of f = 0.15 mm/rev.

The changes in the investigated mechanical characteristics shown in Figures 9 and 10 are explained by the hardening of the drill holes subsurface layers as a result of the cutting process [84,96]. Thus, an increase in the energy characteristics of the drilling process (power P_C and work A_{dcf}), expressed in the form of subsurface layers hardening, prevents the indenter's ability to penetrate these layers. Hence, the cutting process provides pumping a certain part of its energy into the drill holes subsurface layers, generating specific mechanical characteristics in these layers. In turn, the mechanical characteristics of subsurface layers significantly determine the service properties of parts manufactured by the cutting process, as well as their durability and serviceability. Thus, the coincidence of energy characteristics from the drilling process and mechanical characteristics discussed above allows for the required or specified service properties of manufactured parts to be provided by controlling the energy characteristics above.

6. Conclusions

The research object of the present study is to analyze the coincidence between the energy characteristics of the short hole drilling process and the mechanical characteristics of the drill holes' subsurface layers. The energy characteristics of the drilling process are the total drilling power and the cutting work in the tertiary cutting zone. The microhardness of drill holes subsurface layers and the total indenter penetration work, as well as the maximum indenter penetration depth, were used as mechanical characteristics of the subsurface layers.

The cutting power was determined using the measured components of cutting forces. The cutting work in the tertiary cutting zone was calculated by numerical modeling of the short hole drilling process. The validity of the finite element model for the drilling process developed for this purpose was checked by comparing the measured and simulated values of the resulting cutting force. The difference between these values of cutting forces at nominal cutting speed V_C equal to 50 m/min ranged from 10.6 to 18.2% for the drill feed range from 0.05 mm/rev to 0.15 mm/rev. For the nominal cutting speed equal to 100 m/min, the deviation values ranged from 12.1% to 22.7%, and for the cutting speed equal to 150 m/min, from 18.6% to 26%, respectively.

The increase in the energy characteristics of the short hole drilling process causes an increase in the microhardness of the drill holes' subsurface layers. At the same time, this leads to a decrease in the total indenter penetration work and the maximum indenter penetration depth. The reason for this coincidence between the energy characteristics of the drilling process and the mechanical characteristics of the subsurface layers is the prevalence of the phenomenon of strain and strain rate hardening of these subsurface layers over their thermal softening.

The patterns of the identified coincidence between the energy characteristics of the short hole drilling process and the mechanical characteristics of the drill holes' subsurface layers ensure the possibility of controlling the performance properties of the machined parts using the studied drilling process.

Author Contributions: Conceptualization, M.S.; methodology, M.S.; software, M.S.; validation, M.S., L.H., and O.M.; formal analysis, M.S.; investigation, M.S., L.H., O.M. and N.B.; resources, M.S., O.M. and N.B.; data curation, M.S., L.H., and N.B.; writing—original draft preparation, M.S.; writing—review and editing, M.S.; visualization, M.S., L.H., and N.B.; project administration, M.S and O.M.; funding acquisition, M.S. All authors have read and agreed to the published version of the manuscript.

Funding: This study was funded by the German Research Foundation (DFG) in the project HE-1656/153-1, "Development of a Concept for Determining the Mechanical Properties of the Cutting Material in Machining".

Data Availability Statement: The original contributions presented in the study are included in the article, further inquiries can be directed to the corresponding author.

Acknowledgments: The authors would like to thank the German Research Foundation (DFG) for their support, which is highly appreciated.

Conflicts of Interest: The authors declare no conflicts of interest.

References

1. Guo, Y.; Saldana, C.; Compton, W.D.; Chandrasekar, S. Controlling deformation and microstructure on machined surfaces. *Acta Mater.* **2011**, *59*, 4538–4547. [CrossRef]
2. Sagapuram, D.; Udupa, A.; Viswanathan, K.; Mann, J.B.; M'Saoubi, R.; Sugihara, T.; Chandrasekar, S. On the Cutting of Metals: A Mechanics Viewpoint. *ASME J. Manuf. Sci. Eng.* **2020**, *142*, 110808. [CrossRef]
3. Davim, J.P. *Machining of Complex Sculptured Surfaces*; Springer: London, UK, 2012; 258p. [CrossRef]
4. Babichev, D.; Storchak, M. Synthesis of cylindrical gears with optimum rolling fatigue strength. *Prod. Eng. Res. Dev.* **2015**, *9*, 87–97. [CrossRef]
5. Bag, R.; Panda, A.; Sahoo, A.K.; Kumar, R. A Perspective Review on Surface Integrity and Its Machining Behavior of AISI 4340 Hardened Alloy Steel. *Mater. Today Proc.* **2019**, *18 Pt 7*, 3532–3538. [CrossRef]
6. Soori, M.; Arezoo, B. A Review in Machining-Induced Residual Stress. *J. New Technol. Mater.* **2022**, *12*, 64–83. Available online: https://hal.science/hal-03679993 (accessed on 19 April 2019).
7. Malakizadi, A.; Bertolini, R.; Ducobu, F.; Kilic, Z.M.; Magnanini, M.C.; Shokrani, A. Recent advances in modelling and simulation of surface integrity in machining—A review. *Procedia CIRP* **2022**, *115*, 232–240. [CrossRef]
8. Nagaraj, M.; Ezilarasan, C.; Kumar, A.J.P.; Velayudham, A. A review of machining characteristics in mechanical drilling of super alloy. *Int. J. Mech. Prod. Eng. Res. Dev.* **2018**, *8*, 579–588. [CrossRef]

9. Kanaev, A.T.; Ramazanova, Z.M.; Biizhanov, S.K. Study of plasma-hardened wheel steel using nanoindentation. Industrial laboratory. *Diagn. Mater.* **2020**, *86*, 56–60. (In Russian) [CrossRef]
10. Randall, N.X. The current state-of-the-art in scratch testing of coated systems. *Surf. Coat. Technol.* **2019**, *380*, 125092. [CrossRef]
11. Storchak, M. Mechanical Characteristics Generation in the Workpiece Subsurface Layers through Cutting. *Crystals* **2023**, *13*, 761. [CrossRef]
12. Komanduri, R. Machining and Grinding: A Historical Review of the Classical Papers. *Appl. Mech. Rev.* **1993**, *46*, 80–132. [CrossRef]
13. Davim, J.P. *Drilling Technology: Fundamentals and Recent Advances*; De Gruyter Oldenbourg: Boston, MA, USA, 2018; 205p, ISBN 978-3110478631.
14. Krishnaraj, V.; Zitoune, R.; Collombet, F. Comprehensive Review on Drilling of Multimaterial Stacks. *J. Mach. Form. Technol.* **2010**, *2*, 1–32.
15. Kheireddine, A.H.; Ammouri, A.H.; Lu, T.; Jawahir, I.S.; Hamade, R.F. An FEM Analysis with Experimental Validation to Study the Hardness of In-Process Cryogenically Cooled Drilled Holes in Mg AZ31b. *Procedia CIRP* **2013**, *8*, 588–593. [CrossRef]
16. Wegert, R.; Guski, V.; Schmauder, S.; Möhring, H.C. In-process approach for editing the subsurface properties during single-lip deep hole drilling using a sensor-integrated tool. *Prod. Eng. Res. Dev.* **2024**, *18*, 319–337. [CrossRef]
17. Sun, Z.; Geng, D.; Guo, H.; Zhang, Q.; Liu, Y.; Liu, L.; Jiang, X.; Zhang, D. Introducing transversal vibration in twist drilling: Material removal mechanisms and surface integrity. *J. Mater. Process. Technol.* **2024**, *325*, 118296. [CrossRef]
18. Strodick, S.; Schmidt, R.; Donnerbauer, K.; Rozo Vasquez, J.; Zabel, A.; Macias Barrientos, M.; Biermann, D.; Walther, F. Subsurface conditioning in BTA deep hole drilling for improved component performance. *Prod. Eng. Res. Dev.* **2024**, *18*, 299–317. [CrossRef]
19. Ren, X.; Liu, Z. Influence of cutting parameters on work hardening behavior of surface layer during turning superalloy Inconel 718. *Int. J. Adv. Manuf. Technol.* **2016**, *86*, 2319–2327. [CrossRef]
20. Garg, S.; Goyal, R.K. A Study of Surface Roughness in Drilling of AISI H11 Die Steel using Face Centered Design. *Int. J. Innov. Res. Sci. Technol.* **2015**, *1*, 464–474.
21. Lacombe, A.; Landon, Y.; Paredes, M.; Chirol, C.; Benaben, A. Influence of the Hole Surface Integrity on the Fatigue Strength of an Aluminium Drilled Part. In *Advances on Mechanics, Design Engineering and Manufacturing III. JCM 2020*; Roucoules, L., Paredes, M., Eynard, B., Morer Camo, P., Rizzi, C., Eds.; Lecture Notes in Mechanical Engineering; Springer: Cham, Switzerland, 2020. [CrossRef]
22. Farid, A.A.; Sharif, S.; Idris, M.H. Surface integrity study of high-speed drilling of Al–Si alloy using HSS drill. *Proc. Inst. Mech. Eng. Part B J. Eng. Manuf.* **2011**, *225*, 1001–1007. [CrossRef]
23. Barik, T.; Pal, K.; Parimita, S.; Sahoo, P.; Patra, K. Monitoring of hole surface integrity in drilling of bi-directional woven carbon fiber reinforced plastic composites. *Proc. Inst. Mech. Eng. Part C J. Mech. Eng. Sci.* **2020**, *234*, 2432–2458. [CrossRef]
24. Mendas, M.; Benayoun, S.; Miloud, M.H.; Zidane, I. Microhardness model based on geometrically necessary dislocations for heterogeneous material. *J. Mater. Res. Technol.* **2021**, *15*, 2792–2801. [CrossRef]
25. Ameri, A.A.H.; Elewa, N.N.; Ashraf, M.; Escobedo-Diaz, J.P. General methodology to estimate the dislocation density from microhardness measurements. *Mater. Charact.* **2017**, *131*, 324–330. [CrossRef]
26. Varote, N.; Joshi, S.S. Microstructural Analysis of Machined Surface Integrity in Drilling a Titanium Alloy. *J. Mater. Eng. Perform.* **2017**, *26*, 4391–4401. [CrossRef]
27. Haddag, B.; Yameogo, D.; Nouari, M.; Makich, H. Multi-Physics Analysis of Machining Ti-6Al-4V Alloy: Experimental Characterization and a New Material Behavior Modeling. *Metals* **2022**, *12*, 581. [CrossRef]
28. Wang, Z.-Y.; Ren, J.-X.; Zhou, J.-H.; Cai, J. Correlation analysis of microstructure evolution on microhardness and residual stress for cutting Ti-6Al-4V titanium alloy. *Proc. Inst. Mech. Eng. Part B J. Eng. Manuf.* **2023**, *237*, 885–898. [CrossRef]
29. Harun, S.; Burhanuddin, Y.; Ibrahim, G.A. The Effect of Cutting Parameters on Surface Roughness and Morphology of Ti-6Al-4V ELI Titanium Alloy during Turning with Actively Driven Rotary Tools. *J. Manuf. Mater. Process.* **2022**, *6*, 105. [CrossRef]
30. Hussein, R.; Sadek, A.; Elbestawi, M.A.; Attia, M.H. Surface and microstructure characterization of low-frequency vibration-assisted drilling of Ti6Al4V. *Int. J. Adv. Manuf. Technol.* **2019**, *103*, 1443–1457. [CrossRef]
31. Hou, G.; Li, A. Effect of Surface Micro-Hardness Change in Multistep Machining on Friction and Wear Characteristics of Titanium Alloy. *Appl. Sci.* **2021**, *11*, 7471. [CrossRef]
32. Rangasamy, N.; Rakurty, C.S.; Balaji, A.K. A Multiscale Study on Machining Induced Surface Integrity in Ti-6Al-4V Alloy. *Procedia CIRP* **2022**, *108*, 787–792. [CrossRef]
33. Rodriguez, I.; Arrazola, P.J.; Cuesta, M.; Sterle, L.; Pušavec, F. Improving surface integrity when drilling CFRPs and Ti-6Al-4V using sustainable lubricated liquid carbon dioxide. *Chin. J. Aeronaut.* **2023**, *36*, 129–146. [CrossRef]
34. Murad, M.N.; Sharif, S.; Rahim, E.A.; Rival, R. Effect of Drill Point Angle on Surface Integrity when Drilling Titanium Alloy. *In Adv. Mater. Res.* **2013**, *845*, 966–970. [CrossRef]
35. Lotfi, M.; Farid, A.A.; Akbari, J.; Sharif, S.; Mohruni, A.S. Evaluation of surface integrity when drilling Inconel 718 through experimental measurement and finite element analysis. *Int. J. Adv. Manuf. Technol.* **2022**, *119*, 4237–4248. [CrossRef]
36. Dutilh, V.; Dessein, G.; Alexis, J.; Perrin, G. Links Between Machining Parameters and Surface Integrity in Drilling Ni-Superalloy. *Adv. Mater. Res.* **2010**, *112*, 171–178. [CrossRef]

37. Xu, D.; Ding, L.; Liu, Y.; Zhou, J.; Liao, Z. Investigation of the Influence of Tool Rake Angles on Machining of Inconel 718. *J. Manuf. Mater. Process.* **2021**, *5*, 100. [CrossRef]
38. Rahim, E.A.; Sasahara, H. An Analysis of Surface Integrity when Drilling Inconel 718 Using Palm Oil and Synthetic Ester under MQL Condition. *Mach. Sci. Technol.* **2011**, *15*, 76–90. [CrossRef]
39. Karabulut, Y.; Kaynak, Y. Drilling process and resulting surface properties of Inconel 718 alloy fabricated by Selective Laser Melting Additive Manufacturing. *Procedia CIRP* **2020**, *87*, 355–359. [CrossRef]
40. Girinon, M.; Valiorgue, F.; Rech, J.; Feulvarch, E. Development of a Procedure to Characterize Residual Stresses Induced by Drilling. *Procedia CIRP* **2016**, *45*, 79–82. [CrossRef]
41. Nobre, J.P.; Outeiro, J.C. Evaluating Residual Stresses Induced by Drilling of Ti-6Al-4V Alloy by Using an Experimental-numerical Methodology. *Procedia CIRP* **2015**, *31*, 215–220. [CrossRef]
42. Rasti, A.; Sadeghi, M.H.; Farshi, S.S. An analytical study on residual stresses in drilling of hardened steel. *Int. J. Adv. Manuf. Technol.* **2018**, *99*, 2389–2405. [CrossRef]
43. Leveille, T.; Valiorgue, F.; Dumas, M.; Masciantonio, U.; Brosse, A.; Karaouni, H.; Rech, J. 3D numerical modelling of residual stresses induced by reaming. *J. Manuf. Process.* **2024**, *113*, 47–60. [CrossRef]
44. Wu, R.; Huang, S.; Zhang, W.; Zhu, M.; Zheng, Z.; Li, T.; Xue, H. An improved grey wolf optimizer for minimizing drilling deformation and residual stress in AA2024 sheet. *Int. J. Adv. Manuf. Technol.* **2024**, *130*, 4443–4458. [CrossRef]
45. Pharr, G.M. Recent advances in small-scale mechanical property measurement by nanoindentation. *Curr. Opin. Solid State Mater. Sci.* **2015**, *19*, 315–316. [CrossRef]
46. Fischer-Cripps, A.C. Critical review of analysis and interpretation of nanoindentation test data. *Surf. Coat. Technol.* **2006**, *200*, 4153–4165. [CrossRef]
47. Wredenberg, F.; Larsson, P.-L. Scratch testing of metals and polymers: Experiments and numerics. *Wear* **2009**, *266*, 76–83. [CrossRef]
48. Kolawole, O.; Ispas, I. Evaluation of geomechanical properties via scratch tests: Where are we and where do we go from here? *SN Appl. Sci.* **2020**, *2*, 1633. [CrossRef]
49. Fischer-Cripps, A.C. *Nanoindentation*, 2nd ed.; Springer: Berlin/Heidelberg, Germany, 2011; 276p. [CrossRef]
50. Atkins, A.G.; Tabor, D. Plastic Indentation in Metals with Cones. *J. Mech. Phys. Solids* **1965**, *13*, 149–164. [CrossRef]
51. Doerner, M.; Nix, W. A method for interpreting the data from depth-sensing indentation instruments. *J. Mater. Res.* **1986**, *1*, 601–609. [CrossRef]
52. Oliver, W.C.; Pharr, G.M. An improved technique for determining hardness and elastic modulus using load and displacement sensing indentation experiments. *J. Mater. Res.* **1992**, *7*, 1564–1583. [CrossRef]
53. Zivic, F.; Babic, M.; Adamovic, D.; Mitrovic, S.; Todorovic, P.; Favaro, G.; Pantić, M. Influence of the surface roughness on adhesion of chrome coatings on alloy tool steel x165crmov12. *J. Balk. Tribol. Assoc.* **2012**, *18*, 228–237.
54. Bull, S.J. Failure modes in scratch adhesion testing. *Surf. Coat. Technol.* **1991**, *50*, 25–32. [CrossRef]
55. Lin, C.K.; Berndt, C.C. Measurement and analysis of adhesion strength for thermally sprayed coatings. *J. Therm. Spray Technol.* **1994**, *3*, 75–104. [CrossRef]
56. Yildiz, F.; Asaran, A. Multi-pass scratch test behavior of modified layer formed during plasma nitriding. *Tribol. Int.* **2010**, *43*, 1472–1478. [CrossRef]
57. Storchak, M.; Zakiev, I.; Träris, L. Mechanical properties of subsurface layers in the machining of the titanium alloy $Ti_{10}V_2Fe_3Al$. *J. Mech. Sci. Technol.* **2018**, *32*, 315–322. [CrossRef]
58. Sousa, F.J.P.; Tridapalli, D.; Pereira, M.; Flesch, C.A.; Alarcon, O.E. Evaluation of measurement uncertainties for a scratching tester. *Measurement* **2006**, *39*, 594–604. [CrossRef]
59. Fritz, R.; Kiener, D. Development and application of a heated in-situ SEM micro-testing device. *Measurement* **2017**, *110*, 356–366. [CrossRef]
60. Sivaram, S.; Jayasinghe, J.A.S.C.; Bandara, C.S. Qualitative Study on Pile-up Effect on Hardness Test by Nano-Indentation. *Eng. J. Inst. Eng.* **2021**, *54*, 47–55. [CrossRef]
61. Farayibi, P.K.; Hankel, J.; van gen Hassend, F.; Blüm, M.; Weber, S.; Röttger, A. Tribological characteristics of sintered martensitic stainless steels by nano-scratch and nanoindentation tests. *Wear* **2023**, *512–513*, 204547. [CrossRef]
62. Li, Z.; Herrmann, K.; Pohlenz, F. A comparative approach for calibration of the depth measuring system in a nanoindentation instrument. *Measurement* **2006**, *39*, 547–552. [CrossRef]
63. Peng, G.; Xu, F.; Chen, J.; Hu, Y.; Wang, H.; Zhang, T. A cost-effective voice coil motor-based portable micro-indentation device for in situ testing. *Measurement* **2020**, *165*, 108105. [CrossRef]
64. Ding, K.; Zhang, Y.; Birnbaum, A.J.; Michopoulos, J.G.; McDowell, D.L.; Zhu, T. Strain gradient plasticity modeling of nanoindentation of additively manufactured stainless steel. *Extrem. Mech. Lett.* **2021**, *49*, 101503. [CrossRef]
65. Vargas, A.L.M.; Blando, E.; Hübler, R. Elasto—Plastic materials behavior evaluation according to different models applied in indentation hardness tests. *Measurement* **2019**, *139*, 134–139. [CrossRef]
66. Kang, J.J.; Becker, A.A.; Wen, W.; Sun, W. Extracting elastic-plastic properties from experimental loading-unloading indentation curves using different optimization techniques. *Int. J. Mech. Sci.* **2018**, *144*, 102–109. [CrossRef]

67. Tsybenko, H.; Farzam, F.; Dehm, G.; Brinckmann, S. Scratch hardness at a small scale: Experimental methods and correlation to nanoindentation hardness. *Tribol. Int.* **2021**, *163*, 107168. [CrossRef]
68. England, J.; Uddin, M.J.; Ramirez-Cedillo, E.; Karunarathne, D.; Nasrazadani, S.; Golden, T.D.; Siller, H.R. Nanoindentation Hardness and Corrosion Studies of Additively Manufactured 316L Stainless Steel. *J. Mater. Eng. Perform.* **2022**, *31*, 6795–6805. [CrossRef]
69. Moon, J.; Kim, S.; Jang, J.; Lee, J.; Lee, C. Orowan strengthening effect on the nanoindentation hardness of the ferrite matrix in microalloyed steels. *Mater. Sci. Eng. A* **2008**, *487*, 552–557. [CrossRef]
70. Li, C.; Zhao, H.; Sun, L.; Yu, X. In situ nanoindentation method for characterizing tensile properties of AISI 1045 steel based on mesomechanical analysis. *Adv. Mech. Eng.* **2019**, *11*, 1687814019862919. [CrossRef]
71. Paul, V.; Ameyama, K.; Ota-Kawabata, M.; Ohmura, T. Evaluation of Deformation and Fracture Behavior in 304L Austenitic Steel Harmonic Structures through Nanoindentation. *Steel Res. Int.* **2023**, *94*, 2200354. [CrossRef]
72. Zhou, G.; Guo, J.; Zhao, J.; Tang, Q.; Hu, Z. Nanoindentation Properties of 18CrNiMo7-6 Steel after Carburizing and Quenching Determined by Continuous Stiffness Measurement Method. *Metals* **2020**, *10*, 125. [CrossRef]
73. Dean, J.; Aldrich-Smith, G.; Clyne, T.W. Use of nanoindentation to measure residual stresses in surface layers. *Acta Mater.* **2011**, *59*, 2749–2761. [CrossRef]
74. Wang, H.; Zhu, L.; Xu, B. *Residual Stresses and Nanoindentation Testing of Films and Coatings*; Springer Nature Singapore Pte Ltd.: Singapore; Science Press: Beijing, China, 2018; 207p, ISBN 978-981-10-7840-8. [CrossRef]
75. Zhang, W.; Wang, X.; Hu, Y.; Wang, S. Predictive modelling of microstructure changes, micro-hardness and residual stress in machining of 304 austenitic stainless steel. *Int. J. Mach. Tools Manuf.* **2018**, *130–131*, 36–48. [CrossRef]
76. Fan, P.; Katiyar, N.K.; Zhou, X.; Goel, S. Uniaxial pulling and nano-scratching of a newly synthesised high entropy alloy. *APL Mater.* **2022**, *10*, 111118. [CrossRef]
77. Pratap, A.; Divse, V.; Goel, S.; Joshi, S.S. Understanding the surface generation mechanism during micro-scratching of Ti-6Al-4V. *J. Manuf. Process.* **2022**, *82*, 543–558. [CrossRef]
78. Liu, H.; Xu, X.; Zhang, J.; Liu, Z.; He, Y.; Zhao, W.; Liu, Z. The state of the art for numerical simulations of the effect of the microstructure and its evolution in the metal-cutting processes. *Int. J. Mach. Tools Manuf.* **2022**, *177*, 103890. [CrossRef]
79. Bezyazychnyy, V.F.; Prokofev, M.A.; Vinogradova, N.V. Research of the Influence of Technological Machining Conditions on the Accumulation of Latent Energy Deformation in the surface Parts. *Bull. PNIPU Aerosp. Eng.* **2015**, *43*, 131–144. [CrossRef]
80. Yamamoto, M.; Tanaka, M.; Furukimi, O. Hardness–Deformation Energy Relationship in Metals and Alloys: A Comparative Evaluation Based on Nanoindentation Testing and Thermodynamic Consideration. *Materials* **2021**, *14*, 7217. [CrossRef] [PubMed]
81. Storchak, M.; Stehle, T.; Möhring, H.-C. Numerical Modeling of Cutting Characteristics during Short Hole Drilling: Modeling of Kinetic Characteristics. *J. Manuf. Mater. Process.* **2023**, *7*, 195. [CrossRef]
82. Material Property Data. Available online: http://www.matweb.com (accessed on 19 April 2019).
83. Storchak, M.; Stehle, T.; Möhring, H.-C. Determination of thermal material properties for the numerical simulation of cutting processes. *Int. J. Adv. Manuf.* **2021**, *118*, 1941–1956. [CrossRef]
84. Oxley, P.L.B. Development and Application of a Predictive Machining Theory. *Mach. Sci. Technol.* **1998**, *2*, 165–189. [CrossRef]
85. Heisel, U.; Kushner, V.; Storchak, M. Effect of machining conditions on specific tangential forces. *Prod. Eng.* **2012**, *6*, 621–629. [CrossRef]
86. Tsekhanov, J.; Storchak, M. Development of analytical model for orthogonal cutting. *Prod. Eng. Res. Dev.* **2015**, *9*, 247–255. [CrossRef]
87. Heisel, U.; Krivoruchko, D.V.; Zaloha, W.A.; Storchak, M.; Stehle, T. Thermomechanical material models in the modeling of cutting processes. *ZWF Z. Fuer Wirtsch. Fabr.* **2009**, *104*, 482–491. [CrossRef]
88. Storchak, M.; Jiang, L.; Xu, Y.; Li, X. Finite element modeling for the cutting process of the titanium alloy $Ti_{10}V_2Fe_3Al$. *Prod. Eng. Res. Dev.* **2016**, *10*, 509–517. [CrossRef]
89. Johnson, G.R.; Cook, W.H. A constitutive model and data for metals subjected to large strains, high strain and high temperatures. In Proceedings of the 7th International Symposium on Ballistics, The Hague, The Netherlands, 19–21 April 1983; pp. 541–547.
90. Heisel, U.; Krivoruchko, D.V.; Zaloha, W.A.; Storchak, M.; Stehle, T. Thermomechanical exchange effects in machining. *ZWF Z. Fuer Wirtsch. Fabr.* **2009**, *104*, 263–272. [CrossRef]
91. Storchak, M.; Möhring, H.-C.; Stehle, T. Improving the friction model for the simulation of cutting processes. *Tribol. Int.* **2022**, *167*, 107376. [CrossRef]
92. Fluhrer, J. *Deform-User Manual Deform V12.0*; SFTC: Columbus, OH, USA, 2019.
93. Heisel, U.; Krivoruchko, D.V.; Zaloha, W.A.; Storchak, M.; Stehle, T. Breakage models for the modeling of cutting processes. *ZWF Z. Fuer Wirtsch. Fabr.* **2009**, *104*, 330–339. [CrossRef]
94. Storchak, M.; Kushner, V.; Möhring, H.-C.; Stehle, T. Refinement of temperature determination in cutting zones. *J. Mech. Sci. Technol.* **2021**, *35*, 3659–3673. [CrossRef]
95. Kushner, V.; Storchak, M. Determining mechanical characteristics of material resistance to deformation in machining. *Prod. Eng. Res. Dev.* **2014**, *8*, 679–688. [CrossRef]

96. Kushner, V.; Storchak, M. Determination of Material Resistance Characteristics in Cutting. *Procedia CIRP* **2017**, *58*, 293–298. [CrossRef]
97. Heisel, U.; Storchak, M.; Krivoruchko, D.V. Thermal effects in orthogonal cutting. *Prod. Eng. Res. Dev.* **2013**, *7*, 203–211. [CrossRef]

metals

MDPI

Article

Investigation into the Hot-Forming Limit for 22MnB5 Hot-Forming Steel under a Stamping Process

Wenwu He [1], Bin Yang [1], Xuezhong Zhang [1], Min Li [2], Shuli Sun [1], Bao Wang [1,*] and Qingxian Ma [3]

[1] School of Materials Science & Engineering, Taiyuan University of Science and Technology, Taiyuan 030024, China
[2] Weichai Power Co., Ltd., Weifang 261061, China
[3] Department of Mechanical Engineering, Tsinghua University, Beijing 100084, China
* Correspondence: baowang@tyust.edu.cn; Tel.: +86-035-1216-1126

Abstract: Hot-forming technology for 22MnB5 hot-forming steel (22MnB5 HF steel) plates has been widely used in the automobile manufacturing industry in recent years. Physical simulation and numerical modeling were carried out in order to determine the forming limit of a 22MnB5 steel plate for the stamping process. The deformation experiments were performed in a temperature range of 600~900 °C and a strain rate range of 0.1~10 s^{-1}. In the uniaxial tensile tests, it was found that at the forming temperature of 600 °C, the condition of dynamic recrystallization was not fully reached, and thus the corresponding tensile strength was much larger than that at other deformation temperatures. In the numerical simulation of bulging experiments, it was found that 22MnB5 steel had good formability when the initial deformation temperature was high and the forming speed was low by using the instability criterion, combining the maximum punch force and strain path transition. The forming limit diagram of 22MnB5 steel at a temperature of 700 °C and tool speed of 25 mm/s was obtained by means of simulation and a hot stamping experiment. The establishment of the forming limit of the 22MnB5 steel plate can provide theoretical and technical guidance for the hot-forming process.

Keywords: 22MnB5 hot-forming steel; forming limit diagram; fracture criterion; numerical simulation

1. Introduction

With the development of social economy and the improvement in per capita income, the automobile manufacturing industry has developed rapidly. At the same time, the automobile manufacturing industry has also put forward higher requirements for material resources and environmental protection [1–3]. The body steel plate accounts for more than 20% of the total weight of the vehicle. Experiments have shown that if the vehicle mass is reduced by 100 kg, the fuel consumption for driving 100 km can be reduced by 0.3–0.6 L. Fuel consumption can be reduced by 0.7% for every 1% reduction in vehicle weight [4,5]. According to statistics, at present, 90% of the steel plates used in automobiles are made of HF steel. The use of HF steel plates as automobile body manufacturing materials has reduced the total mass of automobiles by about 30% compared with the use of ordinary steel plates. This 22MnB5 steel has the advantages of high strength and excellent collision resistance. The use of HF steel plates can meet the performance requirements and achieve lightweight automobiles [6–8]. Therefore, it has become the preferred material for automobile structural parts of many automobile enterprises [9].

However, the excellent strength also causes many problems and defects in the stamping process of HF steel at room temperature [10,11]. The hot stamping technology of HF steel can avoid the disadvantages of easy wrinkling and cracking of materials at room temperature, the poor forming quality of workpieces, high requirements for equipment and the easy damage of molds [12–14]. Therefore, the hot-forming technology of the HF steel plate has become a hot research topic [15,16]. Germany and Sweden have an absolute leading position and voice in the R&D field of hot stamping. Melak et al. studied the influence of plate thickness and cooling medium on the hot stamping quality of the 22MnB5 steel plate by combining experimental and simulation methods. It was found that increasing the plate thickness and changing the cooling medium can improve the strength and toughness of the parts [17]. Daniel et al. studied the transformation temperature and critical cooling rate of the 22MnB5 steel plate with Zn–Fe coating during hot forming by means of microscopic characterization [18]. Venturato et al. established the generalized incremental stress-state-dependent damage model (GISSMO) to predict the formability in 22MnB5 hot stamping by means of experiment and numerical simulation [19]. Zhuang et al. conducted X-ray diffraction and microstructure investigations to study the phase composition and micro defects, respectively, established a damage-coupled pre-forming constitutive model, and embedded it into ABAQUS 2016 through the VUMAT interface for simulation [20]. Chen et al. proposed a theoretical prediction model for forming a limit prediction model for the high temperature forming process of 22MnB5 based on the Norton–Hoff constitutive model, BBC 2005 yield criterion, and M-K model [21]. Zhou et al. used the combination of theoretical research, finite element simulation, and an experimental test to realize the three-field coupling of thermal mechanical damage in the hot stamping process by using the numerical simulation method [22]. Guo and Zhang et al. introduced the Zener Hollomon parameter to improve the Cockcroft–Latham criterion, Rice–Tracey criterion, and Oyane criterion to predict the initial value of damage and fracture of relevant materials [23,24]. But for the 22MnB5 steel sheet, there is a lack of forming limit research combining simulation and hot stamping experiments. The plasticity of the material in the high-temperature forming process will be affected by the temperature and deformation speed. The steel plate and the heat exchange with the mold will also affect the final experimental results, so it is difficult to obtain a more accurate thermal forming limit.

In the present study, through hot tensile tests and the numerical simulation of bulging tests on 22MnB5 steel, the microstructural changes of 22MnB5 steel and the thermoplasticity of the material under different deformation conditions were analyzed. The fracture criterion and hot-forming limit diagram were established to predict the fracture failure behavior of 22MnB5 steel.

2. Materials and Methods

2.1. Numerical Simulation of Stamping Forming

The material used in the present study was a 22MnB5 steel plate with a thickness of 1.4 mm, and the chemical composition is summarized in Table 1.

Table 1. Chemical composition of the studied 22MnB5 steel (Mass fraction, at. %).

Element	C	Mn	P	S	Si	Ti + Cr	B	Fe
Content (at. %)	0.22	1.2	0.019	0.003	0.25	0.229	0.003	Bal.

The hot tensile test of 22MnB5 steel was carried out on the Gleeble1500D thermal simulation test machine (Poestenkill, NY, USA). The tensile sample width was 14.5 mm and the length was 60 mm. After the sample was broken, the part near the fracture section was taken for metallographic observation. The sample was polished with 240–1000 mesh sandpaper and then mechanically polished to obtain the metallographic sample. The geometric model of the bulging experiment was established as shown in Figure 1. The plate thickness of 22MnB5 steel was set to 1.4 mm and the length to 180 mm. To change the strain path of the 22MnB5 steel plate, 9 different plate widths were set at 20 mm intervals between 20 mm and 180 mm, and the punch diameter was set to 100 mm. The material model required by 22MnB5 steel plate and die was obtained, and DYNAFORM 5.9.4 finite element simulation software was used to simulate the bulging test of the 22MnB5 steel plate. The initial temperature of the 22MnB5 steel and the stamping deformation speed were changed, appropriate instability criteria were selected, and the influence of deformation conditions on the formed parts was analyzed. The no. 106 material model was selected for this simulation. The model requires input material density, elastic modulus, Poisson's ratio, viscosity parameters C and P, thermal expansion coefficient, and stress–strain data, and the material parameters required for 22MnB5 steel by performing a tensile test and using JMatPro V7.0 simulation software. The voce hardening law saturation stress hardening model was used to perform the flow stress curves, corresponding to different temperatures of the 22MnB5 steel line fitting. The hot material parameters of 22MnB5 steel and H13 hot-work die steel were obtained using JMatPro simulation software.

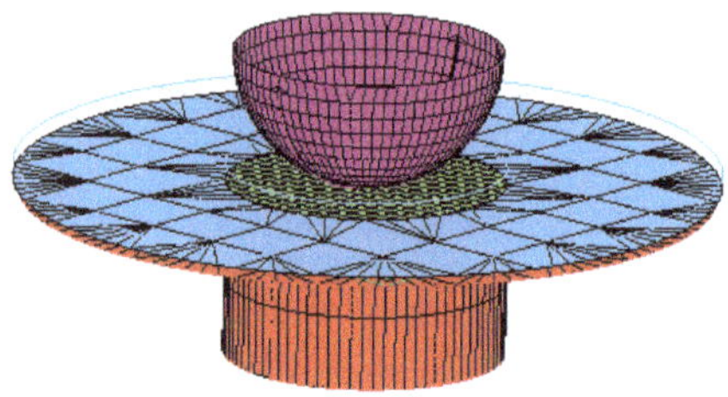

Figure 1. Geometric model of experimental die for Nakazima bulging.

2.2. Selection of Instability Criteria

The 22MnB5 steel is greatly affected by the value of the strain rate sensitivity index in the hot-forming process, and the error of forming limit diagram (FLD) function directly provided by the software was very large. Therefore, the instability criterion was needed to determine whether necking fracture occurred and to find out the fracture time. At present, the maximum punch force criterion and the strain path transition criterion are commonly used in the post-processing analysis of 22MnB5 steel numerical simulation.

Figure 2 shows the punch force–time curve corresponding to the samples taken from 22MnB5 steel plates with widths of 20 mm (Figure 2a), 80 mm (Figure 2b), 100 mm (Figure 2c), and 180 mm (Figure 2d). Through the cup bulging simulation of 22MnB5 steel with different plate widths, it was found that the maximum punch force criterion was only applicable to the small 22MnB5 steel width (20~80 mm). When the 22MnB5 steel width was relatively large (100~180 mm), the specimen would show a double tension state, and the punch force would increase with the growth of displacement without peak value.

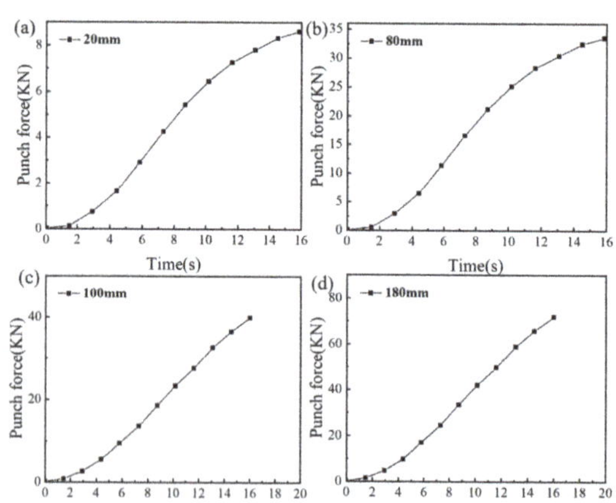

Figure 2. Punch force–time curve corresponding to samples with different widths: (**a**) 20 mm; (**b**) 80 mm; (**c**) 100 mm; (**d**) 180 mm.

Figure 3 presents the transition diagram of the maximum principal strain path corresponding to the specimen when the plate widths are 20 mm and 180 mm. It can be observed that when the plate width is small (less than 100 mm), there will be no inflection point in the strain path of the maximum principal strain, as shown in Figure 3a. When the second principal strain value is almost unchanged, but the maximum principal strain value increases sharply in a vertical shape, the stress state changes to a plane stress state, and the strain corresponding to the inflection point is the limit strain value, as shown in Figure 3b.

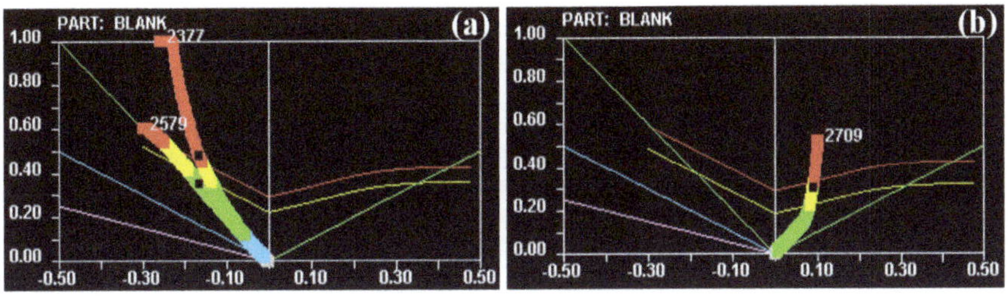

Figure 3. Transition diagram of maximum principal strain path corresponding to samples with different widths: (**a**) 20 mm; (**b**) 180 mm.

2.3. Hot-Stamping-Forming Experiment

On a THP-500T isothermal forging press (Genertec Tianduan, Tianjin, China), an experimental platform for hot stamping was independently set up, including a set of hot stamping dies according to the Nakazima bulging test standard, as well as displacement and force-measurement systems, as shown in Figure 4. The material of the hot stamping die was H13 steel. The hot stamping process parameters referred to the numerical simulation analysis results. Combined with the time–temperature transformation (TTT) curve of 22MnB5 provided by scholars [25], the steel that achieves austenitizing above 900 °C has better formability. The mold was first heated to 650 °C, and then the plate was heated to 930 °C by the resistance furnace, and the heat was maintained for 5 to 10 min. When the plate temperature was 700 °C, the plate was quickly transferred to the mold, and the experiment was carried out on the THP-500T isothermal forging press.

Figure 4. Hot-stamping-forming experiment equipment.

3. Results and Discussion

3.1. Mechanical Performance of the Studied 22MnB5 Steel Plate

The material was heated at a constant rate of 15 °C/s, the heated to 930 °C for 2 min, and the material was completely austenitized. Then, it was cooled to 600 °C~900 °C for the tensile test. The tensile stress–strain curves of 22MnB5 steel under different experimental conditions are presented in Figure 5. The tensile strength of 22MnB5 increased with the decrease in deformation temperature and rise in strain rate. The phenomenon of the strength obtained at temperature of 600 °C being the largest compared to the strength obtained at the other temperatures could be understood through the thermodynamic equilibrium transformation law of the studied steel, as is shown in Figure 6. Figure 6a shows the equilibrium transition diagram of the studied steel calculated through J-mat Pro software, and it could be seen that the A3, A1 temperatures were similarly determined to be 818 °C and 697 °C, and it could be further seen that the tensile temperature zones were mainly distributed in the dual-phase region consisting of ferrite and austenite. Figure 6b shows the continuous cooling transformation curves (CCT) of the studied steel calculated through J-mat Pro software. In accordance with the CCT, it could be seen that the undercooling austenite of the studied steel began to transform to pearlite when the tensile temperature decreased to 600 °C and 1 °C/s was selected as the cooling rate, and the microstructure was mainly composed of proeutectoid ferrite and pearlite. When the tensile temperature was increased to a certain temperature above ~650 °C, the microstructure of the studied steel was composed of proeutectoid ferrite and austenite in terms of the CCT, and it could be further deduced that the strength obtained at the tensile temperature of 600 °C was higher compared to the strength obtained at a tensile temperature above 650 °C due to the presence of the hardened phase of pearlite. According to the flow stress curve of 22MnB5 steel (Figure 5), the tensile strength and fracture strain of different tensile temperatures and strain rates could be obtained, as shown in Table 2. The fracture strain of the material increased with the increase in deformation temperature, and the fracture strain increased first and then decreased with the increase in strain rate (0.1–10 s^{-1}). Therefore, it can be preliminarily judged that the hot-stamping-forming interval of the 22MnB5 steel plate can be selected to be between 700 and 900 °C.

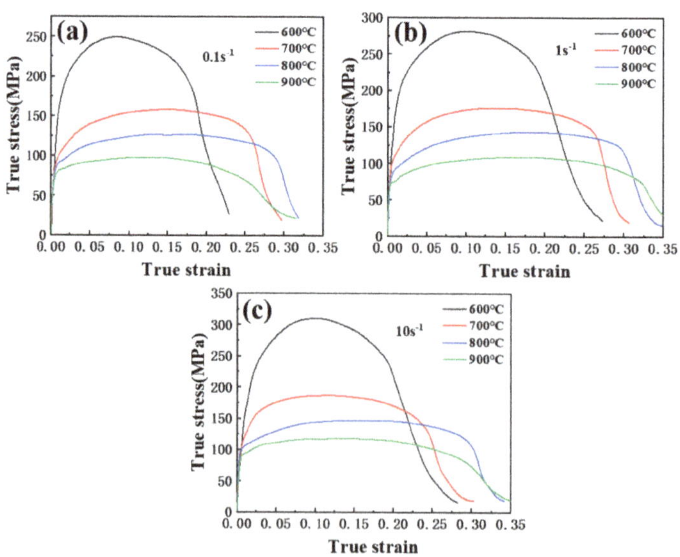

Figure 5. The 22MnB5 steel stress–strain curves under different experimental conditions: (**a**) strain rate = 0.1 s^{-1}; (**b**) strain rate = 1 s^{-1}; (**c**) strain rate = 10 s^{-1}.

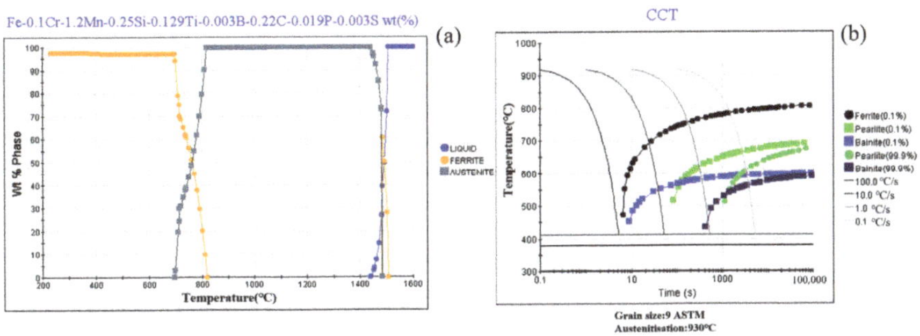

Figure 6. Equilibrium transition diagram and CCT of 22MnB5 steel. (**a**) The equilibrium transition diagram; (**b**) CCT curve.

Table 2. Tensile strength and fracture strain of 22MnB5 steel under different experimental conditions.

Temperature (°C)	**600**			**700**			**800**			**900**		
Strain rate (s^{-1})	0.1	1	10	0.1	1	10	0.1	1	10	0.1	1	10
Tensile strength (MPa)	250	284	310	158	177	192	127	143	148	98	109	121
Fracture strain	0.228	0.243	0.232	0.271	0.283	0.256	0.299	0.316	0.301	0.301	0.313	0.311

3.2. Influence of Deformation Conditions on Simulation Results under Stamping Experiment

For 180 × 180 mm 22MnB5 steel, the initial deformation temperature of 22MnB5 steel was set as 600 °C, 700 °C, 800 °C, or 900 °C, and the forming speed was 2.5 mm/s, 25 mm/s or 250 mm/s, respectively. The maximum principal strain unit was found, and the number of steps corresponding to the critical rupture time of the maximum principal strain unit was found using the strain path transformation criterion. The different deformations at critical rupture time were compared and analyzed under the conditions of 22MnB5 steel temperature field, pressure displacement, and thinning rate.

Figure 7 is the cloud chart of the equivalent plastic strain rate corresponding to different forming speeds. It can be found that when the forming speed is 2.5 mm/s, the equivalent plastic strain rate is particularly small, with a maximum of 0.062 s^{-1}. Although the strain rate nephogram is unevenly distributed, the small value does not affect the uniformity of the deformation area. With the increase in forming speed, the equivalent plastic strain rate increases continuously, and the difference in equivalent plastic strain rate corresponding to different deformation regions becomes larger, resulting in uneven deformation; fracture occurs more easily in the region with large equivalent strain rate.

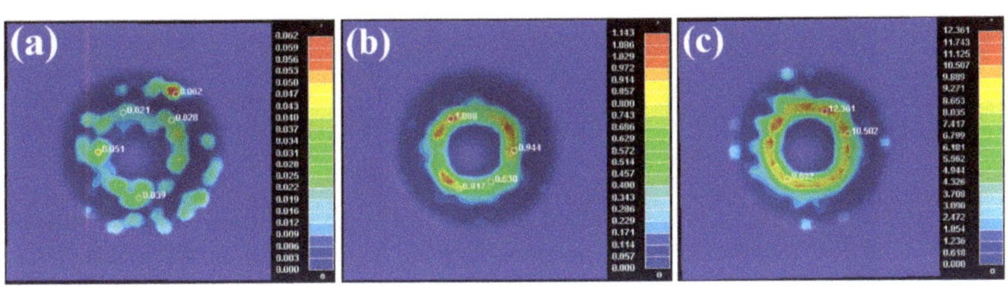

Figure 7. Nephogram of equivalent plastic strain rates under different forming speeds: (**a**) 2.5 mm/s; (**b**) 25 mm/s; (**c**) 250 mm/s.

Figure 8 is the nephogram of the plate-pressing displacement corresponding to the critical fracture time at different forming speeds. The maximum downforce displacements corresponding to the critical fracture time at different forming speeds were 27.308 mm,

23.506 mm, and 21.903 mm, respectively. By increasing the stamping speed, the heat loss becomes lower, the corresponding temperature is higher when the plate breaks, the strain hardening index is smaller, and the uniform deformation ability of the plate under load is weak. Therefore, the maximum downforce displacement of the plate decreases gradually with the increase in the stamping speed. In the forming process, the forming speed should be as small as possible. When the material meets the forming performance, the speed can be appropriately increased in order to save time and cost.

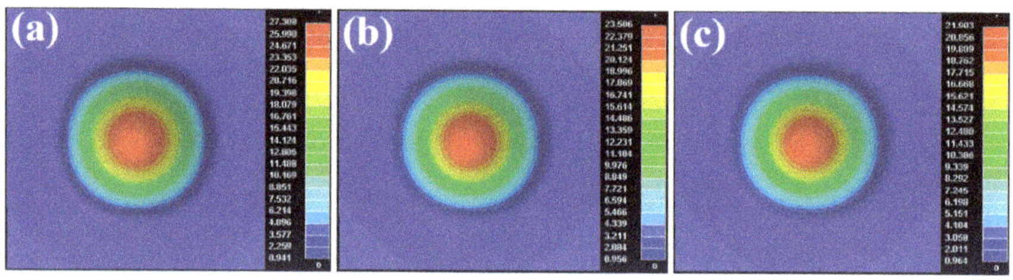

Figure 8. Nephogram of displacement corresponding to 22MnB5 steel under different forming speeds: (**a**) 2.5 mm/s; (**b**) 25 mm/s; (**c**) 250 mm/s.

Figure 9 is the nephogram of thinning rate corresponding to critical fracture time at different forming speeds. The maximum thinning rates corresponding to the critical breaking time at different forming speeds are 28.615%, 25.598%, and 19.992%, respectively. With the increase in forming speed, the pressing displacement corresponding to the critical breaking time of 22MnB5 steel gradually decreases, the load that can be borne in the thickness direction of 22MnB5 steel decreases, and the corresponding thinning rate decreases. Under the same thickness, the faster the forming speed, the easier the workpiece is to crack, and the greater the material thinning rate. Under the same specification, the workpiece obtained is light in weight.

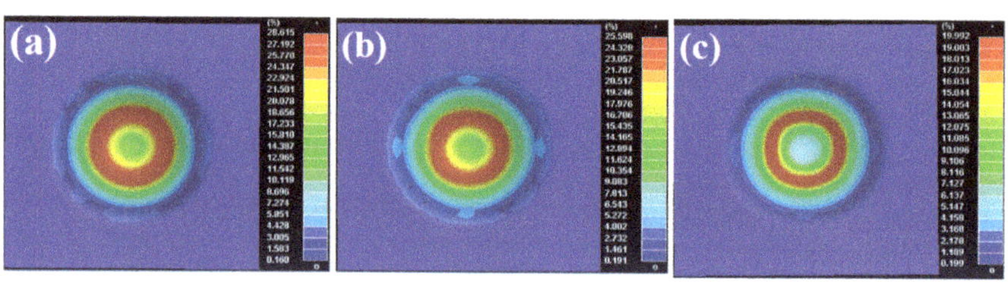

Figure 9. Nephogram of thinning rate corresponding to 22MnB5 steel under different forming speeds: (**a**) 2.5 mm/s; (**b**) 25 mm/s; (**c**) 250 mm/s.

Figure 10 is the cloud diagram of the temperature field corresponding to the critical fracture time of different initial deformation temperatures. It can be observed that when the initial forming temperature of the 22MnB5 steel is higher, the higher the temperature corresponding to the critical fracture time. However, due to the heat transfer and heat conduction between the 22MnB5 steel and the die and the cooling system, the temperature field corresponding to the 22MnB5 steel at the critical fracture time is not different. The temperature in the center of the contact part between the 22MnB5 steel and the punch is the highest, and the further away from the center, the lower the temperature. The temperature distribution near the center is uneven.

Figure 11 is the nephogram of displacement corresponding to the critical fracture time of the 22MnB5 steel under different initial deformation temperatures. It could be found that the maximum downforce displacement of the plate with different initial deformation temperatures at the critical fracture time was 27.219 mm, 29.037 mm, 25.422 mm or 24.822 mm, respectively. The maximum downforce displacement increased with the increase in the initial deformation temperature, reached the maximum at 700 °C, and then decreased with the increase in the initial deformation temperature.

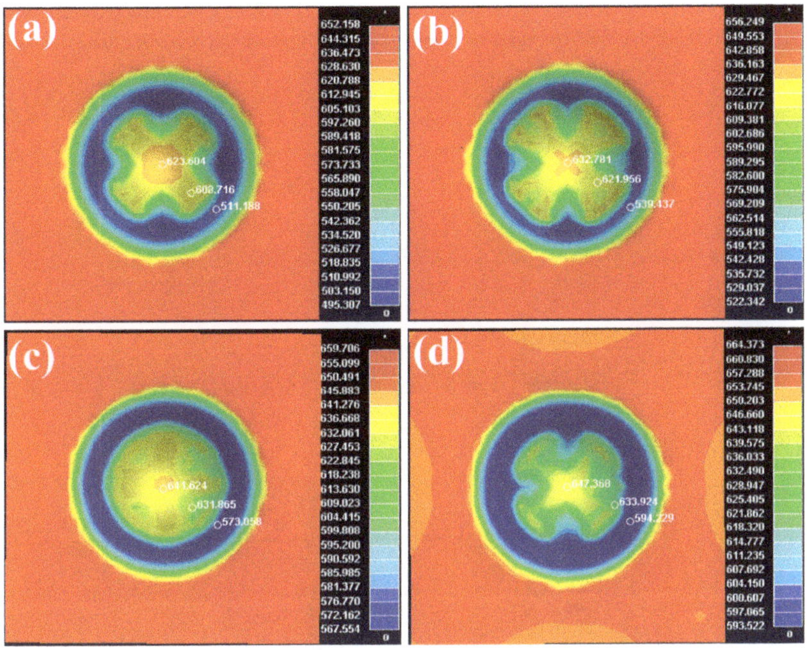

Figure 10. Nephogram of the temperature field corresponding to 22MnB5 steel under different initial deformation temperatures: (**a**) 600 °C; (**b**) 700 °C; (**c**) 800 °C; (**d**) 900 °C.

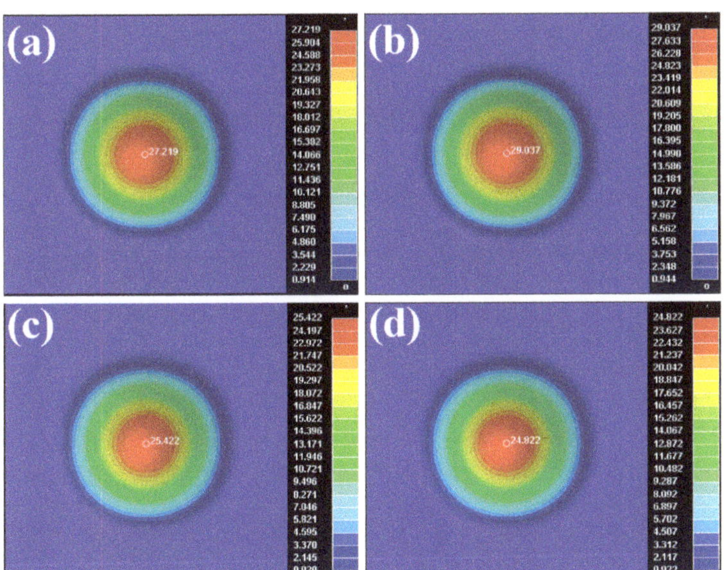

Figure 11. Nephogram of displacement of 22MnB5 steel at different initial forming temperatures: (**a**) 600 °C; (**b**) 700 °C; (**c**) 800 °C; (**d**) 900 °C.

Figure 12 is the cloud chart of the thinning rate corresponding to the critical fracture time at different initial deformation temperatures. It can be seen that the maximum thinning rate corresponding to different initial forming temperatures at the critical fracture time was 27.061%, 31.240%, 28.231% and 24.752%, respectively, which was the same as the law of downforce displacement. It increased with the increase in the initial deformation temperature of the plate. The greater the downforce displacement, the greater the thinning rate. The thinning rate reached the maximum at 700 °C. Thereafter, the thinning rate decreased with the increase in the initial deformation temperature of the plate. The area with the largest thinning rate was located around the contact center between the punch and the plate. Combined with the analysis of the temperature field nephogram, it was found that the area with the most serious thinning was located in the area with uneven temperature distribution.

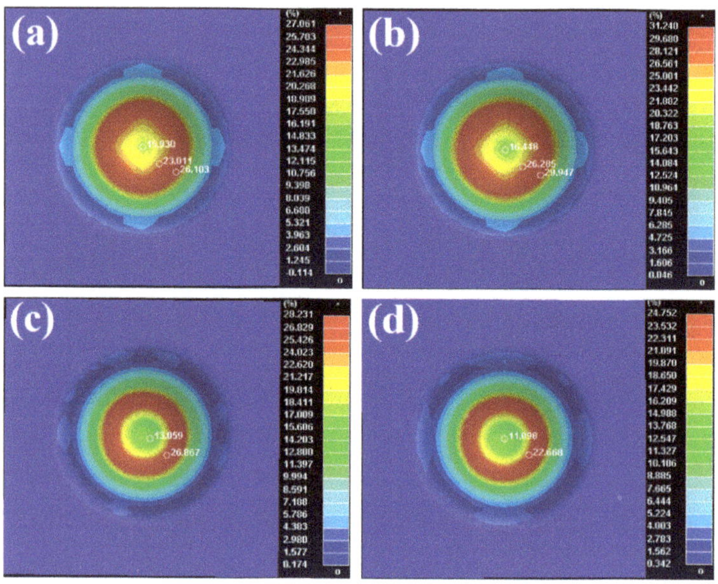

Figure 12. Nephogram of thinning rate of 22MnB5 steel under different initial deformation temperatures: (**a**) 600 °C; (**b**) 700 °C; (**c**) 800 °C; (**d**) 900 °C.

3.3. *Establishment of Hot-Forming Limit Diagram for 22MnB5 Steel*

To study the forming limit of 22MnB5 steel, we set the forming temperature to 700 °C, and the forming speed was set to 25 mm/s to perform the cup bulging simulation on the 22MnB5 steel plate. To analyze the post-processing results and identify the critical instability moments for different plate widths, the criterion combining the maximum punch force with the strain path transition was utilized, and the critical instability moments for different plate widths were identified. Figures 13 and 14 show the axial and radial strain nephograms, respectively, at the critical rupture moments for plate widths of 20 mm and 180 mm.

The analysis and extraction of the axial and radial strain points under different strain paths are depicted in Figures 13 and 14, and their values are shown in Table 3. The hot-forming limit diagram of the 22MnB5 steel plate can be obtained by fitting (as shown in Figure 15). It can be found that the primary and secondary strains on the left side of the FLD curve are linear when the secondary strain is less than 0. For the part where the secondary strain is greater than 0, the right side of the FLD curve shows a parabolic shape by quadratic fitting.

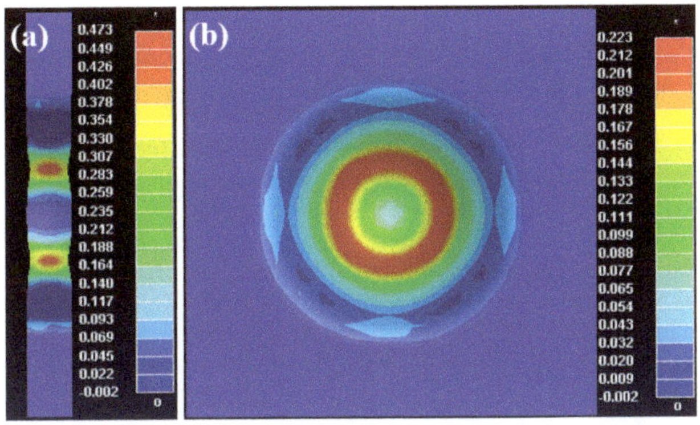

Figure 13. Nephogram of axial strain of critical fracture when plate widths are 20 mm and 180 mm. (**a**) 20 mm, (**b**) 180 mm.

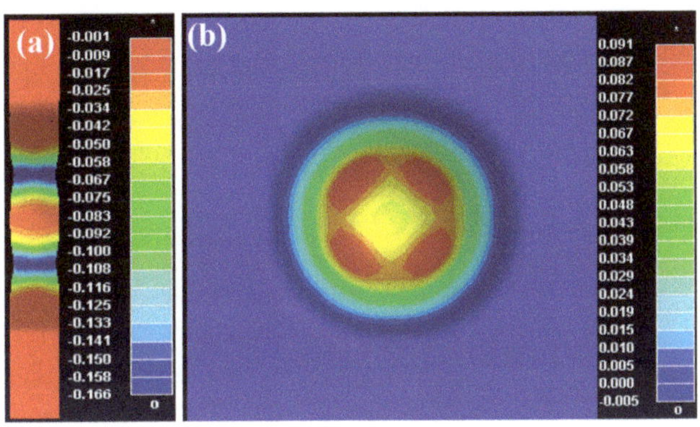

Figure 14. Nephogram of radial strain of critical fracture when plate widths are 20 mm and 180 mm. (**a**) 20 mm, (**b**)180 mm.

Table 3. Axial strain and radial strain points under different strain paths.

Radial strain	−0.169	−0.172	−0.186	−0.146	−0.009	−0.05	0.002	0.048	0.103	0.131	0.153
Axial strain	0.441	0.46	0.456	0.422	0.337	0.347	0.311	0.332	0.349	0.366	0.387

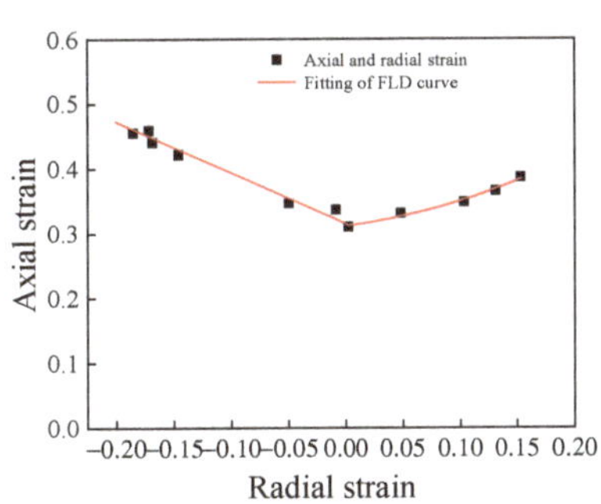

Figure 15. The hot-forming limit diagram of the 22MnB5 high strength steel plate at 700 °C.

3.4. Hot-Stamping Experiment of the 22MnB5 High-Strength Steel Plate

The 22MnB5 high strength steel with the same specifications as the simulation was selected for the hot stamping bulging experiment. Based on the simulation analysis results, the hot stamping forming temperature of 700 °C and the pressing speed of 2.5 mm/s were selected to carry out the hot stamping bulging experiment. In order to keep the temperature field relatively uniform during forming, the hot stamping expansion experiment flow was as follows: (1) Heat the stamping die to 650 °C; (2) heat the 22MnB5 high strength steel plate to 930 °C using a resistance furnace and maintain at this temperature for 10 min. The plate is then transferred into the stamping die, to wait for the hot stamping experiment when the plate was cooled to 700 °C; (3) obtain force and displacement data from the data collector. Figure 16 shows the force–displacement curve during stamping when the plate width is 20 mm. Through curve analysis, record the corresponding pressing depth value and maximum punch force for plates with different widths after stamping, as shown in Table 4. It can be found that in addition to the 20 mm wide plate, the stamping depth value is relatively similar. The maximum punch force increases with the increase in plate width.

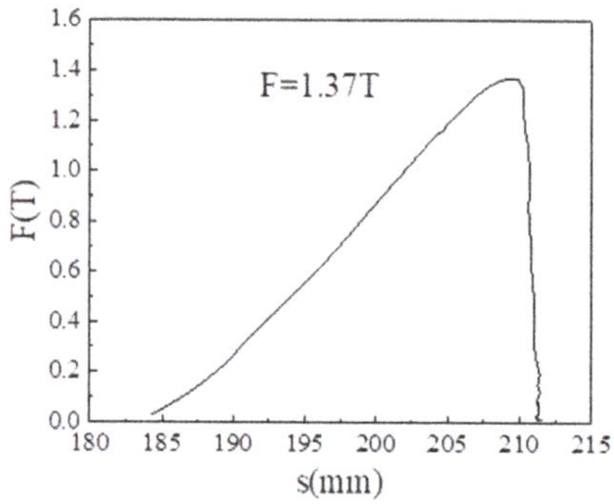

Figure 16. Force–displacement curve during stamping for the plate of 20 mm width.

Table 4. Pressing depth and maximum punch force after stamping with different plate widths.

Plate width (mm)	20	40	60	80	100	120	140	160	180
Stamping displacement (mm)	34.0	25.9	27.6	26.7	27.1	25.2	26.3	27.4	27.6
Maximum punch force (T)	1.37	1.98	2.42	3.25	3.83	4.28	5.62	5.78	6.54

The hot stamping experiment was repeated twice, each with a width of 20–180 mm at 700 °C. Figure 17 shows the maximum principal strain diagram of the hot-stamping-forming simulation and experiment with 22MnB5 steel plate widths of 60 mm and 180 mm, respectively. It could be found that under the same stamping displacement, for the corresponding area to the maximum principal strain, the simulation results were almost the same as the 22MnB5 steel fracture area after the experiment; Table 5 shows the punch force corresponding to the experimental and simulation results under the same stamping displacement. It can be seen that the results of numerical modelling were very close to the experimental results, indicating that the results obtained by the simulation method were reliable.

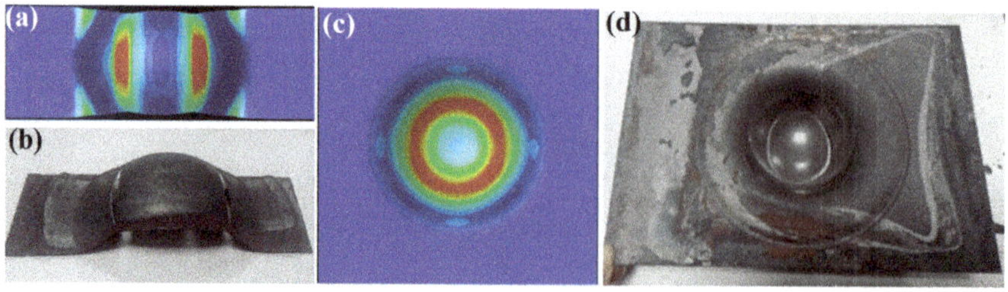

Figure 17. Comparison chart of the maximum principal strain nephogram and plate state after experiments with different plate widths: (**a**) 60 mm; (**b**) 60 mm; (**c**) 180 mm; (**d**) 180 mm.

Table 5. Punch force of the experimental and simulation results under the same downforce displacement.

Plate width (mm)	20	40	60	80	100	120	140	160	180
Stamping displacement (mm)	30	23	25	25	25	23	24	25	25
Punch force of experiment (T)	0.91	1.8	2.4	3.2	3.8	4.2	5.6	5.6	6.5
Punch force of simulation (T)	0.88	1.5	2.6	3.0	3.75	4.0	5.8	5.5	6.3

4. Conclusions

The applicability of the instability criterion was analyzed by simulating different sheet widths of 22MnB5 steel. The influence of forming speed and initial deformation temperature on the formability of the 22MnB5 steel plate was studied, and the forming limit diagram of 22MnB5 steel at 700 °C was investigated. The following conclusions can be drawn:

1. In the numerical simulation of bulging experiments, the maximum punch force criterion was applicable to a plate width of 20~80 mm, and the maximum strain path transition criterion was applicable to 100~180 mm.
2. The formability of 22MnB5 steel decreases with the increase in deformation speed and increases with the rise in initial deformation temperature of 22MnB5 steel, reaching the maximum when the initial temperature of 22MnB5 steel is 700 °C.
3. The limit strain point is determined by the instability criterion, and the hot-forming limit diagram of the 22MnB5 HF steel plate at 700 °C and 25 mm/s is obtained by fitting. It is found that the axial strain and radial strains are linear when the radial strain is less than 0, and the parts where the radial strain is greater than 0 can be fitted by binomial. Through experimental verification, it is found that the simulation results obtained by DYNAFORM finite element simulation software are reliable.

Author Contributions: Conceptualization, W.H.; Methodology, B.Y.; Software, X.Z. and M.L.; Formal analysis, S.S. and Q.M.; Writing—review & editing, B.W. All authors have read and agreed to the published version of the manuscript.

Funding: The authors acknowledge the financial support from the National Key Research and Development Program of China (No. 2023YFB3405404) and the National Nature Science Foundation of China (No. 52304409).

Data Availability Statement: The original contributions presented in the study are included in the article, and further inquiries can be directed to the corresponding author.

Conflicts of Interest: Min Li was employed by the Weichai Power Co., Ltd. The remaining authors declare that the research was conducted in the absence of any commercial or financial relationships that could be construed as a potential conflict of interest.

References

1. Bao, L.; Liu, W.J.; Wang, B.; Li, H.P.; You, X.P.; Zhou, Q.; Liu, M.; Gao, S.R. Experimental investigation on partition controllable induction heating-hot stamping process of high-strength boron alloyed steel plates with designable temperature patterns. *J. Mater. Res. Technol.* **2020**, *9*, 13963–13976. [CrossRef]
2. Guo, J.; Liu, H.L.; Li, X.D.; Yang, T.Y. Fracture behavior of the hot-stamped PHS2000 steel based on GISSMO failure model. *Metals* **2023**, *13*, 1360. [CrossRef]
3. Liang, L.Z.; Hui, L.J.; Sheng, S.Y.; Ze, Y.C. automobile manufacturing. *J. Mech. Sci. Technol.* **2009**, *23*, 1924–1931. [CrossRef]
4. Hu, Y.J.; Wang, Z.L.; Wang, Y.Q. Selection of distributed manufacturing environments based on genetic algorithm for automobile gears. *Mater. Manuf. Process.* **2013**, *28*, 1268–1275. [CrossRef]
5. Quan, G.Z.; Zhan, Z.Y.; Zhang, L.; Wu, D.S.; Luo, G.C.; Xia, Y.F. A study on the multi-phase transformation kinetics of ultra-high strength steel and application in thermal-mechanical-phase coupling simulation of hot stamping process. *Mat. Sci. Eng. A* **2016**, *673*, 24–38. [CrossRef]
6. Sun, H.; Yao, S.J.; Tang, Q.H.; Rong, R.L.; Chu, G.N.; Auer, P.; Domitner, J. Short-wave infrared heating of 22MnB5 steel with adjusted surface emissivity for press hardening of tailored-strength components. *J. Manuf. Process.* **2024**, *120*, 74–85. [CrossRef]
7. Simonetto, E.; Ghiotti, A.; Bruschi, S. Numerical modelling of direct hot tube rotary draw bending of 22MnB5 high strength steel. *CIRP J. Manuf. Sci. Technol.* **2022**, *37*, 547–558. [CrossRef]

8. Li, T.T.; Xie, X.L.; Xu, J.F.; Li, R.F.; Qi, K.; Zhang, X.Q.; Yue, H.Y.; Zhao, Y.; Qiao, L. Research on AZ31 Mg alloy/22MnB5 steel pinless friction stir spot welding process and interfacial temperature field simulation. *J. Mater. Res. Technol.* **2023**, *26*, 3710–3725. [CrossRef]
9. Yao, S.J.; Feng, L.; Yang, D.L.; Han, D.X.; Liu, Y.; Li, Q.; Chao, B.J. A potential hot stamping process for microstructure optimization of 22MnB5 steels characterized by asymmetric pre-rolling and one-or two-step pre-heating. *J. Mater. Process. Technol.* **2018**, *254*, 100–107. [CrossRef]
10. Chen, W.J.; Song, H.W.; Lazarescu, L.; Xu, Y.; Banabic, D. Formability analysis of hot-rolled dual-phase steel during the multistage stamping process of wheel disc. *Int. J. Adv. Manuf. Technol.* **2020**, *110*, 1563–1573. [CrossRef]
11. Huang, D.; Kodur, V.; Wang, W.Y. Temperature-dependent properties of high-strength steel for evaluating the fire resistance of structures. *Adv. Struct. Eng.* **2023**, *26*, 2265–2281. [CrossRef]
12. Hart-Rawug, T.; Buhl, J.; Horn, A.; Bambach, M.; Merklein, M. A unified model for isothermal and non-isothermal phase transformation in hot stamping of 22MnB5 steel. *J. Mater. Process. Technol.* **2023**, *313*, 117856. [CrossRef]
13. Singh, A.; Bhattacharya, A.; Butcher, C. Experimental artefacts affecting characterization of the evolving interfacial heat transfer coefficient in hot stamping of Al-Si coated 22MnB5 steel. *Appl. Thermal. Eng.* **2024**, *24*, 121604. [CrossRef]
14. Liu, H.P.; Lu, X.W.; Jin, X.J.; Dong, H.; Shi, J. Enhanced mechanical properties of a hot stamped advanced high-strength steel treated by quenching and partitioning process. *Scr. Mater.* **2011**, *64*, 749–752. [CrossRef]
15. Couto, C.P.; Revilla, R.I.; Politano, R. Influence of austenitisation temperatures during hot stamping on the local electrochemical behaviour of 22MnB5 steel coated with hot-dip Al-Si. *Corros. Sci.* **2021**, *190*, 109673. [CrossRef]
16. Lkeda, Y.; Ni, H.C.; Chakraborty, A. Segregation-induced grain-boundary precipitation during early stages of liquid-metal em brittlement of an advanced high-strength steel. *Acta Mater.* **2023**, *259*, 1359–6454.
17. Naderi, M.; Uthaisangsuk, V.; Prahl, U. A Numerical and experimental investigation into hot stamping of boron alloyed heat-treated steels. *Steel Res. Int.* **2009**, *79*, 77–84. [CrossRef]
18. Dabuel, A.; Fabio, M. Phase transformation temperatures and Fe enrichment of a 22MnB5 Zn-Fe coated steel under hot stamping conditions. *J. Mater. Res. Technol.* **2019**, *9*, 629–635.
19. Venturato, G.; Ghiotti, A.; Bruschi, S. Stress-state dependent formability modelling in hot stamping. *Prod. Eng.* **2020**, *14*, 105–114. [CrossRef]
20. Zhuang, W.; Wang, P.; Xie, D.; Shi, H. Experimental study and a damage model approach to determine the effect of hot forming deformation on the service performance of 22MnB5 steel. *J. Manuf. Process.* **2019**, *47*, 10–21. [CrossRef]
21. Chen, J.; Zheng, X.; Liu, Y. Theoretical prediction of high strength steel 22MnB5 forming limit in high temperature based on M-K model. *Procedia Eng.* **2017**, *207*, 632–639. [CrossRef]
22. Zhou, J. *Research on Thermal Simulation Experiment and Damage Evolution Modeling and Simulation of Boron Steel Hot Stamping*; Beijing University of Science and Technology: Beijing, China, 2015.
23. Guo, J.; Zhan, M.; Wang, X. Ductile fracture criterions for hot forming process of Ti-6Al-2Zr-1Mo-1V alloy sheet. *J. Cent. South Univ.* **2017**, *48*, 2294–2300.
24. Zhang, X.; Zeng, W.; Shu, Y.; Zhao, Y.; Yu, H.; Zhou, Y. Study on cracking criterion of Ti40 flame retardant alloy based on Zener Hollomon factor. *Rare Met. Mater. Eng.* **2008**, *4*, 604–608.
25. Neumann, R.; Schuster, S.; Gibmeier, J. Two-scale simulation of the hot stamping process based on a Hashin–Shtrikman type mean field model. *J. Mater. Process. Technol.* **2019**, *267*, 124–140. [CrossRef]

metals

Article

Study on Deformation Behavior and Mechanical Properties of 42CrMo High-Strength Steel with Multi-Station Warm Upsetting

Zhiling Xiao [1,2,*], Hao Wang [1], Jinhua Liu [3], Junjie Jiang [1], Liming Yu [2,*] and Yuhao Zhang [1]

1 College of Mechanical and Electrical Engineering, Zhengzhou University of Light Industry, Zhengzhou 450002, China; 18247695198@163.com (H.W.); violet5257@163.com (J.J.); vous200408@163.com (Y.Z.)
2 School of Materials Science and Engineering, Tianjin University, Tianjin 300350, China
3 Sijin Intelligent Forming Machinery Company Limited Co., Ltd., Ningbo 453499, China; zqdgx1@126.com
* Correspondence: xzl03223@126.com (Z.X.); lmyu@tju.edu.cn (L.Y.)

Abstract: In order to find out the deformation behavior and mechanical properties of 42CrMo steel under warm upsetting conditions, the Gleeble-3500 thermal simulation testing machine was used to carry out a warm upsetting physical simulation experiment on 42CrMo steel. By controlling deformation temperature, strain rate, and constant temperature deformation pass, the microstructure evolution rule under different warm upsetting conditions was analyzed, and its hardness value was measured. Then, the simulation experiment is carried out based on the Deform-3D finite element platform. The results show that, with the increase in deformation temperature, 42CrMo steel has a temperature rise softening effect, which significantly reduces the peak value of rheological stress. At 650 °C, the maximum peak value of rheological stress is only 45.3% of that of cold upsetting deformation at room temperature, and the stress-strain curve tends to be gentle at the plastic deformation stage, which is the most suitable temperature for warm upsetting deformation. The maximum peak flow stress of 42CrMo steel increases with the increase in strain rate, but the number of deformation channels has little influence on the stress-strain curve. The warm, upsetting deformation can refine the internal grain structure significantly, and the grain refinement mechanism is mechanical crushing. When the temperature is slightly higher, the broken grain will recover, and the grain size will grow. During the process of warm upsetting, the strain rate has a great influence on the microhardness of the sample. The deformation pass has little influence on the hardness, and the hardness increases slightly with the increase in the deformation pass. Through the Deform-3D simulation, the correlation coefficient R and the average absolute relative error (AARE) between the simulation value and the experimental value were calculated, and the correlation coefficient R-value was 0.9948, and the average absolute relative error (AARE) was 2.05%, indicating that the simulation can accurately reflect the relationship between displacement and applied load.

Keywords: high strength steel; warm upsetting deformation; deformation pass; organizational evolution; peak flow stress

1. Introduction

Fasteners are widely used in machinery, electrical, automotive, transportation, aerospace, and other fields. With the professional and large-scale development of fastener manufacturing, higher requirements are also put forward for the manufacture of fasteners. At present, the most common manufacturing method for fasteners is the cold upsetting process [1,2]. However, due to the greater resistance to plastic deformation of metal materials at room temperature and the characteristics of cold work hardening, the deformation requires greater force, and the upsetting capability of equipment is limited in the manufacture of bolts and nuts with large specifications and deformation [3]. In order to make up for the shortcomings of the cold upsetting process, the hot upsetting process is also applied in a certain range [4,5], while the traditional hot upsetting process also has some problems in the production practice, such as that the billet temperature is not easy to control, the heating is not uniform, the heating time is long, and it is easy to cause over-burning, oxidation, and decarbonization, which is the main reason for the poor quality of the formed parts [6]. Due to the above reasons, the application of warm upsetting technology is promoted.

Warm upsetting technology is applied in fastener production, which solves the manufacturing quality problem of bolts and nuts that are difficult to deform. 42CrMo steel is one of the high-strength steels that is difficult to deform. Due to its high strength, high hardenability, fracture toughness, and wear resistance, it is widely used in bearings, flanges, high-strength bolts, and other fields [7–9]. Upsetting and heat treatment are key processing steps. In recent years, the compression studies of 42CrMo steel mostly focus on hot compression and rolling with the recrystallization temperature above 890 °C [10–14], but there are few studies on warm upsetting forming. Ji et al. [15] carried out high-temperature equal-diameter angle extrusion (ECAP) on 42CrMo steel above the austenite transformation temperature. The microstructure and material strength of the samples at different radial positions were characterized, and the test results showed that DRX of different degrees caused by the uneven distribution of strain and strain rate on the samples was the main reason for the gradient distribution of the radial microstructure and material strength. Y.C. Lin et al. [16] carried out hot compression experiments with different temperatures, deformation rates, and deformation degrees in order to conduct a numerical simulation of the forging of 42CrMo steel and establish its hot-forming process parameters. They also studied the effects of process parameters such as strain rate, forming temperature, and degree of deformation on the microstructure of 42CrMo steel. The results showed that the average grain size of 42CrMo steel increases with an increase in forming temperature and decreases with an increase in deformation degree and strain rate. Qian Dongsheng et al. [17] studied the grain evolution law of 42CrMo ring parts through simulation and experimentation in the processes of blanking, billet making, hot rolling, and residual heat quenching and concluded that in the process of residual heat quenching, the grain refinement of the central part of the ring is obvious, and the forming force of 42CrMo steel in the process of hot forming is small, but the surface is easy to produce an oxide layer, which needs further treatment.

In multi-pass forming, with the increase in deformation passes, the forming force can be significantly reduced each time, and large variable forming can be accumulated. In recent years, many scholars have studied multi-pass compression [18–21], among which the research on 42CrMo steel's multi-pass deformation mainly focuses on the thermal compression experiment above the recrystallization temperature. Fangcheng Qin et al. [22] conducted an isothermal–non-isothermal multi-pass compression test on centrifugally cast 42CrMo steel at a temperature above recrystallization, studied the effects of compression pass and final rolling temperature on deformation behavior and microstructure evolution, and found that the flow stress did not change significantly with the increase in pass. However, controlled-temperature cumulative static recrystallization can make the grains fine and evenly distributed. Wanhui Huang et al. [23] studied the effects of high-temperature multi-direction multi-pass forging and annealing treatment on the microstructure and mechanical properties of 42CrMo steel and found that the average austenite grain size decreased significantly, and dislocation density increased after multi-pass forging. Qingfeng Zhu et al. [24] studied the effect of forging passes on the refining effect of high-purity aluminum during multi-directional forging at room temperature. The results showed that with the increase in forging passes, the fine crystal region in the center of the sample gradually expanded, and the size of coarse grains near the surface continuously decreased with the increase in forging passes to 24 passes.

In order to investigate the deformation behavior and mechanical properties of 42CrMo steel under the condition of warm upsetting, the Gleeble-3500 thermal simulation testing machine was used in this study to carry out the warm upsetting physical simulation experiments on 42CrMo steel at different temperatures, strain rates, and deformation passes. The study on the change in rheological behavior, microstructure, and mechanical properties of 42CrMo steel with warm upsetting parameters provides a theoretical basis for the multi-station warm upsetting deformation of 42CrMo steel.

2. Experimental Conditions and Numerical Simulation Conditions

42CrMo high-strength steel with slow cooling after hot rolling was selected as the material of this experiment, and the material was made into a cylindrical specimen with a diameter of 8 mm and a height of 12 mm by wire cutting, The 42CrMo obtained by this treatment has good comprehensive mechanical properties and is suitable for a warm upsetting process. 42CrMo steel is a kind of medium carbon low alloy steel. Its main chemical composition and content are 0.4 wt% C, 0.23 wt% Si, 0.6 wt% Mn, 0.98 wt% Cr, 0.18 wt% Mo, 0.019 wt% P, 0.012 wt% S, 0.02 wt% Cu.

In order to obtain the mechanical properties and microstructure of warm upsetting forming under different temperatures, strain rates and deformation passes, the constant rate cold upsetting and warm upsetting comparison test, the deformation evolution test, and the multi-pass warm upsetting test were carried out on the Gleeble-3500 thermal simulation test machine of DSI Company (Figure 1a).

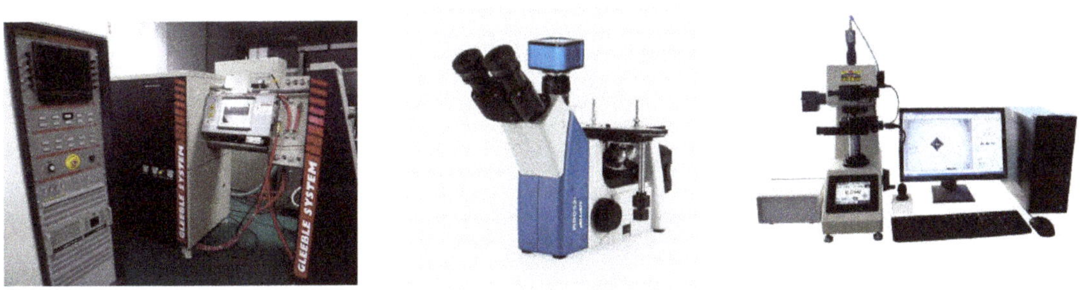

(a) (b) (c)

Figure 1. Experimental equipment and flow chart: (**a**) Gleeble-3500 thermal simulation testing machine; (**b**) IE500M metallographic microscope; (**c**) 430SVA Vickers hardness tester.

The comparison test conditions of cold upsetting and warm upsetting at constant rates were as follows: under the condition of a heating rate of 10 °C/s and a strain rate of 1 s^{-1} dynamic load, the samples were heated to 20 °C, 500 °C, 650 °C, and 800 °C, respectively, for compression experiments, and the total deformation was 60%. The test conditions for the deformation evolution of warm upsetting were as follows: the sample was heated to 650 °C at a heating rate of 10 °C/s, and compression was carried out under the conditions of strain rates of 1, 10, and 15 s^{-1}, respectively, and the total deformation was 60%. The experimental scheme is shown in Table 1. Multi-pass compression conditions are as follows: heating to 650 °C at a heating rate of 10 °C/s; compression tests for two and three passes are carried out at a deformation rate of 10 s^{-1}. During the compression process, the temperature is maintained at not less than 650 °C, and force unloading and loading processes are carried out between each pass. The experimental scheme is shown in Table 2. Air cooling treatment after deformation is completed.

Table 1. Comparison test scheme of cold upsetting and warm upsetting.

Deformation Temperature (°C)	Strain Rate (s^{-1})	Deformation Amount (%)
20	1	60
500	1	60
650	1	60
800	1	60
650	10	60
650	15	60

Table 2. Multi-pass warm upsetting test scheme.

Total Number of Deformation	Number of Deformation	Deformation Amount (mm)	Strain Rate (s^{-1})	Deformation Temperature (°C)
2	1	3.6	10	650
	2	3.6	10	650
3	1	3.6	10	650
	2	2.94	10	650
	3	2.184	10	650

As shown in Figure 2a, the deformed specimen is cut parallel to the longitudinal compression axis. The section was ground and polished, and the surface was corroded with a 4% nitrate alcohol solution. The microstructure of the central area of the section was observed with an IE500M metallographic microscope, as shown in Figure 2b. Then, the average grain size is obtained by the number of grains per unit area.

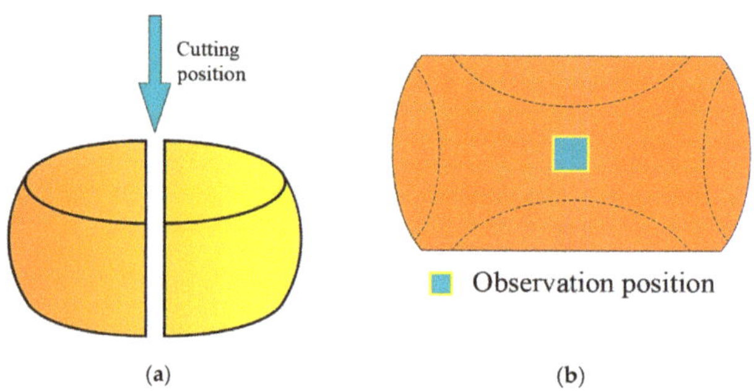

Figure 2. Cutting position and metallographic observation position of the sample: (**a**) cutting position diagram; (**b**) observation position diagram.

Wolpert 430SVA Vickers hardness tester was used to measure Vickers hardness in the central region of the sample with different experimental parameters at room temperature. The test surface is polished with 200- to 2000-mesh metallographic abrasive paper and then polished with polishing paste to avoid flaring ash and debris. The load used for measurement was 10 kgf, the loading time was 20 s, the loading time was 15 s, and the unloading time was 10 s. Finally, the indentation diagonal of the positive pyramid indenter is measured, and the average hardness value of each sample is obtained by measuring at least three times for the same position. The calculation method is shown in Equation (1).

$$HV = 1.8544\frac{p}{d^2} \quad (1)$$

where HV is Vickers hardness (kgf/mm^2), p is the applied load (kgf), d is the average of the diagonal lengths of the two indentations (mm), and 1.8544 is a constant.

Deform-3D Ver.11.0 finite element analysis software was used to simulate the deformation process of 42CrMo steel parameterized. First, the true stress-strain curve obtained from the compression experiment was imported into the software, and then the billet was set as a plastic body with a temperature of 650 °C. At the same time, the die temperature is set to 20 °C and defined as a rigid body; the contact factor between the billet and the die was set to warm upsetting 0.25, and the thermal conductivity was set to 11 N/s/mm/C. The die speed was set at 1 s^{-1}, 10 s^{-1}, and 15 s^{-1}, respectively, and the billet was divided into 35,000 tetrahedral finite elements for analysis.

3. Analysis of Experimental Results of Warm Upsetting

3.1. Single-Station Warm Upsetting Forming Analysis

3.1.1. Influence of Deformation Temperature on Stress-Strain Curve of Single Station Upsetting Experiment

Figure 3 shows the true stress–strain curve of 42CrMo steel at 20~800 °C, a strain rate of 1 s^{-1}, and a compression deformation of 60%. It can be seen from Figure 3 that the peak value of flow stress significantly decreases with the increase in temperature during the compression process, and in the plastic deformation stage, the stress–strain curve of cold upsetting forging at room temperature gradually decreases with the increase in deformation, resulting in rheological softening. When the blue brittleness temperature is exceeded, the curve tends to be gentle when the forging deformation is warm, and when the compression deformation is 800 °C, the curve gradually rises, resulting in deformation hardening. It can be seen from Figure 3 that the variation trend of the stress–strain curve with deformation temperature is consistent with the transformation trend of the stress-strain curve of carbon steel with cold deformation, warm deformation, and hot deformation. This is because temperature is one of the important factors affecting the stress-strain curve. With the increase in temperature, the kinetic energy between atoms increases, the binding force decreases, and the dislocation is more likely to overcome the pinning effect and move, resulting in a temperature rise softening effect, thus gradually reducing the peak metal flow stress [25]. When the temperature is between 20 °C and 650 °C, the softening effect of temperature rise is greater than that of work hardening, so the deformation is mainly softening, and the stress in the plastic deformation stage decreases with the increase in strain [26]. When the blue brittleness temperature is exceeded, the curve tends to flatten and rise at the plastic deformation stage, especially at 800 °C, which is close to the trend of the stress–strain curve of hot upsetting. It can be seen that when 42CrMo steel deforms at about 650 °C, the maximum peak flow stress is only 45.3% of the cold upsetting forging deformation at room temperature, and the stress–strain curve tends to be gentle in the plastic deformation stage, and it is in the best state of warm deformation.

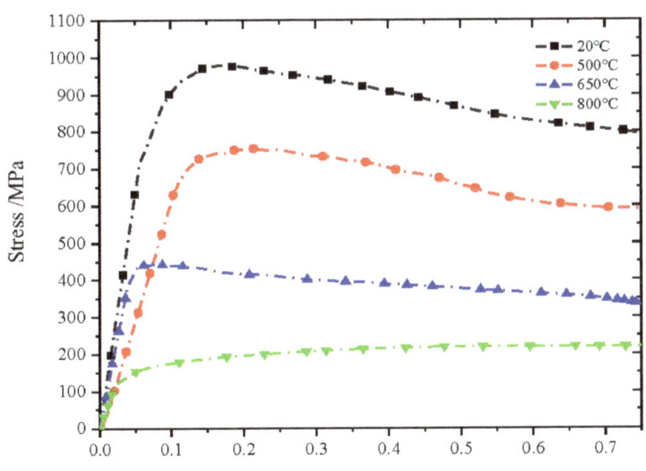

Figure 3. Stress–strain curves at different deformation temperatures with a strain rate of 1 s^{-1}.

3.1.2. Influence of Strain Rate on the Stress–Strain Curve of the Single Station Warm Upsetting Forming Experiment

In order to physically simulate the effect of strain rate on the stress-strain curve of 42CrMo steel in the temperature range of warm upsetting forging, the upsetting test was carried out at 650 °C, with a strain rate ranging from 1 to 15 s^{-1} and compression deformation of 60%. The deformation characteristics are shown in Figure 4.

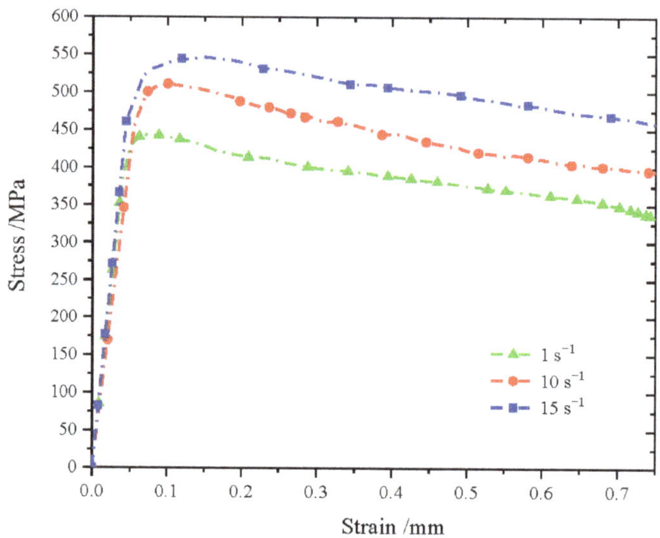

Figure 4. Stress–strain curves at different strain rates at 650 °C.

As can be seen from Figure 4, the sample was deformed under upsetting at a warm forming temperature, and the stress-strain curve of 42CrMo steel showed a trend of first increasing to the stress peak and then slowly decreasing, and with the increase in strain rate, the stress-strain curve increased. This is because, with the increase in deformation rate, dislocation density and dislocation proliferation speed increase rapidly in a short period of time, resulting in an increase in peak flow stress [27]. With the increase in compression deformation, the peak flow stress reaches 443.9 MPa, 511.0 MPa, and 546.5 MPa at the deformation rates of 1 s^{-1}, 10 s^{-1}, and 15 s^{-1}, respectively. It can be seen that under the condition of warm deformation, the peak value of flow stress increases with the increase in deformation rate, and in the plastic stage, the stress value slowly decreases with the increase in strain. This change trend is similar to that of the flow curve under cold deformation, but the decreasing trend tends to be gentle. This shows that the softening effect caused by the rise in deformation temperature is gradually reduced under the condition of warm forming.

3.1.3. Influence of Upsetting Parameters on Microstructure

Figure 5a–e shows the optical microstructure of the original 42CrMo sample and strain rate 1 s^{-1}, deformation temperature of 20 °C, 500 °C, 650 °C, and 800 °C, deformation amount of 60%, respectively. The compression direction is from top to bottom. It can be seen from the figure that the grain size varies significantly with the change in temperature. From Figure 5a, it can be clearly seen that the grains of the original sample have relatively evenly distributed ferrite and bainite structures. After compression at room temperature (20 °C), the grain size was finer than that of the original billet, and the grain size was significantly reduced, as shown in Figure 5b. which is due to the upset force that causes the larger grains to break up and form fine grains [28]. The grain of the sample at 500 °C is further stretched and broken compared with that at 20 °C. However, the sample with a temperature of 650 °C has strip and block grains again, which is due to the upsetting deformation at 650 °C coupled with the deformation temperature rise, and the action of thermal activation makes some dislocation defects disappear or recombine, resulting in the recovery of fine broken grains and some growth. When the sample is compressed and deformed at 800 °C, the grains inside the sample are the smallest and most uniform. This is because the sample is heated to 800 °C, the austenite region above the AC3 temperature of 42CrMo is subjected to rapid compression and deformation, and a fine grain structure is formed after air cooling. However, as can be seen from Figure 5e, due to the phase transformation of the steel during the cooling process after the warm upsetting deformation, the ferrite preferentially extends in the dendrite segregation and non-metallic inclusion and precipitates out in a band along the metal flow direction, resulting in a local band structure [29]. It can be seen that the grain can be refined by upsetting deformation at 500–650 °C, and the grain refinement mechanism is mechanical crushing and refining. When the temperature is slightly higher, the broken grain will recover, and the size will grow.

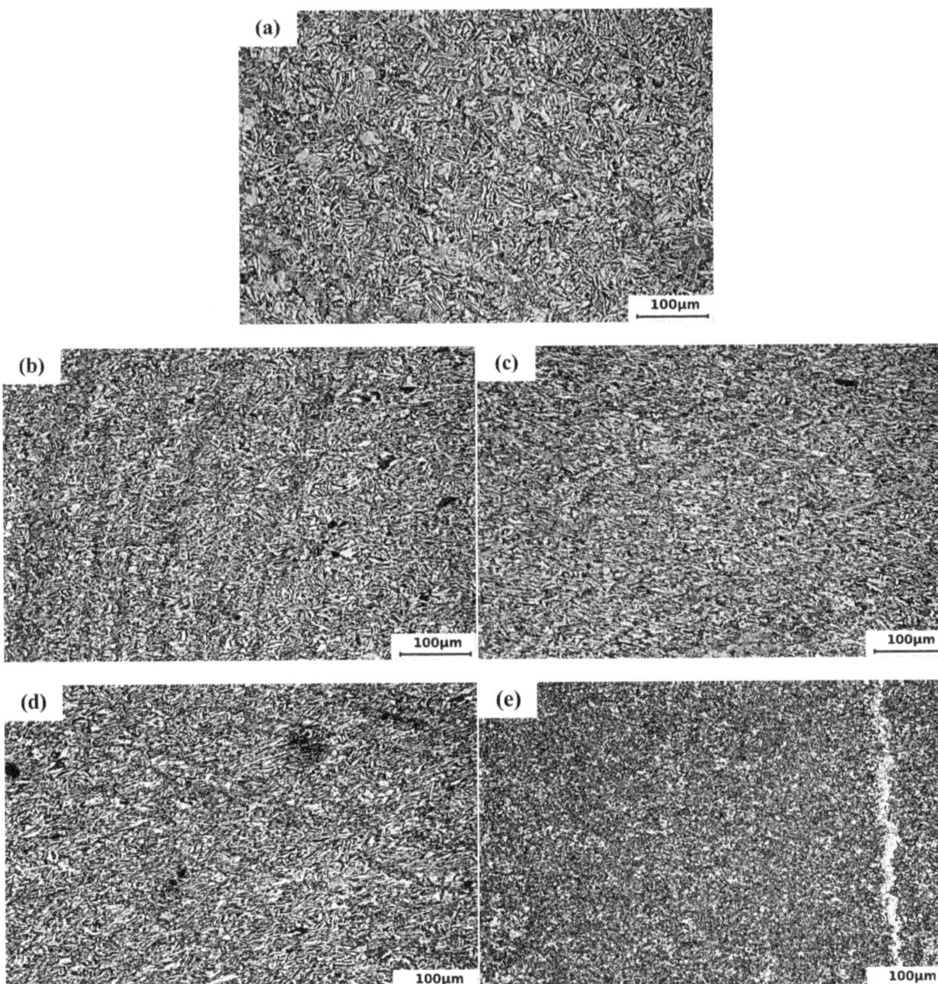

Figure 5. Grain morphology at different temperatures: (**a**) original sample; (**b**) 20 °C; (**c**) 500 °C; (**d**) 650 °C; (**e**) 800 °C.

Figures 5d and 6a,b show the 42CrMo sample at 650 °C; the deformation rate of warm upsetting is 1 s^{-1}, 10 s^{-1}, and 15 s^{-1}, respectively, and the deformation amount is 60%. As can be seen from the figure, the grain size is obviously refined with the increase in strain rate, mainly because the deformation rate is too large. The effect of the mechanical driving force on grain breakage is enhanced. When the strain rate increases, the short deformation time greatly restricts the movement of dislocation, resulting in a rapid increase in dislocation density and further refinement of grains [30].

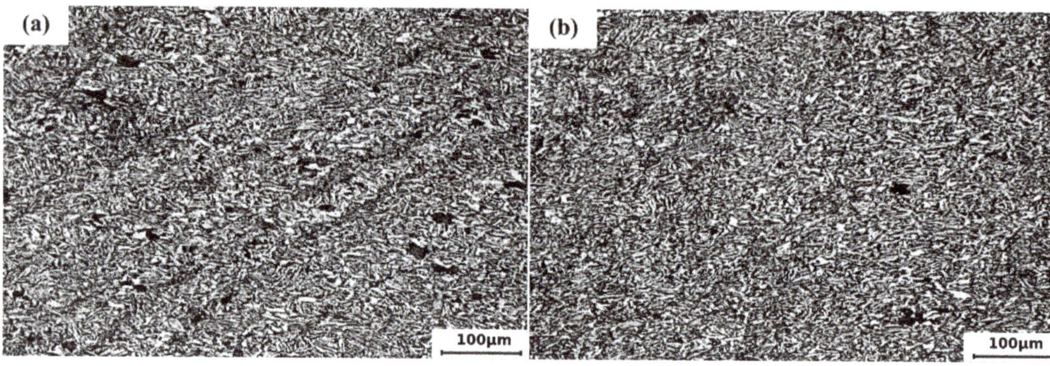

Figure 6. Grain morphology at different strain rates: (**a**) 10 s^{-1}; (**b**) 15 s^{-1}.

3.1.4. Influence of Strain Rate on Hardness

As shown in Figure 7, Vickers hardness values at 650 °C and different strain rates show that when the temperature is 650 °C, and the strain rate is 1 s^{-1} and 10 s^{-1}, the hardness values of the samples have little difference, while the hardness values of the samples with a deformation rate of 15 s^{-1} increase, reaching 286.9 HV, indicating that a higher strain rate can increase the hardness.

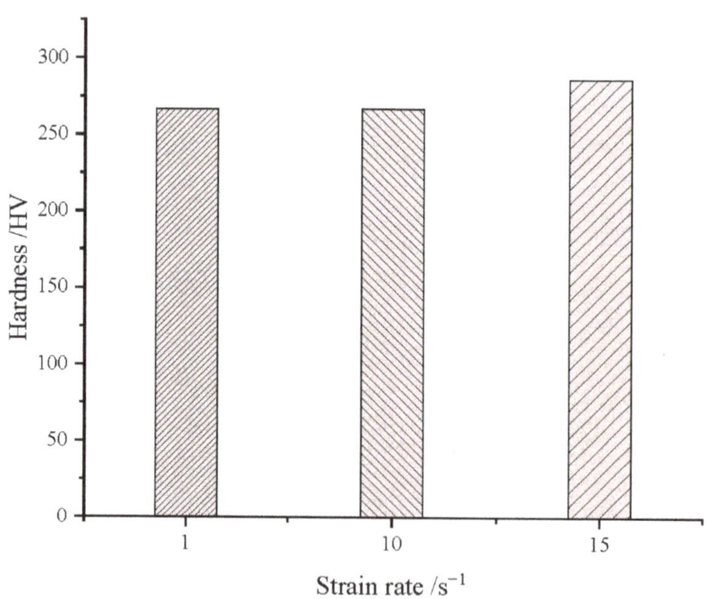

Figure 7. Vickers hardness at different strain rates.

3.2. Analysis of Multi-Station Warm Upsetting Experiment Results

3.2.1. Influence of Passage Number on Stress-Strain Curve of Warm Upsetting Deformation

In order to physically simulate multi-station warm upsetting, a multi-pass compression test was carried out on the bar material. The temperature was set at 650 °C for constant temperature compression, the strain rate was 10 s^{-1}, and the compression deformation of each pass was 30% (3.6 mm) for two passes and 3.6 mm, 2.94 mm, and 2.184 mm for each pass of three passes, respectively. The true stress–strain curve is shown in Figure 8, where the rapid rise and fall of stress is intended to simulate the loading and unloading process between passes. As can be seen from Figure 8, the stress increases first with the increase in strain and then decreases rapidly after reaching the peak stress. In addition, the stress–strain curve transition is relatively smooth when unloading and loading processes are excluded between each pass. It can be seen that the number of deformation channels has little influence on the stress–strain curve of 42CrMo steel under the condition of warm deformation. However, compared with single-pass upsetting with the same deformation conditions, the deformation temperature rise, and softening effect of two—pass and three-pass upsetting are more obvious in the plastic deformation stage, which is because the multi-pass sample is compressive deformation under the state of heat preservation. During the deformation, the temperature of the sample changes little, and the metal deformation resistance does not increase due to the decrease in temperature, so the softening degree of the sample is relatively good [31].

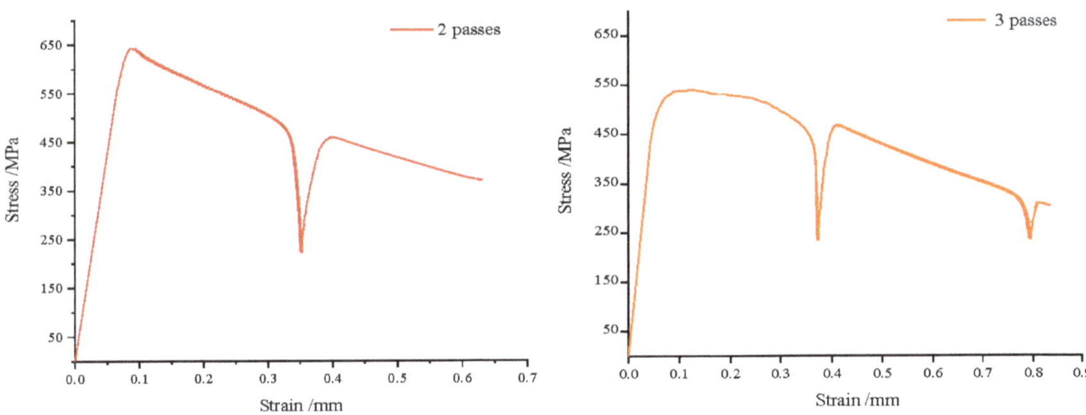

(**a**) (**b**)

Figure 8. Multi-pass stress-strain curve: (**a**) two passes; (**b**) three passes.

3.2.2. Microstructure Analysis of Multi-Station Warm Upsetting and Single-Station Warm Upsetting

Figure 9a shows the microstructure of a 42CrMo sample at 650 °C with a deformation rate of 10 s^{-1} and a deformation amount of 60% for two-pass samples, and Figure 9b shows the internal microstructure of a three-pass sample with a deformation amount of 75%. It can be seen from the figure that the grain size of the two-pass sample is slightly refined compared with that of the single-pass (Figure 6a) sample, but the distribution of ferrite is relatively uniform, indicating that the reduction in fewer passes has less influence on the grain structure. After three passes of compression, the grain structure is obviously refined, mainly because the deformation is highly uneven in the lower temperature range. After multiple upsettings, strong shear deformation of local strip grains is caused, forming fine grains.

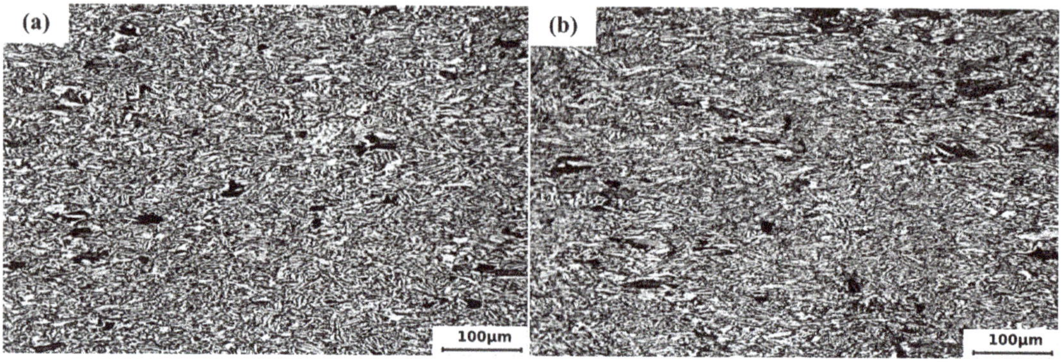

Figure 9. Grain morphology with different pass times: (**a**) two passes; (**b**) three passes.

3.2.3. Analysis of Microhardness of Multi-Station and Single-Station Warm Upsetting

Figure 10 shows the Vickers hardness values at 650 °C and the 10 s^{-1} strain rate under different deformation passes. It can be seen from the figure that deformation passes also have a certain influence on the hardness values. Compared with single-pass, the hardness values of two passes increase slightly, and the hardness values of three passes are the highest, mainly due to the effect of grain refinement, which is consistent with the grain distribution in Figures 6a and 9.

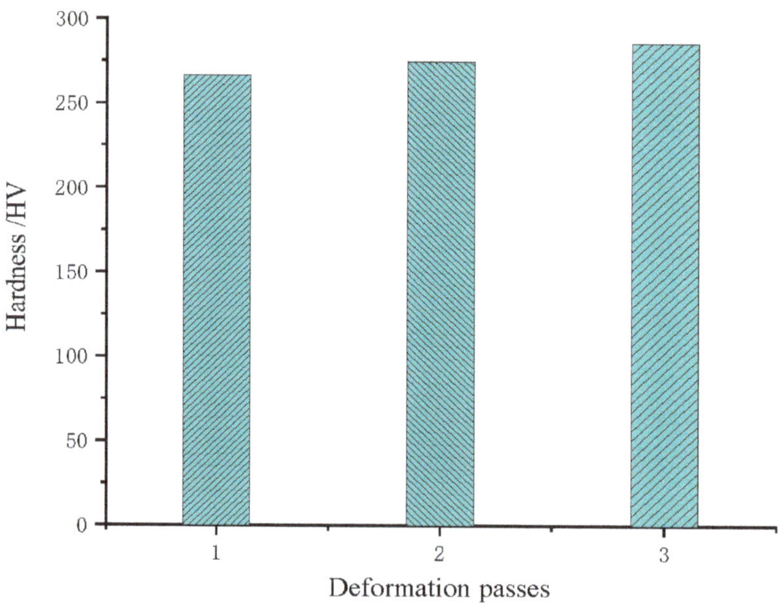

Figure 10. Vickers hardness of different deformation passes.

4. Numerical Simulation Analysis of the Warm Upsetting Forming Process of Typical Parts

4.1. Comparative Analysis between Numerical Simulation and Experiment of Load-Displacement Curve

Figure 11 shows the stroke-load curve of the simulation experiment at a temperature of 650 °C and a strain rate of 1 s^{-1}, 10 s^{-1}, and 15 s^{-1}. Compared with Figure 11, it is found that the stroke-load curve values obtained by the Deform-3D simulation software are in good agreement with the experimental values. Both experimental data and simulation data show that the load increases sharply at the initial stage of displacement and continues to increase as the displacement continues to increase.

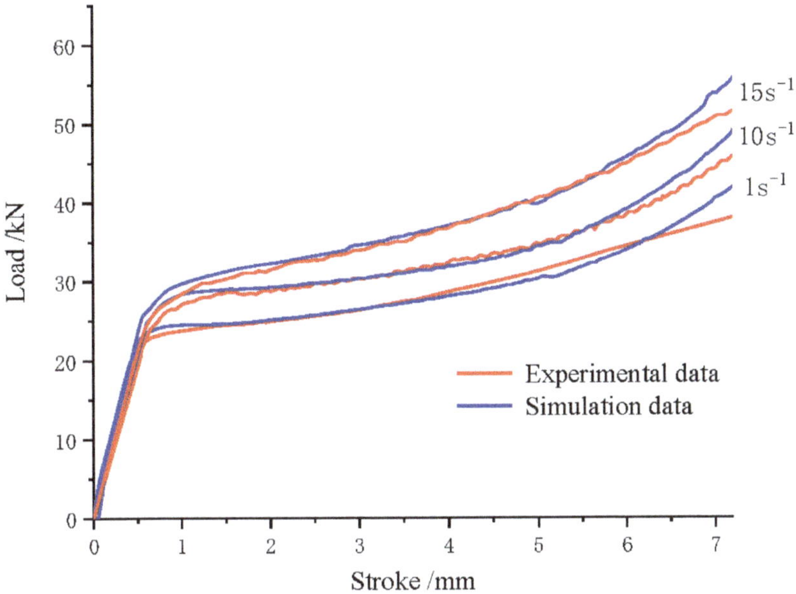

Figure 11. Stroke–load curve at 650 °C.

In order to further quantitatively evaluate the accuracy of the simulation value, the correlation coefficient R (Equation (2) between the simulation value and the experimental value and the absolute mean relative error ($AARE$) (Equation (3) is calculated. The calculated correlation coefficient R is 0.9948, and the average absolute relative error ($AARE$) is 2.05%. Therefore, the relationship between displacement and applied load can be accurately reflected through simulation.

$$R = \frac{\sum_{i=1}^{N} (C_i - \overline{C})(T_i - \overline{T})}{\sqrt{\sum_{i=1}^{N} (C_i - \overline{C})^2} \sqrt{\sum_{i=1}^{N} (T_i - \overline{T})^2}} \tag{2}$$

$$AARE = \frac{1}{N} \sum_{i=1}^{N} |C_i - T_i| / C_i \tag{3}$$

where C_i and $\overline{C}$ are the load experimental values and average values, respectively; T_i and $\overline{T}$ are the simulated load values and average values, respectively.

4.2. Comparative Analysis of Numerical Simulation of Multi-Station Warm Upsetting and Single-Station Warm Upsetting

Figure 12 shows the simulation comparison of stroke-load curves for single-pass and multi-pass at a temperature of 650 °C and a strain rate of 1 s^{-1}. It is found that the stroke–load curves of a single pass are in good agreement with the stroke–load curves of two passes and three passes, and the load difference is small.

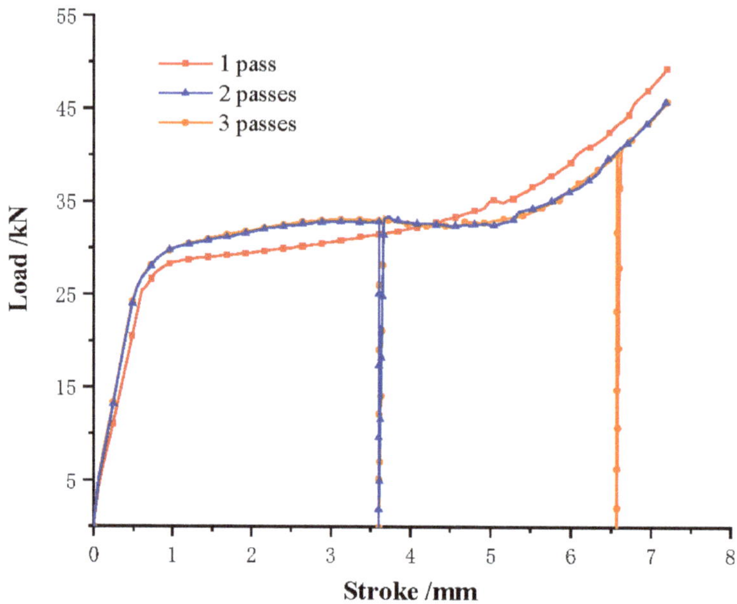

Figure 12. Simulation comparison of single-pass and multi-pass stroke–load curves.

4.3. *Warm Upsetting Process Design and Numerical Simulation Conditions for Typical Parts*

The process scheme of warm upsetting for automobile tire bolts is shown in Figure 13: The first station is necking; the second station has a hexagonal head and flange pre-upset forming; and the third station flange is finally upset forming. According to the deformation degree equation, the maximum deformation degree of each station is 38.3%, 0.17%, and 2.78%; all of them are less than the allowable deformation degree of 42CrMo material under cold forming, which meets the design requirements.

$$\varepsilon = \frac{H_0 - H_1}{H_0} \times 100\% \tag{4}$$

where ε is the deformation amount, H_0 is the original height of the blank before upsetting, H_1 is the height of the part after upsetting.

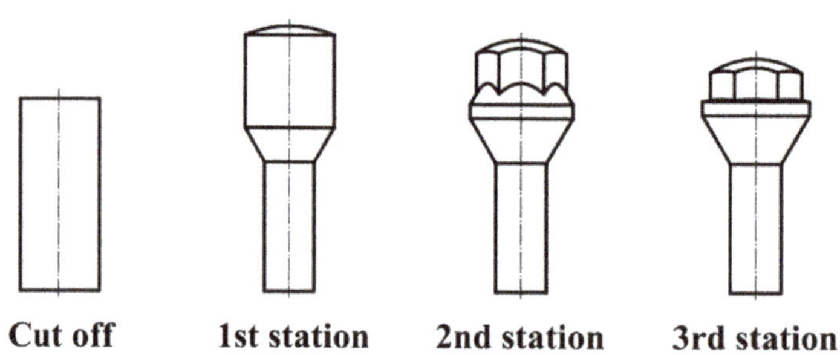

Figure 13. Typical bolt warm upsetting process scheme.

The process of the above design was verified by finite element numerical simulation. The rigid plastic finite element method was used to design the molds of the three stations into an integral part, ignoring the deformation of the molds and the ejector rod. The forming of the typical part requires three sets of dynamic molds and fixed molds, and the dynamic molds and fixed molds are set to have rigid contact. The type of selected element is a tetrahedral element. With the progress of numerical simulation, the mesh is undergoing adaptive repartition. The selected material is 42CrMo, the temperature is 650 °C, and the dynamic and fixed modes are set as rigid bodies without grid division. The moving die moves down at a speed of 84 mm/s, and the fixed die remains static. After completing the formation of one station, the die is replaced, and the formation of subsequent stations is continued until the typical part is fully formed.

4.4. Analysis of Stroke-Load Curve of Typical Parts under Warm Upsetting

As can be seen from the stroke–load curve in Figure 14, there are three load peaks in the entire deformation process, which increase with the increase in the degree of deformation. The initial load of each station is relatively gentle because, in the initial loading phase, the local deformation increases, while most of the rest do not produce large deformation. With the progress of the loading stage, the load–stroke curve appears to have an upward inflection point, the contact area between the blank and the mold cavity increases, the material enters the stage of compression upsetting, and when the maximum load of the station is reached, the material enters the maximum plastic deformation state of the station, and the forming of the station is finally completed. It can be seen from the figure that the load of the first station is the largest, reaching, followed by the third station, mainly because the deformation of the first station is the largest.

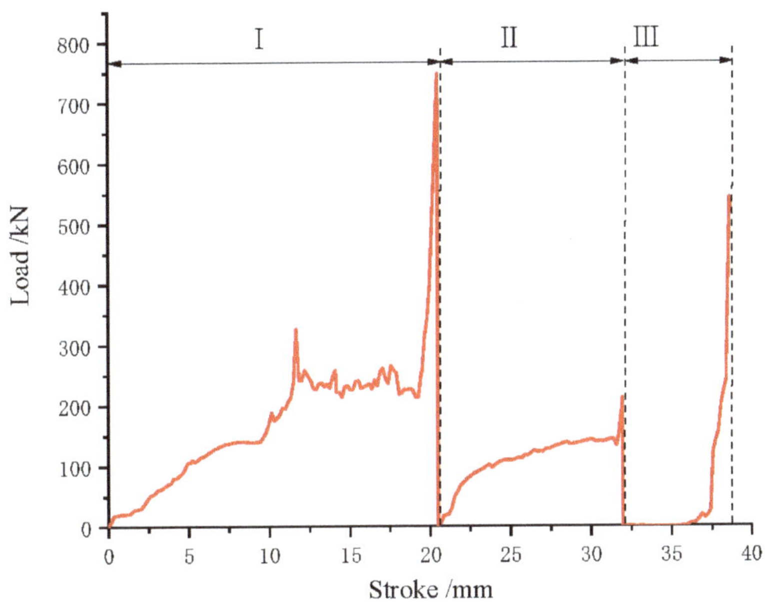

Figure 14. Stroke–load curve of each station.

4.5. Stress Nephogram Analysis of the Warm Upsetting of Typical Parts

Figure 15 shows the distribution diagram of equivalent strain and stress after forming at each station. It can be seen from the diagram that the maximum equivalent strain at each station is mainly distributed at the maximum deformation of each station, among which the first station is mainly reduced in diameter, the material at the lower end of the billet is squeezed, the metal flows to the reduced mold cavity, the deformation is large, the contact area between the upper end of the billet and the mold increases, and the friction also increases due to the restriction of the forging mold cavity. Therefore, the deformation stress in the P1 and P2 regions at the upper end of the billet is large, as shown in Figure 15b. The second station is mainly formed for the hexagonal head and flange pre-upsetting.

Because the lower end of the hexagonal head and the upper end of the flange are not bound by the mold, the strain is larger. The final forming part of the second station is the arc position of the top of the hexagonal head, and the arc of the top is smaller than that of the first station; the maximum stress distribution position of the top of the hexagonal head is a ring, and the maximum stress point is P3. The third station is the final upsetting stage, and each position has different degrees of deformation, among which the largest deformation position is the flange, and the equivalent strain value is also the largest. The whole part is completely covered by the mold, which makes the overall stress larger, and the maximum stress point is in the P4 area at the top of the hexagonal head. Due to the relatively complicated deformation of the hexagonal head, there is a flying edge around the hexagonal head when it forms at the third station. Subsequent trimming is required.

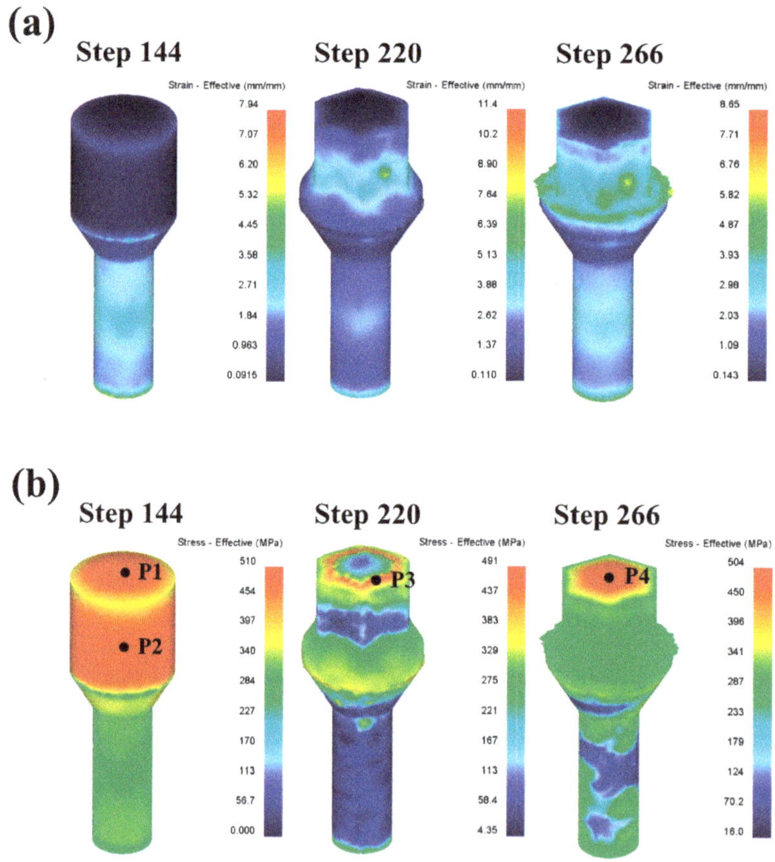

Figure 15. Equivalent stress and strain distribution: (**a**) equivalent strain; (**b**) equivalent stress.

5. Conclusions

In this paper, 42CrMo high-strength steel was used as the material. Compression tests with different parameters were carried out on the Gleeble-3500 thermal simulation machine for 42CrMo steel. The influences of different temperatures, strain rates, and deformation passes on microstructure evolution and mechanical properties were studied in detail, and the warm upsetting forming of typical parts was simulated.

(1) In rapid upsetting deformation, with the increase in deformation temperature, the kinetic energy between atoms increases, the binding force decreases, and the dislocation is easier to overcome the pinning effect and move, resulting in a temperature

rise softening effect that significantly reduces the peak flow stress. In addition, the maximum peak flow stress of 42CrMo steel is only 45.3% of the cold upsetting deformation at room temperature when it deforms at about 650 °C. In the plastic deformation stage, the stress-strain curve tends to be gentle, and it is in the best state of warm deformation;

(2) 42CrMo steel at 500–650 °C warm upsetting deformation conditions, the internal grain structure is significantly refined, and the grain refinement mechanism is mechanical crushing refinement. At slightly higher temperatures, the broken grain recovery and grain size grow;

(3) The correlation coefficient R and the absolute value of average relative error (AARE) between the simulated value and the experimental value were calculated through Deform. The obtained correlation coefficient R is 0.9948, and the average absolute relative error (AARE) is 2.05%, indicating that the simulation can accurately reflect the relationship between displacement and applied load.

Author Contributions: Z.X. carried out the concept, methodology, and writing the first draft. H.W. carried out the formal analysis, wrote the manuscript, and performed the data analysis. J.L. carried out a formal analysis and provided experimental instruments and data processing. J.J. mainly performed the investigation and verification. L.Y. carried out the concept, methodology, and guide on experimental design. Y.Z. undertook originality and form analysis. All authors have read and agreed to the published version of the manuscript.

Funding: This work was supported by the Henan Province Science and Technology Research Project (222102230118) and the Key Research Projects of Higher Education Institutions in Henan Province (23A460016).

Data Availability Statement: The raw data supporting the conclusions of this article will be made available by the authors on request.

Conflicts of Interest: Author Jinhua Liu was employed by the Sijin Intelligent Forming Machinery Company Limited Co., Ltd. The remaining authors declare that the research was conducted in the absence of any commercial or financial relationships that could be construed as a potential conflict of interest.

References

1. Khanawapee, U.; Butdee, S. A study of barreling and DEFORM 3D simulation in cold upsetting of bi-material. *Mater. Today* **2020**, *26*, 1262–1270. [CrossRef]
2. Filipov, A.A.; Pachurin, G.V.; Goncharova, D.A.; Nuzhdina, T.V.; Mukhina, M.V.; Katkova, O.V.; Matveev, U.I.; Tsapina, T.N. Structural and mechanical maintenance of quality of the rolled stock for cold upsetting of metal articles. *IOP Conf. Ser. Mater. Sci. Eng.* **2019**, *632*, 012010. [CrossRef]
3. Pachurin, G.V.; Shevchenko, S.M.; Filippov, A.A.; Mukhina, M.V.; Kuzmin, N.A. March. Defining rolled metal performance for cold bolt upsetting (bolt head). *IOP Conf. Ser. Mater. Sci. Eng.* **2018**, *327*, 032040. [CrossRef]
4. Sweet, G.A.W.; Williams, B.W.; Taylor, A.; Hexemer, R.L.; Donaldson, I.W.; Bishop, D.P. A microstructural and mechanical property investigation of a hot upset forged 2xxx series aluminum powder metallurgy alloy reinforced with AlN. *J. Mater. Process Technol.* **2020**, *284*, 116742. [CrossRef]
5. Li, S.; Ji, H.; Wang, B.; Mu, Y. Numerical analysis and experimental validation of conjunction gear via hot forging-upsetting finishing-radial extrusion. Arch. *Civ. Mech. Eng.* **2019**, *19*, 391–404. [CrossRef]
6. Arulmani, L.; Shridharmurthy, H.N.; Selvan, M.C.P.; Madara, S.R. Hot powder forging behavior analysis of sintered AISI 8740 PM steels for automotive application. *Mater. Today* **2020**, *28*, 1068–1072. [CrossRef]
7. Lu, Y.; Zhu, Z.; Li, D.; Xie, Q. Constitutive model of 42CrMo steel under a wide range of strain rates based on crystal plasticity theory. Mater. *Sci. Eng. A* **2017**, *679*, 215–222. [CrossRef]
8. Zhang, J.; Liu, Z.; Sun, J.; Zhao, H.; Shi, Q.; Ma, D. Microstructure and mechanical property of electropulsing tempered ultrafine grained 42CrMo steel. *Mater. Sci. Eng. A* **2020**, *782*, 139213. [CrossRef]
9. Qi, K.; Yang, Y.; Hu, G.; Lu, X.; Li, J. Thermal expansion control of composite coatings on 42CrMo by laser cladding. *Surf. Coat. Technol.* **2020**, *397*, 125983. [CrossRef]
10. Ji, H.; Duan, H.; Li, Y.; Li, W.; Huang, X.; Pei, W.; Lu, Y. Optimization the working parameters of as-forged 42CrMo steel by constitutive equation-dynamic recrystallization equation and processing maps. *J. Mater. Sci. Technol.* **2020**, *9*, 7210–7224. [CrossRef]

11. Liu, H.; Cheng, Z.; Yu, W.; Wang, G.; Zhou, J.; Cai, Q. Deformation behavior and constitutive equation of 42crmo steel at high temperature. *Metals* **2021**, *11*, 1614. [CrossRef]
12. Zhu, S.; Peng, W.; Shu, X. Effect of tempering on bonding characteristics of cross wedge rolling 42CrMo/Q235 laminated shafts. *J. Iron Steel Res.* **2020**, *27*, 1170–1178. [CrossRef]
13. Qin, F.; Qi, H.; Liu, C.; Qi, H.; Meng, Z. Constitutive Characteristics, Microstructure, and Texture Evolution of As-Cast 42CrMo Alloy in Nonisothermal Multipass Compression. *Adv. Mater. Sci. Eng.* **2021**, *2021*, 6638505. [CrossRef]
14. Wu, Z.; Peng, W.; Shu, X. Influence of rolling temperature on interface properties of the cross wedge rolling of 42CrMo/Q235 laminated shaft. *Int. J. Adv. Manuf. Technol.* **2017**, *91*, 517–526. [CrossRef]
15. Ji, H.; Wang, J.; Wang, Z.; Li, Y.; Cheng, X. The effect of high-temperature ECAP on dynamic recrystallization behavior and material strength of 42CrMo steel. *Mater. Sci. Eng. A* **2023**, *887*, 145732. [CrossRef]
16. Lin, Y.; Chen, M.; Zhong, J. Effect of temperature and strain rate on the compressive deformation behavior of 42CrMo steel. *J. Mater. Process. Technol.* **2008**, *205*, 308–315. [CrossRef]
17. Qian, D.; MA, B.; Deng, J. Numerical simulation and experimental study on grain evolution in whole process of forging and rolling forming-waste heat quenching of 42CrMo ring. *J. Plast. Eng.* **2022**, *29*, 110–117.
18. Wang, B.; Shi, J.; Ye, P.; Deng, L.; Wang, C.; Chen, J.; Li, Q. In-situ investigation on nucleation and propagation of {10–12} twins during uniaxial multi-pass compression in an extruded AZ31 Mg alloy. *Mater. Sci. Eng. A* **2018**, *731*, 71–79. [CrossRef]
19. Hu, Z.; Zhou, X.; Nie, X.; Zhao, S.; Liu, H.; Yi, D.; Zhang, X. Finer subgrain microstructure induced by multi-pass compression in $\alpha + \beta$ phase region in a near-β Ti-5Al-5Mo-5V-1Cr-1Fe alloy. *J. Alloys Compd.* **2019**, *788*, 136–147. [CrossRef]
20. Yan, Z.; Liu, H.; Dai, X.; Li, L.; Zhang, Z.; Wang, Q.; Xue, Y. Effect of multi-pass cooling compression and subsequent heat treatment on microstructural and mechanical evolution of TC4 alloys. *J. Mater. Sci. Technol.* **2023**, *23*, 3137–3150. [CrossRef]
21. Hou, H.; Zhao, G.; Yu, J.; Wei, D. Nonlinear unloading-reloading behavior and macro/micro models during multi-pass cyclic hot compression of aluminum alloys. *Mater. Sci. Eng. A* **2023**, *882*, 145447. [CrossRef]
22. Qin, F.; Li, Y.; Qi, H.; Lv, Z. Deformation behavior and microstructure evolution of as-cast 42CrMo alloy in isothermal and non-isothermal compression. *J. Mater. Eng. Perform.* **2016**, *25*, 5040–5048. [CrossRef]
23. Huang, W.; Zhong, H.; Lei, L.; Fang, G. Microstructure and mechanical properties of multi-pass forged and annealed 42CrMo steel. *Mater. Sci. Eng. A* **2022**, *831*, 142191. [CrossRef]
24. Zhu, Q.; Lei, L.; Ban, C.; Zhao, Z.; Zuo, Y.; Cui, J. Structure uniformity and limits of grain refinement of high purity aluminum during multi-directional forging process at room temperature. *Trans. Nonferrous Met. Soc. China.* **2014**, *24*, 1301–1306. [CrossRef]
25. Liu, Z. *Study on Microstructures, Mechanical Properties and Cold Forming of TC16 Titanium Alloy Bars*; Jiangsu University: Zhenjiang, China, 2021.
26. Zhai, C.; Feng, C.; Chai, L.; Lu, Z.; Nie, Z.; Chen, Z. Study on X2A66 rheological deformation behavior of aluminum-lithium alloy during isothermal compression. *Rare Met. Mater. Eng.* **2017**, *46*, 90–96.
27. Chen, Y.; Qi, H.; Li, Y.; Pang, X. Modification of constitutive model and evolution of activation energy for forged 42CrMo steel during high temperature deformation process. *Forg. Stamp. Technol.* **2021**, *46*, 260–269.
28. Xiao, Z. *Research on Mechanism of Multi-Stage Cold Forging and Optimum Structure of Machine*; China Academy of Machinery Science and Technology: Beijing, China, 2015.
29. Yan, P.; Yan, Z. Improvement method of defect structure in forging steel. *Phys. Testing Chem. Anal. Part A (Phy Test)* **2023**, *59*, 28–31.
30. Huang, F.; Zhong, F.; Li, P.; Liang, Y. Influence factors on grain deformation and refinement mechanisms during plastic deformation. *Foundry Technol.* **2011**, *32*, 1602–1605.
31. Hu, B.; Zhang, B.; Wei, C.; Zhang, Z. Influence of die preheating temperature and punch working speed on quality of roller formed cylinder. *J. Plast. Eng.* **2020**, *27*, 47–52.

metals

MDPI

Article

Aerodynamics Optimization of Multi-Blade Centrifugal Fan Based on Extreme Learning Machine Surrogate Model and Particle Swarm Optimization Algorithm

Fannian Meng [1], Liujie Wang [1], Wuyi Ming [1,2,*] and Hongxiang Zhang [1]

[1] Mechanical and Electrical Engineering Institute, Zhengzhou University of Light Industry, Zhengzhou 450002, China
[2] Guangdong Provincial Key Laboratory of Digital Manufacturing Equipment, Guangdong HUST Industrial Technology Research Institute, Dongguan 523808, China
* Correspondence: mingwuyi@zzuli.edu.cn

Abstract: The centrifugal fan is widely used in converting mechanical energy to aerodynamic energy. To improve the pressure of the multi-blade centrifugal fan used in an air purifier, an optimization process was proposed based on extreme learning machine (ELM) combined with particle swarm optimization (PSO). The blade definition position parameter and blade definition radian parameter were designed using the full-factor simulation experimental method. The steady numerical simulation of each experimental point was carried out using ANSYS CFX software. The total pressure of the multi-blade centrifugal fan was selected as the optimization response. The optimized ELM combined with the PSO algorithm considering the total pressure response value and the two multi-blade centrifugal fan parameters were built. The PSO algorithm was used to optimize the approximation blade profile to obtain the optimum parameters of the multi-blade centrifugal fan. The total pressure was improved from 140.6 Pa to 151 Pa through simulation experiment design and improved surrogate optimization. The method used in the article is meant for improving multi-blade centrifugal total pressure. The coupling optimization of impellers, volutes, and air intakes should be comprehensively considered to further improve the performance of centrifugal fans.

Keywords: multi-blade centrifugal fan; ELM surrogate model; PSO algorithm; optimization

1. Introduction

The multi-blade centrifugal fan is widely used in household appliances such as air conditioners and range hoods; its advantages are its small size, large gas flow, low operating noise, and high pressure coefficient. It is sufficient to complete ventilation work while reducing the impact on the surrounding environment. With the development of industrial technology, multi-blade centrifugal fans are also widely used in industrial equipment such as air supply equipment for metallurgical boilers in metallurgical manufacturing plants and special element extraction equipment for reaction furnaces in nuclear power plants. Improving the performance of multi-blade centrifugal fans is of great significance for energy conservation and emission reduction.

In order to improve the performance of the centrifugal fan, scholars have conducted extensive research on multi-blade centrifugal fan optimization. Xu et al. [1] optimized the blade tip bionic groove parameters based on the flow control mechanism research of blade tip bionic grooves. Ye et al. [2] improved the multi-blade centrifugal fan performance by changing inlet structure parameters based on a bevel-cutting design method and computer simulation technology. Li et al. [3] studied the vortex characteristics of a multi-blade centrifugal fan near the volute outlet region. Wei et al. [4] researched the effect of clearance, the inclination angle, and the volute tongue on a multi-blade centrifugal fan's aerodynamic performance. Zhou et al. [5] adapted the SCT function parameterization method, RBF surrogate model, and NSGA-II algorithm to optimize the multi-blade centrifugal fan blade parameters. Zhou et al. [6] optimized the multi-blade centrifugal fan total pressure efficiency and flow rate to improve the aerodynamic performance based on a modified Hicks–Henne function and Kriging agent model. Zhou et al. [7] proposed an optimal design method for multi-blade centrifugal fan using the Latin hypercube sampling design combined with the Kriging agent model. Alla et al. [8] optimized a squirrel cage on seven design parameters using the Latin hypercube sampling method and OpenFoam software (OpenFOAM foundation, open source, UK). Furthermore, Alla et al. [9] adapted the same method to optimize thirteen squirrel cage design parameters. Heo et al. [10] used the Latin hypercube sampling method, Kriging surrogate model, and genetic algorithm-II to optimize the impeller parameters. Le et al. [11] used ANSYS 15.0 software (ANSYS Inc., USA) to research the optimal operating conditions for the multi-blade centrifugal fan performance and flow characteristics. Kim et al. [12] used the numerical analysis and response surface optimization method to optimize the location of cutoff, radius of cutoff, width of the impeller, and flow coefficient. Table 1 shows the above optimization research on the multi-blade centrifugal fan, including optimization parameters and specific research methods.

Table 1. The optimization research on the multi-blade centrifugal fan.

Reference	Optimization Parameter	Method
Xu et al. [1]	Groove width, groove depth, groove center distance, groove number, different groove shapes	CFD simulation design method
Ye et al. [2]	Inner diameter, bevel-cutting blade	CFD technology, test
Li et al. [3]	Vortex characteristics	CFD technology
Wei et al. [4]	Clearance, the inclination angle, the volute tongue	CFD technology, test
Zhou et al. [5–7]	Multi-blade centrifugal fan blade airfoil parameter	1. CST function parameterization 2. Latin hypercube sampling 3. RBF model 4. Non-dominant sorting genetic algorithm-II
Alla et al. [8,9]	Multi-blade centrifugal fan impeller and wheel parameter	1. Latin hypercube sampling 2. Response surface approximation model 3. EGO approach
Heo et al. [10]	The scroll cut-off angle, the scroll diffuser expansion angle, the diameter ratio of the impeller, the blade exit angle	1. Latin hypercube sampling 2. Kriging surrogate model 3. Algorithm-II
Le et al. [11]	Research operating condition on the centrifugal fan performance and flow characteristics, obtaining the topology optimization design	ANSYS software
Kim et al. [12]	Location of cutoff, radius of cutoff, width of impeller, flow coefficient	Numerical analysis, response surface optimization method

The main research methods for rotating machinery are based on CFD and the surrogate model. For example, among the above literature about multi-blade centrifugal fans, references [1–3,11] are based on CFD research and references [5–7,12] are based on CFD and surrogate model research.

For the rotating machinery of a centrifugal fan, the main research method is also based on the surrogate model. For example, Zhao et al. [13] adapted the reverse design method to optimize the geometric configuration of the centrifugal fan hood using the back-propagation neural networks surrogate model. Soheil et al. [14] used the Kriging surrogate model and MIGA algorithm to optimize the suction and shaft power of a vacuum cleaner centrifugal fan. Zhang et al. [15] used the DOE study, the RSM surrogate model, and the multi-objective genetic algorithm to optimize the fan's 13 design parameters. Lvo et al. [16] adapted the surrogate parameter optimization method to optimize the 3D vane shape.

Jung et al. [17] adapted the DOE method and the response surface methodology to optimize the diffuser of a small high-speed centrifugal fan. Zoran et al. [18] used the 3D parameterization method and the surrogate model to optimize the shape of fan vanes.

For the rotating machinery optimization of a compressor, the research method is also based on the CFD and surrogate model. Xu et al. [19] proposed the local remodeling surrogate model method used in the fluid analysis and structure optimization region of an impeller. Xu et al. [20] also optimized the impeller of a high-speed magnetic drive pump using the optimal Latin hypercube sampling method combined with the response surface methodology surrogate model. Tüchler et al. [21] introduced an aerodynamic shape optimization method for a micro-wave rotor turbine using the CFD model and hybrid algorithm. Furthermore, references [22,23] also optimized the compressor using the surrogate model.

Moreover, Zheng et al. [24] researched the internal flow characteristics of a centrifugal fan from changing edge curvatures of the impeller front disk using the CFD and test method. Engin, T. [25] researched the tip clearance effects for centrifugal fans using the CFD and test methods. References [26,27] optimized the centrifugal fan to obtain the minimize noise. Kruyt et al. [28] adapted the inverse method to design a centrifugal fan. Marian et al. [29] summarized the research method to improve operating efficiency and to reduce noise emissions for fans. Kudela et al. [30] summarized the surrogate model application advances for finite element computation.

The previously reported work mainly focuses on rotating machinery optimization using CFD technology, the surrogate model, and the intelligence algorithm. It can be proven that CFD technology combined with the surrogate model has good advantages in terms of rotating machinery optimization, this article adapted the frequently used method in optimizing the multi-blade centrifugal fan. In this article, the PSO algorithm is used to optimize the ELM surrogate model to improve the surrogate model fitting accuracy. The simulation experimental design method and improved surrogate optimization are used to improve the total pressure of the multi-blade centrifugal fan. The optimization has a wide range of areas and a large range of variables, which is conducive to achieving global optimization.

2. Optimization Problem Definition and Description

2.1. Optimization Object

The research object is one particular type of multi-blade centrifugal fan, which is typically constructed with a built-in motor and is easy to disassemble. The structural parameters of this type of centrifugal fan are shown in Table 2, and its shape is shown in Figure 1.

Setting the blade length as 15 mm, width as 1 mm, and height as 43 mm, and using the plane perpendicular to the width of the blade and over the midpoint of the width as the reference plane (the reference plane is the same as the shape of the blade surface), it is necessary to ensure that the length of the reference line in each simulation model is equal.

The measurement criteria for different blade positions are different values for the included angle between the reference plane and the vertical plane (unit is defined as °), and the measurement criteria for different blade shapes are different values for the included angle between the 5 mm side line at both ends of the reference line and the reference line (unit is also defined as °). The blade profile position definition is shown in Figure 2, where each plane is shown from the top view.

Table 2. Structural parameters of the multi-blade centrifugal fan.

Structural Parameters	Value (mm)
Height	180
Width	61
Length	171
Diameter	133
Inlet diameter	110
Outlet parameter	59 × 71

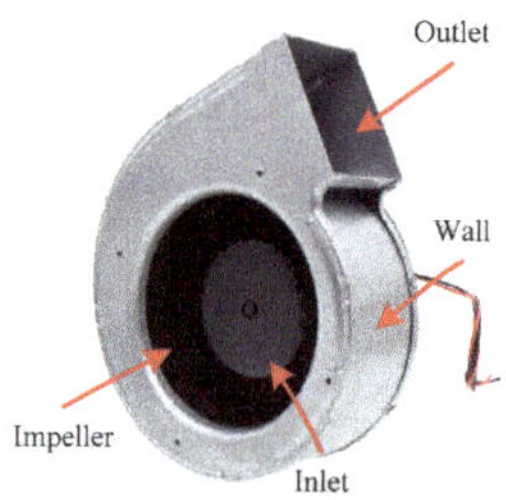

Figure 1. Structural model of the multi-blade centrifugal fan.

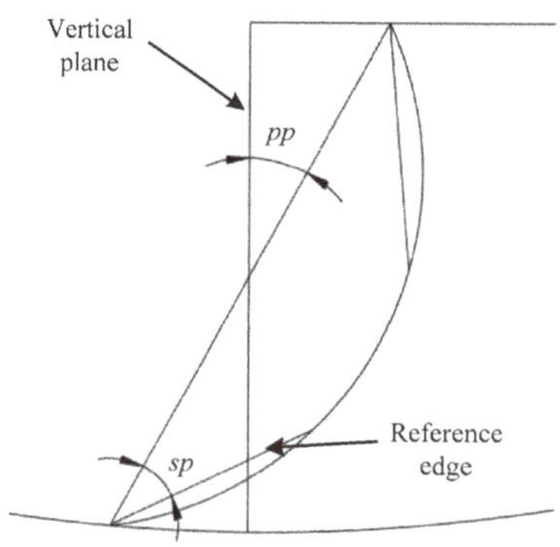

Figure 2. The blade profile position definition diagram.

2.2. Optimization Model

The air volume and total pressure are used to ensure the gas delivery for the multi-blade centrifugal fan. The total pressure (*Tp*) value is set as an optimization target; the multi-blade centrifugal fan position parameter (*pp*) and the shape parameter (*sp*) are optimization parameters. The multi-blade centrifugal fan optimization model is described as follows[5]:

$$\begin{aligned} &\text{find } \boldsymbol{x} = (pp, sp) \\ &\max \widehat{Tp}(\boldsymbol{x}) \\ &\text{s.t. } pp^{LB} \le pp \le pp^{UB} \\ &\quad\ sp^{LB} \le sp \le sp^{UB} \end{aligned} \tag{1}$$

where *pp* is the position parameter for the blade, *sp* is the radian parameter for the blade, and $\widehat{Tp}$ is the surrogate model approximate response value for total pressure, and this value is calculated using the ELM surrogate model optimized by the PSO algorithm.

2.3. Flowchart of Blade Parameter Optimization

Firstly, the *pp* and *sp* parameters are fully factorial simulation experimental design. Secondly, the optimal blade position parameters and shape parameters are obtained according to the process of Figure 3.

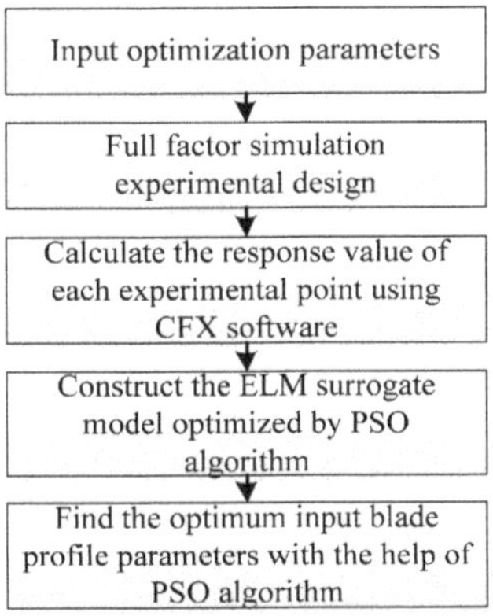

Figure 3. The optimization process flowchart.

3. The Surrogate Model and Optimization Algorithm

3.1. ELM Surrogate Model

Machine learning methods like neural networks are used in many fields [31–33]. Compared with the traditional neural networks, the ELM surrogate model has a faster learning speed and better generalization ability. The training model for the ELM method is shown in Figure 4.

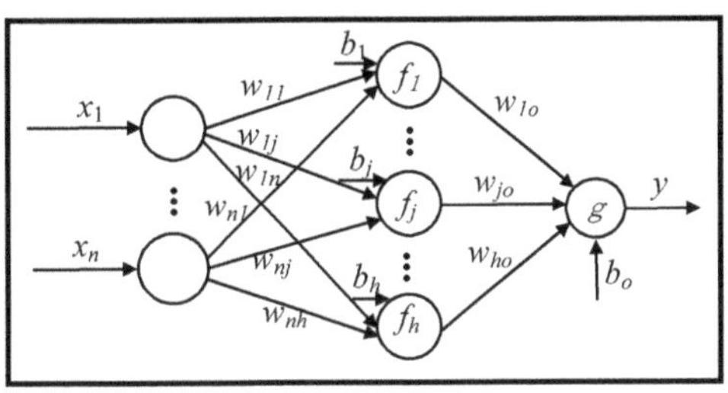

Figure 4. Extreme learning machine network architecture.

In Figure 4, n is the number of the input variables, h is the number of the hidden layer neurons, x_i is the input variable, $i = 1, 2, 3, \cdots, n$, ω_{ij} denotes the connection weights between the input variables and the hidden layer neurons j, w_{jo} denotes the connection weights between the hidden layer neurons j and the output layer, b_j denotes the threshold of the neurons in the hidden layer, $j = 1, 2, 3, \cdots, h$, b_0 denotes the threshold of neurons in the output layer, $f_j(\cdot)$ denotes the activation function of the neurons in the hidden layer, and $g(\cdot)$ denotes the activation function of the neurons in the output layer.

Its mathematical model as follows [34]:

$$\begin{cases} y = g(b_o + \sum\limits_{j=1}^{h} w_{jo} v_j) \\ v_j = f_j(b_j + \sum\limits_{j=1}^{n} w_{ij} x_j) \end{cases} \tag{2}$$

where v_j represents the output of a neuron j in the output layer.

If there are N valid samples, when the output threshold is equal to 0 and the output neuron activation function is a linear function, then Formula (2) can be rewritten as follows:

$$y = (w_o^T v)^T \tag{3}$$

where $y = [y(1), y(2), \cdots, y(N)]^{\mathrm{T}}$ is the network output vector, $w_o = [w_{1o}, w_{2o}, \cdots, w_{ho}]^{\mathrm{T}}$ is the output weight vector, and v is the output matrix of the hidden layer neurons, where the input weight and the threshold matrix w are generated randomly.

$$\begin{cases} v = \begin{bmatrix} v_1(1) & v_1(2) & \cdots & v_1(N) \\ v_2(1) & v_2(2) & \cdots & v_2(N) \\ \vdots & \vdots & \cdots & \vdots \\ v_h(1) & v_h(2) & \cdots & v_h(N) \end{bmatrix} \\ w = \begin{bmatrix} b_1 & b_2 & \cdots & b_h \\ w_{l1} & w_{l2} & \cdots & w_{lh} \\ \vdots & \vdots & \cdots & \vdots \\ w_{n1} & w_{n2} & \cdots & w_{nh} \end{bmatrix} \end{cases} \tag{4}$$

The output weight vector is expressed as

$$w_o = v^* y_e \tag{5}$$

where v^* denotes the Moore–Penrose generalized inverse for output layer matrix v and $y_e = [y_e(1)\ y_e(2) \cdots y_e(N)]^{\mathrm{T}}$ is the expected outputs. If there are $v \in R^{N \times h}$, $N \geq h$, and $\mathrm{rank}(v) = h$, the Moore–Penrose generalized inverse v^* for v can be expressed as

$$v^* = \left(v^T v\right)^{-1} v^T \tag{6}$$

Formula (6) is introduced into Formula (5):

$$w_o = \left(v^T v\right)^{-1} v^T y_e \tag{7}$$

The activation function of the hidden layer used is the sigmoid function shown below:

$$f(a, b, x) = \frac{1}{1 + e^{-ax-b}} \tag{8}$$

Because the extreme learning machine avoids the disadvantages of traditional networks such as learning rate, termination conditions, and being easily trapped in a local optimum, this method can construct a surrogate model more effectively than other feed-forward neural networks. However, an extreme learning machine may need more hidden layer neurons than a traditional neural network, and it may lead to ill-conditioned problems because of randomly assigning input weights and thresholds. To solve this problem, particle swarm optimization (PSO) is used to optimize the ELM and apply it to the multi-blade centrifugal fan optimization process.

3.2. Particle Swarm Optimization

PSO was proposed by Eberhart and Kennedy in 1995 based on artificial life and evolutionary computing theory [35]. The basic idea is derived from the foraging behavior of birds. In PSO, the solution of each optimization problem is a bird in the search space, which is called a particle. Each particle has an initialization speed and position, and the fitness value is determined by the fitness function. Each particle is given a memory function to remember the best location to be searched. In addition, the speed of each particle determines the direction and distance of its flight so that the particle can search in the optimal solution space. In the iteration, the particle updates its speed and position by comparing fitness values and two extremes: the optimal solution found by the particle itself (individual extreme value) and the optimal solution currently found by the whole population (global extreme value):

$$v_i(t+1) = \omega v_i(t) + c_1 R_1 [R_i^b(t) - x_i(t)] + c_2 R_2 [R_g^b(t) - x_i(t)] \tag{9}$$

$$x_i(t+1) = x_i(t) + \phi v_i(t+1) \quad (10)$$

where t is the number of the iterations, $v_i(t)$ represents the speed of the t iteration for particle i, ω is the inertia weight, c_1, c_2 is the cognitive coefficient, R_1, R_2 is the uniformly distributed random number, $R_i^b(t)$ is the optimum position of individual history for particle i, $R_g^b(t)$ is the optimum position of group history, $x_i(t)$ is the position of the particle in the t iteration, and ϕ is a contraction factor used to keep the speed within a certain range.

As a random search and parallel optimization algorithm, the PSO algorithm has the characteristics of simplicity, robustness, easy realization, and fast convergence, and can find the global optimal solution of the problem with a relatively high probability.

3.3. Extreme Learning Machine Based on PSO

According to the basic principles of ELM, we can see that the input layer weight matrix w and the hidden layer threshold matrix b of ELM are generated randomly. Therefore, when the ELM model is fixed, the prediction error of ELM will be very large. PSO has good global optimization ability. PSO can find the optimal initial w and b for the ELM model, and then obtain the optimal ELM model. The main steps of the PSO-ELM algorithm are as follows:

1. Preprocessing of experimental data. The experimental data are divided into training sets and test sets, and then normalized.
2. The parameters of particle swarm optimization are transformed into ELM parameters, and then the input and output samples are sent to ELM for prediction and the fitness is calculated.
3. The individual and global extremums for all particles are calculated.
4. According to the formula of particle renewal speed and position, the particle velocity, position, and weight are updated.
5. When the number of iterations reaches the maximum number of iterations set or the fitness meets the requirements, the search is stopped; otherwise, the iterations are repeated until the condition of stopping iterations is reached.
6. The PSO-ELM model is completed. The particle position corresponding to the optimal individual is the optimal weights and thresholds of the ELM network, which are substituted into the ELM model.

The flow chart of the PSO-ELM algorithm is shown in Figure 5.

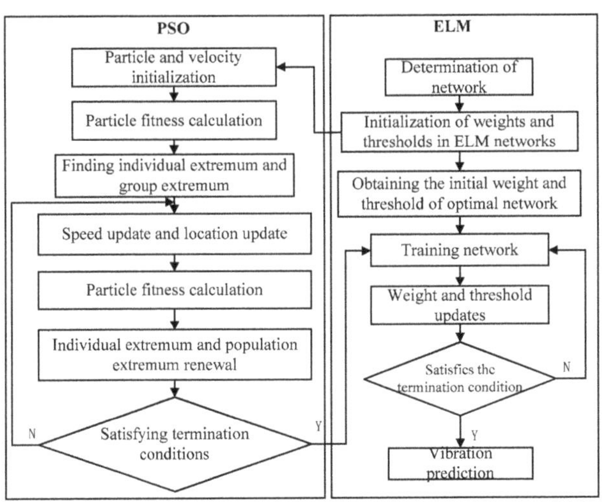

Figure 5. Algorithmic flow of PSO-ELM.

4. Concrete Optimization Process and Results

4.1. Optimization Structural Model

To ensure that the simulation results of the multi-blade centrifugal fan are consistent with the actual value, it is important to make the simulation model as close as possible to the actual working conditions. However, if the simulation model is too complex, the simulation time will be too long to bear. Therefore, it is necessary to simplify the simulation model of the multi-blade centrifugal fan appropriately. In addition, in combination with the requirements of the ANSYS CFX simulation, multiple boundaries need to be defined in the simulation model of the multi-blade centrifugal fan. Furthermore, to ensure that the simulation model meets the simulation requirements, it is essential to establish a standardized approach to follow during the simulation. The specific design principles for the simulation model are as follows: the bolt, nut, gasket, protective net, and other sealing mechanisms of the whole multi-blade centrifugal fan have the least impact on the simulation results during the simulation process and should be deleted from the simulation model. The fillets and grooves on the inner and outer shells of the multi-blade centrifugal fan have little influence on the simulation results and can be omitted. For some arc surfaces on the shell, the connecting arc between the upper shell and the impeller, the built-in motor chip, and the coil can be simplified with regular geometry.

4.2. CFX Simulation

The aerodynamics simulation involves meshing the inlet part, impeller, and volute. The grid division software used is ICEM. The generated grids are then checked using the Check-mesh function in ICEM software (ANSYS Inc., Canonsburg, PA, USA), and then the grids are optimized by the Smooth-Mesh-Globally function in ICEM software. The minimum angle of the grid is required to be greater than 14° to ensure grid quality. The impeller and volute simulation model that completes the grid division is shown in Figure 6.

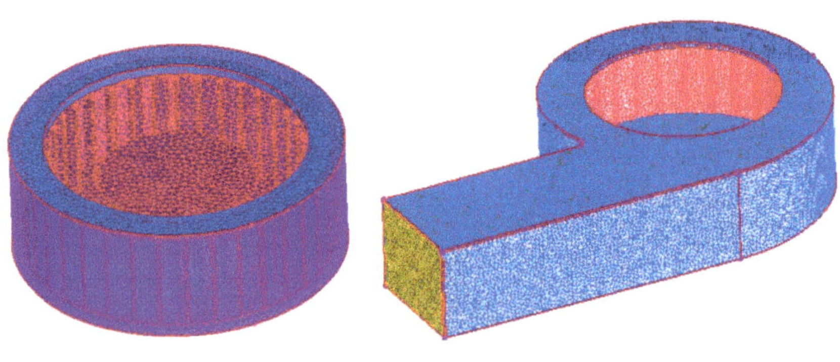

(a) (b)

Figure 6. The generated grid for the multi-blade centrifugal fan. (**a**) Impeller, (**b**) volute.

In addition, in the working process of a multi-blade centrifugal fan, due to the fact that the number of blades is large, the blade distribution is dense, and the space between blades is short, the blade channel is narrow, so an internal secondary flow exists in the volute, forming a relatively complex three-dimensional asymmetric flow. For three-dimensional asymmetric flow, the calculation accuracy will be affected by the turbulence model used. Three-dimensional asymmetric flow often uses the Spalart–Allmaras, standard, and SST turbulence models. It is found that the standard model and SST model can better capture the vortex position and velocity gradient under the condition of small flow through existing research. In addition, due to the thickness of the blade and the fact that the standard model does not consider the wall shear force, there is an error between the numerical value obtained through the CFD simulation and the experimental value. Therefore, in the CFD simulation process, the SST model is used as the turbulence model to improve the accuracy of the simulation.

The author used the same CFD simulation technology to simulate another type of centrifugal fan in reference [36], and it was verified that the CFD simulation results used in the article are in good agreement with the experimental results. In this article's CFD simulation, the same CFD simulation technology was used. Through comparative analysis, the CFD simulation technology used in this paper is reliable and can be used for the optimization analysis of multi-blade centrifugal fans.

4.3. Design of Simulation Experiment

Due to the substantial cost of simulation calculation, the optimized ELM surrogate model is introduced to replace the implicit relationship between the blade parameters and the total pressure responses for the multi-blade centrifugal fan, then the optimized ELM surrogate model is settled as the fitness function used in the PSO optimization process. The multi-blade centrifugal fan parameters *pp* and *sp* are in the design space. The ranges of the parameters *pp* and *sp* are set from 0° to 40°, there is full factor simulation experimental design for parameters *pp* and *sp*, the input parameters and response parameters are shown in Table 3, and the total pressure response values corresponding to each of the input parameters *pp* and *sp* are calculated using CFX software (ANSYS Inc., Canonsburg, PA, USA).

Table 3. Input parameter and response parameter settings.

Sequence	Position Parameter/°	Shape Parameter/°	Total Pressure/Pa
1	0	0	49.73
2	0	10	69.41
3	0	20	91.23
4	0	30	106.10
5	0	40	99.23
6	10	0	50.32
7	10	10	77.76
8	10	20	110.20
9	10	30	129.80
10	10	40	140.60
11	20	0	49.05
12	20	10	71.45
13	20	20	119.20
14	20	30	134.20
15	20	40	150.10
16	30	0	43.04
17	30	10	70.02
18	30	20	103.91
19	30	30	109.7
20	30	40	130.90
21	40	0	30.27
22	40	10	74.29
23	40	20	101.10
24	40	30	86.80
25	40	40	80.35

The simulation results table will be summarized based on the blade position. The blade position data and corresponding fan total pressure data for different blade shapes will be summarized. These summarized data will then be used to plot the corresponding images, as shown in Figure 7.

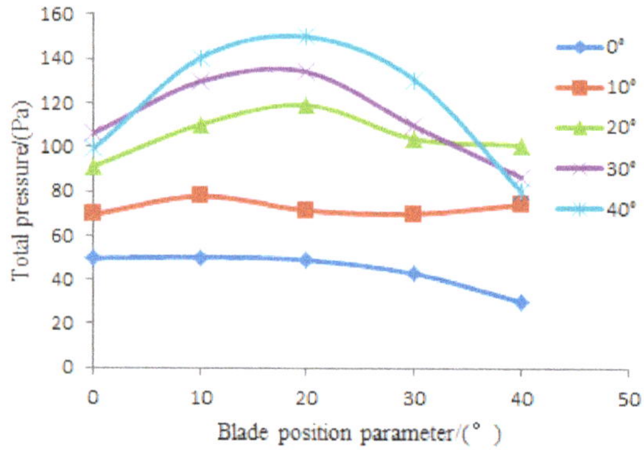

Figure 7. The relationship between the blade position parameters and total pressure.

As can be seen from the above figure, when the blade shape parameter is equal to 0°, the overall total pressure monotonically decreases with an increase in the blade position parameter. Furthermore, the decreasing slope of the blade increases with an increase in the blade position angle. When the blade shape angle ranges from 10° to 50°, the overall total pressure initially increases and then decreases as the blade position angle increases. Within the range of 20° to 40° for the blade shape angle, the maximum value of the overall wind total pressure is observed near the blade shape angle of 20°. It is also noteworthy that the total pressure of the multi-blade centrifugal fan increases as the blade angle increases at a constant blade position angle. The blade position parameters significantly affect the total pressure of a multi-blade centrifugal fan, mainly because the position of the blades is closely related to the airflow inside the impeller. Appropriate position parameters can reduce vortex loss and improve flow efficiency.

Figure 8 illustrates the relationship between blade shape parameters and the total pressure of a multi-blade centrifugal fan for different blade position parameters.

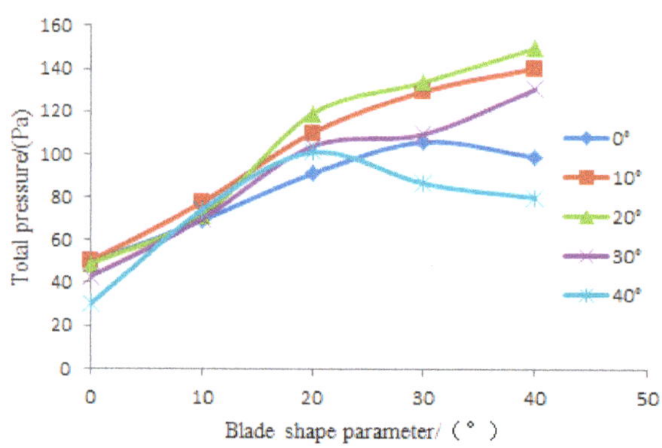

Figure 8. The relationship between the blade shape parameters and total pressure.

From Figure 8, it is evident that the total pressure increases monotonically as the blade shape angle increases within the range of 10° to 30°. When the blade position angles are between 0°and 40°, the total pressure initially increases and then decreases as the blade shape angle increases. When the blade position angle range is from 0° to 20°, the total pressure of the multi-blade centrifugal fan increases gradually for each corresponding blade shape, and when the blade position angle is from 20° to 40°, the total pressure gradually decreases for each blade shape angle. When the blade position angle is set to 20°, the total pressure of the multi-blade centrifugal fan corresponding to each blade shape angle is generally greater than the total pressure corresponding to other blade shape angles. This provides further reference to increase the value of the shape for the multi-blade centrifugal fan and the total pressure required for subsequent designs.

To obtain the blade parameters *pp* and *sp*, corresponding to the maximum total pressure response value, the fitness function is needed in the optimization process of the PSO algorithm. Here, the PSO-ELM surrogate model is settled as a fitness function of the PSO optimization process, and the error index R^2 is used to evaluate the accuracy between the PSO-ELM and ELM methods. The coefficient R^2 is shown below:

$$R^2 = 1 - \frac{\sum_{i=1}^{m} (y_i - \widehat{y}_i)^2}{\sum_{i=1}^{m} (y_i - \overline{y})^2} \quad (11)$$

where R^2 represents the global relative error of the surrogate model. m represents the number of test samples used to evaluate the accuracy of the surrogate model. y_i represents the true value and $\widehat{y}_i$ represents the predicted value through the surrogate model. $\overline{y}$ represents the average value of all true values. The accuracy evaluation values for the R^2 coefficient between the PSO-ELM and ELM surrogate model are shown in Figure 9 and Table 4.

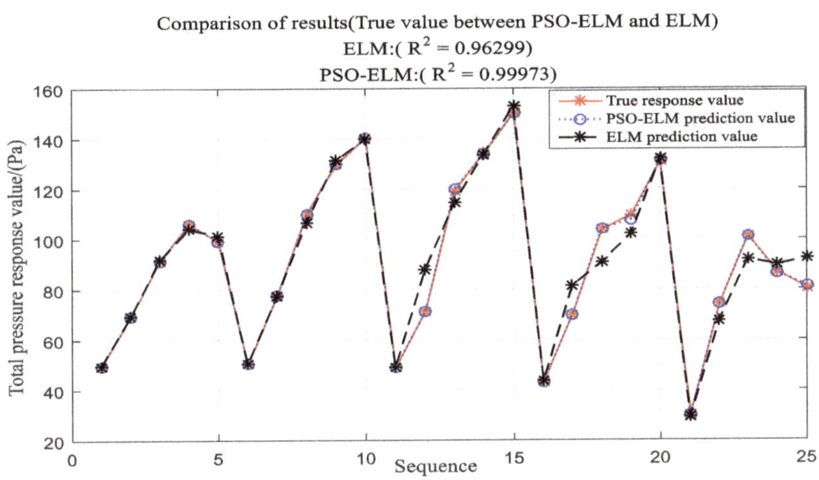

Figure 9. Fitting results evaluation between PSO-ELM and ELM method.

Table 4. Accuracy evaluation results for PSO-ELM and ELM.

Evaluation Index	PSO-ELM	ELM
R^2	0.99973	0.96299

The results in Figure 9 show that the PSO-ELM model is better than the ELM model in the fitting process. Moreover, the evaluation coefficient R^2 value is equal to 0.96299 for to the ELM model, and the R^2 value increases to 0.99973 for the PSO-ELM model.

The ELM model, Kriging model, and RBF model are discussed in the work [37] of Meng. In this study, the ELM model has the highest prediction accuracy compared with the Kriging model and RBF model, so the ELM model is used to predict multi-blade performance in the optimization process.

The article utilizes Matlab (MathWorks Inc., Natick, MA, USA) as the fitting software. In the prediction analysis conducted using Matlab, the ELM intelligent algorithm library function is employed, and the software statement "[IW, B, LW, TF, TYPE] = elm_train (P_train, T_train, N, TF, TYPE);" is used, where "N" represents the number of hidden neurons, "TF" denotes the type of hidden layer activation function, and "TYPE" takes a value of 0 for classification and 1 for surrogate model prediction. After data training, the software statement 'Tsim_2 = elm_predict (P_test, IW, B, LW, TF, TYPE);' is used for proxy prediction.

To obtain the optimal blade parameters *pp* and *sp*, the PSO algorithm is employed. Figure 10 illustrates the optimization process with the PSO-ELM surrogate model set as the fitness function.

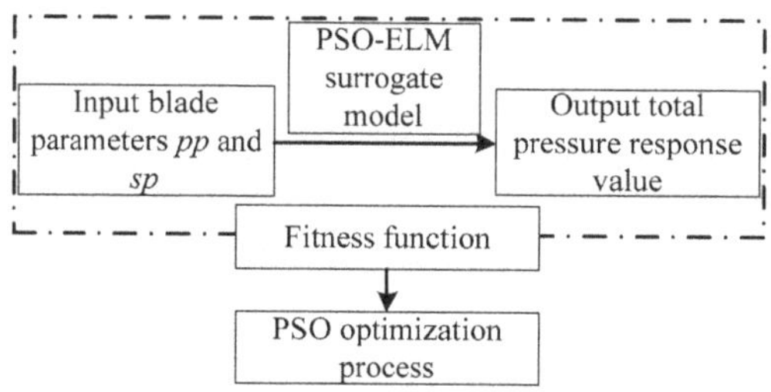

Figure 10. PSO optimization process.

Table 5 shows the optimization results for the maximum total pressure value optimized using the method shown in Figure 10.

Table 5. Total pressure value and blade parameters before and after optimization.

Centrifugal Fan	Blade Parameters [*pp sp*]/(°)	Total Pressure/(Pa)
Before optimization	(10.0, 40.0)	140.6
After optimization	(18.0, 39.0)	151

Table 5 illustrates the comparison of total pressure value before and after optimization. It is evident that the total pressure of the optimized fan has increased from 140.6 Pa to 150 Pa, indicating that the performance of the multi-blade centrifugal fan is superior to that of the original centrifugal fan.

5. Results and Discussion

Overall, with the increase in blade position parameters, the total pressure exhibits a trend of initially increasing and then decreasing. As the blade shape parameter increases, the overall total pressure shows an upward trend.

The method utilized in this article can enhance the total pressure of multi-blade centrifugal fans.

For future research on centrifugal fans, it is crucial to comprehensively consider the coupling optimization of impellers, volutes, and air intakes. This approach will lead to further improvements in the performance of centrifugal fans.

6. Conclusions

In this article, the PSO algorithm is used to improve the accuracy of the ELM model, and the PSO-ELM surrogate model is established, the PSO algorithm is used for optimization, and a blade profile optimization design process considering the aerodynamic performance of the multi-blade centrifugal fan is proposed.

1. Using the CFD method to solve aerodynamic performance is complex and time-consuming. The PSO-ELM surrogate model set as the fitness function is used in the PSO optimization process. From the evaluation index R^2, the R^2 value is equal to

0.99973 for the PSO-ELM model and the R^2 value is equal to 0.96299 for the ELM model; the accuracy of the PSO-ELM surrogate model has increased.
2. The total pressure is improved from 140.6 Pa to 151 Pa after optimization using simulation experiment design and the PSO algorithm.
3. This paper applied the modified ELM model to the optimal design of the multi-blade centrifugal fan and verifies the feasibility of coupling the modified ELM model and PSO algorithm to the blade parametric design. It provides new ideas for the optimal design of multi-blade centrifugal fans and other centrifugal fans. The entire optimization design has certain engineering use value.

Author Contributions: Conceptualization, F.M. and W.M.; methodology, F.M.; software, F.M.; validation, L.W., W.M. and H.Z.; formal analysis, H.Z.; investigation, W.M.; resources, H.Z.; data curation, H.Z.; writing—original draft preparation, L.W.; writing—review and editing, F.M.; supervision, W.M.; project administration, W.M.; funding acquisition, F.M. All authors have read and agreed to the published version of the manuscript.

Funding: This paper was supported by the key science and technology research project of Henan Province (No. 212102210071), and was also supported by the Local Innovative and Research Teams Project of the Guang-dong Pearl River Talents Program (No. 2017BT01G167).

Data Availability Statement: No new data were created or analyzed in this study. Data sharing is not applicable to this article.

Conflicts of Interest: The authors declare no conflict of interest.

References

1. Xu, Z.; Liu, X.; Liu, Y.; Qin, W.; Xi, G. Flow Control Mechanism of Blade Tip Bionic Grooves and Their Influence on Aerodynamic Performance and Noise of Multi-Blade Centrifugal Fan. *Energies* **2022**, *15*, 3431. [CrossRef]
2. Ye, J.; Liu, W.; Duan, P.; Huang, X.; Shao, J.; Zhang, Y. Investigation of the Performance and Flow Behaviors of the Multi-blade Centrifugal Fan Based on the Computer Simulation Technology. *Wirel. Pers. Commun.* **2018**, *103*, 563–574. [CrossRef]
3. Li, Z.; Ye, X.; Wei, Y. Investigation on Vortex Characteristics of a Multi-Blade Centrifugal Fan near Volute Outlet Region. *Processes* **2020**, *8*, 1240. [CrossRef]
4. Wei, Y.; Wang, J.; Xu, J.; Wang, Z.; Luo, J.; Yang, H.; Zhu, Z.; Zhang, W. Effects of Inclined Volute Tongue Structure on the Internal Complex Flow and Aerodynamic Performance of the Multi-Blade Centrifugal Fan. *J. Appl. Fluid Mech.* **2022**, *15*, 901–916. [CrossRef]
5. Zhou, S.; Yang, K.; Zhang, W.; Zhang, K.; Wang, C.; Jin, W. Optimization of Multi-Blade Centrifugal Fan Blade Design for Ventilation and Air-Conditioning System Based on Disturbance CST Function. *Appl. Sci.* **2021**, *11*, 7784. [CrossRef]
6. Zhou, S.; Zhou, H.; Yang, K.; Dong, H.; Gao, Z. Research on blade design method of multi-blade centrifugal fan for building efficient ventilation based on Hicks-Henne function. *Sustain. Energy Technol. Assess.* **2021**, *43*, 100971. [CrossRef]
7. Zhou, S.; Dong, H.; Zhang, K.; Zhou, H.; Jin, W.; Wang, C. Optimal design of multi-blade centrifugal fan based on partial coherence analysis. *Proc. Inst. Mech. Eng. Part C J. Mech. Eng. Sci.* **2022**, *236*, 894–907. [CrossRef]
8. Le Hocine, A.E.B.; Poncet, S.; Fellouah, H. CFD modeling and optimization by metamodels of a squirrel cage fan using OpenFoam and Dakota: Ventilation applications. *Build. Environ.* **2021**, *205*, 108145. [CrossRef]
9. Le Hocine, A.E.B.; Poncet, S.; Fellouah, H. Optimization of a double-intake squirrel cage fan using OpenFoam and metamodels. *Int. J. Heat Fluid Flow* **2023**, *101*, 109129. [CrossRef]
10. Heo, M.-W.; Kim, J.-H.; Seo, T.-W.; Kim, K.-Y. Aerodynamic and aeroacoustic optimization for design of a forward-curved blades centrifugal fan. *Proc. Inst. Mech. Eng. Part A J. Power Energy* **2016**, *230*, 154–174. [CrossRef]
11. Le, T.-L.; Nghia, T.T.; Thong, H.D.; Son, M.H.K. Numerical study of aerodynamic performance and flow characteristics of a centrifugal blower. *Int. J. Intell. Unmanned Syst.* **2022**. *ahead-of-print*. [CrossRef]
12. Kim, K.-Y.; Seo, S.-J. Application of Numerical Optimization Technique to Design of Forward-Curved Blades Centrifugal Fan. *JSME Int. J. Ser. B* **2006**, *49*, 152–158. [CrossRef]
13. Zhao, D.; You, X.-Y. Inverse design optimisation of a novel range hood based on intelligent algorithms and computational fluid dynamics simulations. *Adv. Powder Technol.* **2020**, *31*, 730–745. [CrossRef]
14. Almasi, S.; Ghorani, M.M.; Haghighi, M.H.S.; Mirghavami, S.M.; Riasi, A. Optimization of a vacuum cleaner fan suction and shaft power using Kriging surrogate model and MIGA. *Proc. Inst. Mech. Eng. Part A J. Power Energy* **2022**, *236*, 519–537. [CrossRef]
15. Zhang, J.; Zangeneh, M. Multi-Point, Multi-Objective Optimisation of Centrifugal Fans by 3D Inverse Design Method. *Int. J. Turbomachinery, Propuls. Power* **2023**, *8*, 8. [CrossRef]
16. Lvo, M.; Damir, V. 3D shape 3D shape optimization of fan vanes for multiple operating regimes subject to efficiency and noiserelated excellence criteria and constraints. *Eng. Appl. Comput. Fluid Mech.* **2016**, *10*, 209–277.

17. Jung, U.-H.; Kim, J.-H.; Kim, J.-H.; Park, C.-H.; Jun, S.-O.; Choi, Y.-S. Optimum design of diffuser in a small high-speed centrifugal fan using CFD & DOE. *J. Mech. Sci. Technol.* **2016**, *30*, 1171–1184. [CrossRef]
18. Zoran, M.; Damir, V.; Ivo, M. Multi-Regime Shape Optimization of Fan Vanes for Energy Conversion Efficiency using CFD, 3D Optical Scanning and Parameterization. *Eng. Appl. Comput. Fluid Mech.* **2014**, *8*, 407–421.
19. Xu, H.; Wei, W.; He, H.; Yang, X. Fluid Analysis and Structure Optimization of Impeller Based on Surrogate Model. *Comput. Model. Eng. Sci.* **2022**, *132*, 173–199. [CrossRef]
20. Xu, Z.; Kong, F.; Zhang, K.; Zhu, H.; Wang, J.; Qiu, N. Optimization of the impeller for hydraulic performance improvement of a high-speed magnetic drive pump. *Adv. Mech. Eng.* **2022**, *14*, 16878132221104576. [CrossRef]
21. Tuchler, S.; Copeland, C. Numerical Optimization of a Micro-wave Rotor Turbine Using a Quasi-two-dimensional CFD Model and a Hybrid Algorithm. *Shock Waves* **2021**, *31*, 271–300. [CrossRef]
22. Hosseinimaab, S.; Tousi, A. Optimizing the performance of a single-shaft micro gas turbine engine by modifying its centrifugal compressor design. *Energy Convers. Manag.* **2022**, *271*, 116245. [CrossRef]
23. Chen, J.; Wu, G. Kriging-assisted design optimization of the impeller geometry for an automotive torque converter. *Struct. Multidiscip. Optim.* **2018**, *57*, 2503–2514. [CrossRef]
24. Zheng, S.; Shao, Z.; Liu, J.; Chen, J.; Li, Y.; Chai, M. Flow characteristics of a centrifugal fan with optimized edge curvatures of an impeller front disk. *Proc. Inst. Mech. Eng. Part A J. Power Energy* **2023**. [CrossRef]
25. Engin, T. Study of tip clearance effects in centrifugal fans with unshrouded impellers using computational fluid dynamics. *Proc. Inst. Mech. Eng. Part A J. Power Energy* **2006**, *220*, 599–610. [CrossRef]
26. Jeon, W.-H.; Lee, D.-J.; Rhee, H. An Application of the Acoustic Similarity Law to the Numerical Analysis of Centrifugal Fan Noise. *JSME Int. J. Ser. C* **2004**, *47*, 845–851. [CrossRef]
27. Rebecca, S.; Martin, B. Validation of the Lattice Boltzmann Method for Simulation of Aerodynamics and Aeroacoustics in a Centrifugal Fan. *Acoustics* **2020**, *2*, 735–752.
28. Kruyt, N.P.; Westra, R.W. On the inverse problem of blade design for centrifugal pumps and fans. *Inverse Probl.* **2014**, *30*, 065003. [CrossRef]
29. Piwowarski, M.; Jakowski, D. Areas of Fan Research—A Review of the Literature in Terms of Improving Operating Efficiency and Reducing Noise Emissions. *Energies* **2023**, *16*, 1042. [CrossRef]
30. Kudela, J.; Matousek, R. Recent advances and applications of surrogate models for finite element method computations: A review. *Soft Comput.* **2022**, *26*, 13709–13733. [CrossRef]
31. Zhang, Z.; Qiu, W.; Zhang, G.; Liu, D.; Wang, P. Progress in applications of shockwave induced by short pulsed laser on surface processing. *Opt. Laser Technol.* **2023**, *157*, 108760. [CrossRef]
32. Ming, W.; Sun, P.; Zhang, Z.; Qiu, W.; Du, J.; Li, X.; Zhang, Y.; Zhang, G.; Liu, K.; Wang, Y.; et al. A systematic review of machine learning methods applied to fuel cells in performance evaluation, durability prediction, and application monitoring. *Int. J. Hydrogen Energy* **2023**, *48*, 5197–5228. [CrossRef]
33. Wang, P.; Zhang, Z.; Hao, B.; Wei, S.; Huang, Y.; Zhang, G. Investigation on heat transfer and ablation mechanism of CFRP by different laser scanning directions. *Compos. Part B Eng.* **2023**, *262*, 110827. [CrossRef]
34. Mustafa, A.; Ahmad, T.; Rana, H. Swarm-Based Extreme Learning Machine Models for Global Optimization. *Comput. Mater. Contin.* **2022**, *70*, 6339–6362.
35. Kennedy, J.; Eberhart, R. Particle swarm optimization. In Proceedings of the ICNN'95-International Conference on Neural Networks, Perth, WA, Australia, 27 November–1 December 1995; Volume 4, pp. 1942–1948.
36. Meng, F.; Zhang, Z.; Su, X.; Li, L. Centrifugal fan Blade Optimization Based on Kriging Surrogate model. *Chin. Hydraul. Pneum.* **2023**, *47*, 85–91.
37. Meng, F.; Gong, X.; Du, W.; Wang, L.; Zhao, F.; Li, L. Vibration performance prediction and reliability analysis for rolling bearing. *J. Vibroeng.* **2021**, *23*, 327–346. [CrossRef]

metals

MDPI

Article

Study of Heat Transfer Strategy of Metal Heating/Conduction Plates for Energy Efficiency of Large-Sized Automotive Glass Molding Process

Yanyan Chen [1], Shengfei Zhang [2], Shunchang Hu [2], Yangjing Zhao [2], Guojun Zhang [3], Yang Cao [2] and Wuyi Ming [2,*]

1 College of Mechanical Engineering, Yellow River Conservancy Technical Institute, Kaifeng 475000, China; chyy1106@126.com

2 Mechanical and Electrical Engineering Institute, Zhengzhou University of Light Industry, Zhengzhou 450002, China; zhangshengfei2021@163.com (S.Z.); hushunchang2022@gmail.com (S.H.); zhaoyj0529@163.com (Y.Z.); caoyang8786@zzuli.edu.cn (Y.C.)

3 Guangdong Provincial Key Laboratory of Digital Manufacturing Equipment, Guangdong HUST Industrial Technology Research Institute, Dongguan 523808, China; guojun_zhang_stu@163.com

* Correspondence: mingwuyi@zzuli.edu.cn

Abstract: In recent years, as an important functional material, glass has been widely used in architecture, electronics, optics, and other fields. As an emerging glass processing technology, the glass molding process (GMP) has received widespread attention and research in recent years. In this paper, we study the modeling and analysis of different heat transfer strategies for the energy efficiency of large-sized automotive instrument glass. The heat transfer model of the metal heating plate–conducting plate mold is established, the thermal energy efficiency in the forming process of large automobile glass is analyzed, and the energy efficiency of the mold in the heating stage is compared. The energy consumption per piece generated by the GMP heating device is reduced from 4865.2 to 4668.5 kJ, a reduction of 4.04%. By optimizing the heat flow density, the energy consumption per piece generated by the GMP heating device was reduced from 4865.2 to 4625.5 kJ, a reduction of 4.92%, meeting the sustainable manufacturing requirements.

Keywords: large-sized glass components; molding; metal heat conduction; energy analysis; green manufacturing

1. Introduction

The glass molding process (GMP), as a substitute for the traditional manufacturing process of glass materials, has the advantages of high forming accuracy, a short production cycle, low cost, and no pollution. GMP techniques have become revolutionary techniques in which glass preforms are placed into a mold and subjected to specific temperature and pressure loading conditions to produce glass parts that meet specific requirements [1–3]. The GMP process consists of four main steps: heating, pressing, annealing, and cooling [4,5]. It not only effectively replaces traditional glass processing methods but has also found wide application in the production of optical glass. Compared to traditional techniques, GMP has advantages such as high precision, high efficiency, and a short cycle time. As conventional techniques in glass processing lead to the production of glass components with various defects [6,7] and extremely high energy consumption, the new GMP stands out for its high precision, efficiency, and short cycle time.

The high thermal coupling and ultra-high temperature (700 °C) of GMP technology has hindered its development [8–10]. For example, when using GMP technology to process certain glass components, in order to meet the requirements, the heating system must provide a maximum power of 100 KW to ensure stable temperature control of the mold and glass [11]. Additionally, the consumption of nitrogen gas and the waste rate of graphite molds can lead to increased energy consumption in glass molding pressing [12].

Studies have shown that the configuration of the heating system and the mold losses are the primary factors influencing the energy efficiency and sustainability of glass molding pressing equipment [13]. To improve heat transfer efficiencies, Li et al. [14] proposed a low-energy consumption, short-cycle, and high-precision temperature control heating system for hot press forming to improve temperature control accuracy and process quality. Ming et al. [15–17] simulated and tested the energy use of different heating methods for smartphone casing. Under optimal process parametric conditions, the model decreased the expected energy consumption from 614 kJ to 594.4 kJ, a decrease of 3.19%. At the same time, the glass forming time was shortened from 148.8 s to 139.2 s, with a rate of reduction of 6.45%. Yang et al. [2] established a finite element model for typical surface forming defects, effectively improving the production efficiency and glass quality. Zhang's team [18,19] attempted to analyze the surface integrity of ultra-thin glass using laser and thermal bending techniques. With the results of simulation and experimental studies, the glass forming temperature had a significant effect on the fracture rate, microbubble, and water ripple density. The fracture rate decreased from 67% to 0% at the optimum molding temperature and pressure.

There has been a new demand for oversized-shaped glass in the current market, but the existing technology is not mature. Many studies have focused on the hot bending forming of small-sized glass components, while there has been relatively little research on the hot bending forming of large-sized, shaped glass components. Processing large-sized irregular glass components has problems, such as low yield, high processing energy consumption, and unsustainable manufacturing [20]. At the same time, irregularly sized glass has the characteristics of irregularity and an excessive length–thickness ratio (1430 mm: 1 mm), making it difficult to process. In order to improve the processing yield and quality, reduce the energy consumption, and carry out sustainable green manufacturing of large-scale special-shaped glass, this paper models and analyzes the forming process of large-scale dashboard glass using different heating methods based on the concept of green manufacturing. By establishing a numerical heat transfer model, the effects of energy flow, energy consumption, and the heating strategy of the GMP on performance were studied. In order to better evaluate the performance of the mold structure and its parameters, various simulation experiments were used in this paper to minimize the energy consumption of the GMP.

2. Heat Transfer Model

2.1. Heat Conduction Model and Its Initial Parameters

The thermal equilibrium problem between the glass and the mold is an important element affecting glass forming quality [21]. By analyzing the thermodynamic behavior of the metal heating plates, the metal heat conductors, and a mold, a thermodynamic simulation that can be used for glass molding pressing production was constructed.

Thermodynamic simulations that can be used for GMP production were constructed by analyzing the thermodynamic behavior of metal-heated plates, conductors, and dielectrics. According to Figure 1a,b, the size of the metal heating conductor was 1600 mm × 300 mm× 80 mm, and the size of the metal heating plate was 1600 × 300 × 50 mm. The diameters of the holes on the upper and lower metal heating plates were 28 mm and 25 mm, respectively. Figure 1c clearly shows the three-dimensional structures of the metal heating plate, conductor, and mold, as well as their main dimensions. In this model, the influence of the glass cover plate located inside the mold was ignored. Fifteen 1.0 kW·h heating pipes were installed at the top of the mold, while fourteen 1.2 kW·h heating pipes were installed at the bottom.

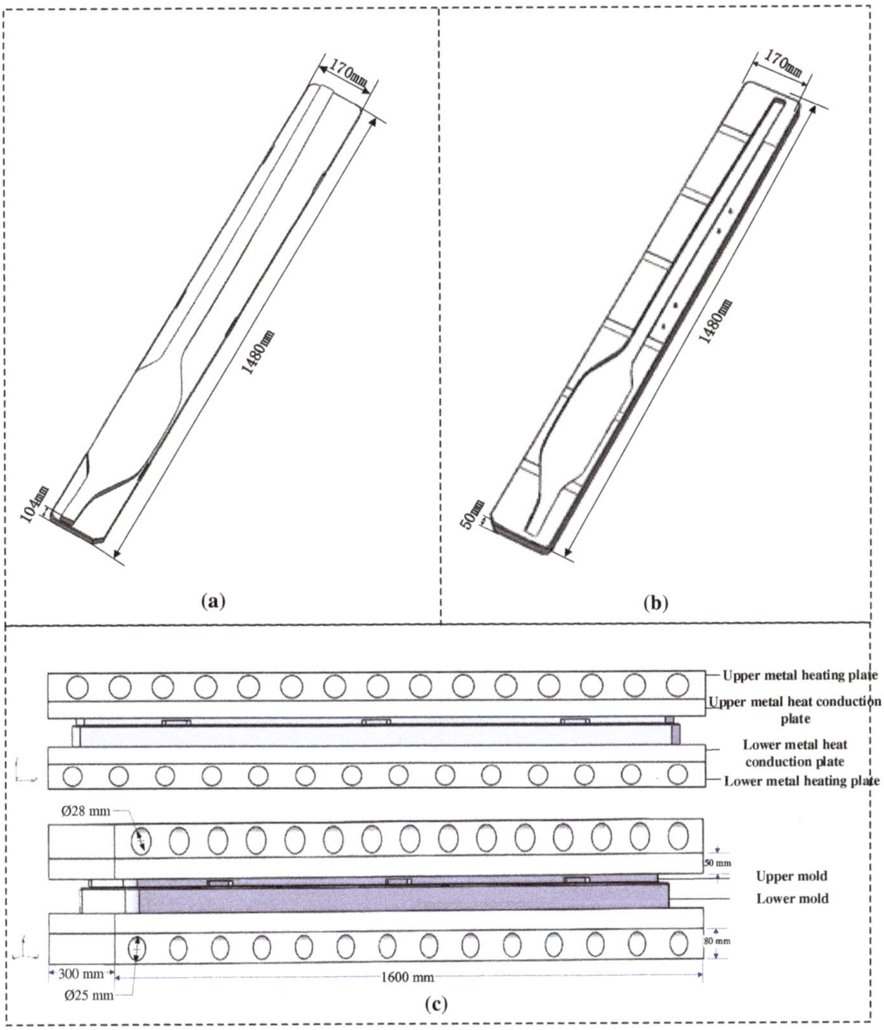

Figure 1. Model of metal heat conduction plates for large-sized automotive instrument glass molding process; (**a**) upper mold, (**b**) lower mold, (**c**) three-dimensional model and dimensions of the metal heating plate and heat-conducting plate (heating conductor).

The heating tube model will not be repeated here, assuming that each heating tube fully absorbs the heat and that it is generated without any loss. As a result, the heating rate of the heating tube can be replaced by the heat flux density, yielding a more efficient heating system. Three different materials—graphite as the mold, tungsten carbide steel (WC) as the heat conductor, and nickel–chromium austenitic stainless steel (SUS 310S) as the heating plate—were used in this study to accomplish this purpose. High-quality austenitic stainless steel SUS 310S was manufactured in Japan. It has excellent corrosion resistance and high-temperature resistance, and could meet the needs of various application scenarios. WC steel is widely used in molds, wear-resistant parts, rolling mills, and other fields, as shown in Table 1.

Table 1. Thermal and mechanical properties of SUS310S and WC materials [2].

Materials	SUS310S	WC
Young's modulus E (GPa)	193	570
Poisson's ratio υ	0.3	0.22
Density $\rho(g/cm^3)$	7.9	14.65
Heat conductivity K (W/m °C)	18.5	63
Specific heat capacity C_P (J/kg °C)	500	314
Coefficient of heat expansion (°C)	18.2×10^{-6}	4.9×10^{-6}

2.2. Boundary Conditions

Numerical modeling and simulation techniques are widely used in the field of machining, and our team has studied them on the basis of MSC Marc and published articles about these problems [2,12,15]. The 3D simulation in this study was implemented by mature software. Figure 2a shows a 3D finite element mesh model of the glass molding process, and Figure 2b displays the cross-sectional view of the model. The thermal coupling analysis model was used to simulate the heating course of the metal heating plates, metal heat conductors, and mold. The tetrahedron was selected as the target family volume, and Patran was used for meshing and modeling. Mesh refinement was performed for the contact surfaces of the metal heating plate, heating tubes, and the inner surface of the mold. Finally, the meshes of the molds (including the upper and lower), heat tubes, and metal heat conductors were accurately refined to 116,851, 23,471, and 105,923 units, respectively. According to the data in Table 2, the heat transfer efficiency can be improved by adjusting the contact area between the metal heating plates, metal heat conductors (heat conducting plates), mold, and nitrogen.

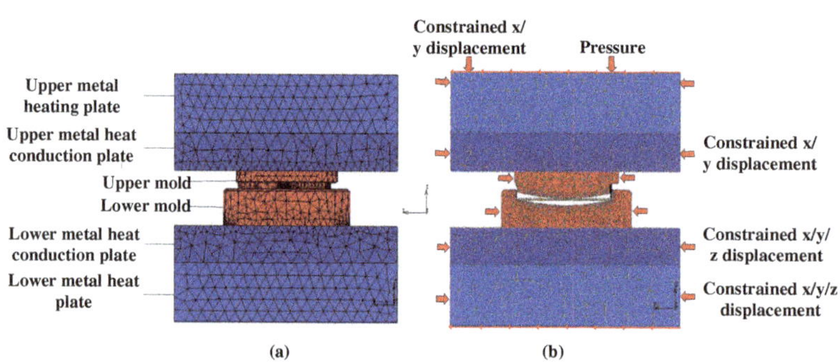

Figure 2. Three-dimensional finite element (FEM) mesh model (using MSC Marc) and its cross-sectional view of GMP. (**a**) 3D finite element mesh model of the glass molding process, (**b**) the cross-sectional view of the model.

Table 2. Constrained displacement, loading, and initial temperature for the FEM model.

FEM Model	Constrained Displacement	Loading (MPa)	Initial Temperature (°C)
Upper heating plate	x/y	0.3	780
Upper heat conduction plate	x/y	-	780
Mold	x/y	-	20
Lower heat conduction plate	x/y/z	-	780
Lower heating plate	x/y/z	-	780

By applying the viscous–sliding friction model [22], the transition from viscous to sliding friction during the thermal expansion and deformation of an object can be effectively simulated. This process can be expressed by Equation (1), as shown in Figure 3 [22].

$$F_t \leq \mu F_n, Static; F_t \leq \alpha \mu F_n, Kinetic \tag{1}$$

where F_t is the shear stress; μ is the coefficient of friction (0.1); F_n is the normal force; α is the coefficient of friction with a default value of 1.05. The energy balance equation allows for the temperature distribution profiles of the mold, the metal heat conductor, and the metal heating plate. The equations are specified as follows [23,24]:

$$\rho_m C_{pm}\left(\frac{\partial T_1}{\partial t}\right) = k_m\left(\frac{\partial^2 T_1}{\partial x^2}\right) + k_m\left(\frac{\partial^2 T_1}{\partial y^2}\right) + k_m\left(\frac{\partial^2 T_1}{\partial z^2}\right) \tag{2}$$

$$\rho_c C_{pc}\left(\frac{\partial T_2}{\partial t}\right) = k_c\left(\frac{\partial^2 T_2}{\partial x^2}\right) + k_c\left(\frac{\partial^2 T_2}{\partial y^2}\right) + k_c\left(\frac{\partial^2 T_2}{\partial z^2}\right) \tag{3}$$

$$\rho_h C_{ph}\left(\frac{\partial T_2}{\partial t}\right) = k_h\left(\frac{\partial^2 T_3}{\partial x^2}\right) + k_h\left(\frac{\partial^2 T_3}{\partial y^2}\right) + k_h\left(\frac{\partial^2 T_3}{\partial z^2}\right) \tag{4}$$

In the above Equations, ρ_h is the metal heating plate; ρ_c is the metal heat-conducting plate; ρ_m is the density (kg/m^3) of the mold; k, C_p are the thermal conductivity (W/m °C) and the specific heat capacity (J/kg °C), respectively; T_1, T_2, and T_3 are the temperature (°C) of the mold, metal heat conductor, and metal heating plate, respectively. The contact thermal conductivity value of the conductor with the heating plate was 2500 W/m^2 °C, and the contact thermal conductivity value of the conductor with the mold was 2800 W/m^2 °C.

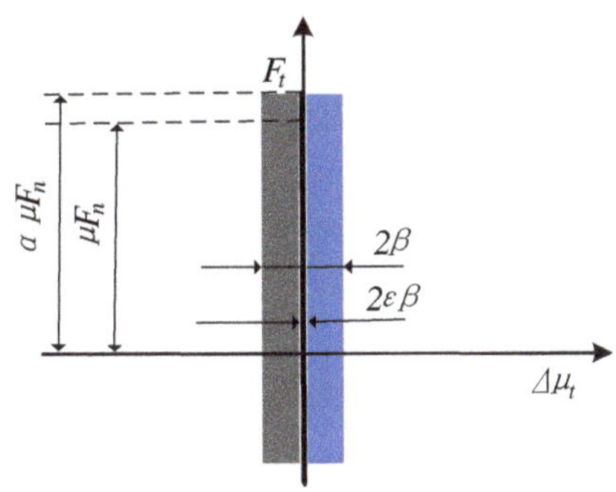

Figure 3. GMP viscous–sliding friction model.

2.3. Calculation of Production Energy Consumption

Figure 4 depicts the energy efficiency calculation flow of the GMP system, which contains the heat transfer between the metal heating plate and the heat conductor, as well as the heating tubes. To better simulate the actual situation, the simulation model can be used with multiple temperature measurement points. T_1 and $T_1{}'$ indicate the average temperatures of the initial and final status of the heating stage, respectively. Then, Equation (5) can be adopted to estimate the effect of heat transfer on the energy inside the metal heating plate, so that the heat loss ($E_1{}''$) can be estimated more accurately [22].

$$E_1{}'' = E_1 - E_1{}' = m_1 c_1 (T_1 - T_1{}') \tag{5}$$

where E_1 is the initial energy of the metal heating plate; $E_1{}'$ is the final energy of the metal heating plate; m represents the mass; and c represents the specific heat capacity. Similarly, the internal energy loss of the metal heat conductor ($E_2{}''$) is obtained from Equation (6) [22]:

$$E_2{}'' = E_2 - E_2{}' = m_2 c_2 (T_2 - T_2{}') \tag{6}$$

Additionally, the energy produced by the heating tubes is determined by Equation (7) [22]:

$$E_3 = n\overline{q}_f St \tag{7}$$

where t, n, and S are the heating time, the number of heating tubes, and the surface area of heating tubes (mm^2), respectively. Therefore, Equation (8) can be used to determine the energy produced by the heating device [22].

$$E = E_1{}'' + E_2{}'' + E_3 \tag{8}$$

The energy absorbed by the mold is depicted by Equation (9) [22]:

$$E' = E_0{}' - E_0 = m_3 c_3 (T_3{}' - T_3) + m_4 c_3 (T_4{}' - T_4) \tag{9}$$

where E_0 is the initial energy of the mold and $E_0{}'$ is the final energy of the mold; T_3 is the upper mold's initial temperature; T_4 is the lower mold's initial temperature; $T_3{}'$ is the upper mold's final temperature; $T_4{}'$ is the lower mold's final temperature; c_3 is the mold's specific heat capacity; m_3 represents the upper mold's mass; and m_4 represents the lower mold's mass.

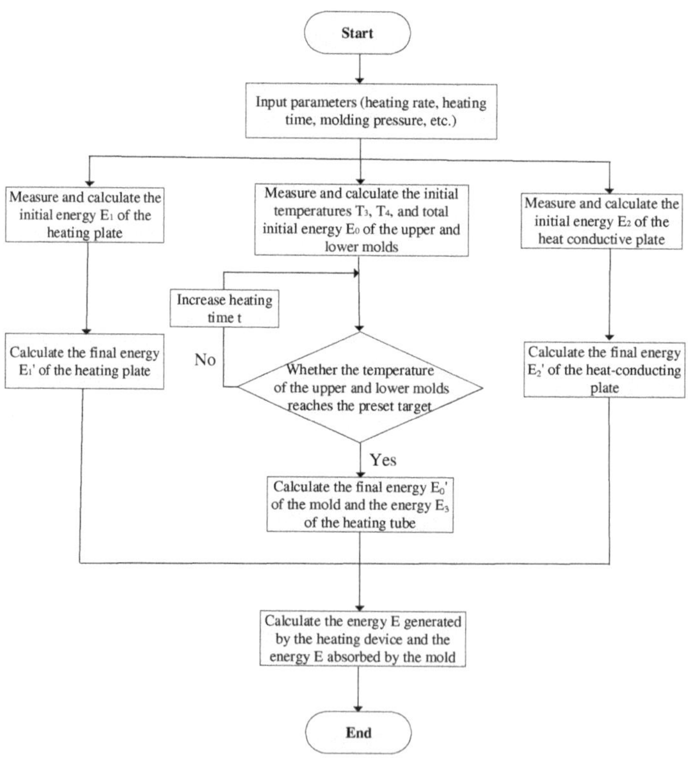

Figure 4. Energy consumption calculation process of GMP system during heating.

3. Results and Analysis

3.1. Impact of Heating Rate on Energy Efficiency

The GMP is affected by the heating rate during the heating phase, which in turn has an effect on energy consumption. In this study, four different strategies for the heating rate were used to investigate this issue. Figure 5 shows the four different heating methods, where the heating rate of the original heating strategy I (0–160 s) was 35 mJ/(mm^2·s), which effectively improved the forming effect of glass. Throughout the heating process, the value of heating rate strategy II remained at 30 mJ/(mm^2·s), the value of strategy III remained at 40 mJ/(mm^2·s), and finally, the value of strategy IV remained at 45 mJ/(mm^2·s).

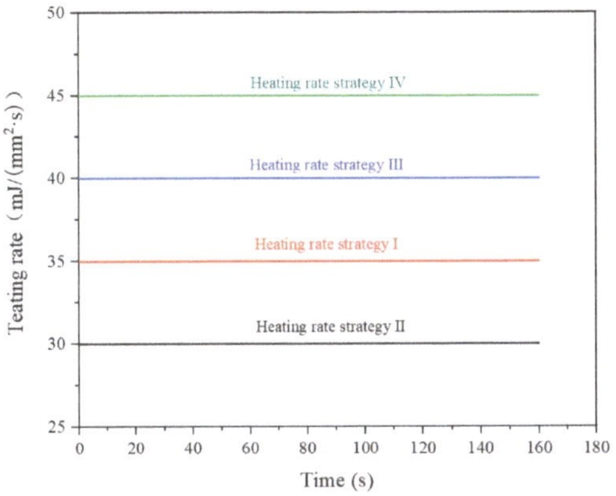

Figure 5. Schematic diagram of different heating rate strategies during glass molding process.

We performed a temperature simulation and analysis on the glass molds under four different heating rate strategies. As shown in Figure 6, observations were made at four points in the middle of the inner surface of the upper/lower mold, and the temperatures at points A, B, C, and D were compared under different strategies. The results of the simulation for the central temperature inside the glass mold are shown in Figure 7. As can be seen in Figure 7, the temperature of the glass mold also increased with the increase in the heating rate, and the temperature inside the mold increased from point A to D. Comparing the temperature changes of the two molds at 160 s, it is clear that the temperature difference of the upper mold was smaller than that of the lower mold. Taking points A_1 and A_2 as examples, the final temperatures of point A_1 under the conditions of increasing the heating rate from low to high were 695.2 °C, 698.4 °C, 702.8 °C, and 706.3 °C, with a maximum temperature difference of 8 °C (in Figure 8). The final temperatures of point A_2 at 160 s under the conditions of increasing the heating rate from low to high were 649.8 °C, 665.4 °C, 675.8 °C, and 677.2 °C, with a maximum temperature difference of 28 °C (in Figure 8). It is evident that the lower mold was more affected than the upper mold under the different heating rate strategies. This was due to the fact that the larger mold was heated, and the upper mold contained an additional heating tube than the lower mold, resulting in a more uniform finish.

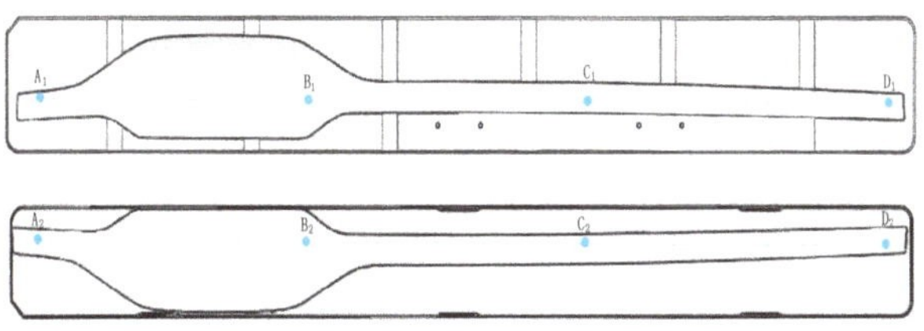

Figure 6. Schematic diagram of temperature measurement points in the upper and lower mold cavities.

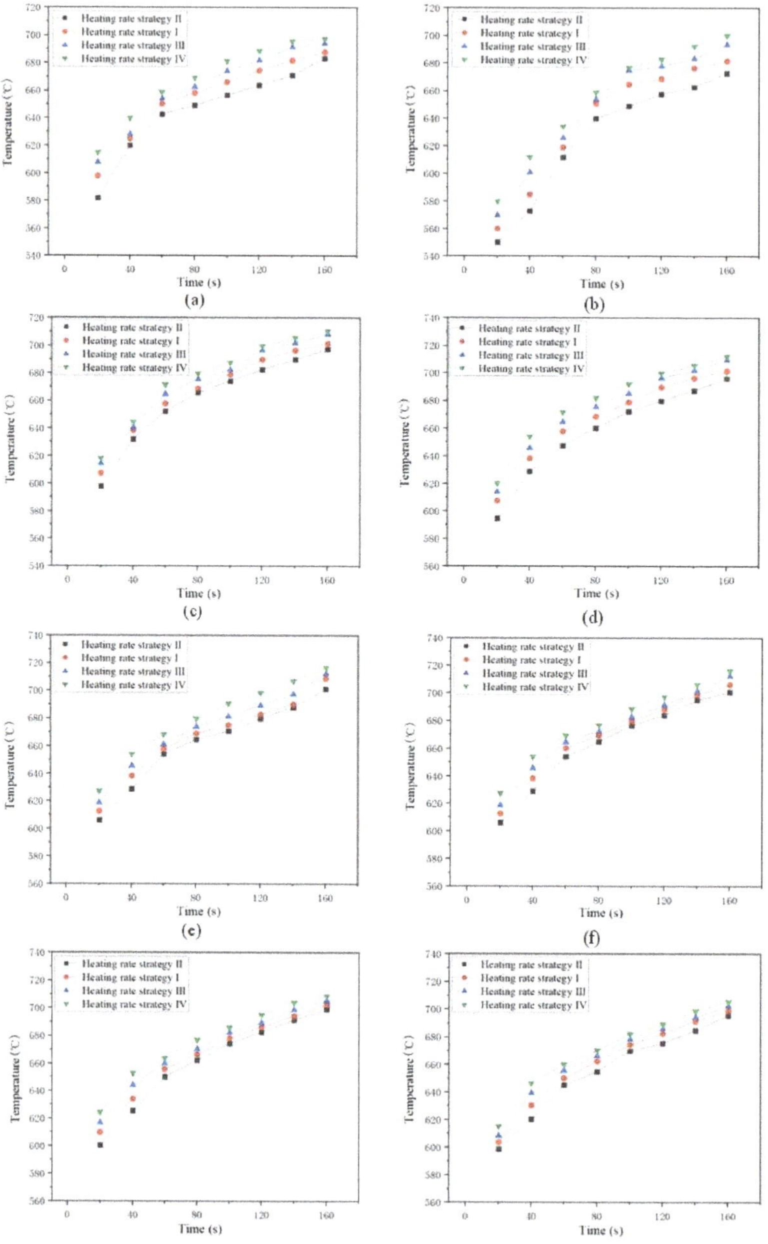

Figure 7. Simulation results of mold temperature under four different heating rate strategies. Upper and lower mold temperature measurement points: **(a)** A_1; **(b)** A_2; **(c)** B_1; **(d)** B_2; **(e)** C_1; **(f)** C_2; **(g)** D_1; **(h)** D_2.

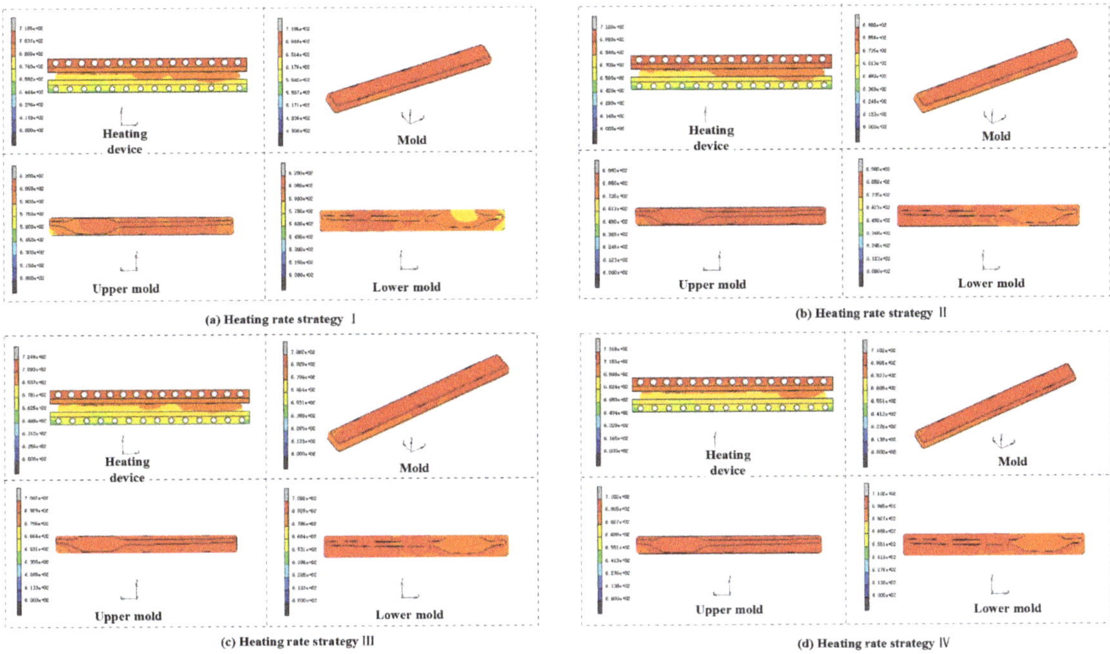

Figure 8. Temperature distribution in the final stage of heating (t = 160 s) using four different heating rates of the heating devices and molds.

Figure 8 shows the temperature distribution in the final stage of heating (t = 160 s) using four different heating rates for the heating devices and molds. It can be observed that their temperature variation trends were also very similar. As the heating rate continued to increase, the maximum temperature of the heating equipment and mold also rose. Conversely, if a slower heating rate is used, the surface temperature will decrease as the heating plate and its transmission parts cool down. After comprehensive consideration, adopting heating strategy IV (45 mJ/(mm^2·s)) can significantly shorten the processing time and effectively promote energy efficiency.

By observing Figure 9, it can be seen that there was a significant difference in the energy output between the heating tubes and heating devices in the four different heating strategies. As shown in Figure 9, the heat output values of the heating tubes for heating rate strategies I~IV were 3287.1, 2886.5, 3316.4, and 3545.6 kJ, respectively, while the energy values produced by the heating devices for heating rate strategies I~IV were 4865.2, 4971.3, 4751.6, and 4668.5 kJ, respectively. It was found that the heating rate strategies with lower energy output from the heating tubes had higher energy output from the heating devices, while the heating rate strategies with higher energy output from the heating tubes had lower energy output from the heating devices. This is because using a fast heating rate strategy can shorten the heating time and reduce energy loss. In addition, using heating rate strategy IV reduced the energy output of the heating device by 4.04% compared to heating rate strategy I, and the heating time was reduced by 7.06%. As the heating rate increases, negative effects may occur, such as excessive internal material stress and difficulty controlling the dimensional accuracy. To find a suitable balance between the processing performance and energy consumption, heating rate strategy IV was adopted in this study to reduce the energy consumption. In addition, under these process parameters, the nitrogen consumption was also reduced due to the shortened heating time.

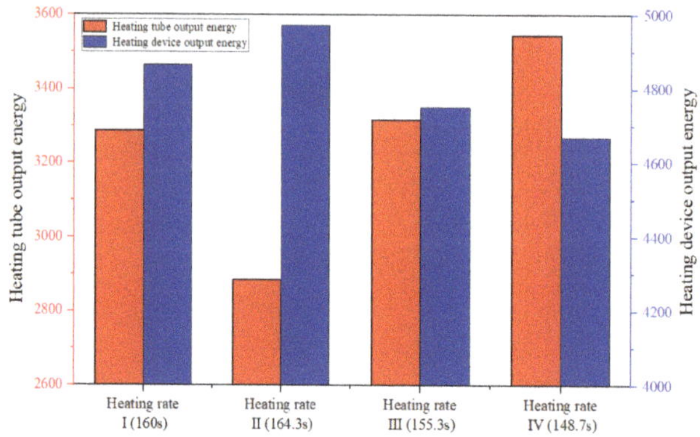

Figure 9. Comparison of output energy of heating tube and heating devices under four different heating rate strategies.

3.2. Impact of Heat Flux Density on Energy Efficacy

In the heating stage, the GMP is mainly heated by heating tubes. To explore the impact of the internal heat flux density of the heating tubes on the energy efficiency of the device, five different heat flow configuration strategies were designed, as shown in Figure 10. Fifteen and fourteen heating tubes were placed inside the upper and lower metal heating plates, respectively, and arranged into five groups. Three conductive pipes were installed in each group of the upper metal heating plate, while the number of conductive pipes in the lower metal heating plate was arranged as 3-3-2-2-3. In Figure 10, it can be seen that the five different strategies of heating heat flux density were applied to the glass molding process. In this section, the effects of five different strategies for the heat flux configuration of the heating tube on the temperature distribution of the heating device and the mold were studied. The impact of different configurations on energy consumption and the optimal energy-saving strategy were determined.

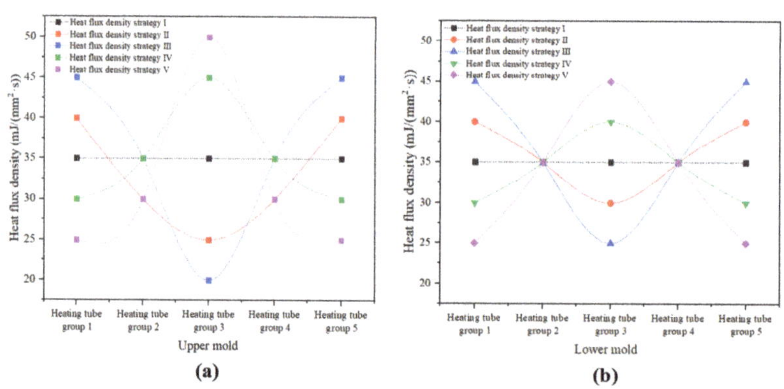

Figure 10. Schematic diagram of five different heat flux density strategies during the glass molding process; (**a**) upper mold heat flux density distribution strategies, (**b**) lower mold heat flux density distribution strategies.

Five different heat flux density strategies were performed in simulation experiments, and the temperature at multiple observation points on the mold surface was observed. The simulation results of the different configurations of heat flux density in the final heating stage (t = 160 s) are shown in Figure 11. Based on the data in Figures 8 and 11, whether using uniform or non-uniform heat flux density, the temperature variation in the heating equipment and the mold it produced was small. When using uniform heat flux density, the lowest temperature can be set to 718.5 °C, while when using non-uniform heat flux density (i.e., heat flux density strategy III), the lowest temperature can be set to 729.8 °C. When employing a strategy of higher heat flow density on the sides and lower heat flow density in the middle, the maximum temperature will be higher compared to using uniform heat flow density. Conversely, adopting the opposite approach will result in a lower temperature. However, the temperature distribution is similar to that of different strategies. Adopting the strategy can achieve balanced heat transfer but may reduce the temperature of the metal heating and conductive plates while increasing the mold temperature. Therefore, to improve production efficiency, a strategy with higher heat flow density on both sides and lower heat flow density in the middle can be chosen when selecting the heat flux density.

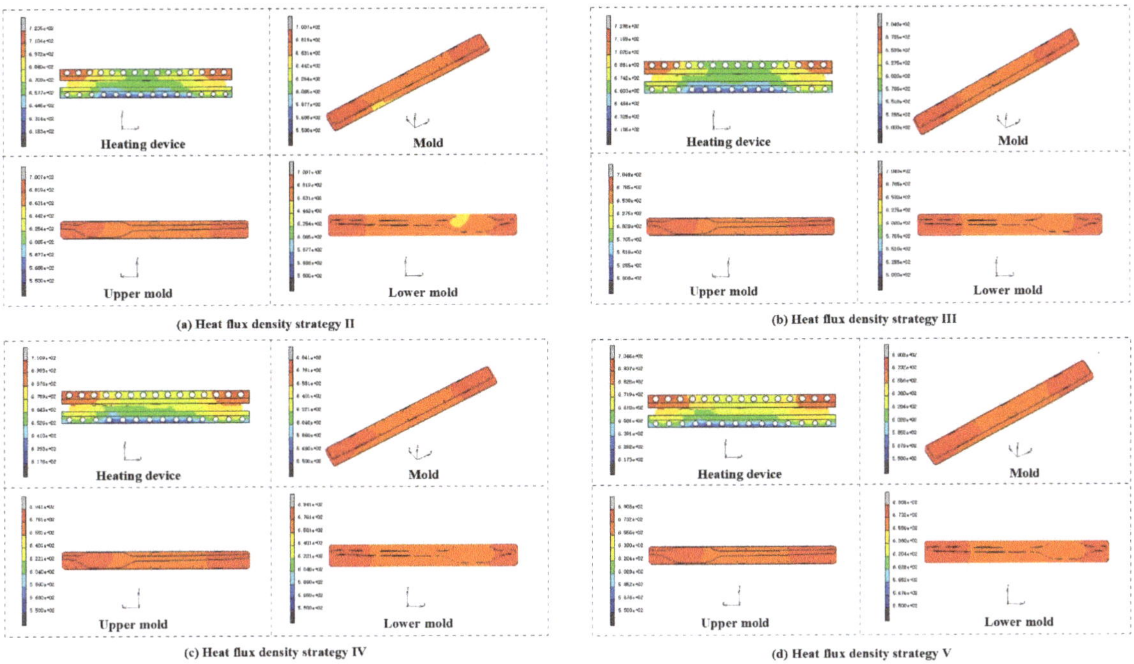

Figure 11. Temperature distribution of heating device and mold under different strategies of heat flux density (t = 160 s).

The use of different heating tube groups with different heat flux density strategies can cause changes in the overall temperature of the mold. According to Figure 12, thirty-three temperature monitoring points (A1-K3) were selected in the lower mold, and the temperature distribution using heat flux density strategies I–III is shown in Figure 13. In Figure 13a, it can be observed that the temperature change inside the lower mold using heat flux density strategy I was not significant, with a temperature change range of up to 5 °C. However, when strategy II was adopted, the highest temperature in the lower mold appeared in the areas around the left and right sides of the mold (A_1-A_3-B_3-B_1 and K_1-K_3-J_3-J_1), and the lowest temperature appeared in the central area of the inner cavity (D_1-D_3-F_3-F_1), as shown in Figure 13b. The temperature distribution using heat flux density strategy III was similar to that of strategy II, and the maximum temperature difference of strategy III was smaller than that of strategy II. A large temperature difference can cause a decrease in the product processing performance (in Figure 13c). Therefore, the final molding performance was the best when using strategy III.

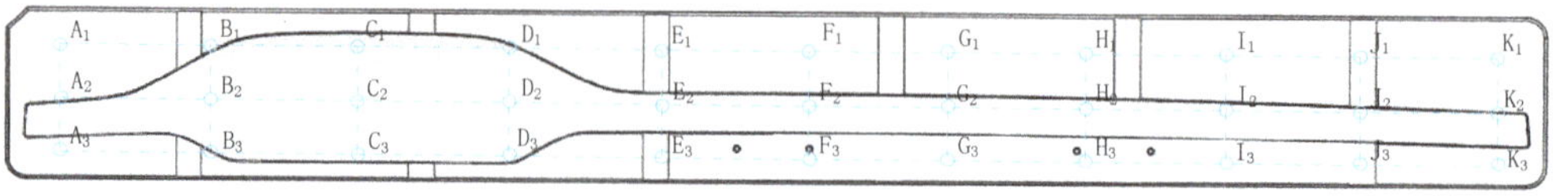

Figure 12. Schematic diagram of the selection of observation points for the lower mold temperature (A1–K3 are the temperature detection points).

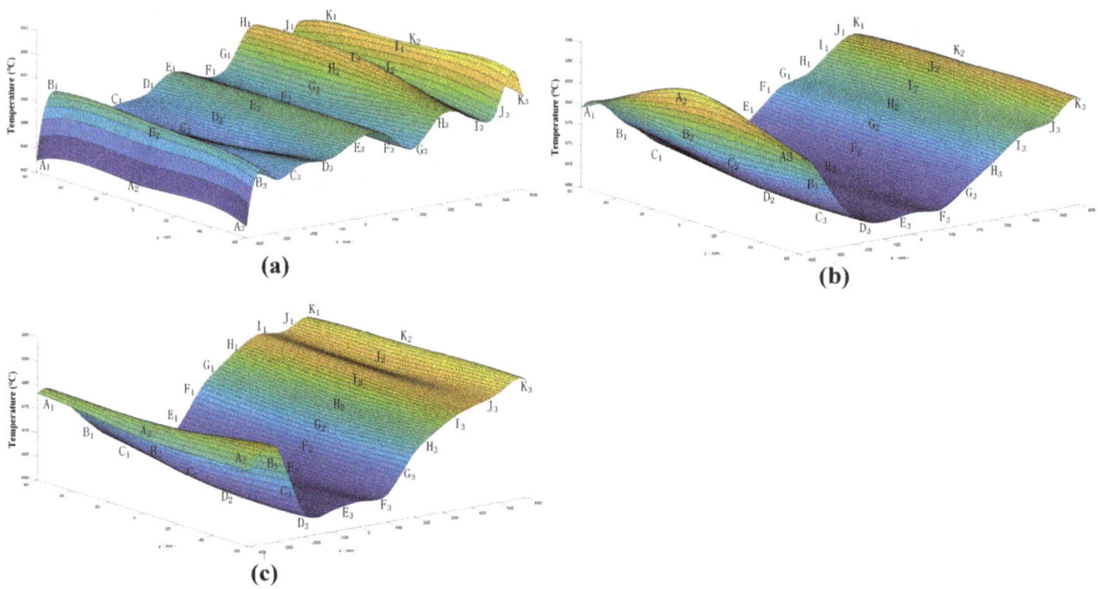

Figure 13. Temperature distribution on the inner surface of the mold under three different strategies of heat flux density; (**a**) heat flux density strategy I, (**b**) heat flux density strategy II, (**c**) heat flux density strategy III.

As shown in Figure 14, by comparing three different strategies, it was observed that in strategies I, II, and III, the output power of the heating tubes reached 3287.1, 3258.6, and 3141.5 kJ, respectively. In the same three strategies, it consumed 4865.2, 4762.3, and 4625.5 kJ for the output power of the heating device. Therefore, it can be concluded that the effectiveness of all three strategies improved. After applying strategy III of non-uniform heat flux, the output energy of the heating tubes and the heating device was lower, and the maximum temperatures of the heating device and the mold were both increased. This is because a non-uniform strategy in heat flux density was used, which shortened the heating time and reduced energy loss. Compared to strategy I, the output energy using strategies II and III was reduced by 4.04% and 4.92%, respectively. In comparison, strategy I completed heating within 160 s, while strategy II completed it in 156.2 s, and strategy III completed it in 150.3 s. By comparison, the heating efficiency using strategies II and III was significantly improved, with the former's reduction reaching 2.37%, and the latter's reduction reaching 6.06%.

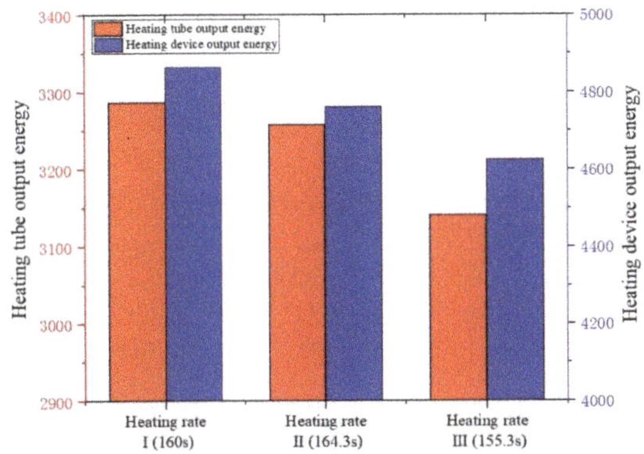

Figure 14. Comparison of the output energy of the heating tubes and the heating device under three different heat flux density strategies.

4. Discussion

As a major contributor to the world economy, China is the world's largest producer and consumer of coal and, naturally, the world's largest emitter of carbon [25–28]. Therefore, improving the efficiency of energy use and reducing carbon emissions are necessary parts of our production and processing [29–31]. During the process of forming large-scale automotive instrument glass, the heating apparatus requires electrical energy, the main source of which is coal used for electricity generation. The nitrogen required to cool the system also consumes a lot of electrical energy, which has an overall negative impact on the environment. Figure 15 shows the percentage of CO_2 emissions from the automotive glass molding process. In the whole process, direct CO_2 emissions account for about 92–97%, and indirect CO_2 emissions account for about 3–6%. In the heating process, heating the mold and manufacturing nitrogen are the main components of direct CO_2 emissions, accounting for 18–20% and 74–77%, respectively. In addition, the service life of graphite molds is significantly shortened in high-temperature environments; therefore, both graphite molds and glass raw materials are consumables, and the indirect emissions of CO_2 are mainly from the process of manufacturing graphite molds (1–3%) and glass raw materials (2–3%). It was found that the direct emissions of CO_2 are more significant than their indirect emissions.

Currently, coal-fired power generation is widely used in developing countries, such as China, due to its irreplaceability. According to the analysis of the heat transfer results for the different strategies, heating rate strategy IV and heat flux strategy III were the optimal strategies for the heat transfer model. The thermal energy output values of the heating device for these two strategies were 4668.5 kJ and 4625.5 kJ, respectively, compared to 4865.2 kJ for the original heat transfer strategy. It can be concluded that the CO_2 emissions of heating rate strategy IV, heat flow density strategy III, and the original heat transfer strategy were 1.0374 kg, 1.0278 kg, and 1.0811 kg, respectively. This means that heat transfer strategy IV reduced carbon emissions by 4.04% compared to the original strategy, and heat transfer density strategy III reduced carbon emissions by 4.92%. Thereafter, the heat transfer model with heat flow density strategy III was the best strategy to reduce energy consumption and improve energy efficiency.

In our previous study [15], we studied the energy flow, energy consumption, and their relationships with the heating rate and heat flux in the heating stage of the GMP and analyzed the impact of the GMP on the environment from the perspective of carbon emissions. However, in this study, we chose glass with a large size and irregular shape as the research object, which is rarely studied by scholars and is also our innovation point. However, in terms of sustainable manufacturing, processing large-size glass using the GMP method has several disadvantages, including low efficiency, uneven heating, large temperature gradients, and low optical properties of the glass components after processing. By combining advanced technologies, such as ultrasonic-assisted GMP [31–33] and laser-assisted GMP [18,34,35], the efficiency and effectiveness of the hot bending process can be improved, resulting in higher-quality and more accurate glass elements. In a follow-up study, we will conduct further research on these two auxiliary processing technologies.

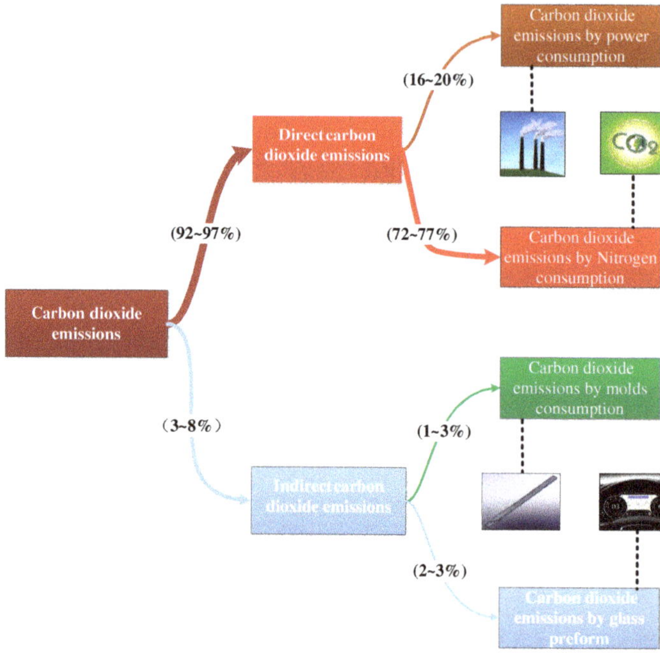

Figure 15. CO_2 emission ratio of glass molding process for large-sized automotive instrument glass.

5. Conclusion and Future Work

In this survey, numerical simulations were performed for large-scale automotive instrument glass, and numerical and experimental models were constructed to explore the energy efficiency issues of different strategies during the heating phase of the glass molding process and to analyze the environmental impact of carbon emissions from the GMP. The following important conclusions can be drawn.

(1) After systematic theoretical analysis and experimental tests, a heat transfer model of a metal heating plate, a metal heat conductor, and a glass model was developed, which can accurately model the action of heat flow, and thus, predict the temperature changes in the glass mold for large-sized automotive instrument glass.
(2) The analysis of the simulation results shows that different heating rate strategies had an impact on the energy efficiency of the GMP. Under heating rate strategy IV, the output energy of the heating device was lower than that under strategy I by 4.04%, and the heating time was reduced by 7.06%. Therefore, using heating rate strategy IV is the most ideal option.
(3) The analysis of the numerical results shows that different heat flux strategies affected the energy consumption of the heating device. The results show that the heat flow density strategy III effect was the best. The output energy of the heating equipment was reduced by 4.92%, and the heating time was reduced by 6.06%.

In the future, we will actively engage in extensive research to introduce new process technologies, such as ultrasound-assisted and laser-assisted techniques. The application of these emerging technologies will further enhance work efficiency and facilitate effective energy conservation.

Author Contributions: Y.C. (Yanyan Chen), Y.C. (Yang Cao) and W.M. carried out the concept and methodology. Y.C. (Yanyan Chen) and S.Z. wrote the first draft. S.Z., S.H., Y.Z., Y.C. (Yanyan Chen) and G.Z. carried out the formal analysis and data analysis. S.H., Y.Z., G.Z. and W.M. mainly carried out the investigation. G.Z. and W.M. undertook funding acquisition. Y.C. (Yanyan Chen) and S.Z. are the first co-authors. All authors have read and agreed to the published version of the manuscript.

Funding: This work was supported by the National Natural Science Foundation of China (No. 51975228), the Henan Province Key Science and Technology Project (Nos. 222102220011, 232102220099), and the Local Innovative and Research Team Project of Guangdong Pearl River Talents Program (No. 2017BT01G167).

Data Availability Statement: Not applicable.

Conflicts of Interest: The authors declare no conflict of interest.

References

1. Akhtar, A.; Ruan, H.H. Review on thin film coatings for precision glass molding. *Surf. Interfaces* **2022**, *30*, 101903. [CrossRef]
2. Yang, W.; Zhang, Z.; Ming, W.Y.; Yin, L.; Zhang, G.J. Study on shape deviation and crack of ultra-thin glass molding process for curved surface. *Ceram. Int.* **2022**, *48*, 6767–6779. [CrossRef]
3. Shu, C.S.; Guo, X.P.; Yin, S.H.; Huang, S.; Mao, Z.Z. Fabrication of curved aspheric compound eye microlens array with high surface quality by precision glass molding. *Precis. Eng.* **2023**, *82*, 129–139. [CrossRef]
4. Yi, A.Y.; Jain, A. Compression Molding of Aspherical Glass Lenses-A Combined Experimental and Numerical Analysis. *J. Am. Ceram. Soc.* **2005**, *88*, 579–586. [CrossRef]
5. He, W.B.; Chen, Z.J.; Ming, W.Y.; Du, J.G.; Cao, Y.; Ma, J.; Wei, A.Y. Multi-objective optimization of glass multi-station bending machining for smartphone curved screen. *J. Braz. Soc. Mech. Sci.* **2019**, *41*, 476. [CrossRef]
6. Ming, W.Y.; Zhang, S.F.; Liu, X.W.; Liu, K.; Yuan, J.; Xie, Z.B.; Sun, P.Y.; Guo, X.D. Survey of Mura Defect Detection in Liquid Crystal Displays Based on Machine Vision. *Crystals* **2021**, *11*, 1444. [CrossRef]
7. Ming, W.Y.; Cao, C.; Zhang, G.J.; Zhang, H.M.; Zhang, F.; Jiang, Z.W.; Yuan, J. Review: Application of Convolutional Neural Network in Defect Detection of 3C Products. *IEEE Access* **2021**, *9*, 135657–135674. [CrossRef]
8. Disfani, M.M.; Arulrajah, A.; Bo, M.W.; Sivakugan, N. Environmental risks of using recycled crushed glass in road applications. *J. Clean. Prod.* **2012**, *20*, 170–179. [CrossRef]
9. Ashie, I.N.A.; Smith, J.P.; Simpson, B.K.; Haard, N.F. Spoilage and shelf-life extension of fresh fish and shellfish. *Crit. Rev. Food Sci. Nutr.* **1996**, *36*, 87–121. [CrossRef]
10. Poonia, A.; Jha, A.; Sharma, R.; Singh, H.B.; Rai, A.K.; Sharma, N. Detection of adulteration in milk: A review. *Int. J. Dairy Technol.* **2017**, *70*, 23–42. [CrossRef]
11. Vu, A.T.; Kreilkamp, H.; Dambon, O.; Klocke, F. Nonisothermal glass molding for the cost-efficient production of precision freeform optics. *Opt. Eng.* **2016**, *55*, 071207. [CrossRef]
12. Zhang, Z.; Ming, W.Y.; Zhang, Y.; Yin, L.; Xue, T.; Yu, H.S.; Chen, Z.J.; Liao, D.M.; Zhang, G.J. Analyzing sustainable performance on high-precision molding process of 3d ultra- thin glass for smart phone. *J. Clean. Prod.* **2020**, *255*, 120196. [CrossRef]
13. Ming, W.Y.; Chen, J.Z.; Du, J.G.; Zhang, Z.; Zhang, G.J.; He, W.B.; Ma, J.; Shen, F. A comprehensive review of theory and technology of glass molding process. *Int. J. Adv. Manuf. Technol.* **2020**, *107*, 2671–2706. [CrossRef]
14. Li, L.H.; Chan, M.K.; Lee, W.B.; Ng, M.C.; Chan, K.L. Modeling and experimental performance analysis of a novel heating system and its application to glass hot embossing technology. *Opt. Lett.* **2018**, *44*, 3454–3457. [CrossRef]
15. Ming, W.Y.; Jiang, Z.W.; Chen, Z.J.; He, W.B.; Li, X.K. Modelling and analysis of energy consumption in glass molding process for smartphone covers using different heating strategies. *Int. J. Adv. Manuf. Technol.* **2023**, *124*, 1491–1512. [CrossRef]
16. Ming, W.Y.; Jia, H.J.; Huang, H.Y.; Zhang, G.J.; Liu, K.; Lu, Y.; Cao, C. Study on Mechanism of Glass Molding Process for Fingerprint Lock Glass Plates. *Crystals* **2021**, *11*, 394. [CrossRef]
17. Ming, W.Y.; Jia, H.J.; Liu, X.; Liu, X.; He, W.B.; Wei, A.Y.; Chen, Z.J. Optimization of process parameters of GMP for 3D ultrathin glass based on cuckoo search algorithm. *IOP Conf. Ser. Mater. Sci. Eng.* **2020**, *768*, 042046. [CrossRef]
18. Wang, P.; Zhang, Z.; Hao, B.; Wei, S.; Huang, Y.; Zhang, G. Investigation on heat transfer and ablation mechanism of CFRP by different laser scanning directions. *Compos Part B Eng.* **2023**, *262*, 110827. [CrossRef]
19. Zhang, Z.; Yang, W.; Ming, W.Y.; Yin, L.; Liao, D.M.; Zhang, G.J. Surface integrity analysis of ultra-thin glass molding process. *Ceram. Int.* **2021**, *47*, 30584–30597. [CrossRef]
20. Ming, W.Y.; Cao, C.; Xie, Z.B.; Liu, X.W.; Xu, Y.J.; Jiang, Z.W.; Li, X.K.; Liu, K.; Guo, X.D.; Yuan, J. Green manufacturing: A comparative study of renewable dielectrics in the EDM process. *J. Braz. Soc. Mech. Sci.* **2022**, *44*, 580. [CrossRef]
21. Zhang, Z.; Qiu, W.Z.; Zhang, G.J.; Liu, D.H.; Wang, P. Progress in applications of shockwave induced by short pulsed laser on surface processing. *Opt. Laser Technol.* **2023**, *157*, 108760. [CrossRef]
22. Chen, H.; Yang, J.; Xue, X. *A New Marc Finite Element Course of Example*; China Machine Press: Beijing, China, 2007; pp. 221–289.
23. Beck, J.V. Transient temperatures in a semi-infinite cylinder heated by a disk heat source. *Int. J. Heat Mass Transf.* **1981**, *24*, 1631–1640. [CrossRef]

24. Storck, K. *Thermal System Analysis: Heat Transfer in Glass Forming and Fluid Temperature-Control Systems;* Linköping University: Linköping, Sweden, 1998.
25. Gu, W.; Zhao, X.H.; Yan, X.B.; Wang, C.; Li, Q. Energy technological progress, energy consumption, and CO_2 emissions: Empirical evidence from China. *J Clean Prod.* **2019**, *236*, 117666. [CrossRef]
26. Ming, W.Y.; Guo, X.D.; Xu, Y.J.; Zhang, G.J.; Jiang, Z.W.; Li, Y.Z.; Li, X.K. Progress in non-traditional machining of amorphous alloys, Ceramics International. *Ceram. Int.* **2023**, *49*, 1585–1604. [CrossRef]
27. Ming, W.Y.; Sun, P.Y.; Zhang, Z.; Qiu, W.Z.; Du, J.G.; Li, X.K.; Zhang, Y.M.; Zhang, G.J.; Liu, K.; Wang, Y.; et al. A systematic review of machine learning methods applied to fuel cells in performance evaluation, durability prediction, and application monitoring. *Int. J. Hydrog. Energy* **2023**, *48*, 5197–5228. [CrossRef]
28. Zhang, J.; Jiang, H.Q.; Liu, G.Y.; Zeng, W.H. A study on the contribution of industrial restructuring to reduction of carbon emissions in China during the five five-year plan periods. *J. Clean Prod.* **2018**, *176*, 629–635. [CrossRef]
29. Ming, W.Y.; Xie, Z.B.; Jiang, Z.W.; Chen, Y.Y.; Zhang, G.J.; Xu, Y.J.; He, W.B. Progress in optical adhesive and lamination process of touch screen in 3C products. *J. Soc. Inf. Display* **2022**, *30*, 851–876. [CrossRef]
30. Ming, W.Y.; Jia, H.J.; Zhang, H.M.; Zhang, Z.; Liu, K.; Du, J.G.; Shen, F.; Zhang, G.J. A comprehensive review of electric discharge machining of advanced ceramics. *Ceram. Int.* **2020**, *46*, 21813–21838. [CrossRef]
31. Yu, J.W.; Luo, H.; Nguyen, T.H.; Huang, L.; Liu, B.; Zhang, Y.Y. Eigenfrequency characterization and tuning of Ti-6Al-4V ultrasonic horn at high temperatures for glass molding. *Ultrasonics* **2020**, *101*, 106002. [CrossRef] [PubMed]
32. Luo, H.; Yu, J.W.; Lou, H.Q.; Huang, K.F.; Hu, J.Z.; Xu, B.C. Thermal/tribological effects of superimposed ultrasonic vibration on viscoelastic responses and mold-filling capacity of optical glass: A comparative study. *Ultrasonics* **2020**, *108*, 106234. [CrossRef]
33. Zhang, Y.; Zhao, H.; Yan, Y.Q.; Tong, X.; Ma, J.; Ke, H.B.; Wang, W.H. Ultrasonic-assisted fabrication of metallic glass composites. *J. Non-Cryst. Solids* **2022**, *597*, 121894. [CrossRef]
34. Li, R.; Li, C.K.; Yan, M.F.; Li, M.; Lin, C.G.; Dai, S.X.; Song, B.A.; Xu, T.F.; Zhang, P.Q. Fabrication of chalcogenide microlens arrays by femtosecond laser writing and precision molding. *Ceram. Int.* **2023**, *49*, 15865–15873. [CrossRef]
35. Gülçür, M.; Romano, J.M.; Penchev, P.; Gough, T.; Brown, E.; Dimov, S.; Whiteside, B. A cost-effective process chain for thermoplastic microneedle manufacture combining laser micro-machining and micro-injection moulding. *CIRP J. Manuf. Sci. Technol.* **2021**, *32*, 311–321. [CrossRef]

metals

MDPI

Article

Evolution of Microstructures, Texture and Mechanical Properties of Al-Mg-Si-Cu Alloy under Different Welding Speeds during Friction Stir Welding

Zhang Luo [1], Youping Sun [1,2,3,*], Wangzhen Li [1,2,3,*], Jiangmei He [1,2], Guojian Luo [1] and Huashen Liu [1]

1 School of Mechanical and Automotive Engineering, Guangxi University of Science and Technology, Liuzhou 545006, China
2 Guangxi Earthmoving Machinery Collaborative Innovation Center, Liuzhou 545006, China
3 Guangxi Key Laboratory of Automobile Components and Vehicle Technology, Liuzhou 545006, China
* Correspondence: syptaiji@126.com (Y.S.); liwangzhen@gxust.edu.cn (W.L.)

Abstract: The effects of different welding speeds on the microstructures and mechanical properties of Al-0.75Mg-0.75Si-0.8Cu alloys were investigated using optical metallographic microscopy (OM), X-ray diffraction (XRD) analysis, an ETM105D electronic universal testing machine and field emission electron microscopy (SEM). The results reveal that during the friction stir welding process, the welded joint forms the base material (BM), heat-affected zone (HAZ), thermomechanically affected zone (TMAZ) and nugget zone (NZ), under the action of shear force and friction heat. The textures present in the BM are mainly C ({001}<100>) recrystallised cubic texture and P({110}<112>) recrystallised texture, Goss (G) texture, brass (B) texture and {112}<110> rotating copper texture. The organisation of the NZ undergoes dynamic recrystallisation, forming fine isometric crystals with large angular grain boundaries, accounting for more than 75% of the total. The geometrically necessary dislocations (GNDs) in the NZ grow as the welding speed rises. Moreover, $\{111\}\langle 1\bar{1}0\rangle$ and $\{111\}\langle \bar{1}10\rangle$ shearing textures, {001}<110> recrystallisation textures and fibre textures are mainly present in NZs. The average grain size in the NZ was the smallest, and the mechanical properties were the best at a welding speed of 125 mm/min. The grain size and the tensile strength and elongation of the NZ were 2.945 μm, 200.7 MPa and 12.7% for the joint at a welding speed of 125 mm/min, respectively.

Keywords: aluminium alloy; EBSD; GND; texture; mechanical properties

1. Introduction

Al-Mg-Si-Cu alloys (6× × × × series aluminium alloys) have the advantages of high specific strength, good welding properties and high corrosion resistance, and are widely used in construction, automotive, aerospace and other fields [1]. The conventional fusion welding process produces defects, such as porosity and thermal cracking, when welding aluminium alloys, which seriously affects the overall performance of the welded joints [2]. Unlike the fusion welding process, friction stir welding (FSW) does not produce defects, such as porosity and cracks, during the welding process and does not produce harmful gases such as fumes, which is a new green and non-polluting solid phase welding technology. Studies have shown that suitable FSW welding process parameters can effectively improve the grain organisation of welded joints and improve the performance of welded joints [3–5].

In stir friction welding, the metal material will accumulate on the advancing side with the shear rotation of the stirring needle, and different regions are subjected to different thermal coupling, which will make the tissue distribution of the welded joint uneven and form different grains and textures, directly affecting the overall performance of the welded joint. Therefore, it is very important to study the microstructure, texture evolution and mechanical properties of welded joints under different welding process parameters.

Jing et al. [6] found that by increasing the welding speed of 6063-T6 aluminium alloy, the mechanical properties of the joint first increased and then decreased, and the fracture location was mostly located in the heat-affected zone of the backwards side. Jin et al. [7] studied the fracture forms of 7055 aluminium alloy joints at different rotational speeds, and the results showed that with increasing rotational speed, the fracture position of the joints first transferred from the HAZ to the NZ, resulting in an "S"-shaped fracture, and then transferred to the thermomechanically affected zone on the advancing side (AS-TMAZ), resulting in fracture. Dong et al. [8] studied the welding speed on the 6005A-T6 aluminium alloy organisation and hardness and found that the original β″ phase in the weld nucleus zone completely entered the α-Al matrix, and the lower welding speed promoted the formation of the GP zone, thereby increasing the hardness of the HZs. The welding speed on the TMAZ hardness was not significant, and the evolution of the precipitation phase was mainly related to the peak temperature. The HAZ due to β″ and Q″ phase aggregation coarsening had the lowest hardness, and the hardness of the HAZ gradually increased with increasing welding speed. Ding et al. [9] investigated the influence of welding speed on the microstructure development and mechanical characteristics of the NZ made of Al-Mg-Mn-Zr-Ti alloy. It was discovered that when the welding speed increased, the DRX grains in the NZ decreased, with no significant texture and low texture strength. The mechanical properties of the welded joints first increased and then decreased. Yuan et al. [10] studied the crystal orientation of the NZ of 5083 aluminium alloy and found that the brass texture and S texture were mainly present in the BM, while these two types of texture were basically absent in the NZ, so the NZ is a weakly oriented organization. Zhang et al. [11] studied the evolution of crystal texture components in different regions of the NZ of 6082-T6 aluminium alloy and found that the {111}< $\bar{1}$10> shear texture and {001}<100> recrystallised cubic texture mainly existed in the backwards side of the NZ. More thermoplastic deformation occurred in the middle of the NZ, thus {111}< $\bar{1}$10> the shear texture became more prominent, and the {001}<100> texture transformed into a {001}<110> rotating cubic texture; on the advancing side of the NZ, the {111}< $\bar{1}$10> texture and {001}<110> texture rotated in the ND direction, forming the {111}< $\bar{1}$10> shear texture and {001}<100> cubic texture. Although a large number of scholars have conducted a lot of research on stir friction welding of aluminium alloys, no one has studied the stir friction welding of the Al-0.75Mg-0.75Si-0.8Cu alloy. Therefore, in the present work, the microstructure, texture evolution and mechanical properties are studied in the NZ of FSW joints made from Al-0.75Mg-0.75Si-0.8Cu alloys at different welding speeds (υ) and a constant rotation speed (ω), i.e., ω = 750 r/min; υ = 75, 100, 125, 150 mm/min, to provide a theoretical basis for the practical application of FSW welding.

2. Experimental Materials and Methods

The material used for the experiments was an Al-0.75Mg-0.75Si-0.8Cu alloy ingot. The industrial, pure aluminium with 99.99% content and the intermediate alloy Al-10Si and Al-10Cu were placed into a resistance melting furnace at 800 °C in a certain proportion. After these materials were melted, the industrial, pure magnesium with 99.99% content in a certain proportion (considering the burning loss of pure magnesium, the content is higher than 10% of the theoretical calculation) was placed into a resistance melting furnace. The resistance furnace was turned off after melting, it was stirred and the switch of the resistance furnace was opened, the temperature was increased to 780 °C for 10 min, then lowered to 720 °C for slagging, and finally poured into a 150 mm × 100 mm × 20 mm Al-0.75Mg-0.75Si-0.8Cu alloy ingot. It was subjected to a 560 °C × 24 h homogenisation process and the ingots were, subsequently, machined into 100 mm × 80 mm × 14 mm plates by a milling machine. The plates were rolled to 4 mm by two hot rolling passes (first pass with 4 mm down and second pass with 6 mm down), both at a holding time of

400 °C/20 min, followed by a solid solution treatment at 510 °C/80 min and an ageing treatment at 190 °C/13 h, and finally cold rolled to 2 mm, both at a strain rate of 10 s^{-1}.

Before welding, all plates were sandpapered to remove oxidation and grease. The shoulder diameter of the stirring head used for welding was 12 mm, the length of the stirring pin was 1.85 mm, the shape was conical, the large diameter of the stirring pin was 3.6 mm, the small diameter was 2.5 mm, and the welding direction (WD) was perpendicular to the rolling direction (TD). The specific welding parameters are shown in Table 1.

Table 1. Welding parameters for the FSW.

Rotational Speed/r·min^{-1}	Welding Speed/mm·min^{-1}
750	75
750	100
750	125
750	150

The size diagram of the tensile sample is shown in Figure 1. An ETM105D electronic universal testing machine was used to test the mechanical properties (tensile rate of 2 mm/min). Three tensile specimens were taken from each group, and the average value was obtained. A SmartLab type X-ray diffractometer was used to examine the texture of the specimens, with specific parameters of 40 KV Cu target rays, a scanning angle of 10–90° and a scanning speed of 2°/min. A Sigma scanning electron microscope was used to observe the fracture morphology of the welded joints. Samples were taken from the welded joints to polish the cross section (ND), and the specimens were electrochemically polished with 70% nitric acid and 30% methanol. The polishing time was 30 s. The EBSD experiments were carried out using a field emission scanning electron microscope (Sigma) with an EBSD probe (Aztec MAX80).

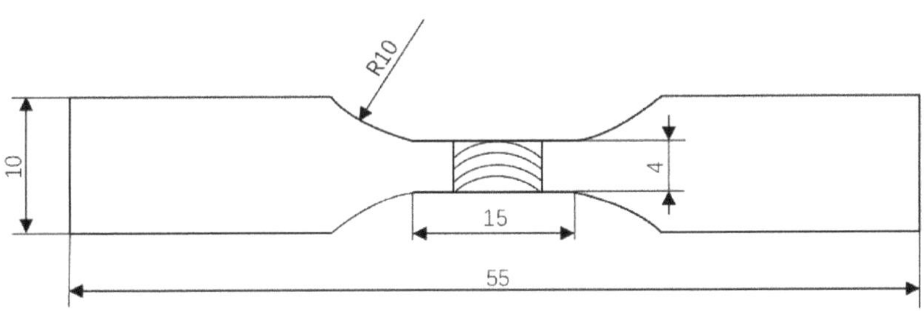

Figure 1. Dimensions of the tensile sample.

3. Experimental Results and Analysis

3.1. Microstructure

The microstructure of each region of the stir friction welded joint is shown in Figure 2. The BM along the rolling direction of the cross-sectional microstructure is shown in Figure 2a. The grains are fibrous and elongated along the rolling direction. The organisation in the NZ is subject to strong thermomechanical coupling, dynamic recrystallisation occurs and fine equiaxed crystals are formed. The grains in the HAZ maintain the original shape of the base material but, in the welding process, the grains become coarser as a result of frictional thermal cycling. As the peak temperature on the advancing side (AS) is higher than that on the retreating side (RS), the grain changes in the AS are more pronounced. The TMAZ undergoes plastic deformation and partial recrystallisation under the action of thermal coupling of the stirring needles, forming a microstructure consisting of curved and elongated grains. Comparing Figure 2c,e, it can be clearly found that the AS-HAZ demarcation line is more obvious, because the tissue of the AS undergoes dynamic recrystallisation under the effect of rotational shear, and the recrystallised microstructure formed is obviously different from the original BM tissue, while the tissue of the RS is mainly subject to extrusion, and the dynamic recrystallisation is limited, so the boundary between it and the NZ is not clear enough [12]. At the same time, in Figure 2e, it can also be observed that the microstructure of the thermally affected zone on the retreating side expands towards the inside of the weld nugget zone, which is due to the good fluidity of the retreating side microstructure. In Figure 2e, the expansion of the tissue in the RS-TMAZ into the inner part of the NZ can also be observed. This is due to the good fluidity of the RS tissue and the combined effect of the compression of the RS tissue by the shaft shoulder and the shearing of the stirring pin, which causes the material in the upper part of the weld to be squeezed into the NZ [13].

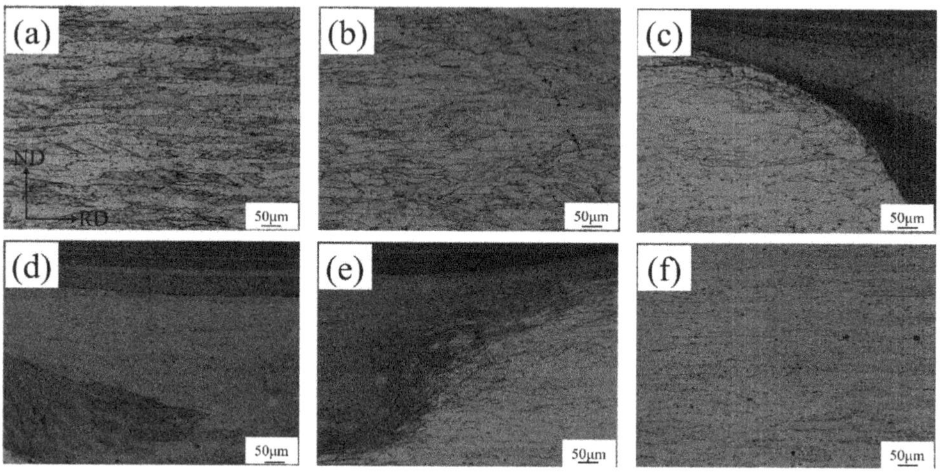

Figure 2. Microstructure of each region of the FSW. (**a**) BM; (**b**) AS-HAZ; (**c**) AS-TMAZ; (**d**) NZ; (**e**) RS-TMAZ; (**f**) TS-HAZ.

Figure 3 shows the EBSD figure, grain size distribution figure and grain boundary orientation angle distribution figure for the NZ at different welding speeds. Figure 3(a1–d1) includes red (2 ~ 5°) thick lines and green (5 ~ 15°) thick lines on behalf of small angle grain boundaries, and black (15 ~ 65°) thick lines on behalf of large angle grain boundaries. The tissue in the NZ is subjected to the compression of the axial shoulder and the stirring action of the stirring needle, and the original tissue undergoes intense thermomechanical coupling, so the grains in the NZ undergo dynamic recrystallisation, producing fine isometric axial grains [14]. Figure 3(a2–d2) shows the grain size distribution map derived through Channel 5 software, while the average grain size of the weld nucleus zone was calculated for the different welding speeds of 3.898 μm, 3.692 μm, 2.945 μm and 3.2 μm. The grain size with the welding speed first increased and then decreased because the welding speed increased, and the peak temperature of the welded joint decreased, thereby reducing the driving force of the grain growth; at the same time, the increase in welding speed reduced the deformation of the welded joint, which caused the grains to grow [15]. Therefore, different welding speeds have different main factors affecting the grain size [16]. At a welding speed of 75 to 125 mm/min, the peak temperature is the main factor affecting the grain size, which decreases; at a welding speed of 125 mm/min to 150 mm/min, deformation is the main factor affecting the grain size, which becomes larger. Figure 3(a3–d3) shows the distribution of the grain boundary orientation angles at different welding speeds. From the figure, the proportion of large angle grain boundaries in the NZ is more than 75%, which is because the thermoplastic deformation during the welding process introduces a large number of dislocations. In the subsequent dynamic response, a small angle grain boundary composed of subgrains will be formed. In the continuous dynamic recrystallisation process, subgrains rotate to the subgrain boundaries through repeated absorption of dislocation growth, evolving into recrystallised grains with large angle grain boundaries [17,18]. As the welding speed increases, the peak temperature of the weld nucleation zone decreases, inhibiting the recrystallisation process and decreasing the percentage of large angle grain boundaries.

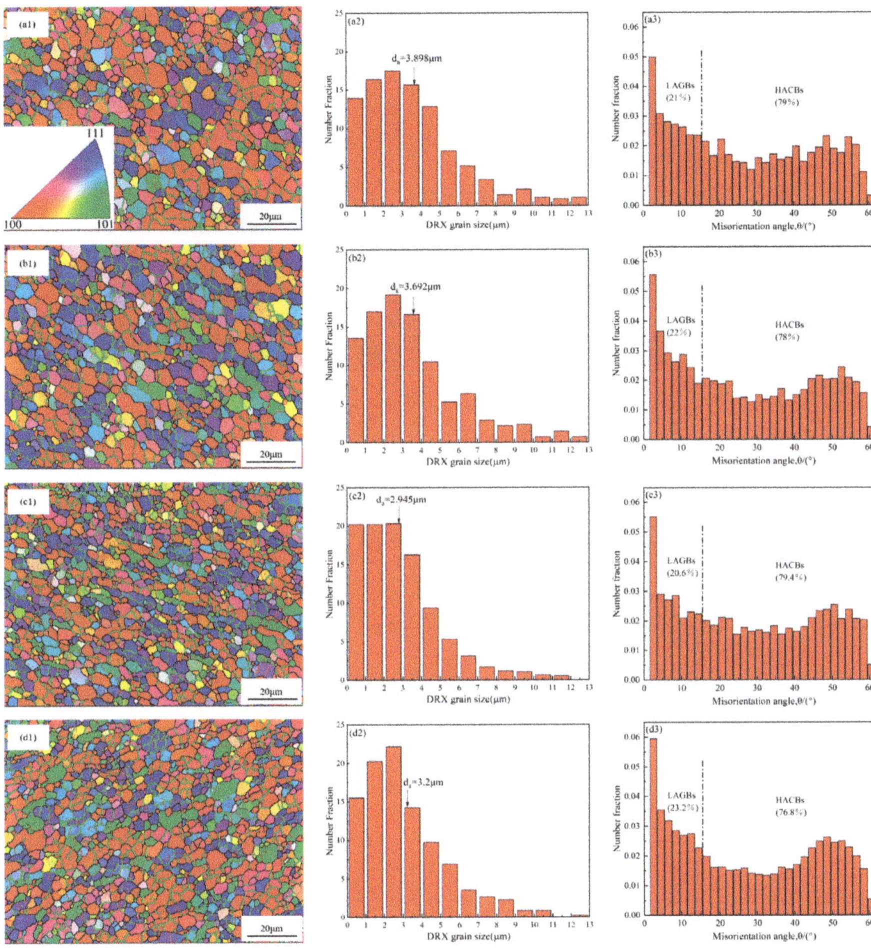

Figure 3. EBSD maps, grain size distribution and misorientation angle distribution of the NZ at different welding speeds. (**a1–a3**) 75 mm/min; (**b1–b3**) 100 mm/min; (**c1–c3**) 125 mm/min; (**d1–d3**) 150 mm/min.

3.2. Geometrically Necessary Dislocations

Figure 4 shows the kernel average misorientation (KAM) figure for different welding speeds, retrieved directly from the EBSD data, which was chosen as a measure for the local misorientations. The KAM quantifies the average misorientation around a measurement point with respect to a defined set of nearest or nearest plus second-nearest neighbour points [19,20]. KAM can reflect the strain energy as well as the magnitude of the dislocation density [21], which is calculated by the software for different welding speeds in the weld kernel zone. The KAM values are 0.42°, 0.43°, 0.42° and 0.43°, which indicates that the welding speed has a small effect on the KAM of the HZ. The smaller KAM values are due to the complete dynamic recrystallisation of the tissue in the NZ.

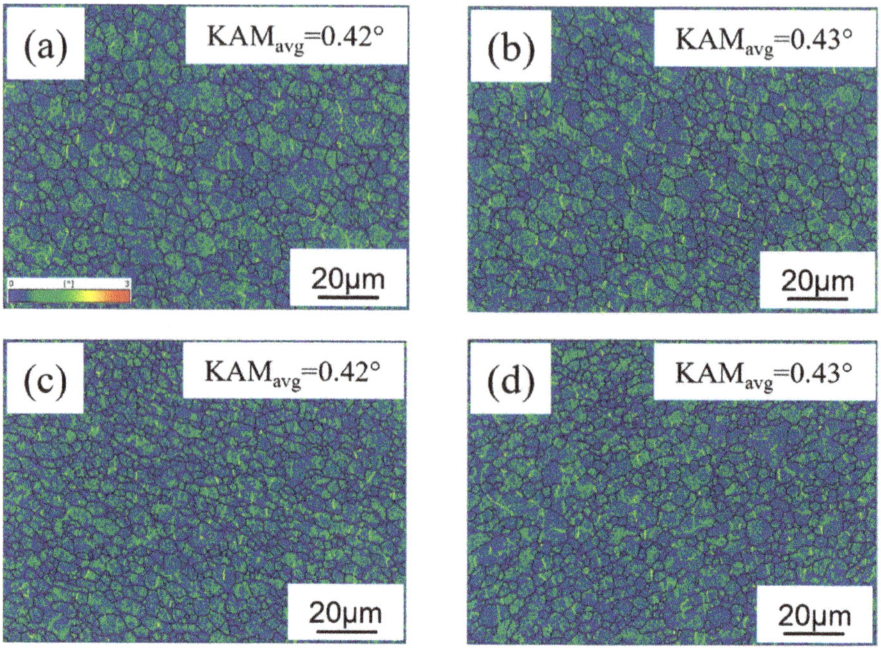

Figure 4. Distribution of the kernel average misorientation in the NZ at different welding speeds. (**a**) 75 mm/min; (**b**) 100 mm/min; (**c**) 125 mm/min; (**d**) 150 mm/min.

The KAM diagram can only qualitatively describe the dislocation density, and the magnitude of the KAM value can only be used to determine the change trend in the dislocation density. In this experiment, the KAM value did not change significantly with the change in welding speed, resulting in the inability to determine the effect of welding speed on dislocation density. Therefore, the mathematical model established by Kubin [22] was introduced to calculate the geometrically necessary dislocations (GNDs) in the microstructure of the NZ at different welding speeds. The model [23] is as follows:

$$\rho^{GND} = \frac{2\Delta\theta_i}{ub} = B\Delta\theta_i \quad (1)$$

Equation (1): ρ^{GND} is the GND density of the measurement point; $\Delta\theta_i$ represents the local orientation difference ($\Delta\theta_i \leq 3°$); u is the EBSD scan step size of 200 nm; b is the Burgers vector; and the Burgers vector of aluminium alloy is 0.286 nm. The columnar distribution of the GND density in the weld nugget region at different welding speeds is shown in Figure 5. At a welding speed of 75 mm/min, the value of the GND was $2.45 \times 1014\ m^{-2}$. Upon increasing the welding speed, the value of the GND reached $2.55 \times 1014\ m^{-2}$, while the dislocation density of the aluminium alloy sheet rolled at room temperature was $1.64536 \times 1015\ m^{-2}$ [24]. Combined with the grain boundary orientation distribution diagram in Figure 6, it can be observed that the dislocations generated during the cold rolling process and the dislocations generated by the deformation of the plate in the FSW process rotated and entangled at high temperatures, which eventually evolved into large angle grain boundaries. The fundamental reason for the minimum GND at a welding speed of 75 mm/min is that more frictional heat is generated in the joint at this parameter, which promotes more dislocations forming large angle grain boundaries.

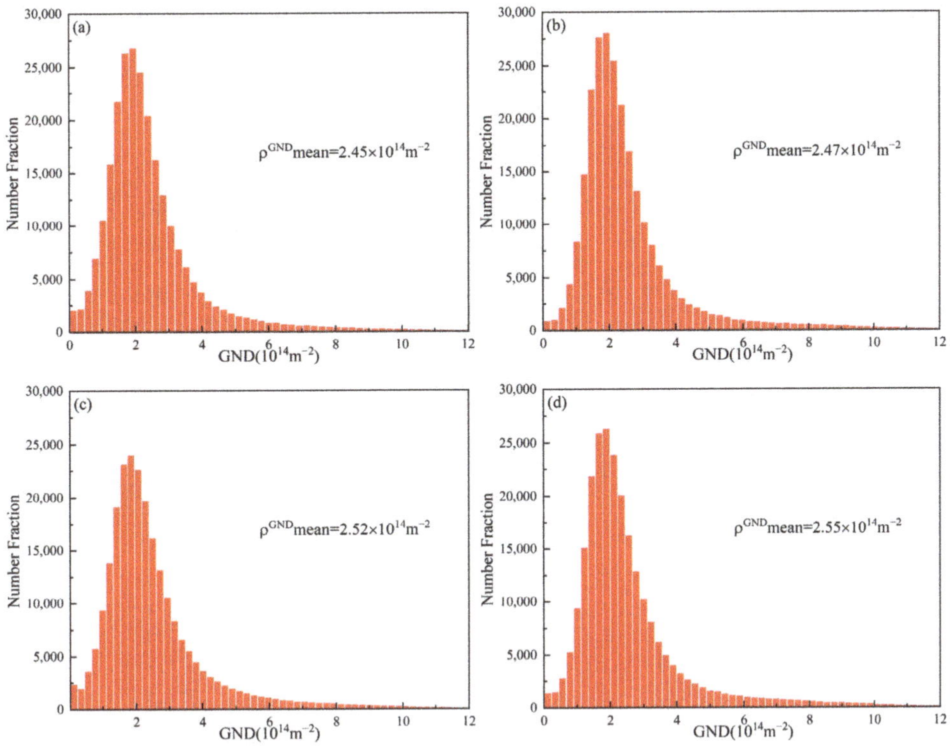

Figure 5. Histogram of the probability of geometric necessary dislocation density in the NZ at different welding speeds. (**a**) 75 mm/min; (**b**) 100 mm/min; (**c**) 125 mm/min; (**d**) 150 mm/min.

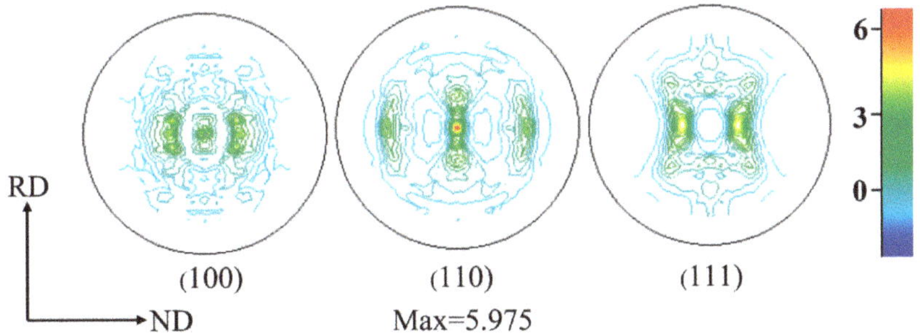

Figure 6. The (001), (110) and (111) pole figures showing the texture of the BM.

3.3. Texture

During the rolling process, the grains of aluminium alloy will rotate, and the rotation of the grains will cause the grain orientation to gather around a certain orientation or certain orientations, thus forming a texture. In the process of hot rolling, two microscopic processes mainly occur in aluminium alloy: one is plastic deformation mainly by dislocation movement, resulting in deformation texture in the metal; the other is dynamic recrystallisation mainly by reversion, nucleation and grain growth, and the dynamic recrystallisation process will cause a recrystallisation texture [25]. The (001), (110) and (111) pole figures of the BM are shown in Figure 6. From Figure 6, the texture in the (110) pole figure is higher than that in the other two pole figures and the max density is 5.975. Figure 7 shows the ODF diagram of the parent material, and it can be determined from the literature [26–28] that the textures present in the BM in the figure are the C ({001}<100>) recrystallised cubic texture and P({110}<112>) recrystallised texture, Goss (G) texture, brass (B) texture and {112}<110> rotating copper texture. The grain orientation during the rolling of aluminium alloys is mainly concentrated on the β-orientation line and α-orientation line, with a small amount of cube texture [29]. The α-fibre texture starts from the Goss orientation {110}<001>, passes through the brass orientation {110}<112> and, finally, reaches the rotating Goss orientation {110}<110>, while the β-fibre texture starts from the copper orientation {112}<111>, passes through the S orientation {123}<634> and, finally, reaches the brass orientation {110}<110>. The G, B textures are common types of textures in aluminium alloys during the rolling process. In aluminium alloys containing large second-phase particles (>1 um), nucleation of the recrystallisation takes place in the deformation zones that form during deformation around these particles. This mechanism is commonly referred to as particle stimulated nucleation (PSN) [30–32]. P textures are typical of recrystallisation orientations that evolve due to PSN and recrystallisation of grains during annealing and rolling [33]. Cube oriented nuclei emerge from band-like structures, which are already present in the as-deformed microstructure, the so-called cube bands. These cube bands are either transition bands that had formed through the splitting of unstable orientations during the preceding deformation or are deformed grains with a retained cube orientation. During the subsequent growth, grains with precise cubic orientation account for the upper fraction, forming the cube texture [30,34,35], and the formation of rotating copper textures may be related to the shear forces on the material.

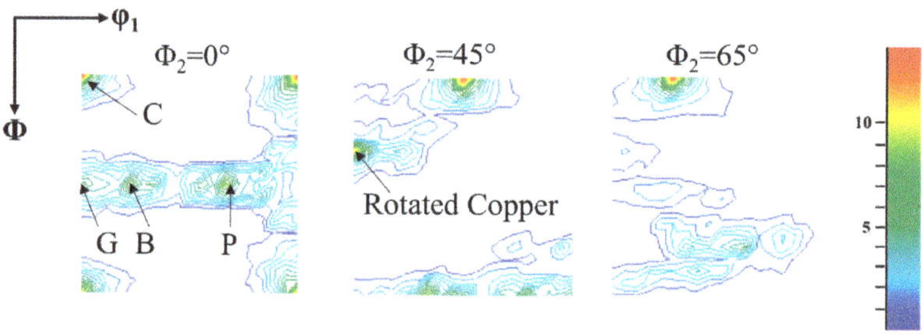

Figure 7. The ODF figures of the BM.

Figures 8 and 9 show different welding speeds for the (100), (110), and (111) pole diagrams and the ODF diagram, respectively. From the figure, it can be seen that at welding speeds of 75 mm/min ~125 mm/min, the texture in the HZ is mainly $\{111\}\langle 1\bar{1}0\rangle$ shear texture and r-cube({001}<110>) texture; at 150 mm/min in the NZ, the texture is mainly $\{111\}\langle \bar{1}10\rangle$ shear texture and r-cube({001}<110>) texture. The (111) slip surface of the aluminium alloy is aligned with the shear surface, and the (110) slip direction is parallel to the shear direction, so the material is subjected to the shear action of the stirring needle, and the $\{111\}\langle \bar{1}10\rangle$ and $\{111\}\langle 1\bar{1}0\rangle$ textures are likely to appear [36]. The {001}<100> cubically oriented deformed substructure has a low dislocation density and simple dislocation grouping [37], so dynamic recovery occurs in the {001}<001> cubically oriented deformed substructure under the action of the welding thermal cycle, becoming the core of the recrystallisation, which in the subsequent recrystallisation process swallows up other non-cubically oriented substructures, for the final formation of the {001}<100> recrystallisation cubic texture [38]. Under the rotational squeezing effect of the shoulder, the {001}<100> texture will rotate along the ND direction and, finally, the {001}<110> rotating cubic recrystallised texture is formed. As shown in Figure 10, the standard {001}<110> texture is deflected by a certain angle from the {001}<110> texture measured in this experiment, which is consistent with the phenomenon observed by Zhang et al. [39].

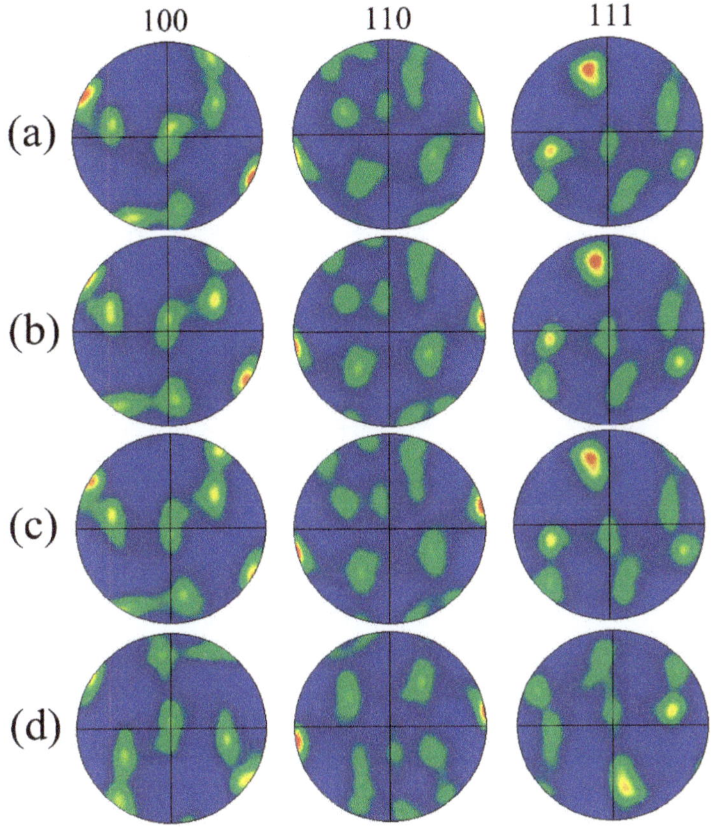

Figure 8. The (001), (110) and (111) pole figures of the NZs at different welding speeds. (**a**) 75 mm/min; (**b**) 100 mm/min; (**c**) 125 mm/min; (**d**) 150 mm/min.

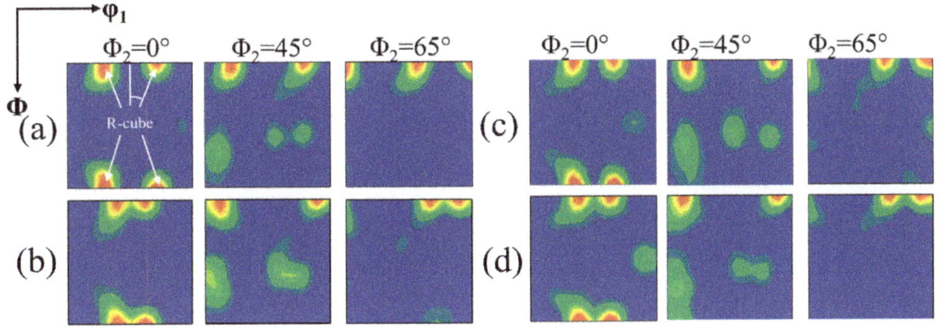

Figure 9. The ODF diagram of the welding core zone at different welding speeds. (**a**) 75 mm/min; (**b**) 100 mm/min; (**c**) 125 mm/min; (**d**) 150 mm/min.

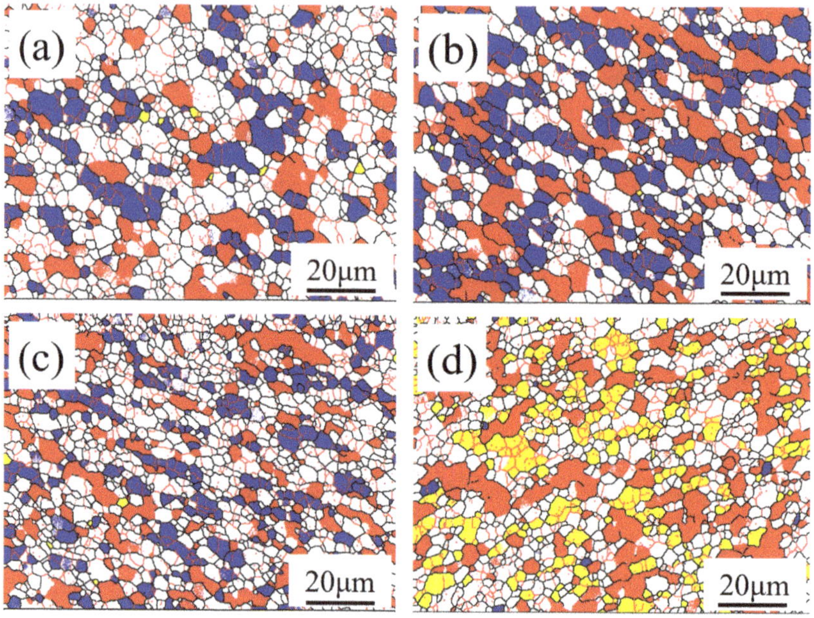

Figure 10. Grain distribution of various textures. (**a**) 75 mm/min; (**b**) 100 mm/min; (**c**) 125 mm/min; (**d**) 150 mm/min.

Table 2 shows the texture structure and content present in the NZ at different welding speeds, and Figure 10 shows the grain colour difference and distribution corresponding to the different textures. At a welding speed of 150 mm/min, the $(111)(\bar{1}10)$ texture content was 22.3%, but with the reduction in welding speed, the $(111)(\bar{1}10)$ texture was transformed into the $(111)(1\bar{1}0)$ texture and gradually increased in content, reaching a maximum at 100 mm/min. Continuing to reduce the welding speed, the $(111)(1\bar{1}0)$ texture again began to decrease. This is because at a welding speed of 100 mm/min, the stirring needle shear force makes the $(111)(\bar{1}10)$ texture rotate in the ND direction because the sliding direction is aligned with the shear direction. The $(111)(\bar{1}10)$ texture will first rotate 45° to form the $(111)(1\bar{1}0)$ texture. At a welding speed of 75 mm/min, the heat input in the HZ becomes large when the grain grows significantly, resulting in the disappearance of a portion of the shear texture [40].

Table 2. Volume fraction of the texture components of NZs at different welding speeds.

Colour Code	Orientation (h k l)(u v w)	75 mm/min	100 mm/min	125 mm/min	150 mm/min
(orange)	(0 0 1)(1 1 0)	16.9	26.6	20.6	26.9
(blue)	$(1\ 1\ 1)(1\ \bar{1}\ 0)$	18.1	31.1	19.2	—
(yellow)	$(1\ 1\ 1)(\bar{1}\ 1\ 0)$	—	—	—	22.3

3.4. *Mechanical Property*

Figure 11 shows the tensile properties of the Al-Mg-Si-Cu alloy at different welding speeds. With increasing welding speed, the tensile strength and elongation are shown to first increase and then decrease because the welding speed and the peak temperature of the welded joint are inversely proportional. When the welding speed is low, the peak temperature is high, grain growth occurs and the impurity phase will gather; when the welding speed is too fast, the peak temperature is low, the material mobility is poor and hole-type defects will occur. Therefore, at a welding speed of 125 mm/min, the joint has the best overall performance, with a tensile strength and elongation of 200.7 MPa and 12.7%, respectively.

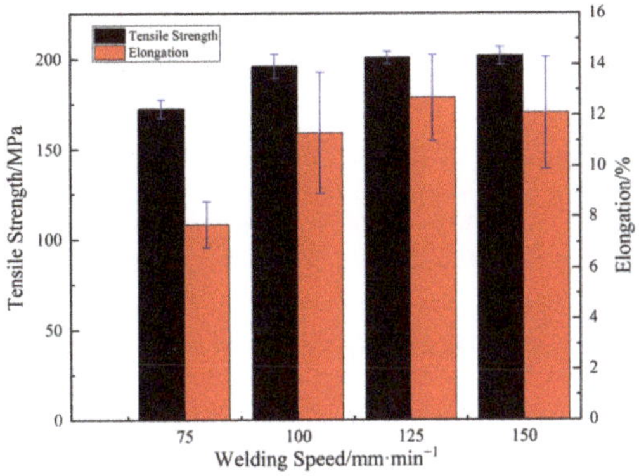

Figure 11. Tensile properties of joints at different welding speeds.

Figure 12 shows the tensile fracture morphology of joints at different welding speeds. As seen from the figure below, there are a large number of small dimples on the tensile fracture surface after FSW, so it can be determined that the fracture form is a ductile fracture. At the lower right of Figure 12a, it can be clearly observed that the distribution of dimples is relatively small and shallow. In the middle of Figure 12a, there are also pits with precipitated phases as the core, so their toughness is poor. Figure 12b shows the most uniform distribution of dimples, but there are parabolic pits, which may be formed by second-phase particles under tensile stress. In Figure 12c, in addition to the dimples, there are also unevenly distributed tear edges. Figure 12d is similar to Figure 12b in that there are parabolic holes.

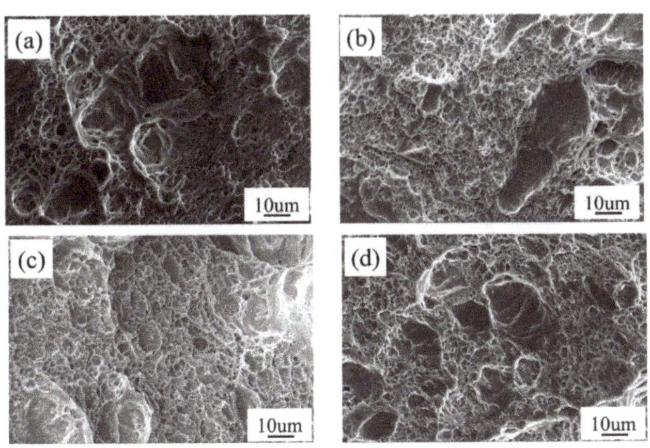

Figure 12. Tensile fracture morphology of the samples at different welding speeds. (**a**) 75 mm/min; (**b**) 100 mm/min; (**c**) 125 mm/min; (**d**) 150 mm/min.

4. Conclusions

(1) During the FSW welding process, the organisation in the HZ underwent thermomechanical coupling, dynamic recrystallisation occurred, fine equiaxed crystals formed and large angular grain boundaries accounted for more than 75% of all the grains. At a welding speed of 125 mm/min, the average grain size is the smallest, 2.945 μm. With an increase in welding speed, the GND density increases;

(2) When the welding speed is between 75 mm/min and 150 mm/min, the main textures in the NZ are $(111)(1\bar{1}0)$ and $(111)(\bar{1}10)$ shear textures, (001)(110) recrystallisation texture, and (110) silk texture. As the welding speed decreases, the $(111)(1\bar{1}0)$ texture will first increase and then decrease;

(3) As the welding speed increases, the tensile strength and elongation of Al-Mg-Si-Cu alloy welded joints show a pattern of increasing and then decreasing. The joint performance is optimal at 125 mm/min, with a tensile strength and elongation of 200.7 MPa and 12.7%, respectively. Fracture of the Al-Mg-Si-Cu alloy joints occurs in the form of ductile fractures.

Author Contributions: Conceptualization, Z.L. and Y.S.; Methodology, Z.L., W.L. and J.H.; Formal analysis, G.L. and H.L.; Data curation, J.H., G.L. and H.L.; Writing—original draft, Z.L.; Funding acquisition, Y.S. and W.L. All authors have read and agreed to the published version of the manuscript.

Funding: Supported by the Guangxi Science and Technology Project (Grant No. AD22080042) and the 2022 Project of Scientific Research Basic Ability Improvement for Young and Middle-aged Teachers in Guangxi Universities (Grant No. 2022ky0351).

Data Availability Statement: The data that support the findings of this study are available from the corresponding author upon reasonable request.

Conflicts of Interest: The authors declare no conflict of interest.

References

1. Zhou, X. *Study on Large Strain Rolling and Mechanical Behavior of AlMgSiCu Alloy*; Guangxi University of Science and Technology: Liuzhou, China, 2017.
2. Sun, J.; Zhu, H.; Yu, H.; Guan, P.; Sun, L. Relationship Between Microstructure Defects and Process Parameters of Aluminum Alloy Friction Stir Welded Joint. *Hot Work Technol.* **2018**, *47*, 45–47, 52.
3. Zhang, L.; Lin, N.; Zou, J.; Xie, R. Research Status on Friction Stir Welding of Aluminum Alloy. *Hot Work Technol.* **2020**, *49*, 1–6.
4. Yi, T.; Liu, S.D.; Fang, C.; Jiang, G.D. Role of oxides in the formation of hole defects in friction stir welded joint of 2519-T87 aluminum alloy. *J. Cent. South Univ.* **2022**, *29*, 3836–3846. [CrossRef]
5. Ming, W.; Sun, P.; Zhang, Z.; Qiu, W.; Du, J.; Li, X.; Guo, X. A systematic review of machine learning methods applied to fuel cells in performance evaluation, durability prediction, and application monitoring. *Int. J. Hydrog. Energy* **2023**, *48*, 5197–5228. [CrossRef]
6. Jing, H.; Feng, Q.; Xu, L.; Zhao, L.; Han, Y. Microstructure and Mechanical Properties of Friction Stir Welds on 6063-T6 Aluminum Alloy. *J. Mech. Eng.* **2020**, *56*, 13–19.
7. Jin, Y.; Huo, R.; Li, C.; Wang, X. Influence of rotational speed on fracture characteristics of 7055 aluminum alloy friction stir welded joints. *Trans. China Weld. Inst.* **2017**, *38*, 10–13, 18.
8. Dong, P.; Li, H.; Sun, D.; Gong, W.; Liu, J. Effects of welding speed on the microstructure and hardness in friction stir welding joints of 6005A-T6 aluminum alloy. *Mater. Des.* **2013**, *45*, 524–531. [CrossRef]
9. Ding, T.; Yan, H.G.; Chen, J.H.; Xia, W.J.; Bin, S.U. Effect of welding speed on microstructure and mechanical properties of Al−Mg−Mn−Zr−Tialloy sheet during friction stir welding. *Trans. Nonferrous Met. Soc. China* **2021**, *31*, 3626–3642. [CrossRef]
10. Yuan, G.C.; Liang, C.L.; Liu, H.; Yuan, Q. Crystal orientation in nugget zone of friction stir welded 5083 aluminum alloy plates. *Trans. Nonferrous Met. Soc. China* **2014**, *35*, 79–82.
11. Zhang, L.; Wang, X.; Wei, X.; Liu, X.; Chai, T. Effect of rotation speed on texture type in friction stir welding joint for 6082-T6 aluminum alloy. *Trans. Nonferrous Met. Soc. China* **2019**, *40*, 128–132.
12. Shi, L.; Dai, X.; Tian, C.; Wu, C. Microstructure and mechanical properties of 2195-T6 Al-Li alloy joint prepared by friction stir welding. *Trans. Nonferrous Met. Soc. China* **2022**, *43*, 25–34.
13. Liu, H.; Hu, Y.; Dou, C.; Sekulic, D.P. An effect of the rotation speed on microstructure and mechanical properties of the friction stir welded 2060-T8 Al-Li alloy. *Mater. Charact.* **2017**, *123*, 9–19. [CrossRef]

14. Yang, H.; Zhao, H.; Xu, X.; Sun, G.; Zhou, L.; Zhao, H.; Liu, H. Microstructure and Properties of 2A14-T4 Aluminum Alloy T Joint by SSFSW. *Mater. Rep.* **2021**, *35*, 20045–20051.
15. Humphreys, F.J.; Hatherly, M. *Recrystallization and Related Annealing Phenomena*, 2nd ed.; Elsevier: Amsterdam, The Netherlands, 2012; pp. 333–378. [CrossRef]
16. Li, Y.; Yan, H.; Chen, J.; Xia, W.; Su, B.; Ding, T.; Li, X. Influences of welding speed on microstructure and mechanical properties of friction stir welded Al–Mg alloy with high Mg content. *Mater. Res. Express* **2020**, *7*, 76506. [CrossRef]
17. Su, J.Q.; Nelson, T.W.; Mishra, R.; Mahoney, M. Microstructural investigation of friction stir welded 7050-T651 aluminium. *Acta Mater.* **2003**, *51*, 713–729. [CrossRef]
18. Li, L.; Zhang, P.; Yi, L.; Wu, S.; Zhou, Q. Effect of Welding Speed on Properties of Friction Stir Welded Joint of Aluminum Alloy. *J. Hunan Univ. Nat. Sci.* **2021**, *48*, 120–128.
19. Cha, J.W.; Jin, S.; Park, S.H. Influence of Ca addition on microstructural characteristics and mechanical properties of Mg–5Bi–3Al alloy extruded at extremely high speed. *Mater. Sci. Eng. A* **2023**, *862*, 144490. [CrossRef]
20. Calcagnotto, M.; Ponge, D.; Demir, E.; Raabe, D. Orientation gradients and geometrically necessary dislocations in ultrafine grained dual-phase steels studied by 2D and 3D EBSD. *Mater. Sci. Eng. A* **2010**, *527*, 2738–2746. [CrossRef]
21. Srivastava, V.C.; Schneider, A.; Uhlenwinkel, V.; Ojha, S.N.; Bauckhage, K. Age-hardening characteristics of Cu–2.4 Ni–0.6 Si alloy produced by the spray forming process. *J. Mater. Process. Technol.* **2004**, *147*, 174–180. [CrossRef]
22. Kubin, L.P.; Mortensen, A. Geometrically necessary dislocations and strain-gradient plasticity: A few critical issues. *Scr. Mater.* **2003**, *48*, 119–125. [CrossRef]
23. Yan, Z.; Wang, D.; He, X.; Wang, W.; Zhang, H.; Dong, P.; Sun, L. Deformation behaviors and cyclic strength assessment of AZ31B magnesium alloy based on steady ratcheting effect. *Mater. Sci. Eng. A* **2018**, *723*, 212–220. [CrossRef]
24. Xie, S.; Sun, Y.; He, J.; Zhu, J.; Fang, D. Effect of cryogenic rolling on microstructure and mechanical properties of 2524 aluminum alloy. *Trans. Mater. Heat Treat.* **2022**, *43*, 58–65.
25. QI, J.; Gong, Z.; Kang, K.; Jin, Z. Effect of Rolling Processes on Texture Evolution in High-strength 7150 Aluminum Alloy Plate. *Shanghai Met.* **2019**, *41*, 59–65.
26. Haase, C.; Barrales-Mora, L.A. Influence of deformation and annealing twinning on the microstructure and texture evolution of face-centered cubic high-entropy alloys. *Acta Mater.* **2018**, *150*, 88–103. [CrossRef]
27. Nezakat, M.; Akhiani, H.; Sabet, S.M.; Szpunar, J. Electron backscatter and X-ray diffraction studies on the deformation and annealing textures of austenitic stainless steel 310S. *Mater. Charact.* **2017**, *123*, 115–127. [CrossRef]
28. Jamaati, R.; Toroghinejad, M.R. Effect of stacking fault energy on deformation texture development of nanostructured materials produced by the ARB process. *Mater. Sci. Eng. A* **2014**, *598*, 263–276. [CrossRef]
29. Yang, Y.; Zhang, Q.; Tian, Q. Mechanical Properties and Texture Evolution Characteristics of 5052 Al Alloy under Large Deformation. *Hot Work. Technol.* **2022**, *51*, 74–78.
30. Engler, O.; Hirsch, J. Control of recrystallisation texture and texture-related properties in industrial production of aluminium sheet. *Int. J. Mater. Res.* **2009**, *100*, 564–575. [CrossRef]
31. Ghosh, A.; Ghosh, M. Microstructure and texture development of 7075 alloy during homogenisation. *Philos. Mag.* **2018**, *98*, 1470–1490. [CrossRef]
32. Ghosh, A.; Roy, A.; Ghosh, A.; Ghosh, M. Influence of temperature on microstructure, crystallographic texture and mechanical properties of EN AW 6016 alloy during plane strain compression. *Mater. Today Commun.* **2021**, *26*, 101808. [CrossRef]
33. Ghosh, A.; Ghosh, M.; Seikh, A.H.; Alharthi, N.H. Phase transformation and dispersoid evolution for Al-Zn-Mg-Cu alloy containing Sn during homogenisation. *J. Mater. Res. Technol.* **2020**, *9*, 1–12. [CrossRef]
34. Ghosh, A.; Ghosh, M.; Kalsar, R. Influence of homogenisation time on evolution of eutectic phases, dispersoid behaviour and crystallographic texture for Al–Zn–Mg–Cu–Ag alloy. *J. Alloys Compd.* **2019**, *802*, 276–289. [CrossRef]
35. Madhavan, R.; Ray, R.K.; Suwas, S. Texture transition in cold-rolled nickel–40 wt.% cobalt alloy. *Acta Mater.* **2014**, *74*, 151–164. [CrossRef]
36. Zhang, C.; Huang, G.; Liu, Q. Quantitative analysis of grain structure and texture evolution of dissimilar AA2024/7075 joints manufactured by friction stir welding. *Mater. Today Commun.* **2021**, *26*, 101920. [CrossRef]
37. Mao, W.M. *Crystallographic Texture and Anisotropy of Metallic Materials*; Science Press China: Beijing, China, 2002.
38. Zhang, L.; Wang, X.; Liu, X. Effect of Dynamic Recrystallization Mode on Texture Type for the Friction-Stir-Welded 6082-T6 Aluminum Alloy. *Mater. Rep.* **2019**, *33*, 665–669.
39. Zhang, L. *Microstructure Evolution of 6082-T6 Aluminum Alloy in Friction Stir Welding Process and Its Influence on Mechanical Properties of Welded Joints*; Lanzhou University of Technology: Lanzhou, China, 2018.
40. Fonda, R.W.; Bingert, J.F. Texture variations in an aluminum friction stir weld. *Scr. Mater.* **2007**, *57*, 1052–1055. [CrossRef]

metals

MDPI

Article

Multi-Objective Lightweight Optimization Design of the Aluminium Alloy Front Subframe of a Vehicle

Xiangchao Meng [1,2], Youping Sun [1,2,3,*], Jiangmei He [3], Wangzhen Li [1,2] and Zhifeng Zhou [4]

1 School of Mechanical and Transportation Engineering, Guangxi University of Science and Technology, Liuzhou 545006, China
2 Guangxi Earthmoving Machinery Collaborative Innovation Center, Liuzhou 545006, China
3 Guangxi Key Laboratory of Automobile Components and Vehicle Technology, Guangxi University of Science and Technology, Liuzhou 545006, China
4 Liuzhou Wuling New Energy Automobile Co., Ltd., Liuzhou 545007, China
* Correspondence: syptaiji@126.com

Abstract: The aluminium alloy front subframe of an automobile was developed through multi-operating condition topology optimization and multi-objective optimization methods. By considering the influences of loads on the strength, static stiffness, and modal of the aluminium alloy front subframe under typical operating conditions, the performance parameters of the aluminium alloy front subframe after topology optimization were obtained. After topology optimization was performed, the parametric model of the aluminium alloy front subframe was established. Based on the Isight optimization platform, sample points were generated with the optimal Latin hypercube test method, and the response surface approximate model was constructed. The minimum mass and maximum first-order frequency were taken as the objectives, the stress under typical working conditions did not exceed the set target value, and the maximum displacement of the installation point was taken as the constraint condition. The multi-objective particle swarm optimization algorithm was used to optimize the aluminium alloy front subframe. The error of the free modal and finite element free modal analysis of the aluminium alloy front subframe samples was less than 15%. The optimized aluminium alloy front subframe was 2.4 kg lighter than the original subframe under the premise of satisfying various performance indices, and the lightweight rate was up to 12%.

Keywords: multi-condition topology optimization; aluminium alloy front subframe; Isight; optimal Latin hypercube; response surface approximate model; multi-objective particle swarm optimization algorithm

1. Introduction

The subframe is an auxiliary device connecting the suspension connector and the body. It is an important part of the chassis. The traditional subframe is made of steel, which undoubtedly increases the weight of the whole vehicle. Studies have shown that for every 100 kg reduction in the vehicle weight, fuel consumption will be reduced by approximately 0.4 L per 100 km, and CO_2 emissions will be reduced by 8 to 11 g [1]. With the rise of new energy vehicles, an increasing number of automobile enterprises have begun to use aluminium alloy subframes. Using aluminium alloy subframes not only improves vehicle handling and stability but also reduces the weight of the vehicle. The production mode of aluminium alloy subframes mainly includes the extruded profile welding type and casting type. In recent years, many scholars have performed research on the development of subframes and have obtained some research results. Li et al. [1] proposed a subframe topology optimization method from a conceptual design to a detailed design, which further reduced the overall subframe weight, met the requirements of stiffness and natural frequency, passed the durability requirements, and, finally, reduced the subframe mass from the initial structure of 82.6 kg to 21.4 kg. Kim et al. [2] proposed a subframe method based on topology optimization, which took the bar section and stiffness value as the design variables to simultaneously optimize the topology and shape of the subframe. Ali and Fraser [3] used the finite element method to analyse a certain subframe of a Chevrolet vehicle, used LS-DYNA commercial software to model the subframe, analysed the subframe under normal driving conditions and vehicle collision conditions, and established the appropriate subframe material by comparison with the performance of the original subframe of the vehicle. Nam et al. [4] proposed a fatigue life evaluation technology for the aluminium subframes of automobiles. Based on the virtual test simulation technology of the nonlinear suspension component model, this technology can effectively predict the fatigue life of the vehicle chassis structure. By testing an actual aluminium subframe, the accuracy of the model was verified. Lee et al. [5] optimized the shape and thickness of the subframe to meet the requirements of multidisciplinary design optimizations (MDOs) and meet the weight, fatigue, crash, NVH, and K & C performance requirements. The proposed method was also suitable for complex vehicle design problems. Oh et al. [6] developed a subframe using a hydraulic forming technology. To improve the stiffness of the suspension and reduce the maximum stress that affects the durability cycle life, they adopted a variety of optimization design techniques, shape, size, and topology optimization, and formed the shape of the optimized rear suspension through a hydraulic forming process. The effectiveness of this design method was proven by finite element commercial software. To improve the NVH performance of vehicles, Park et al. [7] studied the front subframe of the vehicle, optimized the dynamic stiffness of the suspension bushing of the subframe, analysed the advantages and disadvantages of the integral subframe with different total shapes, and proposed structural reinforcement and other methods to improve the overall NVH performance of the vehicle. Hur and Lee [8] designed an integrated method for pre-tests, modelled associations and updated analyses of automobile subframes, evaluated model correlations by combining modal parameters of modal tests and finite element analyses, and analysed changes in the natural frequencies and MAC values of material characteristics based on sensitivity analysis results. The iterative method was used to modify the finite element model, and good results were obtained. Belingardi et al. [9] established the goals of strength, stiffness, and natural frequency of the subframe. The subframe designed using composite materials and optimizing the placement of stiffeners was superior to the steel subframe. Chiu Huang [10] designed a hydraulic moulding mould for the subframe, which could be used to achieve a lightweight subframe design. Han et al. [11] applied composite materials to the chassis component subframe, comparing the natural frequency and damping characteristics of steel and composite test pieces, achieving a certain improvement in stiffness, strength, and NVH performance of the composite subframe compared to the steel subframe, achieving a 50% weight reduction effect. Law et al. [12] proposed a new design method for the subframe after studying the subframe structure and design methods. The subframe structure designed using this method was superior in terms of lightness of weight and durability. Jang et al. [13] conducted a lightweight optimization design for the front subframe structure of a car. After optimization, while ensuring performance indicators, the optimized front subframe mass decreased by 30%, while verifying the practicality of the optimization design method. Fichera et al. [14] conducted theoretical calculation and analysis of the performance and load of the subframe, and then conducted topology optimization on the subframe structure, which resulted in a reasonable subframe structure. Rotondella et al. [15] constructed a subframe welding model and verified the reliability of the model through comparison of experiments and simulations. This method could predict structural dynamic characteristics. Da'Quan et al. [16] used ANSYS to study the topology optimization of the subframe under complex load conditions. The study found that the structural performance of the side member of the subframe was improved, while the weight of the subframe remained unchanged. Price et al. [17] designed a cast aluminium alloy subframe that reduced the weight by 40% while maintaining the same strength as the original subframe by optimizing weight and geometry. Liao et al. [18] designed a new high-strength steel subframe structure based on the stiffness and strength analysis of the subframe under various working conditions. Through comparative analysis of various performances with the original subframe, the feasibility of the new subframe was verified. Hamdi et al. [19] optimized the design of the subframe, which resulted in a significant improvement in NVH performance and a significant reduction in quality. Through verification, this method has made a positive contribution to improving vehicle NVH performance.

In this paper, an aluminium alloy front subframe envelope is first established based on the original steel front subframe, the topology optimization of the aluminium alloy front subframe envelope is carried out by the method of multi-condition topology optimization, and the optimal force path of the subframe is obtained. According to the topology optimization results in CATIA with regards to the detailed mathematical modelling design, its strength, stiffness, and modal performance analysis, the aluminium alloy front subframe performance indicators are obtained. Then, the aluminium alloy front subframe after topology optimization is used to establish a parametric model. Based on the Isight optimization platform, the optimal Latin hypercube test method is used to generate sample points [20], and an approximate response surface model of the aluminium alloy front subframe is constructed. Regarding a former deputy frame optimization design, the multi-objective particle swarm optimization algorithm is adopted for aluminium alloys. Through the analysis of the strength, static stiffness, and modal and fatigue durability, the performance of the aluminium alloy front subframe meets the design requirements, and the modal error of the aluminium alloy front subframe sample and the finite element analysis modal is less than 15%. Moreover, this model successfully passed the bench fatigue durability test.

2. Multi-Condition Topology Optimization of the Aluminium Alloy front Subframe

The aluminium alloy front subframe envelope body is established based on the hard points of each subframe connection. The principle of establishing the aluminium alloy front subframe envelope body should ensure that there is no movement interference between the envelope body and the surrounding parts. The established envelope body is shown in Figure 1. The aluminium alloy material used is a ZL114A cast aluminium alloy with a density of 2.68×10^{-9} t/mm^3, Poisson's ratio of 0.33, elastic modulus of 7.24×10^4 MPa, yield strength of 230 MPa, and tensile strength of 290 MPa.

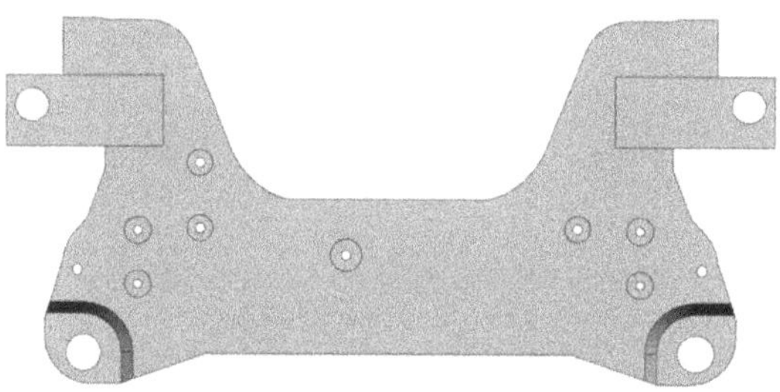

Figure 1. Aluminium alloy front subframe envelope.

The subframe will encounter various working conditions during use, so the influence of the various working conditions on the subframe structure should be comprehensively considered in exploitation and design. Regarding vice frames, the typical working conditions are the forward braking condition, backward braking condition, forward emergency braking condition, backward emergency braking condition, overconvex hull condition, over unilateral hull condition, over one side deep pit condition, ultimate steering condition, steering braking condition, steering drive condition, maximum driving acceleration condition, and diagonal torsion condition. The load information of each typical working condition extracted from the multibody dynamics analysis software Adams/car is summarized and used to check the strength of the aluminium alloy front subframe, as shown in Table 1. Due to space limitations, only part of the load information is shown here.

Table 1. Summary of load information of each typical working condition.

Load Loading Condition	Load Extraction Position	FX/N	FY/N	FZ/N	TX/N.mm	TY/N.mm	TZ/N.mm
Forward braking	Left front mounting point of lower control arm	3284.07	−13,698.91	976.43	−12,565.55	−4972.73	17,758.18
Backward braking		−1938.66	5250.36	440.28	2433.93	7908.52	−17,822.63
Forward emergency braking		3603.93	−12,297.02	503.65	−10,162.87	−6142.74	17,802.89
Backward emergency braking		−2567.50	5922.08	384.72	1116.05	10,045.76	−19,683.54
Overconvex hull		−312.96	−2037.32	330.97	−16,806.04	682.42	−1978.72
Over unilateral hull		−471.74	−3045.79	522.29	−17,978.29	624.39	−2796.88
Over one side deep pit		2486.79	−19,902.80	3368.50	−19,125.35	−2931.95	18,310.30
Ultimate steering		160.90	12,037.52	−1142.56	−14,712.09	−395.00	−8286.05
Steering braking		2172.52	−4666.31	467.00	−13,050.78	−3580.23	14,608.78
Steering drive		−1786.45	13,389.08	933.28	501.83	8327.25	−22,282.73
Maximum driving acceleration		−959.99	3711.35	661.49	11,256.40	5734.13	−10,161.87
Diagonal torsion		−330.45	−2186.29	192.93	−12,220.28	305.09	−2008.30
Forward braking	Left rear mounting point of lower control arm	6415.23	11,730.02	−399.92	−17,445.81	4444.70	1461.34
Backward braking		−3404.68	−6275.54	−788.88	2620.12	−84.12	−2600.15
Forward emergency braking		7662.75	13,728.38	10.44	−13,110.00	3269.03	2080.89
Backward emergency braking		−5073.90	−8907.96	−945.29	248.28	838.25	−2830.12
Overconvex hull		−392.48	−749.21	88.65	−28,138.93	7174.87	−3304.56
Over unilateral hull		−608.28	−1141.71	153.51	−30,776.38	7415.47	−3843.70
Over one side deep pit		3891.64	7412.00	−1043.45	−32,024.11	6871.48	−588.33
Ultimate steering		−355.88	−657.69	56.79	−23,436.62	6460.99	−3803.89
Steering braking		3647.18	7040.50	−319.42	−18,503.06	4927.97	506.97
Steering drive		−4046.88	−7266.67	−699.38	−772.01	983.45	−3228.57
Maximum driving acceleration		−1541.12	−2936.89	−705.71	15,925.78	−3253.31	−2182.21
Diagonal torsion		−439.74	−838.45	42.95	−18,237.10	5292.08	−1926.40
Forward braking	Steering gear left mounting point	−35.71	282.71	−42.47	−5.02	−335.09	−5480.94
Backward braking		−37.87	−61.51	−76.32	179.50	−675.91	1313.98
Forward emergency braking		−481.47	250.38	150.66	164.99	−2733.61	−5023.66
Backward emergency braking		352.29	−63.17	64.88	217.27	1332.79	1586.27
Overconvex hull		−89.30	74.38	9.83	166.08	−829.91	−1352.61
Over unilateral hull		−150.66	−221.96	79.69	−853.46	−1710.82	−1956.74
Over one side deep pit		502.53	2156.57	−778.59	1839.08	6405.35	−4312.94
Ultimate steering		349.17	887.21	−328.71	−229.11	4762.88	2935.18
Steering braking		47.06	633.97	−97.38	−452.64	1676.53	−2247.49
Steering drive		−730.26	−1058.18	−224.87	−1624.26	−8359.98	2036.35
Maximum driving acceleration		−290.88	−45.50	−356.40	43.70	−1885.46	733.59
Diagonal torsion		−98.05	−185.83	−19.56	−1143.84	−1169.12	−1932.29

The topology optimization method based on the compromise programming method in Optistruct software is used to optimize the topology of the aluminium alloy front subframe under multiple working conditions. The compromise programming method solves the optimal solution of the multi-objective optimization problem by approximating the one-by-one objective function. It can not only deal with the relative importance of different objective functions, but also give the sensitivity analysis of the optimal strategy to each constraint condition. It can also weigh multiple objectives, avoiding the disadvantage of not considering other influencing factors in the single objective optimization. The volume percentage and weighted compliance are taken as the optimization response; the goal was to minimize the weighted compliance. The volume percentage is taken as the constraint condition, and the symmetric constraint and the process restriction of the unidirectional drawing die are introduced. Considering the influences of 12 typical working conditions on the topology optimization, the corresponding topology optimization model can be expressed as [11]:

$$\begin{cases} \min C(\rho) = \left\{ \sum_{k=1}^{m} w_k^q \left[\frac{C_k(\rho) - C_k^{\min}}{C_k^{\max} - C_k^{\min}} \right]^q \right\}^{\frac{1}{q}} \\ s.t \frac{V(\rho)}{V_0} \leq volfrac \\ cons = volfrac \\ geo = symmetry \\ \sigma_{\max} < [\sigma] \end{cases} \tag{1}$$

where $C(\rho)$ is the objective function of comprehensive compliance; m is the total number of working conditions; w_k is the weight value in the $k-th$ working condition; $C_k(\rho)$ is the objective function of compliance under the $k-th$ working condition; $C_k^{\min}$ and $C_k^{\max}$ are the minimum and maximum compliance values of the $k-th$ working condition, respectively; $V(\rho)$ is the volume after topology optimization; V_0 is the volume before topology optimization; $volfrac$ is the volume percentage; $cons$ is the constraint condition; geo is the geometric limit; $symmetry$ is the symmetric constraint; $\sigma_{\max}$ is the maximum stress under the ultimate load condition; and $[\sigma]$ is the allowable stress of the material.

By dividing the aluminium alloy front subframe envelope into the design area and the non-design area, the minimum and maximum member sizes are set. The minimum member size should be more than 3 times the average mesh size of the model, and the maximum member size should be more than 2 times the minimum member size, in order to eliminate the excessive concentration of materials. Because the aluminium alloy front subframe envelope mesh size is 5 mm, here the minimum member size is set to 16 and the maximum member size is set to 40. After 21 iterations, the topology optimization results of the aluminium alloy front subframe envelope obtained when the threshold is adjusted to 0.3 are shown in Figure 2.

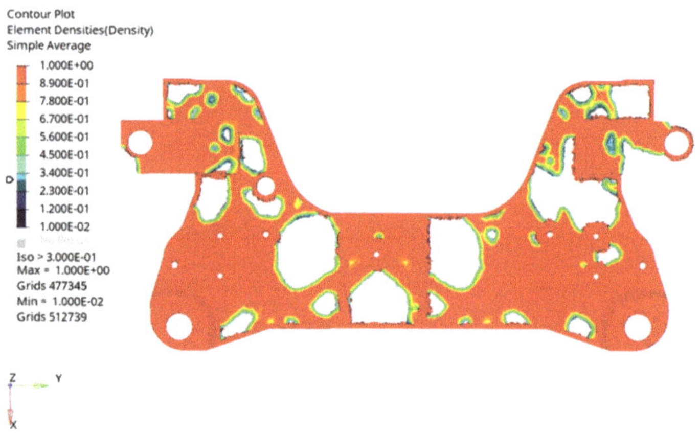

(a)

Figure 2. *Cont.*

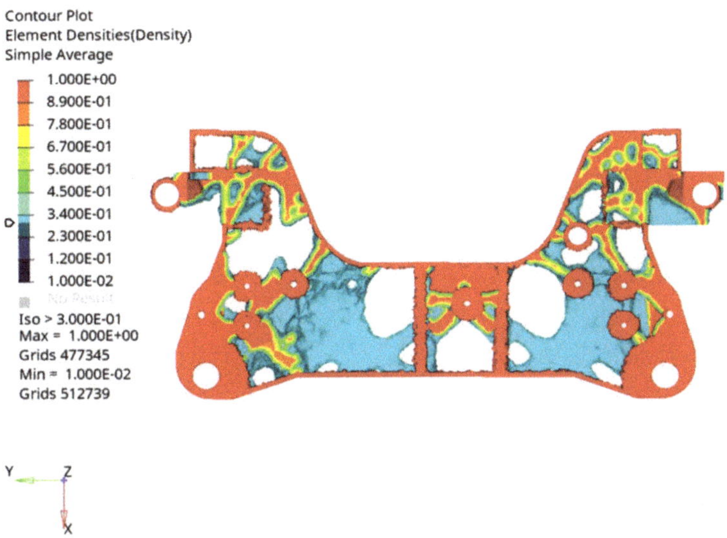

(b)

Figure 2. Topology optimization results of aluminium alloy front subframe envelope under multiple working conditions. (**a**) Positive results of multi-condition topology optimization. (**b**) Rear result of multi-condition topology optimization.

Based on the topology optimization results, according to the stress path of the aluminium alloy front subframe envelope and considering the position of the reinforcing bars, the detailed digital modelling design of the aluminium alloy front subframe is carried out in CATIA. The detailed design digital modelling of the aluminium alloy front subframe is shown in Figure 3. The weighing result of the aluminium alloy front subframe in CATIA is 18.55 kg.

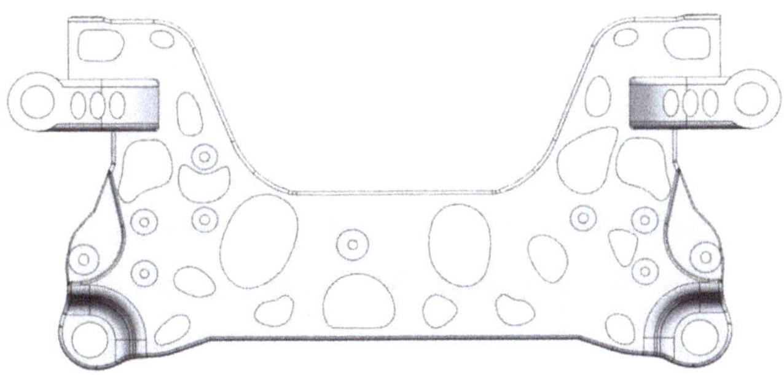

Figure 3. Aluminium alloy front subframe detailed design of digital model.

3. Performance Analysis of the Aluminium Alloy Front Subframe

3.1. Strength Performance Analysis of the Aluminium Alloy Front Subframe

The strength performance of the aluminium alloy front subframe under each typical working condition is analysed according to the typical working condition load extracted by multibody dynamics. The establishment of displacement and load boundary conditions for the aluminium alloy front subframe is shown in Figure 4.

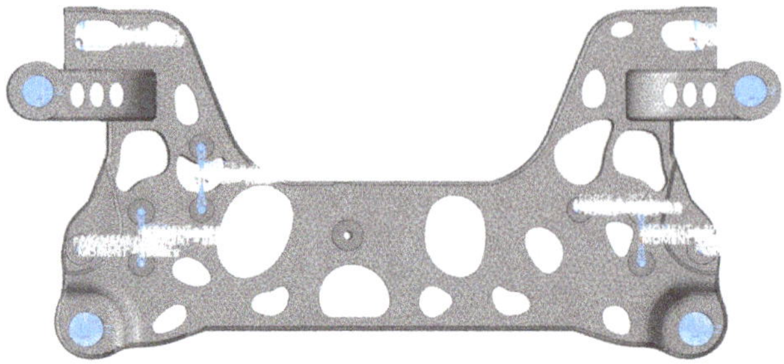

Figure 4. Establishment of displacement and load boundary conditions for aluminium alloy front subframe.

The analysis results are shown in Figure 5. Due to space limitations, only part of the stress analysis results of the aluminium alloy front subframe under typical working conditions are listed. According to the cloud diagram of the stress analysis results, it can be seen that the stress of the aluminium alloy front subframe under each typical working condition is the maximum when the aluminium alloy front subframe is under the one side deep pit condition. Moreover, the stress value is 180.6 MPa, which is far less than the material yield strength of 230 MPa and is within the safe range of the material. This shows that the aluminium alloy front subframe meets the strength performance requirements and has a large space for optimization.

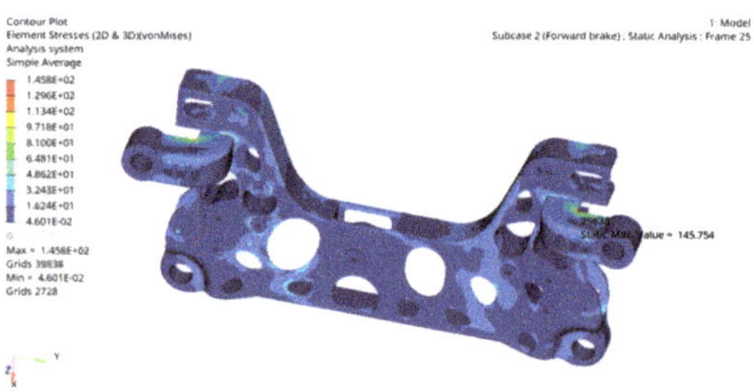

(a)

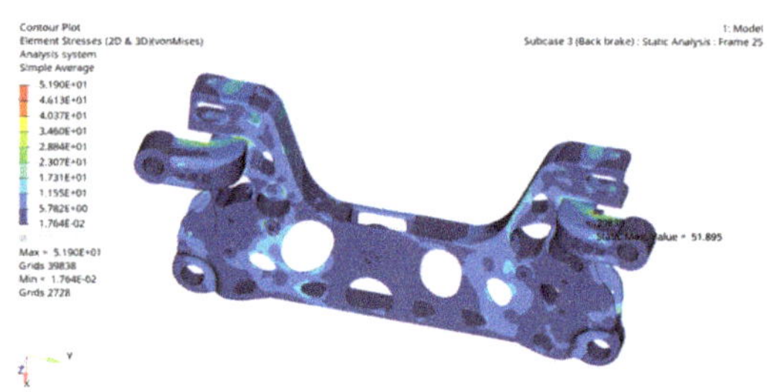

(b)

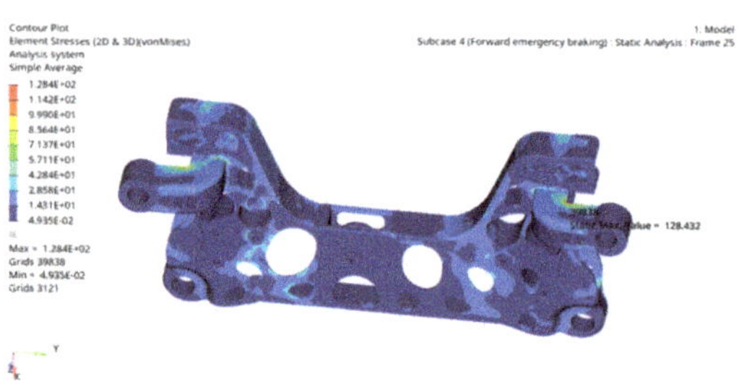

(c)

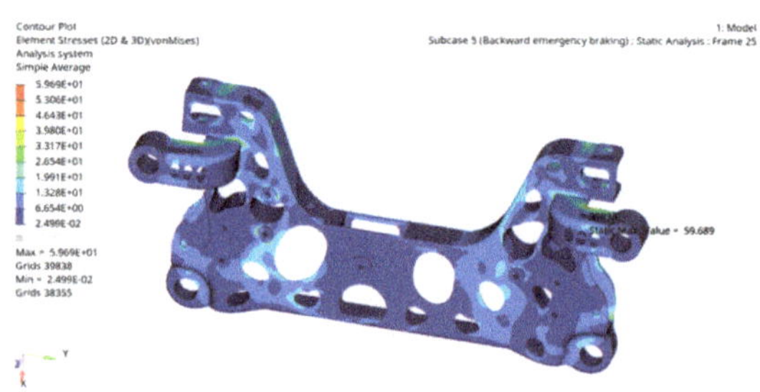

(d)

Figure 5. *Cont.*

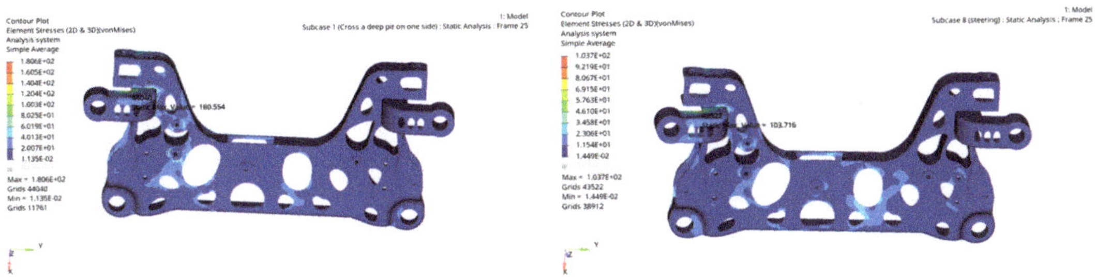

(e) (f)

Figure 5. Cloud diagram of stress results of aluminium alloy front subframe under some typical working conditions. (**a**) Forward braking condition. (**b**) Backward braking condition. (**c**) Forward emergency braking condition. (**d**) Backward emergency braking condition. (**e**) Over one side deep pit condition. (**f**) Ultimate steering condition.

3.2. Static Stiffness Performance Analysis of the Aluminium Alloy Front Subframe

As a bearing part, the aluminium alloy front subframe needs to meet certain static stiffness requirements. Thus, it is necessary to carry out static stiffness analysis on the aluminium alloy front subframe. The static stiffness can be calculated with the ratio of the load of the subframe to the deformation under the corresponding load, as shown in Equation (2):

$$K = \frac{F}{L} \tag{2}$$

where K is the static stiffness value of the subframe, F is the load of the subframe at each connection hard point, and L is the deformation amount under the corresponding load.

In the static stiffness analysis of the aluminium alloy front subframe, the three translational degrees of freedom and three rotational degrees of freedom of the connection position between the aluminium alloy front subframe and the body are constrained. The load condition is a unit force of 1000 N in each of the three translational directions being applied to each installation point of the aluminium alloy front subframe. Table 2 summarizes the static stiffness analysis results at each mounting point of the aluminium alloy front subframe.

Table 2. Static stiffness analysis results of each mounting point of the aluminium alloy front subframe.

Number	Position of Mounting Point	Direction	Static Stiffness Value (N/mm)	Target Static Stiffness Value (N/mm)
1	Lower arm left front point	X Y Z	23,810 25,641 40,000	12,000 16,000 5000
2	Lower swing arm left back point	X Y Z	50,0000 90,909 52,631	20,000 20,000 5000
3	Lower arm right front point	X Y Z	13,889 45,455 45,455	12,000 16,000 5000
4	Lower swing arm right back point	X Y Z	200,000 200,000 62,500	20,000 20,000 5000
5	Steering gear left point	X Y Z	40,000 55,556 22,727	25,000 25,000 8000

Table 2. *Cont.*

Number	Position of Mounting Point	Direction	Static Stiffness Value (N/mm)	Target Static Stiffness Value (N/mm)
6	Steering gear right point	X Y Z	71,429 90,909 28,571	25,000 25,000 8000
7	Stabilizer bar left point	X Y Z	58,824 76,923 40,000	10,000 5000 16,000
8	Stabilizer bar right point	X Y Z	100,000 111,111 47,619	10,000 5000 16,000

According to the static stiffness analysis results of each installation point, the static stiffness of the aluminium alloy front subframe at each installation point reaches and exceeds the target value.

3.3. Free Modal Analysis of the Aluminium Alloy Front Subframe

The natural frequency is a characteristic property of objects. Resonance occurs when externally applied vibrations coincide with the natural frequency of the object. When resonance occurs, vibration and noise will increase dramatically and eventually lead to the destruction of parts [11]. Because the low-order mode has a great influence on the vibration performance of the subframe, usually only the influence of the low-order mode on the subframe is considered. When the vibration frequency of the subframe is close to the excitation frequency, the damage of the parts will be accelerated. Therefore, the frequency of the subframe is close to the excitation frequency, and the frequency of the first-order modal should be increased as much as possible. The modal calculation of the aluminium alloy front subframe is carried out under the OptiStruct module of HyperMesh. The Lanczos algorithm is used to analyse the free modal of the aluminium alloy front subframe. The Lanczos method has a fast convergence speed. The first 6 modals are rigid body modals, so only the modals after the 7th order are considered. To clearly see the modal shape of the subframe, the scaling coefficient is set to 3. The first 6 orders of nonrigid body modal shapes are shown in Figure 6.

The subframe is mainly affected by the ground excitation frequency and engine excitation frequency, while the excitation frequency of general highway pavement is below 25 Hz, and the engine excitation frequency can be calculated by the following formula:

$$\mathrm{f} = \frac{n \cdot z}{120} \tag{3}$$

where n is the idle speed of the engine, r/min, and z is the number of cylinders of the engine.

The idle speed of the engine of the model under study is 1500 r/min, and the engine is a four-cylinder engine. Thus, the engine excitation frequency is calculated to be 50 Hz, and the free modal of the aluminium alloy subframe should be beyond 50 Hz to avoid resonance. The results of the finite element analysis show that the free modal first-order frequency of the aluminium alloy subframe is 97 Hz. It is much higher than the ground and engine excitation frequency, and the difference between each modal frequency is more than 20 Hz. Thus, the aluminium alloy subframe will not resonate with the excitation frequency, and the modal performance of the aluminium alloy subframe meets the design requirements.

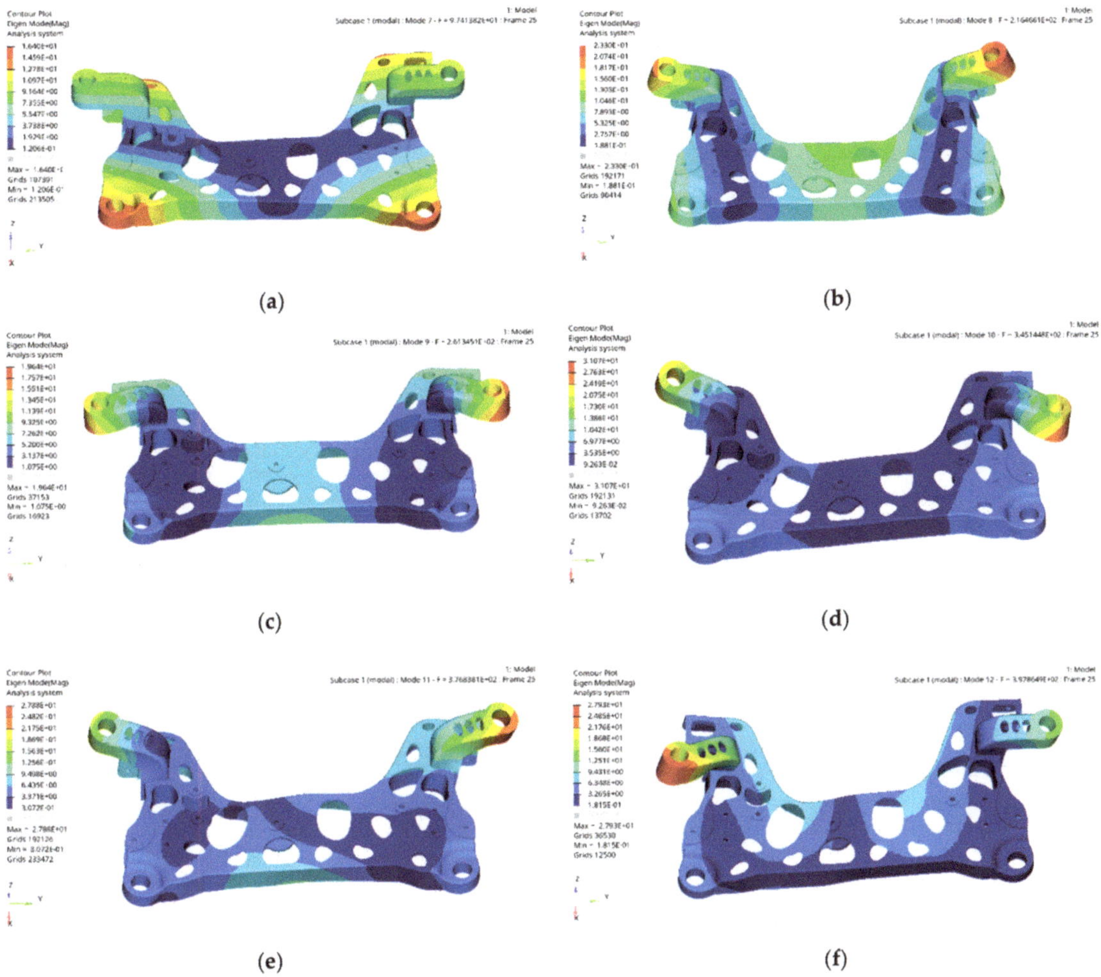

Figure 6. Non-rigid body modal shapes for the front 6th order of aluminium alloy front subframe. (**a**) First order modal shape. (**b**) Second order modal shape. (**c**) Third modal shape. (**d**) Fourth modal shape. (**e**) Fifth modal shape. (**f**) Sixth modal shape.

4. Lightweight Optimization Design Based on Multi-Objective Methods

4.1. Establishment of the Parametric Model of the Aluminium Alloy Front Subframe

The parametric model of the aluminium alloy front subframe is created in the Morph module of Ansa. The Morph module can quickly carry out mesh deformation and create shape design variables. Four positions of the aluminium alloy front subframe are selected to set the parametric variables. The parameterized model and optimized positions are shown in Figure 7. The four positions are named F_01, R_01, LZ_t, and LZ_b. The variation range of the parameterized variables at four positions of the aluminium alloy front subframe is shown in Table 3.

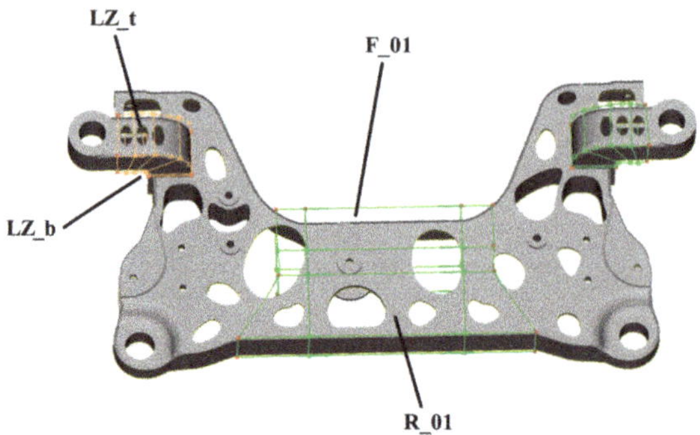

Figure 7. Parameterized model and optimized position of aluminium alloy front subframe.

Table 3. Variation range of design variables of aluminium alloy front subframe.

Design Variable	Initial Value (mm)	Lower Limiting Value (mm)	Upper Limit Value (mm)
F_01	0	−20	10
R_01	0	−30	15
LZ_t	0	−5	5
LZ_b	0	−8	0

4.2. Establishment of the Optimization Mathematical Model

Based on the strength performance obtained in Section 3.1 and the performance of the modal analysis obtained in Section 3.3, the aluminium alloy front subframe can be further optimized, and the mass can be further reduced. The design of the aluminium alloy front subframe should meet various performance requirements at the same time, including the strength, static stiffness, modal performance, and fatigue endurance performance [18], which is a multi-objective optimization problem [21]. The establishment of a multi-objective optimization design mainly includes the determination of the design variables, constraints, and optimization objectives, as well as the selection of an optimization algorithm. In this paper, the multi-objective optimization platform Isight, combined with Ansa, OptiStruct, and Meta, is adopted to conduct the multi-objective optimization of the aluminium alloy front subframe. The parameters of the aluminium alloy front subframe model were established in the morph module of Ansa; the strength, stiffness, and modal were calculated in the OptiStruct module; and the strength, stiffness, and modal values were extracted in the Meta software. The specific mathematical model is as follows:

$$Object : \{Mass_{\min}(X) \le 18, Freq\_1_{\max}(X), Disp_{\min}(X)\} \tag{4}$$

$$s.t. \left\{ \begin{array}{c} 97 \le Freq\_1 \le 200 \\ \mathrm{Max}\_stress1 \le 180 \\ \mathrm{Max}\_stress2 \le 145 \\ \mathrm{Max}\_stress3 \le 128 \\ Disp_{\min}(\mathrm{X}) \le 0.07 \end{array} \right\} \tag{5}$$

$$Variable : \left\{ \begin{array}{c} -20 \le F\_01 \le 10 \\ -8 \le LZ\_b \le 0 \\ -5 \le LZ\_t \le 5 \\ -30 \le R\_01 \le 15 \end{array} \right\} \tag{6}$$

where $Mass_{\min}(X)$ is the minimum mass of the aluminium alloy front subframe; $Freq\_1_{\max}(X)$ is the first order frequency of the nonrigid body modal of the aluminium alloy front subframe; $Disp_{\min}(X)$ is the minimum maximum displacement; $Freq\_1$ is the first order nonrigid body modal of the aluminium alloy front subframe, Max_*stress*1, Max_*stress*2, Max_*stress*3 are the stress values of the over one side deep pit condition, forward braking condition, and forward emergency braking condition, respectively; and *F*_01, *R*_01, *LZ*_*t*, *LZ*_*b* are the four defined optimization design variables.

4.3. Experimental Design and Establishment of the Approximate Model

Through experimental design, key test factors of the aluminium alloy front subframe can be quickly identified, the optimal combination of the variables and parameters can be identified, and the relationship between the variables and optimization objectives can be analysed to reduce the subjectivity of the design and improve the reliability of the design [22]. In this paper, the optimal Latin hypercube experimental design algorithm is adopted [23,24]. It can make the fitting of factors and responses more accurate and make the sampling points more uniform. The distribution of the optimal Latin hypercube sample points is shown in Figure 8. The number of DOE design points is 25. Figure 9 shows the DOE multi-objective optimization platform built based on Isight.

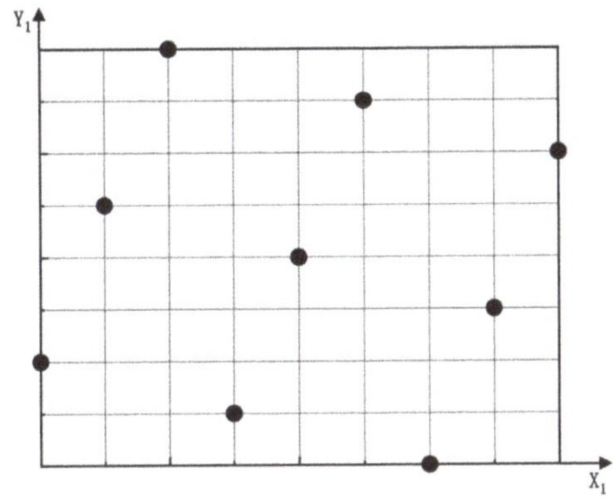

Figure 8. Distribution of optimal Latin hypercube sample points.

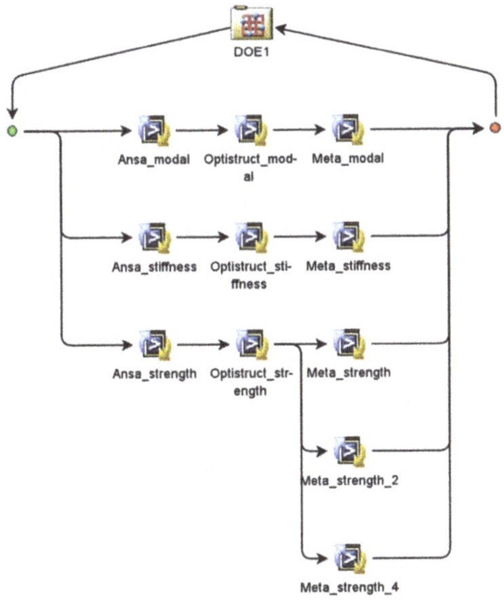

Figure 9. Multi-objective experimental design optimization platform.

DOE1 is the experimental design module; the Ansa_modal branch is the modal optimization branch, the Ansa_stiffness branch is the static stiffness optimization branch, and the Ansa_strength is the strength optimization branch. Regarding the strength optimization, three working conditions with maximum stresses are considered. They are the over one side deep pit condition, forward braking condition, and forward emergency braking condition.

The approximate model method is a method used to approximate a set of input variables and output variables by the mathematical model method. The relationship between the input variables and output response of the approximate model is as follows:

$$y(x) = \widetilde{y}(x) + \varepsilon \tag{7}$$

where $y(x)$ is the actual response value and an unknown function; $\widetilde{y}(x)$ is the response approximation with a known polynomial; and ε is the error between the actual value and the approximate value.

When the approximate model has a sufficient accuracy, the approximate model can be used to replace the actual model for simulation analyses to shorten the calculation time and improve the calculation efficiency. Figure 10 shows the construction process of the aluminium alloy front subframe approximate model.

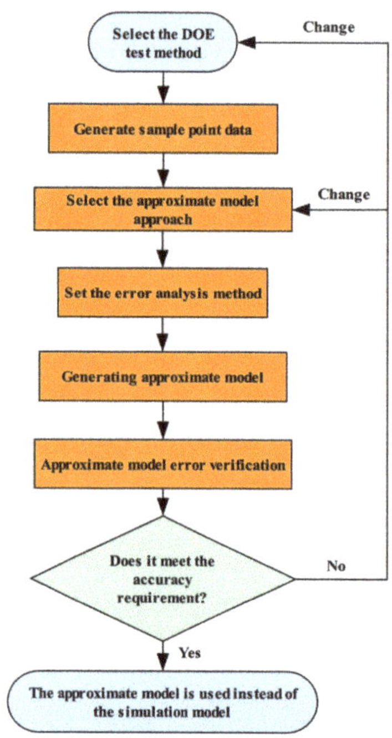

Figure 10. Process of building approximate model of aluminium alloy front subframe.

A response surface model is used to construct approximate models [25–27]. Because four positions of the aluminium alloy front subframe are defined as parametric design variables, a second-order polynomial is selected to establish the response surface approximation model, and the specific formula is as follows [28]:

$$\begin{aligned}\widetilde{y} = \; & \beta_0 + \beta_1 x_1 + \beta_2 x_2 + \cdots + \beta_M x_M + \beta_{M+1} {x_1}^2 + \\ & \beta_{M+2} {x_2}^2 + \cdots + \beta_{2M} {x_{2M}}^2 + \sum_{i \neq j} \beta_{ij} x_i x_j\end{aligned} \tag{8}$$

where $\widetilde{y}$ is the response approximation; $\beta_0, \beta_1, \beta_2 \cdots \beta_M \cdots \beta_{ij}$ are the polynomial coefficients; and $x_1, x_2 \cdots x_M \cdots x_i$ are the components of the design variable.

After the approximate model is constructed, an error test should be conducted on the fitting results of the approximate model, and the R-squared error analysis method should be used to test the fitting accuracy between the approximate model and the actual sample points [29–31]. The closer R-squared is to 1, the higher the accuracy of the approximate model, with a high credibility. Moreover, a fitting accuracy above 0.9 is generally acceptable. The R-squared value of each response of the aluminium alloy front subframe is shown in Table 4, and it can be seen from Table 4 that the fitting accuracy of each response is greater than 0.98, indicating that the fitting effect of each response variable is good. An approximation model can be used in place of the actual model for multi-objective optimization simulations, thus greatly shortening the computing workload and calculation time.

Table 4. Fitted R-squared values for each response.

Responses	Total Error
Freq_1	0.99675
Mass	0.99995
Max_stress1	0.99493
Max_stress2	0.99783
Max_stress3	0.99781
Disp	0.9897

Figure 11 shows the fitting results of the predicted and actual values of each response of the aluminium alloy front subframe. The abscissa is the predicted value, and the ordinate is the actual value. The points on the 45° oblique line indicate that the predicted results are the same as the actual results.

Figure 12 shows the response surface approximation model between the first-order frequency response of the aluminium alloy front subframe and each variable, and the response surface can better express the changing trend of the test point. The corresponding response value of the two variables can be found at any value within the design range. It can be seen from the equal potential surface that the degree of linearity of the fitting is relatively high.

4.4. Multi-Objective Optimization of the Aluminium Alloy Front Subframe Based on the Approximate Model

The multi-objective particle swarm optimization algorithm [32] was used to optimize the aluminium alloy front subframe. Multi objective particle swarm optimization is a population-based search process that uses the principles of evolutionary computation: first, initialize from a group of random populations, and search for the optimal value through updating the population generations. The evolution process depends on the previous population. Compared to other optimization algorithms, the advantage of the Multi Objective Particle Swarm optimization algorithm is its strong versatility, simplicity, and ease of implementation, with fewer parameters to adjust. Figure 13 shows the multi-objective optimization design platform based on the approximate model, in which Optimization is the multi-objective optimization module. Moreover, Approximation is the approximate model obtained after generating sample points through the optimal Latin hypercube experimental design. The parameters of the multi-objective particle swarm optimization algorithm are as follows: the maximum number of iterations is set to 100, in order to obtain more optimization results; the number of particles is set to 10; the inertia weight is set to 0.9; the global increment is set to 0.9; the particle increment is set to 0.9; the maximum velocity is set to 0.1; the max failed runs is set to 5; the failed run penalty value is set to 1.0×10^{30}; the failed run objective value is set to 1.0×10^{30}.

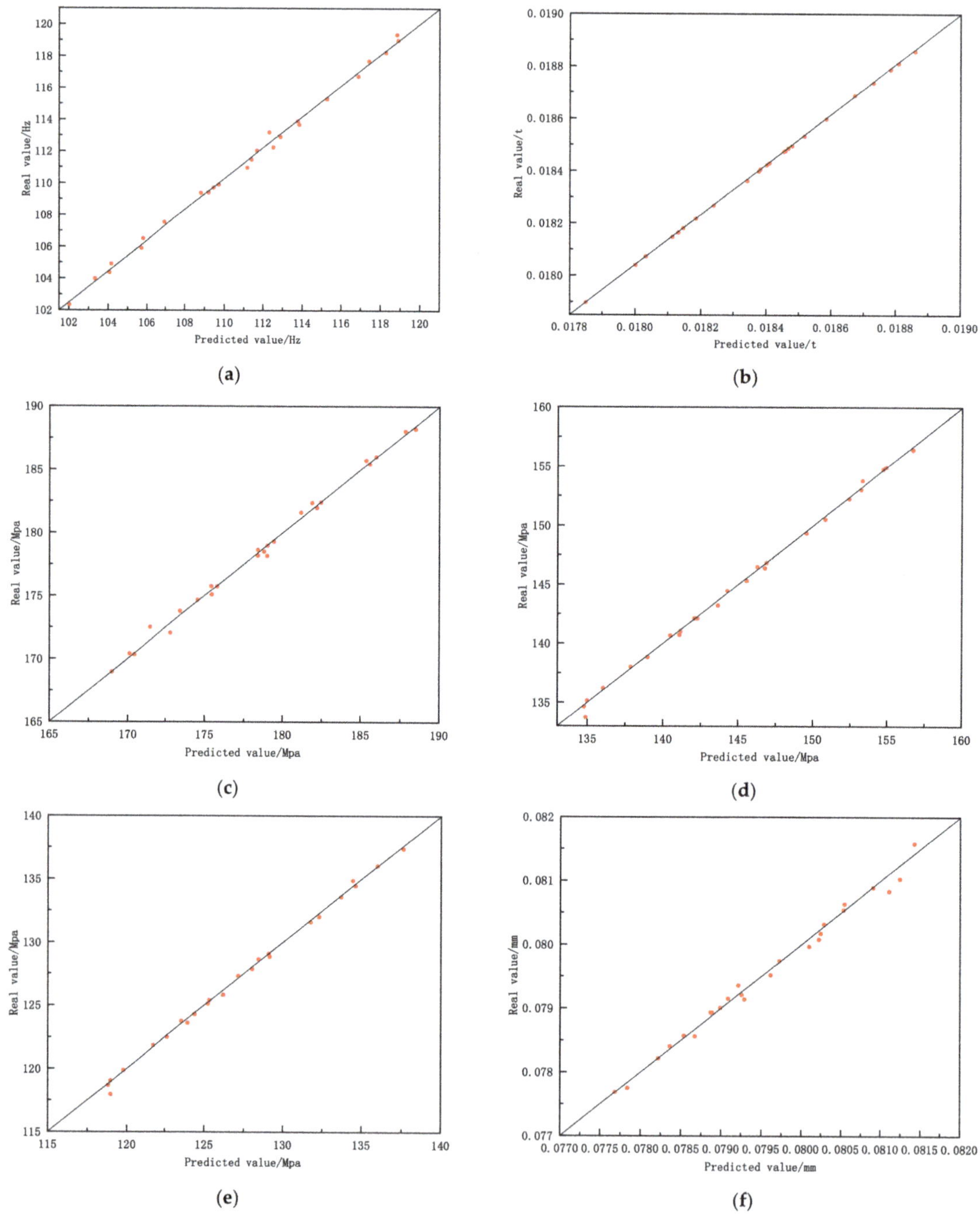

Figure 11. The fitting results of the predicted and actual response values of the aluminium alloy front subframe. (**a**) Results of first-order modal frequency fitting. (**b**) Mass fitting results. (**c**) Stress fitting results of over one side deep pit condition. (**d**) Stress fitting results of forward braking condition. (**e**) Stress fitting results of forward emergency braking condition. (**f**) Fitting results of the maximum displacement of the right front point of the lower arm.

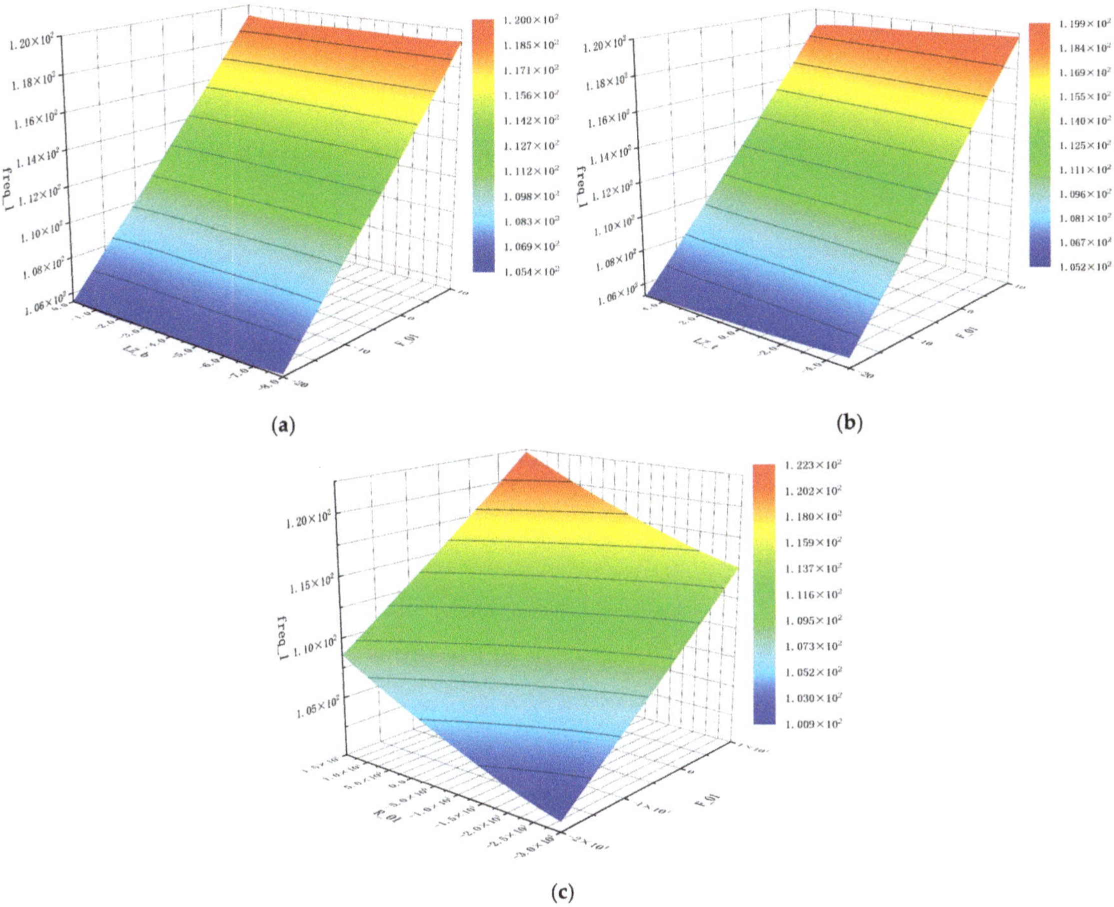

Figure 12. First order frequency response and variable response surface approximation model of the aluminium alloy front subframe. (**a**) Response surface approximation model between the first order frequency of the aluminium alloy front subframe and design variables LZ_b and F_01. (**b**) Response surface approximation model between the first order frequency of the aluminium alloy front subframe and design variables LZ_t and F_01. (**c**) Response surface approximation model between the first-order frequency of the aluminium alloy front subframe and design variables R_01 and F_01.

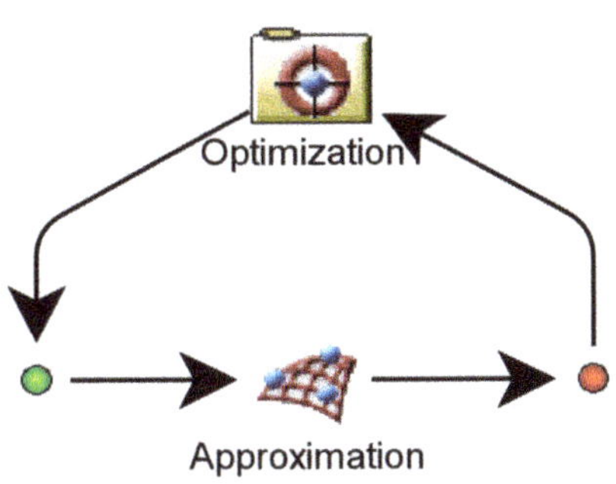

Figure 13. Approximate model multi-objective optimization design platform.

The multi-objective optimization design based on the response surface approximation model iteration was run 1002 times. The local effect diagram can show which had the greatest influence on the target response variable and can find the weak link in the design. To better guide the design, the local effect between each variable and the response is shown in Figure 14. Blue represents a positive effect, and red represents a negative effect. From the local effect diagram, it can be seen that F_01 has the most significant effect on the frequency, R_01 has the most significant effect on the mass, and LZ_b has the most significant effect on the stress over one side deep pit condition and forwards braking condition. The optimal solution set was selected from the 1002 iterations to make it closest to the set target value, and the optimal solution of the four variables was finally determined, as shown in Table 5.

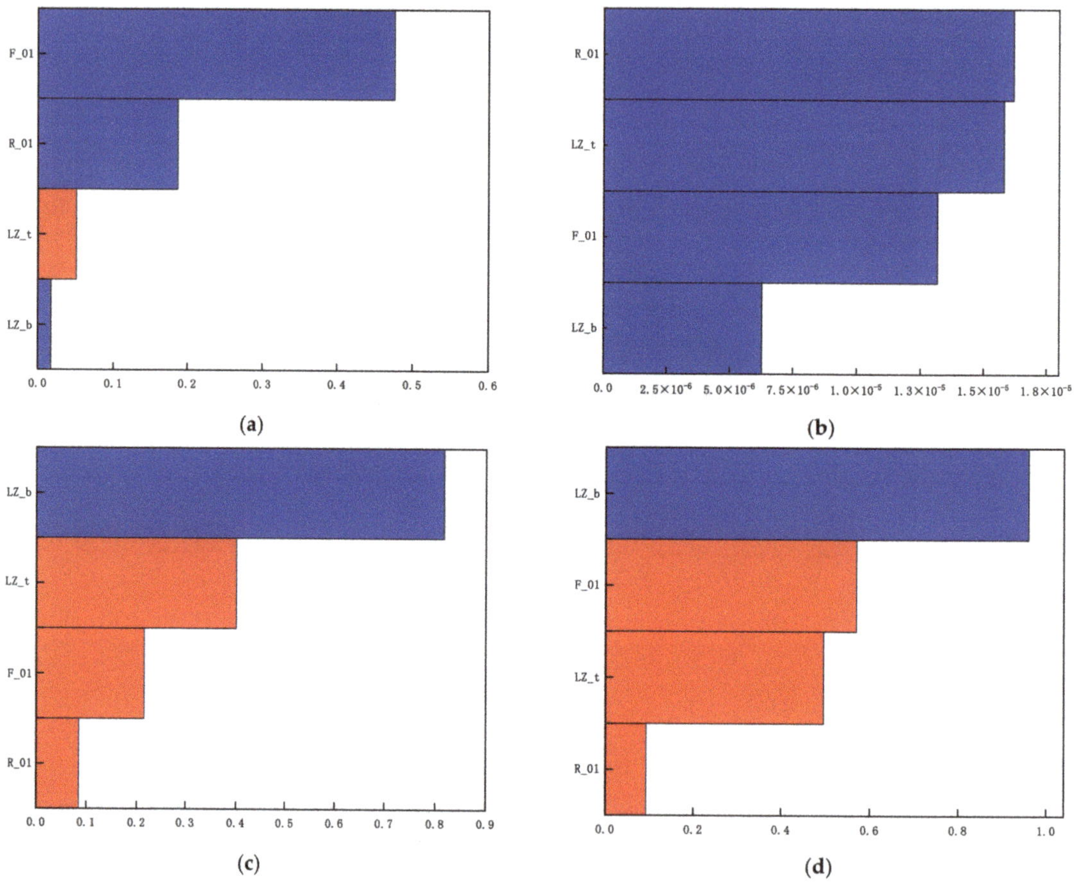

Figure 14. Local effect diagram of each variable and response. (**a**) Frequency local effect diagram. (**b**) Mass Local Effect diagram. (**c**) Local effect diagram of stress in over one side deep pit condition. (**d**) Local effect diagram of forward braking condition.

Figure 15 shows the digital–analogue comparison before and after multi-objective optimization of the aluminium alloy front subframe. After optimization, the weight is reduced from 18.55 kg to 17.6 kg, reducing the weight by 0.95 kg.

Table 5. Comparison of variable data before and after optimization.

Design Variable	Initial Value/mm	Lower Limiting Value/mm	Upper Limit Value/mm	Optimal Value/mm
F_01	0	−20	10	1.5
R_01	0	−30	15	−27
LZ_t	0	−5	5	−4
LZ_b	0	−8	0	−4.5

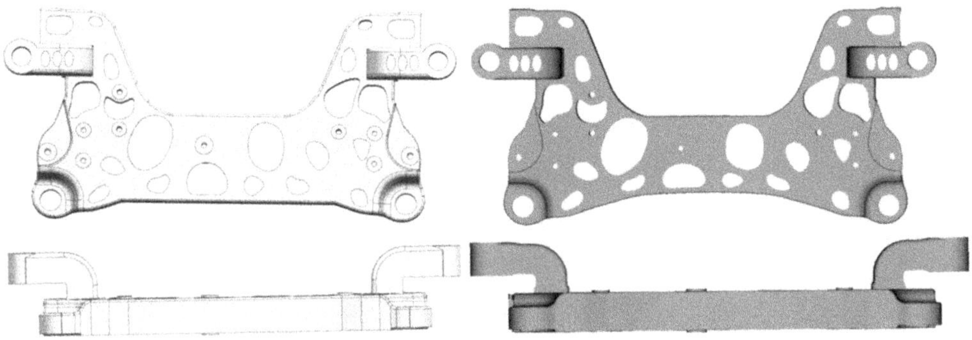

Figure 15. Digital and analogue comparison before and after multi-objective optimization of aluminium alloy front subframe.

5. Performance Analysis and Verification of the Aluminium Alloy Front Subframe after Optimization

5.1. Strength Performance Analysis of the Optimized Aluminium Alloy Front Subframe

The displacement boundary conditions and load boundary conditions from Section 3.1 are used to analyse the strength performance of the optimized aluminium alloy front subframe. Figure 16 shows the cloud diagram of the stress analysis results of the optimized aluminium alloy front subframe under some typical working conditions. According to the cloud diagram of the stress analysis results, the working condition corresponding to the maximum stress of the optimized aluminium alloy front subframe is the limiting condition of the one side deep pit condition, and the stress value is 179.3 MPa, which is 1.3 MPa less than the value of 180.6 MPa before optimization.

Table 6 shows the comparative analysis results of the stress of the aluminium alloy front subframe before and after optimization. After optimization, the stress of the aluminium alloy front subframe is reduced in all typical working conditions, which is less than the material yield strength. After optimization, the aluminium alloy front subframe meets the strength requirements.

Table 6. Comparative analysis results of stress before and after optimization of aluminium alloy front subframe.

Typical Working Conditions	Model Stress before Optimization/MPa	Model Stress after Optimization/MPa
Forward braking condition	145.7	141.7
Backward braking condition	51.9	50.2
Forward emergency braking condition	128.4	124.8
Backward emergency braking condition	59.7	57.8
Over one side deep pit condition	180.6	179.3
Ultimate steering condition	103.7	101.9

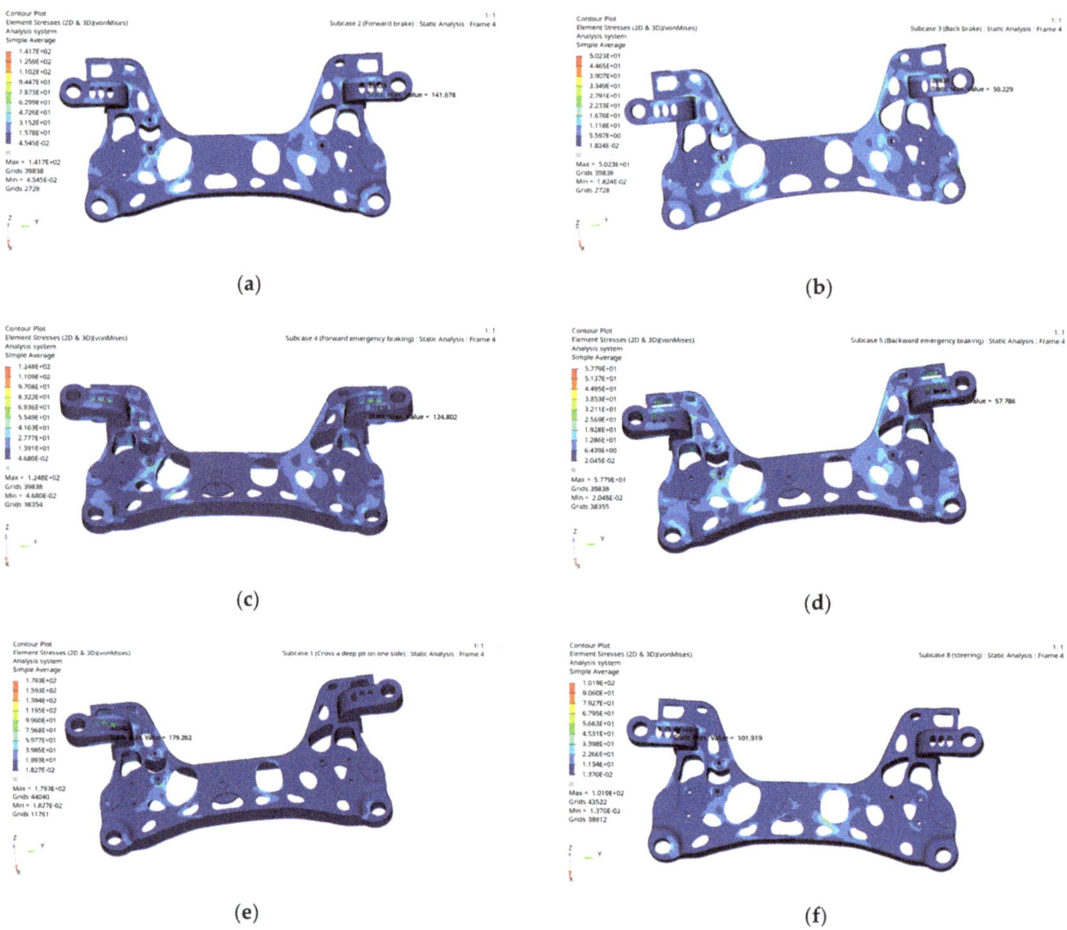

Figure 16. Cloud image of stress analysis results of aluminium alloy front subframe under some typical working conditions after optimization. (**a**) Forward braking condition. (**b**) Backward braking condition. (**c**) Forward emergency braking condition. (**d**) Backward emergency braking condition. (**e**) Over one side deep pit condition. (**f**) Ultimate steering condition.

5.2. Static Stiffness Performance Analysis of the Optimized Aluminium Alloy Front Subframe

Table 7 shows the comparative analysis results of the static stiffness performance of the aluminium alloy front subframe before and after optimization. The results show that the static stiffness values of each hard point of the aluminium alloy front subframe have little change compared with those before optimization, which is within the acceptable range and meets the design requirements.

5.3. Free Modal Performance Analysis of the Optimized Aluminium Alloy Front Subframe

Free modal analysis is carried out on the optimized aluminium alloy front subframe, and the first six rigid body modals are filtered out. The cloud diagram of the free modal analysis results is shown in Figure 17. Table 8 shows the comparison results of the free modal analysis before and after optimization of the aluminium alloy front subframe. After optimization, the first-order modals and fifth-order modals of the aluminium alloy front subframe are slightly reduced, while the other modals are improved. These results are higher than the ground and engine excitation frequencies, and the difference between each modal frequency is more than 20 Hz. The optimized aluminium alloy front subframe meets the modal performance requirements.

Table 7. Comparative analysis results of static stiffness performance of aluminium alloy front subframe before and after optimization.

Number	Position of Mounting Point	Direction	Static Stiffness Value of the Model before Optimization (N/mm)	Static Stiffness Value of Optimized Model (N/mm)
1	Lower arm left front point	X Y Z	23,810 25,641 40,000	24,390 23,809 37,037
2	Lower swing arm left back point	X Y Z	500,000 90,909 52,631	500,000 83,333 50,000
3	Lower arm right front point	X Y Z	13,889 45,455 45,455	13,888 40,000 40,000
4	Lower swing arm right back point	X Y Z	200,000 200,000 62,500	200,000 200,000 62,500
5	Steering gear left point	X Y Z	40,000 55,556 22,727	38,461 52,631 21,739
6	Steering gear right point	X Y Z	71,429 90,909 28,571	71,428 83,333 27,027
7	Stabilizer bar left point	X Y Z	58,824 76,923 40,000	58,823 83,333 38,461
8	Stabilizer bar right point	X Y Z	100,000 111,111 47,619	100,000 111,111 45,454

Table 8. Comparison of free modal analysis results before and after optimization of aluminium alloy front subframe.

Order	Model Frequency before Optimization/Hz	Optimized Model Frequency/Hz
First order modal	97	96
Second order modal	216	222
Third order modal	261	268
Fourth order modal	345	346
Fifth order modal	376	373
Sixth order modal	397	417

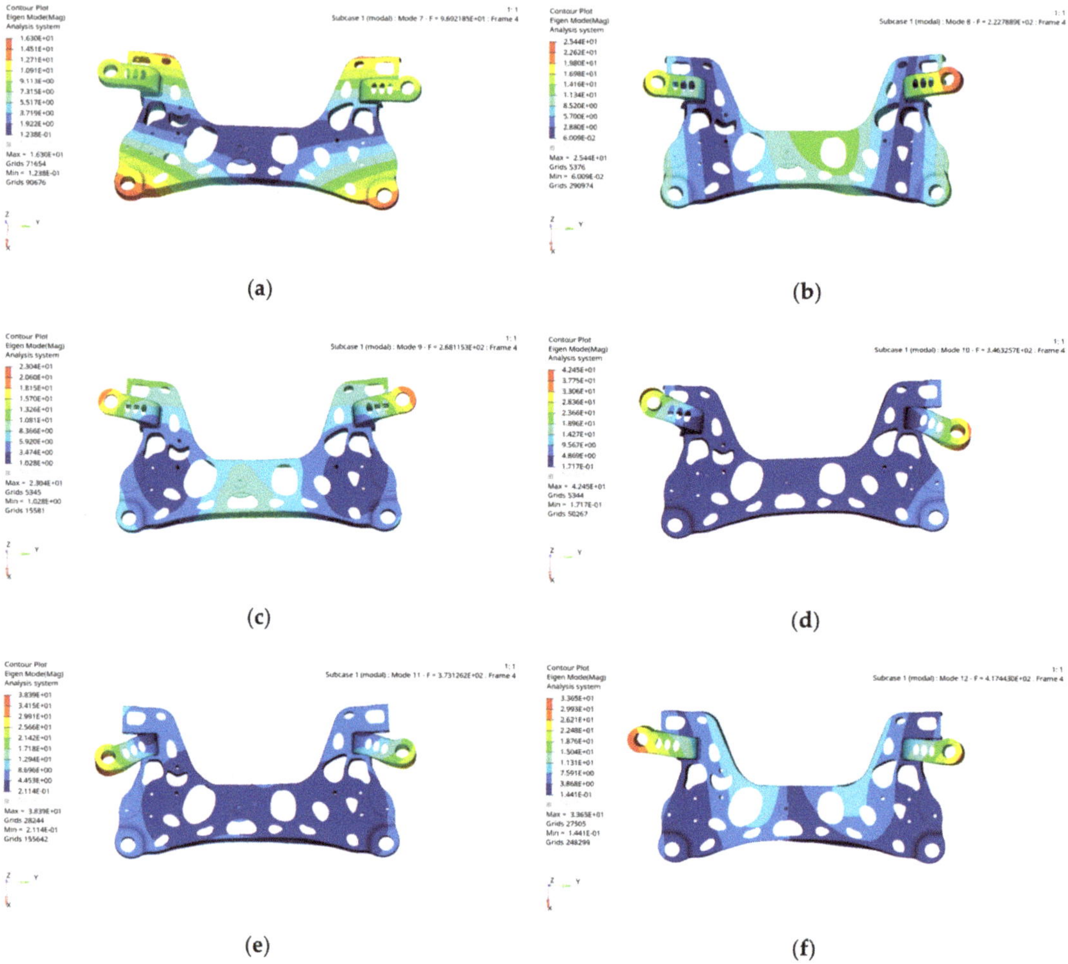

Figure 17. Optimized non-rigid body modal shapes for the front 6th order of the aluminium alloy front subframe. (**a**) First modal shape. (**b**) Second order modal shapes. (**c**) Third order modal shapes. (**d**) Fourth order modal shapes. (**e**) Fifth order modal shapes. (**f**) Sixth order modal shapes.

5.4. Fatigue Life Analysis of the Optimized Aluminium Alloy Front Subframe

Automobile parts will be subjected to various excitation loads during use. Even if the automobile parts meet the strength requirements, they will fracture or fail due to fatigue damage. Therefore, it is imperative to carry out a fatigue analysis of the aluminium alloy front subframe. The aluminium alloy front subframe will be affected by the excitation of the ground and the engine during use because of a high stress alternating frequency; therefore, the analysis method of the high cycle fatigue is adopted, that is, the stress-life (S–N) fatigue analysis method [33–37]. The fatigue life analysis of the aluminium alloy front subframe is carried out in the fatigue process module of the HyperMesh software, and the whole fatigue analysis process is shown in Figure 18.

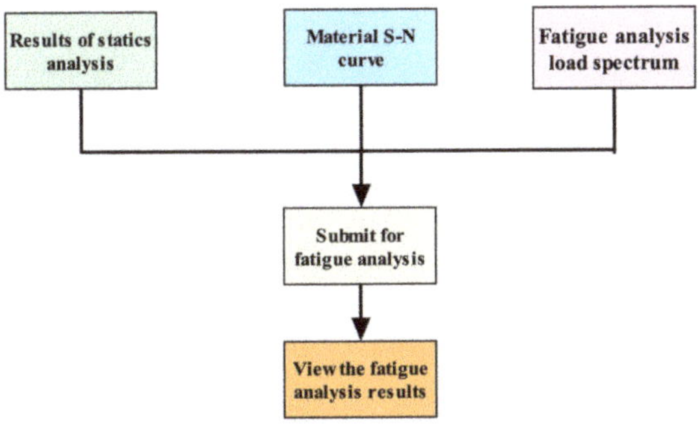

Figure 18. Fatigue analysis flow of aluminium subframe.

After the pre-treatment parameters of the fatigue analysis are set, they are submitted to the solver for calculation. The fatigue life of the aluminium alloy front subframe is analysed under two conditions of high stress: the over one side deep pit condition and forward braking condition. The target value of the fatigue damage is less than 0.25. The fatigue damage cloud diagram and fatigue life cloud diagram of the optimized aluminium alloy front subframe are shown in Figures 19 and 20, respectively, under the one side deep pit conditions.

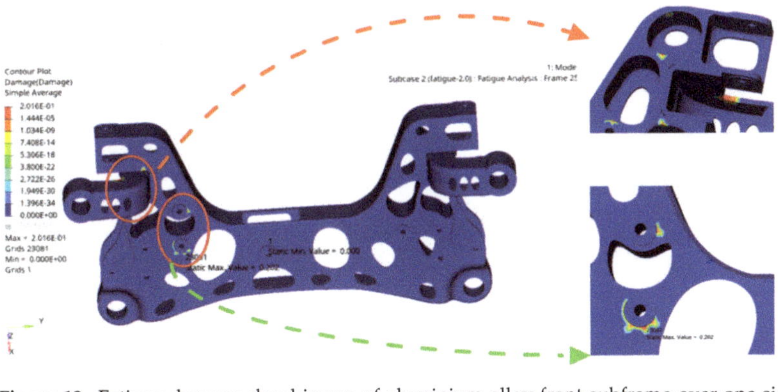

Figure 19. Fatigue damage cloud image of aluminium alloy front subframe over one side deep pit condition.

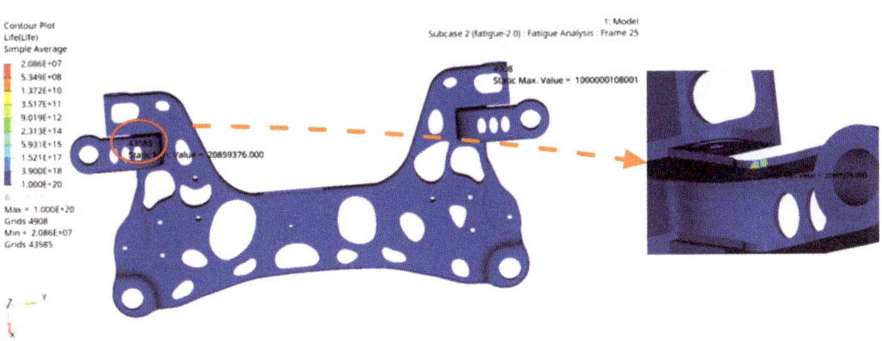

Figure 20. Fatigue life cloud diagram of aluminium alloy front subframe over one side deep pit condition.

It can be seen from Figure 19 that the maximum fatigue damage of the aluminium alloy front subframe is 0.201, which is less than the target value of 0.25; thus, the design requirements are met. It can be seen from Figure 20 that the minimum fatigue life of the aluminium alloy front subframe is 2.08×10^7, which also meets the design requirements.

The fatigue damage cloud diagram and fatigue life cloud diagram of the aluminium alloy front subframe under forwards braking conditions are shown in Figures 21 and 22.

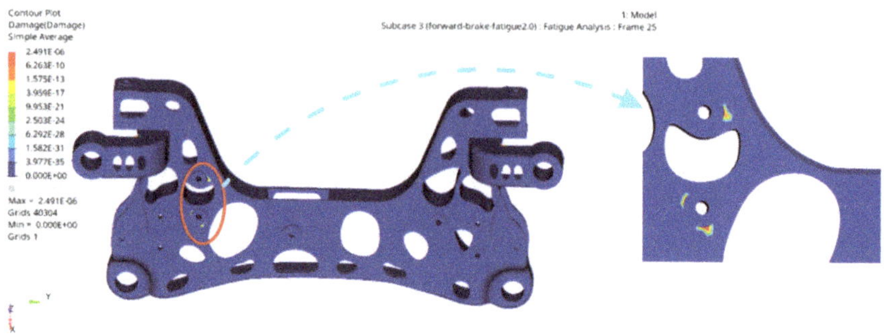

Figure 21. Fatigue damage cloud diagram of aluminium alloy front subframe under forward braking condition.

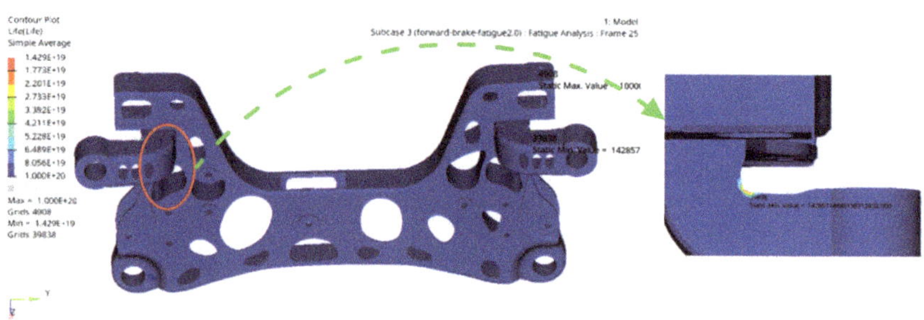

Figure 22. Fatigue life cloud diagram of aluminium alloy front subframe under forward braking condition.

Figure 21 shows that the maximum fatigue damage of the aluminium alloy front subframe in the forward braking condition is 2.49×10^{-6}, which is far less than the target value of 0.25; thus, the design requirements are satisfied. Figure 22 shows that the minimum fatigue life of the aluminium alloy front subframe in the forward braking condition is 1.42×10^{19}, which also meets the design requirements.

6. Test Analysis of the Aluminium Alloy Front Subframe Sample

6.1. Free Modal Test Analysis of the Aluminium Alloy Front Subframe

The aluminium alloy front subframe studied in this paper is produced by an integrated casting method. After casting, the aluminium alloy front subframe sample is obtained by fine machining. To verify the accuracy and reliability of the finite element analysis results, a free modal test of the finished aluminium alloy front subframe was carried out. The modal test equipment used is shown in Figure 23, and the free modal test platform built for the aluminium alloy subframe is shown in Figure 24. The aluminium alloy front subframe sample was suspended on the test bench by a special rubber cord for the modal test so that it was in a free suspension state. The method of force hammer knocking and multipoint picking up was adopted for the modal test [38]. Eight test points were selected in the aluminium alloy front subframe body, and the force hammer knocking position was repeatedly knocked six times to obtain the average value to improve the accuracy of the test results.

Figure 23. Aluminium alloy front subframe modal test equipment.

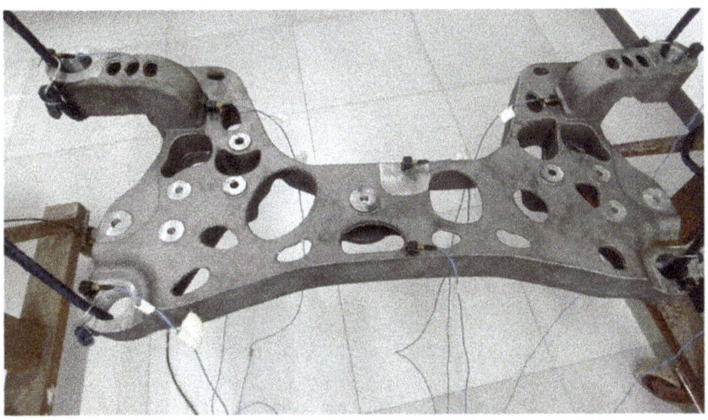

Figure 24. Free modal test platform of aluminium alloy subframe.

The least square frequency domain method is used to identify the modal shapes. Figure 25 shows the comprehensive frequency response function curve of the aluminium alloy front subframe. The results show that the first-order modal of the aluminium alloy front subframe is 84 Hz, the second-order modal is 201 Hz, the third-order modal is 254 Hz, the fourth-order modal is 342 Hz, the fifth-order modal is 389 Hz, and the sixth-order modal is 390 Hz. By comparing the free modal finite element analysis results of the aluminium alloy front subframe with the test analysis results, as shown in Table 9, it can be seen that the maximum error of the free modal finite element results and test results of the first six orders of the aluminium alloy front subframe is 14.29%, and the maximum error is no more than 15%, which verifies the accuracy of the finite element model.

Table 9. The results of free modal finite element analysis of aluminium alloy front subframe are compared with the results of test analysis.

Order Number	Finite Element Analysis Frequency/Hz	Test Analysis Frequency/Hz	Error/%
First order modal	96	84	14.29
Second order modal	222	201	10.45
Third order modal	268	254	5.93
Fourth order modal	346	342	1.47
Fifth order modal	373	389	1.32
Sixth order modal	417	390	6.92

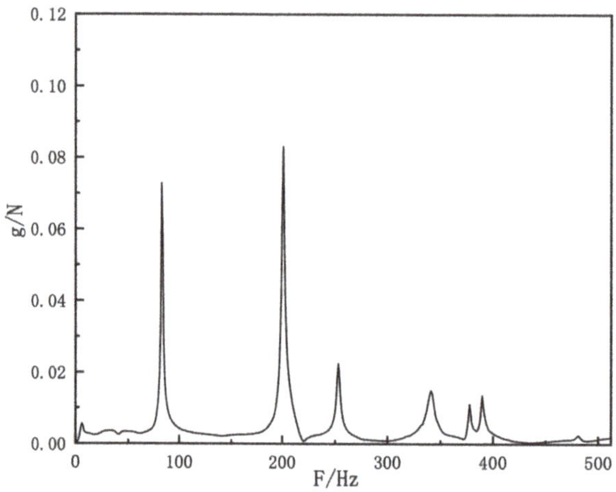

Figure 25. Integrated frequency response function curve of aluminium alloy front subframe.

6.2. Fatigue Endurance Bench Test of the Aluminium Alloy Front Subframe

To verify the fatigue durability of the front subframe of the aluminium alloy, a fatigue durability bench test was carried out on the front subframe of the aluminium alloy. According to the bench test standards of enterprises, the longitudinal durability test and lateral durability test of the front subframe were investigated, and the durability of the front and rear installation points of the swinging arm of the front subframe of the aluminium alloy was checked. The simulated loading of the aluminium alloy front subframe and the body were bolted together. In the longitudinal endurance test analysis, force values ranging from −6 kN to 8.6 kN in the X-direction were loaded on the swing arm ball pin in the form of a sine wave. The test frequency range was 2 Hz, and the test target was 350,000 times without cracking. The built longitudinal fatigue endurance bench test platform is shown in Figure 26. In the analysis of the lateral force durability test, a force in the direction of Y ± 9 kN was loaded on the swing arm ball pin in the form of a sine wave. The test frequency range was also 2 Hz, and the test target was 300,000 times without cracking. The built lateral force fatigue endurance bench test platform is shown in Figure 27.

Figure 26. Longitudinal fatigue endurance test of aluminium alloy front subframe on bench.

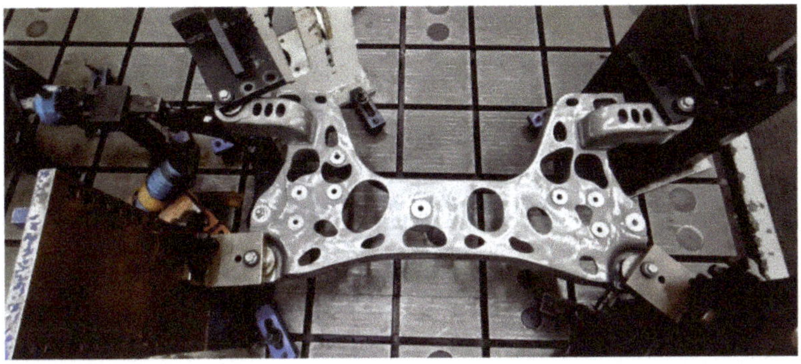

Figure 27. Lateral force fatigue endurance test of aluminium alloy front subframe on bench.

The longitudinal durability test and lateral durability test of the aluminium alloy front subframe verified three samples. It was verified by the test that the aluminium alloy front subframe could reach 350,000 times without cracking in the longitudinal durability test and 300,000 times without cracking in the lateral durability test. These results satisfy the requirements of the bench durability test.

7. Conclusions

In this paper, the influence of typical working conditions on an aluminium alloy front subframe structure was considered by the multi-working condition topology optimization method, and the optimal stress path structure of the aluminium alloy front subframe was obtained. According to the optimal stress path structure, the detailed structural design of the aluminium alloy front subframe and the performance analysis showed that the aluminium alloy front subframe met the requirements for the performance of strength, static stiffness, and modal. Further optimization can be obtained. Then, the response surface approximation model was constructed using the optimal Latin hypercube test method Isight. Meanwhile, the multi-objective particle swarm optimization algorithm was used to optimize the multi-objective design of the aluminium alloy front subframe, which further reduced the weight of the aluminium alloy subframe. Aluminium alloy subframe samples were manufactured based on the optimization results. Free modal test analysis and fatigue endurance bench tests were carried out on the aluminium alloy subframe samples. The following conclusions can be drawn from this study.

- The multi-operating condition topology optimization method is adopted to find the best stress path of the aluminium alloy front subframe, which prevents blindness in the design and development process of the aluminium alloy front subframe and can greatly shorten the development cycle.
- The response surface approximation model was constructed by using the optimal Latin Hypercube test method in the Isight software. Meanwhile, the multi-objective particle swarm optimization algorithm was used to carry out the multi-objective optimization design of the aluminium alloy front subframe. After 1002 iterations, the optimal structure of the aluminium alloy front subframe was obtained. After optimization, the maximum stress of the aluminium alloy front subframe corresponds to the extreme single-side pit condition, and the stress value is 179.3 MPa, which is 1.3 MPa less than that of 180.6 MPa before optimization. After optimization, the static stiffness values of each hard point of the aluminium alloy front subframe have little change, while the static stiffness values of some hard points remain unchanged. Under the premise of satisfying various performance indexes, the aluminium alloy front subframe loses 0.95 kg compared with the topology optimization, and 2.4 kg compared with the original subframe, with a lightweight rate of 12%.

- Free modal test analysis was carried out on the aluminium alloy front subframe sample after multi-objective optimizations. The error between the test results and the finite element free modal analysis results was less than 15%, which verified the accuracy of the finite element model. The fatigue endurance bench test of the aluminium alloy front subframe samples showed that the longitudinal endurance test of the aluminium alloy front subframe could reach the target value of 350,000 times, and the lateral endurance test could reach the target value of 300,000 times, satisfying the bench fatigue durability requirements.

Author Contributions: Conceptualization, Y.S. and X.M.; methodology, X.M., W.L. and Z.Z.; software, X.M.; validation, Y.S., X.M. and Z.Z.; formal analysis, J.H.; writing—original draft preparation, X.M.; writing—review and editing, Y.S.; funding acquisition, Y.S. All authors have read and agreed to the published version of the manuscript.

Funding: This work was supported by the Central Guidance Project of China (ZY21195053) and the Science and Technology Project of Liuzhou (2021CBA0102).

Data Availability Statement: All data used to support the findings of this study are included within the article.

Conflicts of Interest: The authors declare no conflict of interest.

References

1. Li, C.; Kim, I.Y.; Jeswiet, J. Conceptual and detailed design of an automotive engine cradle by using topology, shape, and size optimization. *Struct. Multidiscip. Optim.* **2015**, *51*, 547–564. [CrossRef]
2. Kim, S.I.; Kang, S.W.; Yi, Y.S.; Park, J.; Kim, Y.Y. Topology optimization of vehicle rear suspension mechanisms. *Int. J. Numer. Methods Eng.* **2018**, *113*, 1412–1433. [CrossRef]
3. Ali, U.; Fraser, R.A. *Numerical Modeling of Rear Subframe under Different Loading Conditions*; 2013-01-0571; SAE 2013 World Congress & Exhibition: Detroit, MI, USA, 2013.
4. Nam, J.; Shin, H.; Choi, G. Durability prediction for automobile aluminum front subframe using nonlinear models in virtual test simulations. *Int. J. Automot. Technol.* **2014**, *15*, 593–601. [CrossRef]
5. Lee, G.S.; Park, J.M.; Choi, B.L.; Choi, D.H.; Nam, C.H.; Kim, G.H. Multidisciplinary Design Optimization of Vehicle Front Suspension System Using PIDO Technology. *Trans. Korean Soc. Automot. Eng.* **2012**, *20*, 1–8.
6. Oh, J.H.; Choi, H.H.; Park, S.H. The Optimization of Rear Suspension Using Hydroforming. *Trans. Mater. Process.* **2008**, *17*, 481–485.
7. Park, S.Y.; Park, D.C.; Yoon, K.S.; Lee, M.S. *A Study of Front Subframe System Optimization for Improving Vehicle Nvh Performance*; 2009-01-2097; SAE 2009 Noise and Vibration Conference and Exhibition: St. Charles, IL, USA, 2009.
8. Hur, D.; Lee, K. A Study on the FE Model Updating and Optimization for Vehicle Subframe. *Trans. KSAE* **2003**, *10*, 220–227.
9. Belingardi, G.; Koricho, E.G. Design of a composite engine support sub-frame to achieve lightweight vehicles. *Int. J. Automot. Compos.* **2014**, *1*, 90–111. [CrossRef]
10. Chiu, H.; Cheng, K.; Chang, M.; Chi, S. *The Lightweight Chassis Sub-Frame Design by Applying Hydroforming Technology*; 2010-01-0989; SAE 2010 World Congress & Exhibition: Detroit, MI, USA, 2010.
11. Han, D.-H.; Ha, S. Development of lightweight composite sub-frame in automotive chassis parts considering structure & NVH performance. *Compos. Res.* **2019**, *32*, 21–28.
12. Law, S.E.; Widmer, D.R. *Crossmember Design Process—A Case Study*; 2001-01-3068; SAE 2002 World Congress & Exhibition: Detroit, MI, USA, 2001.
13. Jang, C.-S.; Yoo, Y.-D.; Jeon, J.-W.; Choi, B.-L.; Jung, E.-W. *Design of a Light Weight Suspension Component Using CAE*; 980901; International Congress & Exposition: Detroit, MI, USA, 1998.
14. Fichera, G.; Scionti, M.; Petrone, F. Optimization of the mount system of a vehicle's front subframe. *Shock. Vib. Dig.* **2006**, *38*, 342–343.
15. Rotondella, V.; Merulla, A.; Baldini, A.; Mantovani, S. Dynamic modal correlation of an automotive rear subframe, with particular reference to the modelling of welded joints. *Adv. Acoust. Vib.* **2017**, *2017*, 8572674. [CrossRef]
16. Da'Quan, L.; Zhou, J.; Park, J. The Use of Topology Optimization in Enhancing the Structural Property of an Automotive Front Sub-Frame. In Proceedings of the 2022 ASEE Gulf Southwest Annual Conference, Prairie View, TX, USA, 16–18 March 2022.
17. Price, M. Design Optimization of a Magnesium Subframe. Master's Thesis, Mississippi State University, Starkville, MS, USA, 2018.
18. Liao, J.; Shan, L.; Feng, Y. The Design and Size Optimization of the New FCEV Subframe Based on Hypermesh Platform. *Adv. Mater. Res.* **2011**, *328–330*, 435–440. [CrossRef]

19. Hamdi, A.; Yannou, B.; Landel, E. Design Target Cascading for Vibro-Acoustic Conceptual Design of An Automobile Subframe. In Proceedings of the DS 32, Proceedings of DESIGN 2004, the 8th International Design Conference, Dubrovnik, Croatia, 17–20 May 2004.
20. Huang, L.; Yuan, C. *Physical Drawbead Design and Modeling with ABAQUS/Isight*; 2017-01-0305; WCX™ 17: SAE World Congress Experience: Detroit, MU, USA, 2017.
21. Coello Coello, C.A. Recent Trends in Evolutionary Multiobjective Optimization. In *Evolutionary Multiobjective Optimization*; Springer: London, UK, 2005; pp. 7–32.
22. Chen, X.; Yu, X.; Ji, B. Study of Crankshaft Strength Based on iSIGHT Platform and DOE Methods. In Proceedings of the 2010 International Conference on Measuring Technology and Mechatronics Automation, Changsha, China, 13–14 March 2010; pp. 548–551.
23. Park, J.S. Optimal Latin-hypercube designs for computer experiments. *J. Stat. Plan. Inference* **1994**, *39*, 95–111. [CrossRef]
24. Liefvendahl, M.; Stocki, R. A study on algorithms for optimization of Latin hypercubes. *J. Stat. Plan. Inference* **2005**, *136*, 3231–3247. [CrossRef]
25. Patil, S.A.; Moradi, R.; Lankarani, H.M. Vehicle mass optimization for frontal structure using I-sight and study of weld parameterization for mass improvement. In *ASME International Mechanical Engineering Congress and Exposition*; ASME: New York, NY, USA, 2014; Volume 46613, p. V012T015A011.
26. Fang, H.; Horstemeyer, M.F. Global response approximation with radial basis functions. *Eng. Optim.* **2006**, *38*, 407–424. [CrossRef]
27. Alsharif, F.; McNaughton, J.L., Jr. Optimization Analysis of a V-Twin Motorcycle Engine Using WAVE Cycle Analysis and an iSight Optimization Framework. In *Ricardo Software e-Brochure V3*; Ricardo, Inc.: Shoreham-by-Sea, UK, 2005; Volume 14, p. 15.
28. Smith, M. *Abaqus User's Manual, Version 2020*; Dassault Systèmes Simulia Corp.: Johnston, RI, USA, 2020.
29. Nicholson, J.C.; Arora, J.S.; Goyal, D.; Tinjum, J.M. Multi-objective structural optimization of wind turbine tower and foundation systems using Isight: A process automation and design exploration software. In Proceedings of the 10th World Congress on Structural and Multidisciplinary Optimization, Orlando, FL, USA, 19–24 May 2013; pp. 19–24.
30. Ren, J.; Zhu, H.W.; Wang, H.; Zhao, C.F.; Zhong, J.L. Multi-objective structural optimization of VL seal ring based on Isight. *J. Phys. Conf. Ser.* **2020**, *1622*, 012031. [CrossRef]
31. Wang, D.F.; Xie, C.; Liu, Y.C.; Xu, W.C.; Chen, Q. Multi-objective Collaborative Optimization for the Lightweight Design of an Electric Bus Body Frame. *Automot. Innov.* **2020**, *3*, 250–259. [CrossRef]
32. Richards, Z.D.; Valavanis, K. Particle Swarm trade-off curve analysis for bi-objective optimization. In Proceedings of the IEEE Congress on Evolutionary Computation, Barcelona, Spain, 18–23 July 2010; p. 16.
33. Campbell, F.C. *Fatigue and Fracture: Understanding the Basics*; ASM International: Almere, The Netherlands, 2012.
34. Xu, J.T.; Song, Y.D.; Ding, J.B.; Ding, S.Q. Analysis of a front sub-frame fatigue strength based on Miner theory. In Proceedings of the 2010 International Conference on Computer Application and System Modeling (ICCASM 2010), Taiyuan, China, 22–24 October 2010.
35. Zhang, L.B.; Liu, Y.; Huang, P.C. FEM-based fatigue analysis method and application in automobile industry. *Comput. Aided Eng.* **2006**, *1*, 195–198.
36. Bignonnet, A.; Thomas, J.J. *Fatigue Assessment and Reliability in Automotive Design*; 2001-01-4061; SAE Brasil International Conference on Fatigue: São Paulo, Brasil, 2001.
37. Peiskammer, D.; Dannbauer, H.; Dutzler, E.; Puchner, K. Using fatigue life prediction to increase reliability of automotive structures. *SAE Trans.* **2003**, *112*, 174–180.
38. Yu, Z.Q.; Jia, H.F.; Huang, X.Y. Design of the Lower Control Arm of an Electric SUV Front Suspension Based on Multi-Disciplinary Optimization Technology. *Jordan J. Mech. Ind. Eng.* **2021**, *15*, 7–14.

metals

MDPI

Article

Design Optimization of Chute Structure Based on E-SVR Surrogate Model

Xiaoke Li [1], Qianlong Jiang [1], Yu Long [2], Zhenzhong Chen [3], Wenbo Zhao [4], Wuyi Ming [1], Yang Cao [1] and Jun Ma [1,*]

[1] Henan Key Laboratory of Intelligent Manufacturing of Mechanical Equipment, Zhengzhou University of Light Industry, Zhengzhou 450002, China
[2] Zhejiang PanGood Power Technology Co., Ltd., Jinhua 321100, China
[3] College of Mechanical Engineering, Donghua University, Shanghai 201620, China
[4] Luoyang TiHot Railway Machinery Manufacturing Co., Ltd., Luoyang 471000, China
* Correspondence: majun@zzuli.edu.cn

Abstract: To reduce the wear and damage of the chute caused by long-term impact of coke, a structure parameter optimization model was established in this paper, which takes the minimum impact force as the objective and the coke-conveying speed as the constraint. Furthermore, the ensemble of support vector regression (E-SVR) with different kernel functions was developed to replace the implicit relationship between the conveying speed, the impact force, and the structure parameters. Using the numerical examples, the effectiveness of the E-SVR model was verified. Finally, the optimal chute structure parameters were obtained by using the E-SVR model. After optimization, the maximum impact force was reduced by 17.07% and the maximum conveying speed was reduced by 6.59%, which still falls within the specified range. Therefore, the feasibility of the optimization results and the effectiveness of the E-SVR surrogate model were verified.

Keywords: chute; support vector regression; kernel function; surrogate model; ensemble of surrogates; design optimization

1. Introduction

The steel industry is one of the basic industries of modern industrialized countries and is also an important symbol of a country's degree of development and economic strength. The transportation and delivery of bulk materials (such as coke and iron ore) are the basic production link in the steel industry. As the key equipment in the steel industry, the chute delivers the bulk materials from high to low. However, due to the long-term impact of the bulk materials, the chute is easily worn or damaged, reducing its service life. Once the chute is damaged, it can only be stopped for maintenance, which seriously affects the production schedule. Moreover, maintenance usually consists of padding the steel plate at the worn place, which leads to the unsmooth conveying of bulk materials in the chute and can even result in the plugging phenomenon. Furthermore, the collision to the chute may cause the bulk materials to be broken, producing a large amount of dust or even the phenomenon of the bulk materials falling, resulting in serious safety risks to the on-site staff.

To overcome the above issues, scholars have conducted extensive research on chute structure design. Based on the wear mechanism, new materials have been adopted for the chute, and the wear-resistant lining board has been adhered to the inner shell of the chute [1–3] to increase its service life. On the other hand, to avoid the plugging phenomenon, structural design methods were adopted to improve the conveying smoothness of the materials in the chute [4,5]. The structural parameters, the contour curves of the impact plate, and the contour curves of the diversion plate of the chute were analyzed to reduce chute wear under the premise of conveying smoothness requirements [6–9]. Other methods for improving the working performance and service life of the chute include structural parameters optimization [10–13], baffle settings [14], and motion control [15].

However, the existing structural design methods of the chute still depend on experience and simple verification, which lacks data support. Moreover, the implicit relationship between chute structure parameters and performance responses (such as impact force of coke to the chute and conveying speed of coke) further increases the difficulty of optimization.

To reduce optimization costs, various surrogate models were applied to approximate the implicit relationship between the design parameters and performance responses. Among the surrogate models, the Support Vector Regression (SVR) model was widely used in engineering problems because of its good performance in small samples, including nonlinear, high-dimension, overfitting, and multiple local minima problems [16,17]. SVR is the specific application of a Support Vector Machine (SVM) in the field of functional regression and has been widely used in the field of structural reliability analysis, such as the lightweight design of complex structures [18–22] and process parameters optimization [23,24]. However, there are multiple kernel functions in SVR, and each kernel function has its characteristic. Therefore, for an unknown implicit problem, how to select the optimal kernel function is still a challenge [25,26].

In this paper, the idea of ensemble of surrogates (EoS) is introduced to alleviate dependency on the kernel functions in modeling of performance responses of the chute. The E-SVR model with multiple kernel functions was constructed to replace the implicit relationship between the chute structural parameters and the performance response. Then, the design optimization of the chute structure was carried out to reduce the maximum impact force with the maximum conveying-speed constraint.

2. Simulation of the Material-Conveying Process

In practical production, the service life of the chute and the smoothness of material conveying in the chute are closely related to the structural parameters of the chute. In this paper, the coke-conveying process is simulated by EDEM 2019 software to obtain the maximum impact force and maximum material-conveying-speed data at different structural parameters.

According to the workshop layout of one steel plant, the initial 3D model of chute was built (shown in Figure 1). Then, the model was imported into EDEM software for material-conveying process simulation to obtain the impact force of coke on the chute and the conveying speed of coke in the chute under the current combination of structural parameters.

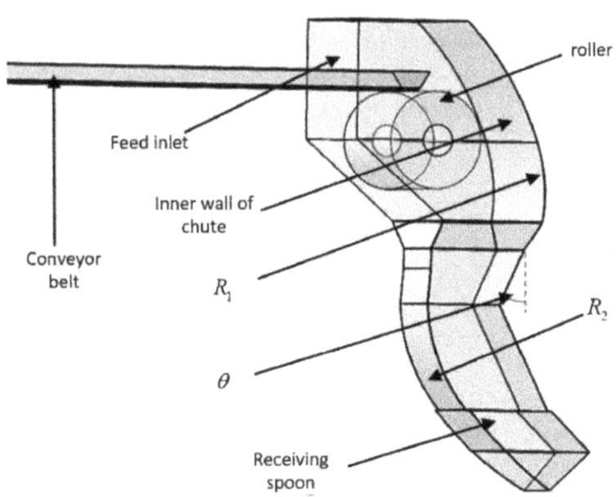

Figure 1. Structure of the chute.

EDEM is a commonly used software to solve the motion analysis of discrete particles. By simulating the motion state of particles during the conveying process, the response results of particle velocity and impact force can be obtained. As shown in Figure 2, the material was conveyed from the input conveyor belt to the feed inlet. Under the influence of inertial force and gravity, the material dashed against the inner wall of chute and then changed direction and moved along the inner wall of the chute. Finally, the material was conveyed to the output conveyor belt through the receiving spoon at the bottom of the chute.

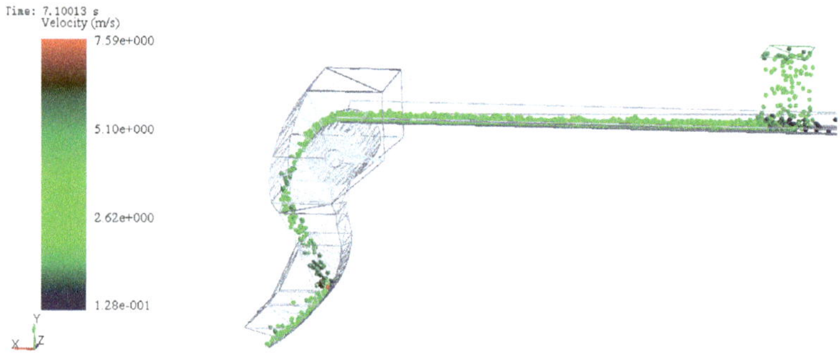

Figure 2. Simulation diagram of coke speed.

The settings of the material-conveying process in EDEM software were as follows: the contact model is the rolling friction model; the direction of gravity acceleration is Z, and the magnitude is -9.81 m/s^2; the direction of the initial moving speed of the material is X, and the magnitude is 1.5 m/s; and the conveying materials is the spherical coke. Specific parameters are shown in Table 1.

Table 1. Material properties and simulation settings.

	Parameters	**Value**
Spherical material property	Diameter (mm)	80
	Poisson ratio	0.28
	Shear modulus (GPa)	1.98
	Density (Kg/m^3)	500
Chute properties	Poisson ratio	0.3
	Shear modulus (GPa)	80
	Density (Kg/m^3)	7850
Collision properties between materials	Coefficient of restitution	0.40
	Static friction coefficient	0.40
	Coefficient of rolling friction	0.05
Collision properties between material and chute	Coefficient of restitution	0.50
	Static friction coefficient	0.30
	Coefficient of rolling friction	0.10
Simulation parameter setting	Number of particles between materials/(s)	5000
	Fixed time step(s)	6.8×10^4

Particles were generated by EDEM's built-in particle factory. The particle position was set to "random," and the weight generation rate was set to 50 kg/s.

In order to obtain the conveying speed of coke and the impact force on the chute, the conveying process of coke in the chute was simulated. The simulation began from the time the coke entered the chute until the batch of materials was output from the receiving spoon. The maximum conveying speed and the maximum impact force of coke under the current structural parameters were obtained, as shown in Figures 2 and 3. At 7.1 s, the coke-conveying speed reached the maximum value of 7.59 m/s. At 17 s, the maximum impact force of coke (2460 N) on the chute occurred.

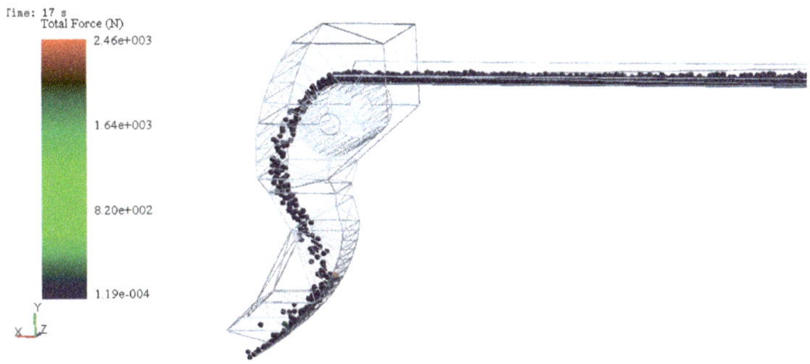

Figure 3. Simulation diagram of impact force.

3. Optimization Model and Flowchart

3.1. Optimization Model

The phenomenon of blocking, dusting, and crushing often occurs in the conveying process of the chute. Moreover, the long-term impact easily leads wear and destruction of the chute. Therefore, the conveying speed of coke in the chute and the impact of coke on the chute should be considered comprehensively in the design of chute structure. According to the practical experience of enterprises, the phenomenon of plugging and high breakage rate will not occur when the maximum conveying speed of coke is 7–7.2 m/s. Therefore, taking the maximum impact force of coke on the chute as the optimization objective and the maximum conveying speed of coke within the specified interval as the constraint, the optimization model of chute structural parameters is constructed, as shown below:

$$\begin{array}{l} find : d \\ \min : f(d) \\ s.t. : 7 \le v \le 7.2 \\ d = (R_1, R_2, \theta) \\ 1700 \le R_1 \le 2500 \\ 1800 \le R_2 \le 2200 \\ 25^\circ \le \theta \le 30^\circ \end{array} \quad (1)$$

In Equation (1), d is the structural parameters of the chute, $f(d)$ is the maximum impact force of materials on the chute (N), v is the maximum conveying speed of materials (m/s), R_1, R_2 is the key structure size of the chute (mm) and θ is the key angle of the chute (°). Figure 1 shows the location of R_1, R_2 and θ. The structure parameters [2100, 2000, 27.5] of the chute previously used in a steel mill are determined as the initial design variables. It is worth noting that the uncertainty of the design variable is not considered in the optimization model (1). In the future, a reliability-based design optimization method will be conducted to obtain the optimal design that meets the probabilistic constraints [27].

3.2. Flowchart of Chute Structure Optimization

The flowchart of the proposed chute structure optimization method is shown in Figure 4. First, the initial value and the range of each design variable are determined. Then, the objective function and constraint function are approximated by the surrogate model to reduce the optimization cost. The E-SVR surrogate model is used here to achieve a good performance for the implicit objective or constraint function. Finally, the gradient-based optimization algorithm is used to obtain the optimal design.

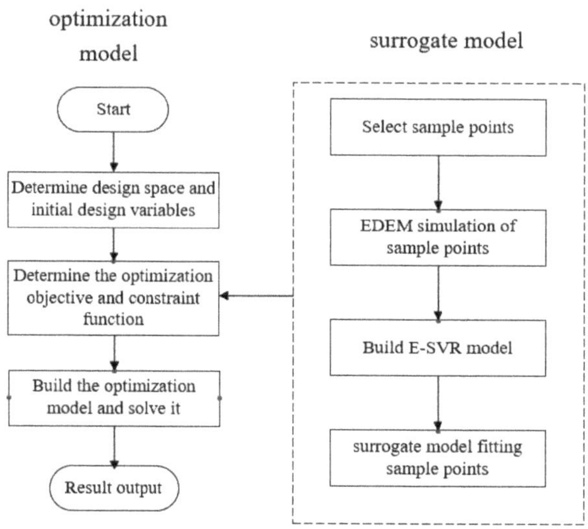

Figure 4. Design optimization of chute structure.

4. E-SVR Model

4.1. SVR Model

Based on the classification model of the Support Vector Machine (SVM), the SVR model is extended to fit the implicit function. The main advantages of SVR are as follows [28,29]: strong adaptability to small sample data, strong generalization ability, and the ability to avoid the "dimension disaster" because the complexity of SVR is determined by the number of support vectors rather than the dimension of design space. The SVR model is expressed as

$$y(x) = \langle w \cdot \Phi(x, x_i) \rangle + b \tag{2}$$

In Equation (2), $\Phi(x, x_i)$ is the kernel function, which is the key to solving nonlinear problems [28]. The weight coefficient w and the deviation b are obtained by solving the following optimization problems:

$$\begin{array}{l} \min: \frac{1}{2}\|w\|^2 + C\sum\limits_{i=1}^{n}(\xi_i^+ + \xi_i^-) \\ s.t.: y_i - \langle w \cdot \Phi(x, x_i) \rangle - b \le \varepsilon + \xi^+ \\ \langle w \cdot \Phi(x, x_i) \rangle + b - y_i \le \varepsilon + \xi^- \\ \xi^+, \xi^- \ge 0 \end{array} \tag{3}$$

In Equation (3), C is the penalty factor and ξ_i^+, ξ_i^- are the relaxation factors.

Great differences in prediction accuracy of SVR can be observed when different kernel functions are used. Therefore, the selection of kernel functions is very important in SVR. Typical kernel functions are shown in Table 2.

Table 2. Kernel functions in SVR.

Kernel Function	Formulas
Linear kernel function	$k(x, x') = x^T x'$
Sigmoid kernel function	$k(x, x') = \tanh(ax * x' + c)$
Radial basis kernel function	$k(x, x') = \exp\left[-\lvert x - x'\rvert^2/(2s^2)\right]$
Polynomial kernel function	$k(x, x') = (x^T x' + h)^d$

For given sample points, various surrogate model can be fitted, but the fitting precision is different, which has a great influence on the final optimization result. Therefore, the error index is used to evaluate the accuracy of the surrogate model. In this paper, commonly used error indices are used:

(1) Multiple correlation coefficient R^2

$$R^2 = 1 - \frac{\sum_{i=1}^{m} (y_i - \hat{y}_i)^2}{\sum_{i=1}^{m} (y_i - \overline{y})^2} = 1 - \frac{MSE}{Var} \tag{4}$$

(2) Relative average absolute error ($RAAE$)

$$RAAE = \frac{\sum_{i=1}^{m} |y_i - \hat{y}_i|}{m \times STD} \tag{5}$$

(3) Relative maximum absolute error ($RMAE$)

$$RMAE = \frac{\max_{i=1,\ldots,m} |y_i - \hat{y}_i|}{STD} \tag{6}$$

In Equations (4)–(6), R^2 and $RAAE$ represent the global relative error of the surrogate model. $RMAE$ represents the local relative error. m represents the number of test samples used to evaluate the accuracy of the surrogate model. y_i represents the true value, and $\hat{y}_i$ represents the predicted value through the surrogate model. y represents the average value of all true values. MSE, Var and STD represent the mean square error, variance of true value, and standard deviation, respectively, which are calculated as follows:

$$MSE = \frac{\sum_{i=1}^{m} (y_i - \hat{y}_i)^2}{m}, Var = \frac{\sum_{i=1}^{m} (y_i - \overline{y})^2}{m}, STD = \sqrt{Var} \tag{7}$$

As seen from the above equations, the smaller MSE is, the larger R^2 and the smaller $RAAE$ are. Therefore, the global error of the surrogate model is small, and the prediction accuracy is high. The local error index $RMAE$ represents the local fitting accuracy of the surrogate model. In a certain region, the smaller $RMAE$ is, the higher the fitting accuracy of the surrogate model is. In this paper, larger R^2, smaller $RAAE$, and smaller $RMAE$ indicate that the surrogate model has higher fitting accuracy.

4.2. E-SVR Surrogate Model

The kernel functions in SVR can be classified as the global kernel function and the local kernel function [30]. In the global kernel function (such as the polynomial kernel function), the farther to the test points, the greater the influence on the kernel function. Therefore, the strong generalization ability and the weak learning ability are observed in the global kernel function. Differently, in the local kernel function (such as the Gaussian radial basis kernel function), the closer to the test points, the greater the influence on the kernel function. Therefore, the strong learning ability and the weak generalization ability are observed in the local kernel function [31]. To comprehensively use the advantages of both the global kernel function and the local kernel function, an ensemble method was developed in this paper.

Among all the kernel functions, the polynomial kernel function and the Gaussian radial basis kernel function are most commonly used. When the samples are linearly or quadratic nonlinearly separable, the linear kernel function has better classification ability. On the other hand, the Gaussian radial basis kernel function has good performance for general nonlinear problems. The polynomial kernel function is a global kernel function. The Gaussian RBF kernel function has better adaptability to data and strong anti-interference ability to noise, so it belongs to local kernel function. To assess the problems of generalization and learning ability of the single-kernel function, the ensemble of SVR models constituted by polynomial kernel function and Gaussian radial basis kernel function was carried out.

The basic principle of establishing the E-SVR surrogate model is as follows. First, the polynomial kernel function-based SVR model (Poly-SVR) and the Gaussian RBF kernel function-based SVR model (RBF-SVR) were established. Then, the weight coefficients of each single-kernel-based SVR model were calculated. Finally, two single-kernel-based SVR models were linearly combined to obtain the E-SVR model. Compared with the single-kernel-based SVR model, the E-SVR model has better performance on the adaptability, robustness, and prediction ability of the implicit problems.

The equation for the E-SVR model is as follows:

$$\hat{y}_{E-SVR}(x) = \sum_{i=1}^{N} w_i \hat{y}_i(x) \tag{8}$$

where, the weight coefficient should satisfy the following equation:

$$\sum_{i=1}^{N} w_i = 1 \tag{9}$$

In Equations (8) and (9), N is the number of the single-kernel SVR models used to construct the final E-SVR model, $\hat{y}_{E-SVR}$ is the predicted value of the E-SVR model at test point x, $\hat{y}_i(x)$ is the predicted value of the i-th single-kernel SVR model at test point x, and w_i represents the corresponding weight coefficient of the i-th single-kernel SVR model.

The weight value quantifies the prediction accuracy of the single-kernel function-based SVR model. In general, among the single-kernel SVR models, the higher the prediction accuracy of the single-kernel SVR model, the larger the weight value that will be assigned to it. Conversely, a smaller weight value will be assigned to the single-kernel SVR model with less accuracy. It is worth noting that, when the weight value of the single-kernel SVR model is assigned as 1 and the weight values of the other single-kernel SVR models are assigned as 0, the E-SVR model will degenerate into the single-kernel-based SVR model.

In this paper, the heuristic weight-solving strategy [32] proposed by Goel and Haftka is used to calculate the weight values of single-kernel function-based SVR models. The heuristic weight-solving strategy is simple and easy to implement, which can better balance the weight values of each single-kernel-based SVR model. The equation of Goel and Haftka's method is as follows:

$$\begin{aligned} E_{avg} &= \frac{1}{N}\sum_{i=1}^{N} E_i \\ \omega_i^* &= (E_i + \alpha E_{avg})^{\beta}, \alpha < 1, \beta < 0 \\ \omega_i &= \omega_i^* / \sum_{i=1}^{N} \omega_i^* \end{aligned} \tag{10}$$

In Equation (10), E_i is the predicted error of the single-kernel SVR model, which can be expressed by global error or local error. N is the number of single-kernel SVR models required to construct the final E-SVR model. α and β are used to control the importance of averaging and importance of individual surrogate, respectively [32]. Smaller α and larger negative β impart higher weight for the single-kernel SVR model with higher prediction accuracy. On the contrary, larger α and smaller β show high confidence in the averaging scheme for all the single-kernel SVR models [32]. According to Goel and Haftka's study, $\alpha = 0.05$ and $\beta = -1$ are used in this paper.

4.3. Numerical Examples

In this section, 2 test functions were introduced to compare the performance of Poly-SVR, RBF-SVR, and E-SVR models proposed in this paper.

(1) Brain-Hoo function [33]:

$$f(x,y) = \left(y - 5.1x^2/4\pi^2 + 5x/\pi - 6\right)^2 + 10(1 - 1/8\pi)\cos(x) + 10, x \in [-5,10], y \in [0,15] \tag{11}$$

This is moderate nonlinear problem with 2 variables. The 3D graph of Brain-Hoo function is shown in Figure 5. In this example, 20 points by Latin hypercube sampling (LHS) were selected as the training samples. The error evaluation index (R^2, $RAAE$, $RMAE$) were calculated through leave-one-out cross-validation [34]. The comparative results of different SVR models for the Brain-Hoo function are shown in Table 3.

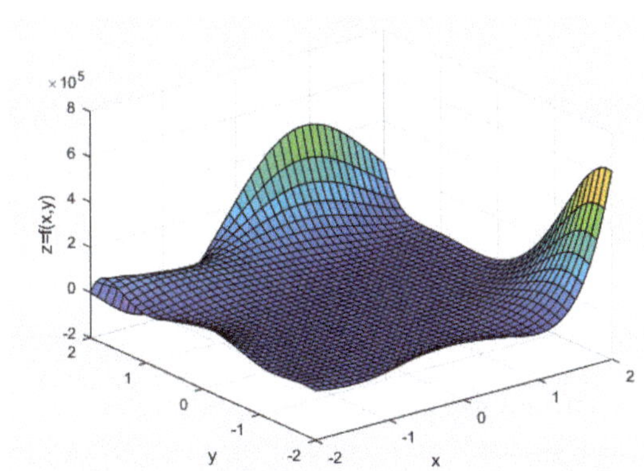

Figure 5. Three-dimensional graph of the Brain-Hoo function.

Table 3. Brain-Hoo function test results.

Evaluation Index	Poly-SVR	RBF-SVR	E-SVR
R^2	**1**	0.9575	**1**
RAAE	**0.0026**	0.1078	0.0038
RMAE	**0.0276**	1.3177	0.031

Based on the results in Table 3, a much smaller global error evaluation index $RAAE$ and local error evaluation index $RMAE$ are observed in the Poly-SVR model compared to the RBF-SVR model. Therefore, the accuracy of the Poly-SVR model is much better than the RBF-SVR model in this example. Differently, E-SVR obtained by the ensemble of two different kernels has similar accuracy to the Poly-SVR model, where the same R^2 and slightly larger $RAAE$, $RMAE$ are observed. Therefore, taking the 2D Brain-Hoo test function with moderate nonlinearity as an example, the E-SVR model has better performance.

(2) Hartman3 function [32]

$$f(x) = -\sum_{i=1}^{4} c_i \exp\left\{-\sum_{j=1}^{n} a_{ij}(x_j - p_{ij})^2\right\} \quad x = (x_1, x_2, \cdots, x_n), x_i \in [0,1] \tag{12}$$

This is a highly nonlinear problem with 3 variables. The corresponding parameters of Hartman3 function are given in Table 4 [32].

Table 4. Parameters of the Hartman function.

i	a_{ij}			c_i	p_{ij}		
1	3.0	10	30	1.0	0.3689	0.1170	0.2673
2	0.1	10	35	1.2	0.4699	0.4387	0.7470
3	3.0	10	30	3.0	0.1091	0.8732	0.5547
4	0.1	10	35	3,2	0.03815	0.5743	0.8828

In this example, 30 points selected by Latin hypercube sampling (LHS) were used as the training samples. The error evaluation indices (R^2, $RAAE$, $RMAE$) were calculated through leave-one-out cross-validation [34]. The comparative results of different SVR models for the Hartman3 function are shown in Table 5.

Table 5. Hartman3 function test results.

Evaluation Index	Poly-SVR	RBF-SVR	E-SVR
R^2	0.2628	0.7461	**0.7581**
RAAE	0.6744	0.3854	**0.3689**
RMAE	3.3907	**1.8152**	2.1837

By comparing the experimental results in Table 5, it can be seen that the performance of the RBF-SVR model is better compared to the Poly-SVR model whether viewed from the global evaluation index (R^2, $RAAE$) or the local evaluation index $RMAE$. Among the 3 SVR models, the global fitting performance of the E-SVR model is better than the Poly-SVR model and the RBF-SVR model because larger R^2 and smaller $RAAE$ are observed. On the other hand, the local error evaluation index $RMAE$ in the E-SVR model is smaller than the Poly-SVR model and slightly larger than the RBF-SVR model. Therefore, taking the Hartman3 test function as an example, the E-SVR model has better performance among the three different surrogate models.

After comprehensive analysis of these 2 examples with different dimension and nonlinearity, the following conclusions can be drawn. For the first problem, the Poly-SVR model has the best performance, which is a little better than the E-SVR model. For the second problem, the E-SVR model has the best performance when using most error indices. In other words, the E-SVR model has high fitting accuracy for both problems. Therefore, for the test functions with different dimension and nonlinearity, the E-SVR model constructed by RBF-kernel and Poly-kernel has good fitting performance.

5. Design Optimization of Chute Structure

5.1. Design of Experiment

In order to obtain the optimal design parameters of the chute, the optimization model (1) can be solved directly by using the EDEM simulation results. However, due to the huge cost of calculation, this is difficult to realize in practice. In this paper, the surrogate model is introduced to replace the implicit relationship between structural parameters and performance responses of the chute, then the surrogate model is directly used in the optimization process to greatly reduce the simulation cost.

To reduce the number of EDEM experiments but generate representative samples, various experiment design methods can be used to select the chute structure parameter combinations (R_1,R_2,θ) in the design space. LHS has the characteristic of uniform stratification and can obtain the representative sample in the case of less sampling, so LHS is more efficient than the ordinary sampling method [35]. Moreover, LHS has the freedom to define the sample number, thereby providing greater flexibility for problem with different dimensionality [36]. Therefore, the LHS method is used to select 30 chute structure parameter combinations (R_1,R_2,θ) in this paper. Specific data are shown in Table 6.

Table 6. Chute structure parameter combinations.

R_1, R_2, θ	R_1, R_2, θ	R_1, R_2, θ	R_1, R_2, θ	R_1, R_2, θ
2168/2034/27.9	2350/2125/29.1	2406/2153/29.4	2042/1971/27.1	2296/2098/28.7
2375/2137/29.2	1877/1888/26.1	1738/1819/25.2	1961/1931/26.6	2338/2119/29.0
1907/1903/26.3	1934/1917/26.5	1979/1939/26.7	2144/2022/27.8	1830/1865/25.8
2052/1976/27.2	1782/1841/25.5	2428/2164/29.6	1763/1832/25.4	2088/1994/27.4
1852/1876/25.9	2205/2052/28.2	2220/2060/28.3	2495/2198/30.0	2010/1955/26.9
1705/1802/25.0	2281/2090/28.6	2460/2180/29.8	2248/2074/28.4	2116/2008/27.6

5.2. Solution and Verification

In this paper, the E-SVR surrogate model was used in the iterative optimization procedure of chute structure, and the Sequential Quadratic Programming (SQP) method was used to calculate the next design point. The optimal chute structure parameters [2259.6, 1813.4, 29.9] were obtained under the condition that the coke velocity was between 7 m/s and 7.2 m/s. All program codes are tested in Matlab R2014b.

In order to verify the effectiveness of the optimization results, a 3D model of the chute was reconstructed according to the optimal structure parameter [2259.6, 1813.4, 29.9]. The 3D model was imported into EDEM software for verification experiments. Moreover, the original chute of the factory was upgraded by using the structural parameters obtained from the optimization design of the chute structure through the E-SVR model. During the actual service of the upgraded chute for 1000 h, there was no blocking phenomenon. Through actual monitoring, the maximum impact force of coke on the chute was reduced to 2047 N, which was basically consistent with the simulation result of 2040 N and proved the effectiveness of using EDEM simulation.

After optimization, the maximum impact force was reduced by 17.07% and the maximum conveying speed was reduced by 6.59%, which still falls within the specified range. The result comparisons of the chute structure before and after optimization are shown in Table 7.

Table 7. Result comparisons of the chute structure before and after optimization.

	Structure Parameter	Maximum Impact Force	Maximum Conveying Speed
Before optimization	2100, 2000, 27.5	2460 N	7.59 m/s
After optimization	2259.6, 1813.4, 29.9	2040 N	7.09 m/s

5.3. Comparison of Single/Ensemble-Kernel-Based SVR

To further verify the effectiveness of the proposed E-SVR model in design optimization of chute structure, the performance of commonly used single-kernel-based SVR models (Poly-SVR and RBF-SVR) are used for comparison.

The 30 chute structure parameter combinations in Section 5.1 to construct the Poly-SVR model and the RBF-SVR model, respectively. According to the Poly-SVR model, the design optimization model of the chute structure was solved, and the optimal parameter combination [2296.7, 1742.4, 29.3] was obtained. Using this parameter combination, a 3D model of the chute structure was reconstructed. Then, the model was imported into EDEM software to conduct simulation experiments.

Similarly, according to the RBF-SVR model, the design optimization model of the chute structure was solved, and the optimal parameter combination [2173.5, 1909.5, 28.2] was obtained. According to this parameter combination, the 3D model of chute structure was reconstructed, and the model was imported into EDEM software for verification experiments.

The comparison results of single/ensemble-kernel SVR models are shown in Table 8. As can be seen, different optimization results were obtained based on different kernel SVR models for the chute structure optimization problem. By comparing the optimization results of 3 single/ensemble-kernel SVR models with that before optimization, the performance of chute structure based on the 3 SVR models improved in different degrees. Moreover, the optimization result of E-SVR model is better than that of Poly-SVR and RBF-SVR models. Therefore, better performance is observed in the parameter optimization of chute structure.

Table 8. Comparison results of single/ensemble-kernel SVR models.

	Structure Parameter	Maximum Impact Force	Maximum Conveying Speed
Poly-SVR	2296.7, 1742.4, 29.3	2420 N	7.14 m/s
RBF-SVR	2173.5, 1909.5, 28.2	2370 N	7.18 m/s
E-SVR	2259.6, 1813.4, 29.9	2040 N	7.09 m/s

6. Conclusions

To avoid blocking and crushing of the coke and reduce wear and destruction of the chute during the conveying process, the design optimization model of structure parameters of the chute was established and solved in this paper. In the proposed method, the conveying speed of coke in the chute and the impact force of coke on the chute were comprehensively considered.

(1) The optimization model of the chute structure parameter was established and solved in this paper. The maximum impact force of coke on the chute and the material crushing rate were effectively reduced within the allowed speed range, which improved the utilization rate of coke and prolonged the service life of the chute.
(2) The E-SVR model was developed to integrate the advantages of both the Poly-SVR model and the RBF-SVR model. The effectiveness of the E-SVR model was verified through numerical examples.
(3) By using the E-SVR surrogate model in the chute structural parameter optimization models to replace the implicit relationship between the maximum impact force, maximum conveying speed, and design variables, the computational cost of design optimization was reduced.
(4) Through comparison with the results by using the Poly-SVR model and results by using the RBF-SVR model, the optimal results obtained by the proposed E-SVR model are accurate and effective, which is of great significance for the design of chute structure.

Author Contributions: Conceptualization, X.L. and J.M.; methodology, X.L.; software, Y.L.; validation, Q.J., Z.C. and W.Z.; formal analysis, W.M.; investigation, Y.C.; resources, Z.C.; data curation, Y.L.; writing—original draft preparation, Q.J.; writing—review and editing, X.L.; visualization, W.M.; supervision, J.M.; project administration, J.M.; funding acquisition, X.L. All authors have read and agreed to the published version of the manuscript.

Funding: This research was funded by the National Natural Science Foundation of China, grant number 51905492.

Institutional Review Board Statement: Not applicable.

Informed Consent Statement: Not applicable.

Data Availability Statement: No new data were created or analyzed in this study. Data sharing is not applicable to this article.

Conflicts of Interest: The authors declare no conflict of interest. The funders had no role in the design of the study; in the collection, analyses, or interpretation of data; in the writing of the manuscript; or in the decision to publish the results.

References

1. Tamoghna, M.; Henrik, S. Simulation of Burden Distribution and Charging in an Ironmaking Blast Furnace. *IFAC Pap. Online* **2015**, *48*, 183–188. [CrossRef]
2. Yu, Y.H.; Wu, R.F.; Ruan, W.S. Design and performance analysis of chute with aluminum foam laminated structure. *J. Mach. Des.* **2017**, *34*, 65–69.

3. Ma, C.; Ren, T. Structure Design and Optimization of Distributing Chutes in Blast Furnace for Wear Resistance. *China Mech. Eng.* **2017**, *28*, 253–257, 266.
4. O'Shaughnessy, C.; Masoero, E.; Gosling, P.D. Topology optimization using the discrete element method. Part 1: Methodology, validation, and geometric nonlinearity. *Meccanica* **2022**, *57*, 1213–1231. [CrossRef]
5. Zhou, L.; Esteban, R.; Bryan, E.; Antonio, M. A smooth contact algorithm for the combined finite discrete element method. *Comput. Part Mech.* **2020**, *7*, 807–821. [CrossRef]
6. Li, X.; Yuan, H.; Sun, G.S. Anti-impact Design of Discharge Chute. *Coal. Mine Mach.* **2020**, *41*, 4–5.
7. Roberts, A.W. Chute Performance and design for rapid flow conditions. *Chem. Eng. Technol.* **2010**, *26*, 163–170. [CrossRef]
8. Ren, X.P.; Jia, Y.W. Value analysis of influence of different factors on wear of distributing chute. *China Metall.* **2020**, *30*, 11–15, 40.
9. Zhang, C.H.; Xu, Y.S.; Qin, Y.L.; Zhao, J.Y.; Li, W.L. Optimization design and application of the material-transfer device based on DEM. *J. Mach. Des.* **2020**, *37*, 112–117.
10. Murao, A.; Kashihara, Y.; Oyama, N.; Sato, M.; Watakabe, S.; Yamamoto, K.; Fukumoto, Y. Development of Control Techniques for Mixing Small Coke at Bell-less Top Blast Furnace. *Tetsu Hagane J. Iron Steel Inst. Jpn.* **2016**, *102*, 614–622. [CrossRef]
11. Xia, R.; Wang, X.W.; Li, B. Abrasion performance of the scraper conveyor chute under complex working conditions. *Proc. Inst. Mech. Eng. Part J J. Eng. Tribol.* **2021**, *235*, 2364–2375. [CrossRef]
12. Guixin, S.; Futang, X.; Yue, H.; Xiaogang, W. BP-ANN Model to Optimize the Structural Parameter Setting of Transfer Point. In Proceedings of the 2020 4th International Workshop on Renewable Energy and Development (Iwred 2020), Sanya, China, 24–26 April 2020; Volume 510. [CrossRef]
13. Li, H.F.; Luo, Z.G.; Zhang, S.C.; Zou, Z.S. Effect of Chute Angle on Process of Mixing Charging. *J. Northeast. Univ. Nat. Sci.* **2012**, *33*, 681–684.
14. Chen, L.S.; Luo, Z.G.; You, Y.; Zou, Z.S. Effects of Flap Angles on the Charging Procedure of Flap Distributors. *J. Northeast. Univ. Nat. Sci.* **2013**, *34*, 971.
15. Huang, W.F.; Sun, X.M. Development and application of the breakup-proof device for the telescopic chute. *Goal Eng.* **2004**, *10*, 69–70.
16. Jha, R.K.; Swami, P.D. Failure prognosis of rolling bearings using maximum variance wavelet subband selection and support vector regression. *J. Braz. Soc. Mech. Sci. Eng.* **2022**, *44*, 13. [CrossRef]
17. Li, Y.X.; Huang, X.Z.; Zhao, C.Y.; Ding, P.F. Stochastic fractal search-optimized multi-support vector regression for remaining useful life prediction of bearings. *J. Braz. Soc. Mech. Sci. Eng.* **2021**, *43*, 18. [CrossRef]
18. Andres-Perez, E.; Gonzalez-Juarez, D.; Martin-Burgos, M.J.; Carro-Calvo, L.; Salcedo-Sanz, S. Influence of the number and location of design parameters in the aerodynamic shape optimization of a transonic aerofoil and a wing through evolutionary algorithms and support vector machines. *Eng. Optimiz.* **2017**, *49*, 181–198. [CrossRef]
19. Duan, L.; Xiao, N.; Li, G.; Cheng, A.; Chen, T. Design optimization of tailor-rolled blank thin-walled structures based on -support vector regression technique and genetic algorithm. *Eng. Optimiz.* **2017**, *49*, 1148–1165. [CrossRef]
20. Ye, Y.F.; Gao, J.B.; Shao, Y.H.; Li, C.N.; Jin, Y.; Hua, X.Y. Robust support vector regression with generic quadratic nonconvex epsilon-insensitive loss. *Appl. Math. Model.* **2020**, *82*, 235–251. [CrossRef]
21. Zhu, P.; Pan, F.; Chen, W.; Zhang, S. Use of support vector regression in structural optimization: Application to vehicle crashworthiness design. *Math. Comput. Simulat.* **2012**, *86*, 21–31. [CrossRef]
22. Gupta, D.; Acharjee, K.; Richhariya, B. Lagrangian twin parametric insensitive support vector regression (LTPISVR). *Neural Comput. Appl.* **2020**, *32*, 5989–6007. [CrossRef]
23. Liu, Y.F.; Yan, C.P.; Ni, H.X. The approach to multi-objective optimization for process parameters of dry hobbing under carbon quota policy. *Int. J. Adv. Manuf. Technol.* **2022**, *121*, 6073–6094. [CrossRef]
24. Xiang, H.Y.; Li, Y.L.; Liao, H.L.; Li, C.J. An adaptive surrogate model based on support vector regression and its application to the optimization of railway wind barriers. *Struct. Multidiscip. Optim.* **2017**, *55*, 701–713. [CrossRef]
25. Tsirikoglou, P.; Abraham, S.; Contino, F.; Lacor, C.; Ghorbaniasl, G. A hyperparameters selection technique for support vector regression models. *Appl. Soft Comput.* **2017**, *10*, 139–148. [CrossRef]
26. Kong, D.D.; Chen, Y.J.; Li, N.; Tan, S.L. Tool wear monitoring based on kernel principal component analysis and v-support vector regression. *Int. J. Adv. Manuf. Tech.* **2017**, *89*, 175–190. [CrossRef]
27. Kaminski, M.; Solecka, M. Optimization of the truss-type structures using the generalized perturbation-based Stochastic Finite Element Method. *Finite Elem. Anal. Des.* **2013**, *63*, 69–79. [CrossRef]
28. Azamathulla, H.M.; Ghani, A.A.; Chang, C.K.; Hasan, Z.A.; Zakaria, N.A. Machine Learning Approach to Predict Sedment Load—A Case Study. *Clean Soil Air Water* **2010**, *38*, 969–976. [CrossRef]
29. Yang, C.Z.; Liu, J.Q.; Zeng, Y.Y.; Xie, G.Y. Prediction of components degradation using support vector regression with optimized parameters. *Energy Procedia* **2017**, *127*, 284–290. [CrossRef]
30. Huang, H.; Ding, S.; Jin, F.; Yu, J.; Han, Y. A Novel Granular Support Vector Machine Based on Mixed Kernel Function. *Int. J. Digit. Content Technol. Its Appl.* **2012**, *6*, 484–492. [CrossRef]
31. Zhu, P.; Zhang, Y.; Chen, G.L. Metamodeling development for reliability-based design optimization of automotive body structure. *Comput. Ind.* **2011**, *62*, 729–741. [CrossRef]
32. Goel, T.; Haftka, R.T.; Shyy, W.; Queipo, N.V. Ensemble of surrogates. *Struct. Multidiscip. Optim.* **2007**, *33*, 199–216. [CrossRef]

33. Viana, F.A.C.; Haftka, R.T.; Steffen, V. Multiple surrogates: How cross-validation errors can help us to obtain the best predictor. *Struct. Multidiscip. Optim.* **2009**, *39*, 439–457. [CrossRef]
34. Shan, S.Q.; Wang, G.G. Survey of modeling and optimization strategies to solve high-dimensional design problems with computationally-expensive black-box functions. *Struct. Multidiscip. Optim.* **2010**, *41*, 219–241. [CrossRef]
35. Li, X.K.; Zhu, H.; Chen, Z.Z.; Ming, W.Y.; Cao, Y.; He, W.B.; Ma, J. Limit state Kriging modeling for reliability-based design optimization through classification uncertainty quantification. *Reliab. Eng. Syst. Saf.* **2022**, *224*, 108539. [CrossRef]
36. Li, X.K.; Han, X.Y.; Chen, Z.Z.; Ming, W.Y.; Cao, Y.; Ma, J. A multi-constraint failure-pursuing sampling method for reliability-based design optimization using adaptive Kriging. *Eng. Comput.* **2022**, *38*, 297–310. [CrossRef]

metals

Review

Application of Artificial Intelligence to Support Design and Analysis of Steel Structures

Sina Sarfarazi, Ida Mascolo *, Mariano Modano and Federico Guarracino *

Department of Structural Engineering, University of Naples "Federico II", 80125 Naples, Italy; sina.srfz@gmail.com (S.S.); modano@unina.it (M.M.)

* Correspondence: ida.mascolo@unina.it (I.M.); fguarrac@unina.it (F.G.)

Abstract: In steel structural engineering, artificial intelligence (AI) and machine learning (ML) are improving accuracy, efficiency, and automation. This review explores AI-driven approaches, emphasizing how AI models improve predictive capabilities, optimize performance, and reduce computational costs compared to traditional methods. Inverse Machine Learning (IML) is a major focus since it helps engineers to minimize reliance on iterative trial-and-error by allowing them to identify ideal material properties and geometric configurations depending on predefined performance targets. Unlike conventional ML models that focus mostly on forward predictions, IML helps data-driven design generation, enabling more adaptive engineering solutions. Furthermore, underlined is Explainable Artificial Intelligence (XAI), which enhances model transparency, interpretability, and trust of AI. The paper categorizes AI applications in steel construction based on their impact on design automation, structural health monitoring, failure prediction and performance evaluation throughout research from 1990 to 2025. The review explores challenges such as data limitations, model generalization, engineering reliability, and the need for physics-informed learning while examining AI's role in bridging research and real-world structural applications. By integrating AI into structural engineering, this work supports the adoption of ML, IML, and XAI in structural analysis and design, paving the way for more reliable and interpretable engineering practices.

Keywords: machine learning; inverse machine learning; explainable AI; AI-based performance evaluation; data-driven design automation; steel structures

1. Introduction

Structural engineering depends on both exact design specification and extensive analysis to ensure that infrastructure and buildings satisfy safety and performance criteria. Although conventional structural engineering depends on analytical calculations, experimental testing, and numerical simulations, modern developments demand a change toward data-driven methods and automation. Like they once moved from hand calculations to computational tools like finite element (FE) analysis, engineers have to adopt fresh approaches to solve problems as technology develops. The increasing application of artificial intelligence (AI) and machine learning (ML) in structural engineering [1–3] presents a similar challenge and demands experts to become familiar with AI-driven tools. These new technologies reinforce decision-making with adaptive, data-supported solutions rather than replacing conventional knowledge; they are instead extensions of engineering intuition [4]. Structural engineering guarantees design and analysis accuracy by means of methodical processes and uniform guidelines. These approaches have been included into software tools over time to automatically complete compliance checks and regular computations. Large-scale steel buildings, however, often call for high-performance workstations and licensed tools since they demand significant computational resources, specialized simulation software, and expert oversight.

Among the deterministic, physics-based models supporting most conventional structural engineering approaches are reliability-based optimization frameworks, Computational Fluid Dynamics (CFD), and the Finite Element Method (FEM). These reliable methods provide engineers with complete awareness of mechanical behavior, stress distribution, failure modes, and safety margins. However, the application of structural systems becomes more challenging and computationally expensive as they get larger and include nonlinearities, material heterogeneity, or advanced loading scenarios. Even with strong simulation tools, each analysis iteration requires manual intervention, time, and expert interpretation—factors that can slow down the design and optimization process. Another paradigm comes from AI—in particular from ML. ML models learn from patterns in past datasets, derived from experiments, simulations, or hybrid sources, instead of solving governing equations repeatedly. Once trained, these models can save computational time, let engineers rapidly investigate large design areas, and offer almost instantaneous predictions for new designs. This capacity underlies early-stage design, parametric sensitivity studies, and multi-objective optimization where speed and adaptability are particularly critical. From a structural perspective, ML techniques have been used to precisely predict material strength, buckling resistance, joint behavior, and global system performance. From the AI side, the use of surrogate modeling, ensemble learning, and neural networks has made it possible to capture highly nonlinear relationships that are difficult to express analytically. Acting as a fast, approximative evaluator, trained ML models enable real-time structural assessment, uncertainty-based optimization, or integration with digital twin systems. This synergy between physics-based rigor and data-driven intelligence marks a potential revolution in how future structures are conceived, assessed, and optimized.

Analyzing a structure and designing it, have different uses and needs for different skill sets in structural engineering. Analysis is mostly concerned with how a structure responds to external forces; design is the process of developing answers that fit performance criteria and practical limitations. These two disciplines sometimes have different logical frameworks. Analysis depends on scientific ideas and deterministic models while design is adaptive, iterative, and affected by outside events including material availability, cost, and safety rules. One challenge of engineering education is that students are mostly taught mathematical and computational methods, which stress strict formulations and simulations. However, when they enter the industry, they must deal with real-world design problems that require creative problem-solving, adaptability, and decision-making under uncertainty. This gap emphasizes the need of bridging analytical precision with design intuition so that engineers may properly move from knowledge of structural behavior to developing practical, efficient, creative designs that satisfy real-world needs [5].

Particularly in inverse design, ML is becoming more popular since it lets engineers directly decide ideal material properties, geometric layouts, and performance criteria depending on predefined structural goals [4,6]. Inverse Machine Learning (IML) allows a data-driven approach to quickly find the best design solutions in steel constructions, not only iteratively improving designs through simulations.

Apart from predictive powers, the growing complexity of ML models has generated questions about interpretability and transparency in structural engineering uses [7]. Explainable machine learning (XML) has become increasingly important in order to meet these difficulties since it guarantees that ML-driven models produce justifiable, clear, and understandable results [8]. By means of explainability techniques including feature importance analysis and SHapley Additive exPlanations (SHAP) [9], engineers can better grasp the contribution of various design parameters in decision-making, rendering ML-based structural analysis more reliable and useful. The identification of possible biases, validation of model reliability, and enhancement of decision-making confidence in safety-critical applications including load-bearing assessments, failure predictions, and material selection also quite heavily depend on XML techniques [10].

Typically addressing different engineering challenges, three main categories define ML techniques: supervised learning, unsupervised learning, and reinforcement learning. Supervised learning, which relies on labeled datasets, is particularly effective for regression tasks such as estimating material properties and classification problems like detecting structural damage [11]. Conversely, unsupervised learning is useful for predictive maintenance and structural design optimization since it finds hidden trends in engineering data without depending on predefined labels [12]. Though less often used, reinforcement learning is very important for decision-making under uncertainty—especially in dynamic structural analysis [13].

Table 1 provides a comparative overview of these three categories—highlighting their underlying logic, typical algorithms, structural applications, and key limitations.

Table 1. Overview of main ML categories.

ML Category	Core Principle	Common Algorithms	Applications in Structural Engineering	Limitations
Supervised Learning	Learns from labeled datasets to predict outcomes	Linear Regression, Random Forest, ANN	Material property estimation, damage classification, load prediction	Requires large, labeled datasets; limited in extrapolation
Unsupervised Learning	Finds hidden patterns or groupings in unlabeled data	K-means, PCA, Autoencoders	Structural health monitoring, design clustering, pattern discovery in sensor data	Results may lack clear interpretation; requires expert analysis
Reinforcement Learning	Learns via trial-and-error interactions to maximize rewards over time	Q-learning, Deep Q-Networks	Real-time structural control, decision-making under uncertainty, adaptive load redistribution	High computational cost; limited adoption due to training complexity

In structural engineering, ML first emerged in the late 1980s when researchers started working on design tasks using artificial neural networks (ANN) [14]. Recent advances in data-driven design methods, ML-driven predictive modeling, and optimization techniques have expanded its opportunities. Although more and more studies on AI in civil and structural engineering are available, there is still a dearth of targeted studies methodically analyzing ML uses in steel constructions. Emphasizing improved accuracy, efficiency, and automation, this review shows how AI is transforming steel structure engineering and synthesizing recent developments from 1994 to 2025. Previous studies have mostly focused on particular AI techniques in structural engineering, including pattern recognition for structural health monitoring, genetic algorithms in optimization problems, ML applications in concrete structures, and steel–concrete composite structures. Although these works offer insightful analysis, their scope is usually more general and they do not particularly address the function of ML in steel structure engineering.

This review paper provides a concentrated study of AI applications in design and analysis of steel structures. It also underlines the increasing relevance of IML, which allows engineers to directly choose optimal material properties and design criteria independent of iterative trial-and-error processes. By learning complicated interactions between input parameters and structural behavior, ML models can accelerate the design process, increase prediction accuracy, and maximize structural performance unlike conventional simulation-based techniques. This review provides a basic resource for researchers and engineers by aggregating and critically evaluating the most recent ML-driven innovations in steel structure engineering, thus bridging the gap between conventional computing methods and modern AI-enhanced approaches. The aim of this work is to enable the acceptance of ML, IML, and XML in steel structure design and analysis so guaranteeing more effective, transparent, and data-driven engineering solutions.

Objectives of This Review

- Providing a Focused Overview of ML in Steel Structure Engineering:

This paper attempts to give a thorough study of the part ML plays in the design and analysis of steel buildings. This work will show how ML-driven approaches are improving structural performance prediction, material selection, load-bearing capacity estimate, and failure assessment by aggregating significant developments. The paper will also discuss how ML is changing conventional design methods, thus enabling structural engineering to be more efficient, data-driven, and optimized.

- Exploring the Role of IML in Steel Structure Design:

This work tries to show a thorough investigation of the part ML does in the design and analysis of steel buildings. IML allows engineers, free from trial-and-error methods, to directly choose ideal design parameters including geometric configurations, material properties, and performance criteria unlike conventional iterative approaches. This review will examine how application of IML is improving steel structure optimization.

- Investigating XML for Transparent Structural Engineering

The growing complexity of ML models raises questions about interpretability and confidence in structural engineering uses. This paper will review XML techniques to show how engineers might better understand and validate AI-driven predictions. By means of better model transparency, XML can help to identify significant influencing factors, guarantee safety compliance, and improve reliability in steel analysis and design.

- Addressing Challenges, Limitations, and Future Directions

Even if ML has great promise for structural engineering, several issues have to be resolved before general acceptance. Important questions including data shortage, model interpretability, generalization, and integration of physics-informed learning to improve engineering accuracy are investigated in this review. Moreover, covered are problems with feature engineering, uncertainty quantification, and regulatory approval of ML-driven models in structural safety assessments. Future research directions are also under discussion, underlining the need of hybrid AI-physics models, industry-driven validation, and responsible AI application in design and analysis processes.

2. Research Data Extraction Process

This research evaluates steel building design and analysis through bibliometric methods and content analysis to investigate the implementation of AI techniques.

Scopus's extensive scholarly coverage enabled this study to compile a comprehensive dataset for examining the impact of AI on the design and analysis of steel structures. The selected papers included applications of ML, IML, deep learning, fuzzy logic, genetic algorithms, optimization, surrogate modeling, and XML across various domains of structural engineering. Special attention was given to steel building components such as cold-formed steel, stainless steel, high-strength steel, beams, columns, plates, trusses, connections, and full steel frames. The dataset also included research on design strategies—such as structural optimization, topology optimization, reliability analysis, uncertainty quantification, and performance-based design—and structural analysis techniques including FEM, numerical simulations, structural health monitoring, load capacity evaluation, fatigue assessment, buckling behavior, seismic performance, wind load effects, and fire resistance.

The following inclusion criteria ensured both scientific quality and thematic relevance: only peer-reviewed journal articles were selected; the search was limited to publications between 1994 and 2025; all chosen papers explicitly focused on the application of artificial intelligence in steel structural engineering, excluding studies centered solely on concrete, timber, or hybrid systems; and the selected studies were required to demonstrate either methodological innovation or practical relevance in structural modeling, performance prediction, or design optimization. A keyword-based search was conducted in Scopus using combinations of terms such as "steel structures", "machine learning", "artificial intelligence", "predictive modeling", and "structural optimization". Duplicate records, non-English publications, editorials, and studies without accessible full texts were excluded.

Overview of Research Contributions in AI-Driven Steel Structure Studies

A two-stage screening process was applied. In the first stage, papers were screened depending on title, abstract, and keywords to ensure relevance. The second stage involved a thorough review to select studies focusing on ML-driven structural analysis and design strategies. The bibliometric analysis identified 2291 English-language publications, as shown in Table 2, highlighting the growing scholarly attention on AI's role in structural engineering. This collection shows a wide international collaboration, with contributions from 1277 researchers affiliated with 1367 institutions across 85 countries. Covering from 1994 to 2025, the dataset captures three decades of increasing interest in AI-driven structural performance evaluation and optimization. Of the collected works published across 159 scientific journals, 2227 are research articles and 64 are review papers. The citation impact is also noteworthy, with a total of 50,893 citations and an average of 22.21 citations per paper. An annual citation rate of 2035.72 further emphasizes the rapid acceptance and industrial relevance of AI-based technologies in steel construction engineering.

Table 2. Core Data Insights.

Detailed Overview	Outcomes
Citations	50,893
Authors	1277
Organization	1367
Countries	85
Journal	159
Documents	2291
Average citations per year	2035.72
Average citations per document	22.21
Time span	1994–2025

The Citation Overview graph (Figure 1) displays the annual publication and citation distribution in the field of AI-driven steel structure research. From 2000 to about 2015, the results show a slow but steady increase in research activity despite rather low numbers of publications and citations. Starting in 2015, however, both measures clearly speed forward, a reflection of structural engineering's increasing integration of AI and data-driven approaches. This trend fits world developments in computational modeling, deep learning, and ML, which have made AI-based solutions more pragmatic and successful for engineering use.

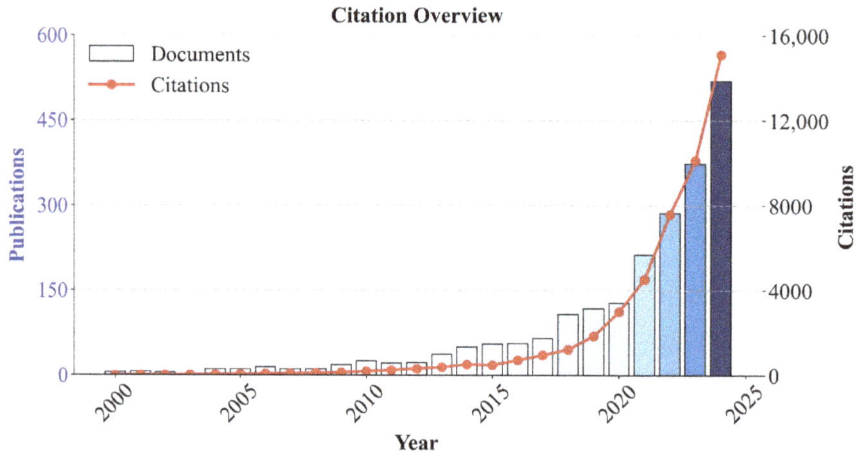

Figure 1. Citation and publication trends over time. Bar color intensity reflects the number of publications (darker = more publications).

Publications pointing a shift toward AI acceptance in the field began to appear around 2018. This development confirms that the research community is placing increasing emphasis on AI applications in structural analysis, optimization, and design. Interest continues to rise, reaching its highest levels between 2022 and 2024. Citation counts follow a similar trend, showing a marked increase from 2015 onward and peaking in 2024. This trend underlines both the growing academic weight of AI-related research and its relevance in addressing challenging engineering problems.

Figure 2 shows the publication trends of AI-driven structural engineering research across several journals from 2000 to 2024. Among the most prominent sources, only the top five journals with the highest number of publications are considered: Engineering Structures, Structures, Automation in Construction, Journal of Building Engineering, and Thin-Walled Structures. These journals are important venues for disseminating innovative research, since their major contributions to the development of AI applications in structural engineering have played a major contribution in advancing the field. The statistics show a slow rise in research activity in the early years followed by a sudden surge in publications after 2015, suggesting an increasing interest in AI applications within this field. Engineering Structures and Structures show the most notable rise among the sources, especially starting from 2020 forward. This trend implies that in steel structure engineering, these publications now serve as main venues for publishing research on AI. Particularly noteworthy is the fast rise in publications within Structures after 2020, which reflects a recent surge in research contributions most likely driven by the growing acceptance of ML, optimization strategies, and computational modeling in structural design.

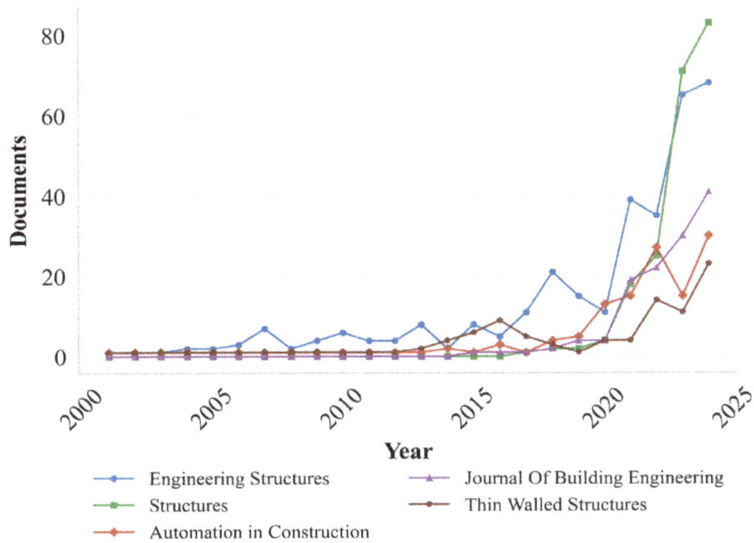

Figure 2. Annual publication trends in AI-driven structural engineering across the top five contributing journals by volume (2000–2024).

Though at a rather slower pace, other journals including Automation in Construction, Journal of Building Engineering, and Thin-Walled Structures—also exhibit a consistent upward trend in AI-related publications. Particularly from 2018 onward, these journals have gained popularity as AI-based approaches are increasingly applied across multiple domains, including structural design, optimization, and performance assessment. This steady rise in publication frequency suggests that AI methods are expanding beyond traditional design principles into broader applications, such as structural resilience evaluation, real-time monitoring, and automated construction.

Figure 3 shows the worldwide distribution of structural engineering research motivated by AI and identifies important contributing nations. Reflecting its strong investment in data-driven approaches and computational technologies, China (23.6%) leads the field, followed by the United States (10.3%), which continues to be a major center for structural analysis and design. Iran (8.8%) and India (5.7%)—both actively advancing structural performance evaluation and optimization—are other major contributors. Similarly, the United Kingdom (5.0%) and South Korea (5.0%) have made significant contributions, with a focus on enhancing the accuracy and efficiency of the engineering process. Apart from these leading countries, Australia (3.3%), Vietnam (3.1%), Canada (2.8%), and Turkey (2.7%) have also played instrumental roles in shaping AI-driven structural engineering research. The "Other Countries" category, accounting for 29.7%, emphasizes the global reach and growing relevance of artificial intelligence applications in structural engineering across many regions.

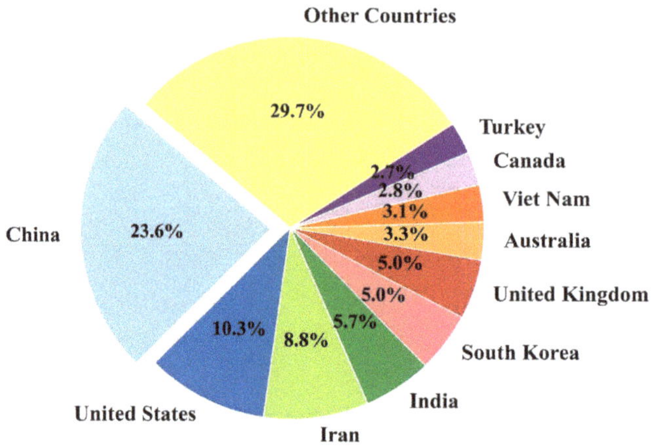

Figure 3. Global Distribution of AI-Driven Structural Engineering Publications by Country.

Visualizing the most often used terms in AI-driven structural engineering research, the keyword co-occurrence network (Figure 4) shows their interconnections and thematic clusters. Comprising several color-coded clusters, each with a different research focus, the network is each keyword's size reflects its importance in the dataset; conversely, the proximity and connecting lines show the strength of terms' relationships. Key words including "machine learning", "prediction", "model", "structure", "analysis", "method", and "genetic algorithm" appear most at the center of the network. This implies that research in this domain heavily emphasizes predictive modeling, structural analysis, and algorithmic optimization. Emphasizing its use in structural performance enhancement, the term "genetic algorithm" is tightly related to "optimization", "frame", and "optimal design". The red cluster is centered around "machine learning", "prediction", and "strength", indicating a focus on AI-driven approaches for capacity estimation, structural behavior modeling, and material property analysis. This cluster includes terms such as "fire resistance", "column", "shear strength", and "capacity prediction", showing the role of AI in evaluating load-bearing capacities and failure mechanisms in steel structures. The blue cluster emphasizes "structural health monitoring", "detection", and "classification", indicating a strong research focus on damage detection, real-time assessment, and digital twin technology. This fits the growing acceptance of computer vision techniques and deep learning for automated structural inspection and reliability assessment. The green cluster—which centers on "genetic algorithm", "frame", and "optimization design"—represents another important area. This points to the focus of research on seismic reliability analysis, structural optimization methods driven by AI, and improvement of steel frame performance. Terms like "finite element simulation", "deep learning method", and "graph neural network" point to the way AI is being combined with cutting-edge computational approaches. In engineering applications, structural reliability, performance evaluation, and uncertainty quantification all depend on this integration. The keyword network (Figure 4) shows generally the increasing importance of AI, deep learning, and optimization methods in structural engineering research. The grouping of related keywords emphasizes how AI is being used in predictive modeling, performance analysis, structural monitoring, and optimization, determining the course of data-driven engineering solutions.

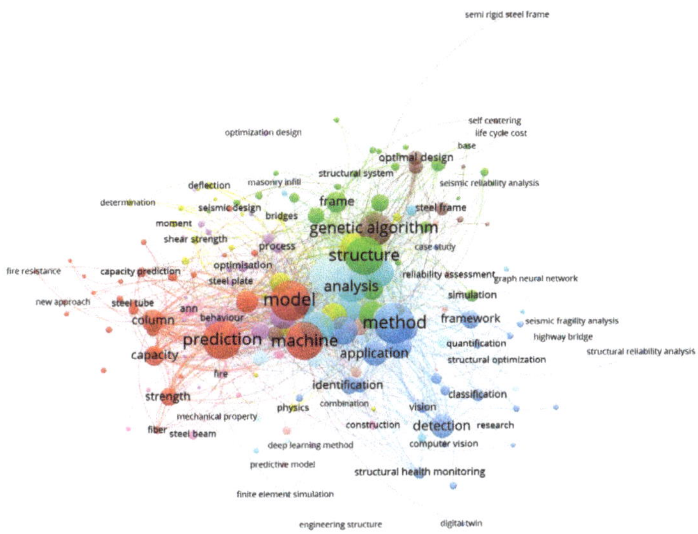

Figure 4. Keyword Co-Occurrence Network in AI-Driven Structural Engineering Research.

The integration of AI in structural engineering has obviously evolved over the last two decades. Early years mostly focused on rule-based systems, genetic algorithms, and shallow neural networks—mostly used for design optimization or basic capacity prediction. As the field matured, researchers gradually shifted toward more sophisticated methods, including support vector machines, ensemble learning models, and metaheuristic optimization frameworks. After 2015, with increasing computational resources and access to larger datasets, deep learning techniques—particularly convolutional and recurrent neural networks—began to dominate studies focused on structural health monitoring, damage detection, and time-series forecasting. Reflecting the need for transparency in safety-critical domains, most recently the interest has grown into Explainable AI (XAI) and hybrid approaches.

Three main applications—ML, IML, and XML—are investigated in our scientometric analysis. Sections 3–5 include representative studies that demonstrate how these approaches tackle complex engineering problems to show an extensive view of AI applications in the field.

3. An Overview of ML Application in Steel Structures

To maintain focus and ensure scientific clarity, this review concentrates on supervised learning algorithms, which represent the most widely applied category in structural engineering tasks. Although other machine learning paradigms such unsupervised and reinforcement learning have great value, a thorough review of all current ML models would exceed the scope and intended length of this paper. Therefore, a curated set of algorithms with a demonstrated applicability to structural prediction, optimization, and evaluation is examined in the sections that follow. To help better navigate the hierarchy and relationships among these supervised learning techniques, Figure 5 offers a graphical summary of their classification.

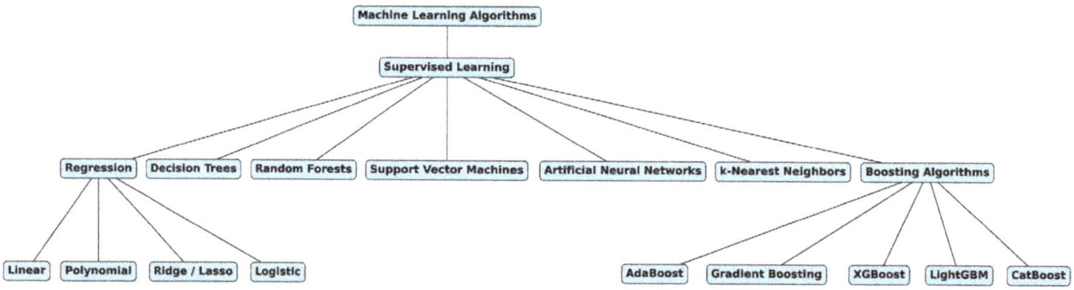

Figure 5. Hierarchical classification of supervised ML algorithms commonly applied in structural engineering.

3.1. Supervised ML Algorithms

3.1.1. Regression Algorithms

One of the simplest regression models, linear regression, uses a straight-line equation to build a direct link between input and output variables. This approach is categorized as simple linear regression (for a single predictor) or multiple linear regression (for many predictors) depending on the count of input variables. Expanding on regression techniques, multivariate regression extends multiple linear regression by concurrently predicting several output variables, thus helping to understand interdependencies between inputs and outputs in complex systems. By including higher-degree polynomial terms in the equation, polynomial regression unlike linear models captures nonlinear relationships. Lasso regression adds an L1 regularization term to handle problems with correlated input features, helping to shrink less relevant coefficients toward zero and so accomplish feature selection. Another type of regularized linear regression, ridge regression, maintains all variables in the model by using L2 regularization to lower the impact of less important features. Although these methods are meant for continuous variable prediction, logistic regression is extensively applied for classification problems by estimating the probability of an outcome and mapping inputs to discrete categories. In structural engineering, each of these regression techniques has unique uses; linear regression is appropriate for simple predictions like load-bearing estimates, polynomial regression is helpful for modeling nonlinear material behavior, and logistic regression helps in binary classification tasks including evaluating structural safety compliance [15,16].

3.1.2. Decision Tree

Considered a generally used ML method with simplicity and interpretability [17], decision tree (DT) analysis is particularly useful for classification and regression activities. This approach methodically divides difficult datasets into smaller pieces. A DT is a nonparametric model that divides the input space into discrete areas, each matching a particular decision result. The structure comprises a root node, which acts as the starting point for decision-making, branches reflecting various conditions, decision nodes performing feature-based tests, and leaf nodes, indicating final predictions. The learning process alternately divides the dataset depending on chosen criteria, such minimizing Mean Squared Error (MSE) in regression problems. With each leaf node serving as a last classification or prediction, the path from root to leaf defines the decision rules within the tree. The aim is to build a compact tree with few decision nodes such that predictive accuracy is maintained. Avoiding overfitting depends on appropriate stopping criteria that limit tree depth or stop too frequent divisions. DTs demand minimal data preparation, are quite flexible, able of managing both numerical and categorical data. Their sensitivity to small data variations and overfitting sensitivity, however, can cause instability and accuracy loss relative to more sophisticated tree-based models like Random Forests and Boosting Algorithms. Regularizing methods—such as pruning or limiting tree complexity—are widely used to improve performance, guaranteeing better generalization to unseen data.

3.1.3. Random Forest

Widely used in both classification and regression tasks, Random Forest (RF) [18] is a potent ensemble learning method. RF improves predictive accuracy by building several DTs and aggregating their outputs, thus lowering the overfitting risk. The basis of RF, the idea of ensemble learning, depends on the combined ability of several models to generate more consistent results. Bootstrapped aggregation (bagging), in which the dataset is randomly split into several subsets and separate classifiers are trained on each subset, is one of the most successful ensemble techniques available within RF. This method reduces variance and enhances model stability by means of some data points appearing in several subsets. Majority voting (for classification) or averaging (for regression) determines the last prediction, thus guaranteeing better generalization on fresh data. RF is unique in that it randomly chooses features during tree building, thus lowering correlation between individual trees and improving model performance. For handling high-dimensional datasets with many input variables—a difficulty that conventional Decision Trees often face—this randomness makes RF especially efficient. While training many trees is computationally efficient, producing predictions from a trained RF model can be time-intensive because the demand to process several trees concurrently. Among RF's main benefits are its lower sensitivity to overfitting and greater resilience than of a single Decision Tree. RF is easily available for users without strong knowledge in ML since it often performs well with default settings even without great fine-tuning.

3.1.4. Support Vector Machines

Support Vector Machines (SVM) are a widely used ML technique known for their effectiveness in classification tasks [19]. Originally presented for handling linearly separable data, SVM has since evolved to accommodate more complex problems including nonlinearly separable datasets, regression (Support Vector Regression, SVR), and clustering (Support Vector Clustering, SVC). Although its main use is still in classification, SVM's flexibility lets it be applied in many fields needing strong accuracy and solid decision limits. Fundamentally, SVM seeks to identify the best separating hyperplane maximizing the margin separating several data classes.

Its orientation and position are much influenced by the support vectors, the important data points closest to the hyperplane. SVR uses a similar idea but concentrates on fitting a function that retains most data points inside a specified margin, thus allowing some flexibility in error tolerance in regression problems. SVM uses kernel functions and penalty parameters to improve its capability in situations when data are not linearly separable. While preserving a wide margin, the penalty parameter adds slack variables that enable controlled misclassifications, thus balancing accuracy with generalization. Conversely, kernel functions create a higher-dimensional space from the original input space where a linear separation is feasible. Commonly used kernels include sigmoid, linear, poisson, radial basis function (RBF), and sigmoid functions, each with different benefits contingent on the dataset.

3.1.5. Artificial Neural Networks

Inspired by the organization and capability of the human brain, artificial neural networks (ANNs) are a fundamental ML method. Originally designed for pattern recognition tasks, Rosenblatt's perceptron model in 1958 helped to form the idea. Computational power improvements over time have allowed ANNs to develop into more advanced designs able to address challenging ML tasks. ANNs have developed into several specialized designs each intended to solve particular ML problems. The most fundamental form is the feed-forward neural network (FFNN), in which data moves just from input to output layers. Incorporating several hidden layers, an advanced form of the Multilayer Perceptron (MLP) improves processing capability. Conversely, the Radial Basis Function Neural Network (RBFNN) is especially fit for specialized uses since it uses radial basis functions as activation mechanisms. Applied in fields including structural engineering, convolutional neural networks (CNNs) are optimized for image processing and pattern recognition, playing a major part in crack detection. Designed to manage sequential data, recurrent neural networks (RNNs) and their sophisticated variant, Long Short-Term Memory (LSTM) networks, shine in capturing long-term dependencies. The Adaptive Neuro-Fuzzy Inference System (ANFIS) is another hybrid method whereby neural networks are combined with fuzzy logic to improve decision-making capacity in uncertain surroundings [20].

Inspired by the human brain, which has linked layers of neurons processing and interpreting data, ANN architecture follows Input, hidden, and output layers. The hidden layers use sophisticated computations to derive significant patterns from unprocessed data. An ANN's efficiency mostly relies on elements like activation functions, which control signal flow across the network. The sigmoid or Rectified Linear Unit (ReLU) controls neuron activations, influencing the learning process. ANNs need hyperparameter fine-tuning of the number of layers, learning rate, and training epochs if they are to reach optimal performance. Correct optimization guarantees against overfitting or underfitting and guarantees the network can generalize to unseen data effectively.

The capacity of ANNs to replicate the learning mechanisms of the human brain defines one of their main advantages. Like neurons and synapses, these networks handle data in a way where experience shapes connections either strengthening or weakening. By means of deep feature extraction made possible by hidden layers, the network can identify complex interactions inside data. In difficult problem-solving activities, high predictive accuracy depends on this layered approach. Modern ML applications depend on ANNs since they offer a strong framework for adaptive learning by modeling biological neural paths [21].

3.1.6. k-Nearest Neighbor

Mostly used for classification and regression problems, the k-Nearest Neighbor (KNN) algorithm [22] is among the simplest yet most often used ML methods. KNN preserves the whole dataset and generates predictions depending on proximity-based decision-making unlike many other models that build an explicit function or discard training data after learning. The method assigns a class label depending on the majority vote among the k-closest data points, sometimes known as "nearest neighbors", when classifying a new instance. KNN averages the target values of the closest neighbors to forecast a continuous value in regression problems. Being a non-parametric method, KNN does not presume any predefined data distribution. Its efficiency relies on choosing a suitable value for k and measuring data point similarity using an appropriate distance metric. When no previous domain knowledge is available, the most often used metric is the Euclidean distance; nevertheless, depending on the dataset properties other metrics, such Manhattan or Minkowski distance, could also be used. KNN is a common choice, especially for applications where model transparency is crucial, because of its simplicity and interpretability.

3.1.7. Boosting Algorithms

Combining several weak learners into a more accurate and robust model helps boosting, a ML method, increase predictive performance. Freund and Schapire [23] first proposed the idea of boosting in the middle of the 1990s, which helped to create AdaBoost, among the first adaptive boosting systems. Many sophisticated boosting techniques have been developed since then to improve computational efficiency as well as accuracy. Among the notable examples are Friedman's Gradient Boosting Machine (GBM), which iteratively improves predictions by minimizing errors via gradient optimization. More recent developments including Extreme Gradient Boosting (XGBoost), Light Gradient Boosting Machine (LightGBM), and Categorical Gradient Boosting (CatBoost) have further optimized boosting techniques by improving speed, lowering memory use, and more effectively handling categorical data.

First presented in 1996 by Freund and Schapire [23], Adaptive Boosting (AdaBoost) was among the first boosting methods to show notable success. AdaBoost's central concept is to prioritize challenging-to-classify samples, sequentially improving weak learners. The method starts with training a base model on the whole dataset and evaluates mistakes afterwards, strengthening the impact of misclassified events. This guarantees that next models give more of their attention on fixing past errors. Many weak learners are trained as the process runs, each tackling particular shortcomings of the previous models. The resultant ensemble model uses a weighted voting system to decide predictions, giving more weight to more accurate models. Although AdaBoost is quite successful in raising prediction accuracy, its dependence on reweighting misclassified data makes it vulnerable to outliers and noisy data, thus compromising general performance. AdaBoost's efficiency and ability to improve simpler models help it to still be extensively used in many applications despite this restriction.

Designed to increase predictive accuracy by building models in a stepwise manner—where each new model seeks to correct the errors of the previous one—gradient boosting (GB) [24] is a ML method. GB, unlike AdaBoost, which changes instance weights, trains each next model on the mistakes left by its predecessor, minimizing the residual errors. The name "gradient" derives from the technique applied to maximize the loss function—each iteration moves in the direction most effectively lowering errors. Flexibility is one of its main advantages since it can be used with different loss functions depending on the kind of the problem. But because of its incremental character, computational cost may be high—especially in relation to big datasets.

Designed especially for speed and scalability when handling vast amounts of data, XGBoost [25] is an advanced application of the gradient boosting framework. It increases prediction accuracy by consecutively training decision trees, in which each new tree fixes the mistakes made by the one before it, same as in conventional GB. Still, XGBoost brings some improvements that boost efficiency. Its regularizing systems are an important aspect, since they stop overfitting and enhance computational efficiency.

Particularly useful for large-scale ML applications, Light Gradient Boosting Machine (LightGBM) [26] is a high-performance gradient boosting method intended for speed and efficiency. Furthermore, unique to LightGBM is its leaf-wise tree growth method, which deviates from standard level-wise methods. This enables the method to create increasingly sophisticated, deeper trees, thus improving predictive accuracy. But especially if the number of leaves is not under control, this rapid tree growth raises the possibility of overfitting. Despite this difficulty, LightGBM is still among the most effective boosting methods, since it can manage vast amounts of high computational-efficient data.

CatBoost [27] is a gradient boosting method that handles categorical features using target-based statistics, instead of one-hot encoding like many other boosting algorithms. This approach helps reduce the risk of data leakage and overfitting. CatBoost uses an ordered boosting method, where each model iteration is trained only on the data that comes before it. This helps produce stable predictions and avoids bias caused by using future data. CatBoost also uses a symmetric tree structure to maximize the search space and speed training while preserving great accuracy. It is a user-friendly option for many ML projects since it requires less hyperparameter tuning when compared to XGBoost and LightGBM. CatBoost is especially suited for uses involving structured data with multiple categorical features since it can effectively manage categorical variables and provide fast, accurate results.

Figure 6 shows the distribution of ML algorithms used in structural engineering research, based on the reviewed papers published between 1990 and 2025. Neural Networks (NN), with the largest share at 53%, remain the dominant choice across a wide range of structural applications. Boosting Algorithms (BA) follow with 18%, clearly indicating their growing presence in recent years. Support Vector Machines (SVM) account for 8% of the total, while Random Forest (RF) and Decision Trees (DT) each contribute 5%. Regression-based Approaches (RA) also represent 5%, and the remaining 6% are grouped under Other methods, which reflects a combination of various ML techniques. This distribution underscores the strong reliance on NN-based models, while also highlighting the increasing role of ensemble methods such as BA, RF, and DT in structural engineering research.

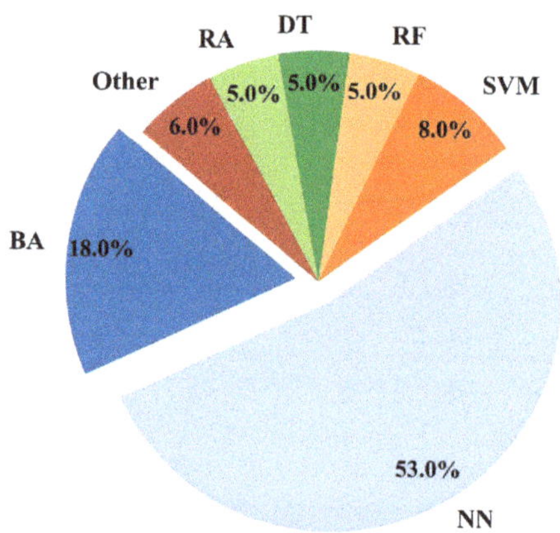

Figure 6. ML techniques applied in the considered research of this review spanning 1990 to 2025.

3.2. *ML for Steel Joints, Connections, and Rotational Stiffness Prediction*

Including ML into steel connections and joints helps to predict structural behavior with much more accuracy and efficiency, thus reducing dependence on expensive experimental testing and conventional empirical formulations [28–33]. Several research have used ML techniques to enhance the prediction of moment-rotation behavior, joint stiffness, and strength of bolted and welded connections.

Paral et al. [34] presented a deep learning-based method for evaluation of semi-rigid joint condition in steel frames. The model efficiently analyzed global vibration response signals from impulse excitation by combining Convolutional Neural Networks (CNN) with Continuous Wavelet Transform (CWT). Emphasizing moment-rotation (M-φ) correlations, Kueh [35] suggested explicit mathematical formulations for steel flush endplate beam–column connection rotational stiffness prediction. Practical formulations were derived depending on geometric and material characteristics using ANNs and MLR. In another study on moment-rotation behavior, Tran [36] examined flush end-plate connections at elevated temperatures using FE simulations and ANNs to forecast the ultimate moment (Mu) and shape parameters of an optimal M-φ model. Expanding the use of ML in bolted connections, Sarothi et al. [37] created a predictive ML framework for structural steel bearing strength estimation of double shear bolted joints. RF showed the best accuracy ($R^2 = 0.88$) above current design formulations. In a similar vein, Jiang et al. [38] investigated ML-based failure load and failure mode predictions for bolted connections of high strength steel. Training eight ML algorithms—including SVM, RF, and XGBoost—the study obtained a failure mode prediction accuracy of 97.2%, higher than the 67.9–85.3% accuracy of traditional design codes.

Beyond ANN models, Multi-Gene Genetic Programming (MGGP) has been investigated for moment-rotation prediction. In modeling semi-rigid connection behavior, Rabbani et al. [39] compared MGGP with ANN-based models and showed that MGGP displayed better accuracy and generalization capacity.

3.3. ML for Buckling and Stability Analysis

Recent developments in ML and AI have produced more exact, data-driven approaches for predicting lateral-torsional buckling (LTB), elastic and inelastic buckling, and stability failure modes in steel buildings. Several studies have also shown how well ANNs might predict LTB resistance. For LTB resistance in slender steel cellular beams, Ferreira et al. [40] created an ANN-based predictive model. The findings revealed notable progress over current analytical models, thus highlighting ML's structural optimization potential.

The existence of web openings in steel beams presents yet another difficulty for buckling resistance estimate since it can greatly affect post-buckling strength. Developing an ANN model for web-post buckling resistance and failure modes in steel beams with elliptically based web openings, Shamass et al. [41] addressed this problem. While web opening height, width, and radius had negative effects, the study found important geometric parameters influencing buckling behavior including beam height and web thickness (which favorably affected resistance). Their results underlined how successfully ANN-based models capture intricate geometric effects on stability performance. Furthermore, difficult in LTB prediction are steel beams with perforated web geometries. De Carvalho et al. [42] investigated the lateral-torsional buckling behavior of I-beams including sinusoidal web openings. The results imply that engineers using perforated I-sections might find useful design tools in ANN-enhanced predictive models. Xing et al. [43] presented an ANN-based model for local buckling prediction in the fire-resistant design of stainless steel beams. The study trained ANN models optimized using Kruppa's criteria and k-Fold cross-valuation by including experimental and FE results into a comprehensive dataset, thus ensuring robust and reliable buckling predictions for stainless steel I-sections under fire exposure. Rossi et al. [44] showed how well ANN-based models predicted LTB strength for steel I-beams. Their FEA parametric study investigated LTB resistance under geometric parameters and loading conditions. The ANN model gave better predictive accuracy than current design equations. Developing an ML framework for high-strength steel I-section columns, Cheng et al. [45] addressed the scattered accuracy of conventional design rules. Seven ML models were trained; Categorical Boosting turned out to be the most accurate method.

3.4. ML for Strength Prediction and Optimization of Cold-Formed Steel Structures

Cold-formed steel (CFS) buildings are rather common because of their great strength-to-weight ratio, economy, and sustainability. Predicting their strength, stability, and failure mechanisms still challenges since traditional design approaches sometimes oversimplify nonlinear interactions, residual stresses, and defects. ML is quickly becoming a great tool for load-bearing predictions, cross-section optimization, and structural reliability increase.

By addressing constraints in conventional design approaches that mostly concentrate on global buckling, Xu et al. [46] helped to predict ML-based bearing capacity for cold-formed stainless steel circular hollow section (CHS) columns. Trained on a database of 280 CHS columns, their ML-based approach showed significant accuracy improvements over current design codes, enabling improved predictive models in cold-formed steel engineering. Nguyen et al. [47] estimated the axial compression capacity of cold-formed steel oval hollow section columns using ANN and Adaptive Neural Fuzzy Inference System (ANFIS) models. Comparisons with three current design codes confirmed even more that ML-based approaches are better in estimating CFS section capacity. Fang et al. [48] developed a deep learning framework to evaluate the web crippling strength of cold-formed stainless steel channels addressing web crippling in perforated steel sections. Their DBN-based predictions exceeded conventional web crippling equations; new design equations were proposed based on the ML results. Lu et al. [49] developed a regression-classification ensemble ML model for predicting loading capacity and buckling modes of cold-formed steel built-up I-section columns. With high accuracy in both capacity estimation and buckling mode classification, XGBoost stood out among the tested models. To solve constraints in traditional fire design methods, Shaheen et al. [50] developed an ML-based predictive model for estimating the mechanical properties of high-strength steel at elevated temperatures. The work trained Deep Neural Networks (DNNs) on a large experimental dataset using temperature and chemical composition as input features to predict ultimate tensile strength, yield strength, 0.2% proof strength, and elastic modulus. Shahin et al. [51] developed hybrid ML methods for cold-formed steel-lipped channels to forecast web crippling capacity. Their work honed forecasts by combining ANN with GA and PSO. The PSO-ANN hybrid model exceeded other methods and offered more consistent strength estimates than accepted design guidelines. Yılmaz et al. [52] developed an ML-based predictive model for the load-bearing capacity of lipped channel sections. By training their model on a dataset of 2240 FE simulations, they found that flange length and section thickness were the most important parameters, underlining the importance of ML in geometry-based strength evaluations.

3.5. ML Applications in Steel Frame Design, Optimization, and Damage Detection

For automated visual inspection of steel frame structures, Kim et al. [53] presented a deep convolutional neural network-based damage locating (DCNN-DL) method. The DCNN-DL approach found and overlaid damage sites on input images, thus enabling exact, real-time steel frame inspections. Using ML models, Truong et al. [54] investigated their applicability for load-carrying capacity prediction of semi-rigid steel structures. Twelve ML techniques were tested. With regard to error metrics and determination coefficients, XGBoost showed the best accuracy among the tested models. Aiming to replicate the decision-making process of an experienced designer, Jahjouh [55] investigated the possibilities of ANNs in the design optimization of steel frames. Optimal structural designs for 2D steel frames were produced using an adaptive harmony search technique, subsequently regarded as training data for ANNs. The trained ANN models showed a 99% accuracy in forecasting appropriate designs during verification. By including real-time model training and parameter tuning, Shan et al. [56] developed an ML-assisted optimization framework for steel frame design, thus improving the efficiency of metaheuristic algorithms. The technique dynamically creates surrogate models to improve convergent accuracy and convergence.

3.6. ML-Based Structural Health Monitoring in Steel Structures

Pal et al. [57] presented a deep-learning method for structural health monitoring of a steel frame construction with bolted connections based on vibrations. Developed to extract discriminative features from time-frequency scalogram images of vibration data, a convolutional neural network (CNN) classified Bolt loosening conditions into fully loose, hand-tight, and completely tight (undamaged) categories. Focusing on connection damage identification using statistical vibration features, Naresh et al. [58] developed an ML-based health monitoring system for steel frame buildings. High accuracy in structural damage detection allowed an SVM model to classify both undamaged and damaged conditions to show Analyzing natural frequency fluctuations, Vu et al. [59] investigated ML-based damage identification in steel beams, offering a substitute for conventional structural health monitoring techniques. FEM generated a dataset of natural frequencies under several damage scenarios; ML models including ANN, XGB, and RF were trained to predict damage location, width, and depth.

Although conventional sensor-based monitoring systems—such as strain gauges, fiber Bragg grating sensors, and accelerometers—have been extensively used for damage detection in steel constructions due to their high precision and long-term stability, they are often limited to localized measurements and demand direct physical installation and continuous maintenance. Conversely, for non-contact, wide-area structural inspection, recent developments in computer vision and deep learning present interesting substitutes. These AI-powered visual techniques, particularly those using convolutional neural networks (CNNs), have shown strong performance in tasks such as surface crack segmentation, corrosion detection, and bolt loosening identification. The primary advantage of vision-based approaches lies in their scalability, cost-effectiveness, and ability to process large amounts of visual data rapidly. However, they also face practical challenges such as sensitivity to lighting, occlusion, camera angle, and the need for high-quality annotated datasets. Several recent studies suggest hybrid approaches that use the strengths of both systems to combine sensor signals with vision-based outputs, thus enabling stronger condition assessment. Integrating these two techniques could provide a more robust and interpretable structure for real-time steel structure monitoring in complex environments as the field develops.

Beyond real-time detection, one of the most exciting frontiers of AI in structural engineering is the prediction of long-term degradation in steel structures including industrial frameworks, bridges, and high-rise buildings. Recurrent neural networks (RNNs), long short-term memory (LSTM), and Bayesian networks have been investigated recently in order to forecast damage accumulation depending on operational load, climate exposure, and historical sensor data. Models have been trained, for example, to learn patterns from multi-year monitoring datasets to forecast fatigue crack propagation, coating degradation, and corrosion depth over time. Though early results are encouraging, such applications still provide practical difficulties. These comprise limited access to high-quality long-term field data, environmental condition variability, and the challenge of documenting rare but critical degradation modes. Recent efforts aiming at overcoming these constraints have concentrated on hybrid models combining AI with physics-based deterioration models, providing better interpretability and generalizability. Transfer learning methods let models developed on one structure be adjusted for others with similar typologies concurrently. These methods, which are still under development, imply that AI has great possibilities for lifecycle assessment of steel infrastructure under actual conditions and predictive maintenance planning.

Especially in the automated damage and failure mode detection in steel constructions, ML techniques have brought significant SHM advances. Deep learning and feature extraction-based models have remarkably faithfully classified bolt loosening, connection damage, and other structural failures. Moreover, ML-driven frequency analysis provides a fast approach to identify damage in steel beams and frames, thus reducing dependency on traditional sensor-based methods and hand inspections.

3.7. ML-Based Prediction of Structural Performance in Steel Components

ML uses in shear strength, bending capacity, moment resistance, and structural stability prediction in steel components is investigated in this part.

Li et al. [60] addressed errors in conventional design approaches by developing a data-driven ML model for shear strength of high-strength friction-grip bolts. Backpropagation Neural Networks (BPNN) obtained the lowest coefficient of variation and the best accuracy (93% goodness of fit) among the several ML techniques the study examined. Using 108 stainless steel lipped channels and 238 carbon steel Lite Steel sections as test data, Dissanayake et al. [61] further used ML models to forecast the shear capacity of steel channel sections. The work evaluated several ML methods and found that SVR gave the most consistent shear capacity predictions. Aiming to overcome conventional design codes, Dai et al. [62] concentrated on moment capacity prediction of cold-formed steel channel beams with edge-stiffened and un-stiffened web holes. A FEM was validated against 12 experimental datasets, and an XGBoost model was trained using a database of 1620 samples, obtaining a R^2 score of ~99%, much above traditional design approaches. Reliability analysis helped to validate the suggested ML-driven equations, guaranteeing adherence to AISI safety standards. Liu et al. [63] expanded this study by developing ML-based models for predicting the ultimate bending moment resistance of high-strength steel (HSS) welded I-section beams. The results confirmed that ML models greatly outperformed both European and American design codes.

To estimate the failure loads of normal and high-strength steel welded I-section beam–columns, Su et al. [64] proposed an ML-based unified design method. Traditional methods were found to understate or scatter failure load predictions when compared to Eurocode and American design rules; the ML approach produced more consistent and dependable results. Utilizing structural mechanics and ANNs to forecast point forces and distributed loads, Tusnin et al. [65] applied ML for load identification in steel structural systems. The suggested approach allows final load estimations, structural decomposition, ML-based projections, and structural integrity monitoring better accuracy in engineering inspections and structural health monitoring.

3.8. ML-Based Seismic Performance Assessment and Optimization

Aiming at seismic collapse behavior using ML, Sediek et al. [66] developed a database of over 900 experimental and numerical results for deep wide flange (W-shaped) columns. The work estimated cumulative inelastic rotation until failure under axial and lateral loading and categorized failure modes (local, global, and coupled). By raising classification accuracy by 30%, the ML models exceeded highly ductile limits and demonstrated the possibilities of data-driven methods in creating future seismic design rules. Using ML models, Imam et al. [67] investigated how best to forecast the seismic performance of steel moment-resisting frames (SMRFs). The study assembled a sizable 29,200 data point collection from 292 structural models, training RF, XGBoost, and ANN. Compared to conventional nonlinear dynamic analysis techniques, the ML-based models showed great predictive accuracy.

Addressing constraints in traditional multivariable regression models, Shin and Kim [68] developed an ML-based framework for estimating the hysteretic behavior of two-sided clamped steel shear walls (TCSSWs). Under cyclic load, the ML-based model shown efficiency in automating and enhancing the hysteretic behavior modeling of steel shear walls. Cho and Han [69] developed a numerical material model to simulate the cyclic behavior of high-strength steel (HSS), thus extending ML applications in seismic engineering. Eliminating the need for expensive LCF tests, empirical equations were proposed to estimate model parameters using just monotonic tensile test data, improving practical applicability.

Hu et al. [70] presented an ML-aided peak and residual displacement-based design (PRDBD) method to use self-centering braces to increase the seismic resilience of SMRFs. Conventional SMRFs dissipate energy through steel yielding, producing significant residual displacements that sometimes call for expensive post-earthquake demolition. Samadian et al. [71] developed surrogate models for seismic and pushover responses of SMRFs, furthering ML applications in seismic engineering. Providing a computationally efficient substitute for traditional nonlinear time history and pushover analyses, CatBoost models emerged as the most effective in the study. Salama [72] investigated ML techniques to maximize the seismic resilience of vertically irregular steel building. Using XGBoost with Owl Search Algorithm (OSA) for hyperparameter tuning produced accurate seismic behavior predictions. The results underlined the great reliance of base shear capacity and general seismic performance on structural irregularities.

3.9. AI Applications in Real-World Steel Structures

In bridge engineering, AI-based inspection and SHM systems are advancing safety, automation, and predictive maintenance. Huang et al. [73] developed a CNN-based deep learning method for structural damage identification (SDI) on steel truss bridges, achieving high detection accuracy using field data. Iacussi et al. [74] explored a drive-by monitoring approach, where smart sensor nodes on moving vehicles predicted bridge deflection shapes with strong agreement to installed sensors. Wang et al. [75] introduced SBDNet, a segmentation network trained on real crack images, enabling pixel-level fatigue crack detection on steel bridge surfaces. Svendsen et al. [76] proposed a hybrid SHM framework combining numerical simulations with experimental data to detect loose bolts and member failure on full-scale steel bridges under environmental fluctuations. In high-rise buildings, Zhou et al. [77] monitored a 300 m twin-tower steel frame where smartphone videos were processed with computer vision to extract vibration patterns. Ghaffari et al. [78] developed an RNN model trained on multi-hazard FE data to predict the dynamic response of tall steel structures, achieving less than 1% error compared to actual displacements.

For offshore and tower structures, AI tools have demonstrated strong practical potential. Wang et al. [79] introduced an ensemble deep learning model (CNN + BiLSTM + SENet) to detect damage in jacket platforms with 95% accuracy, validated by experimental results. Martzikos et al. [80] used field data from scaled floating steel platforms to train ANN models predicting mooring line forces under wave loading. Kouchaki et al. [81] used ML classifiers to detect damage in transmission towers based on vibration data, achieving up to 96% accuracy even under noise-contaminated conditions. Kiyoki et al. [82] predicted bending moments in wind turbine towers using only nacelle sensor data, offering a low-cost alternative to strain-based measurements.

Beyond monitoring, AI is being embedded into design optimization and material reuse. Vlasenko et al. [83] developed an ensemble ML model to non-destructively estimate yield strength of reclaimed steel using magnetic sensor readings, enabling sustainable material reuse with real-time decisions.

Connection and joint evaluation has also benefited from AI integration. Shang et al. [84] created a machine vision system that automatically quantifies bolt loosening by detecting nut rotation angles through image processing.

In the area of maintenance and corrosion monitoring, Huang et al. [85] used an Inception-v3 deep CNN trained on laboratory corrosion images to classify rust severity levels on steel surfaces. The model achieved performance comparable to expert inspectors and enables consistent, rapid corrosion assessment across large steel structures such as bridges, towers, and tanks.

These studies collectively show that AI technologies are now being deployed in real-world settings for steel structure monitoring, inspection, design, reuse, and maintenance. As AI continues to mature, its integration into practical structural workflows will likely expand, making structural engineering more data-informed, efficient, and adaptive. This expanding body of validated applications confirms that AI is no longer limited to simulation environments but is becoming an essential companion to structural engineers in the field.

4. Inverse Machine Learning (IML) for Design Optimization and Performance Enhancement in Steel Structures

IML has become a powerful tool in structural engineering for rapidly investigating complex geometrical designs [86]. Particularly for complicated structural systems, conventional design techniques sometimes involve several iterative changes that can be computationally expensive and time-consuming. Direct mapping design objectives to optimal parameters helps IML simplify this process and enable a greater spectrum of design possibilities by lowering the need for major trial-and-error adjustments [87]. This capacity is especially important for handling complex structural designs where conventional methods might not be able to provide sufficient effective answers. In inverse modeling, one of the main challenges is the ill-posed character of inverse problems, whereby solutions could lack uniqueness or stability [88]. To solve these challenges, several computational techniques have been developed to ensure strong and consistent predictions, thus improving the applicability of IML in useful engineering environments.

In structural engineering, inverse problems mostly concern determining design parameters—such as geometry and material properties—needed to meet specific performance criteria [89]. Conversely, in forward problems structural performance is expected to be dependent on predefined input criteria. Unlike forward analyses, inverse problems are often ill-posed, meaning they can show instability or have several possible answers. Particularly for complex structural designs [90], conventional techniques—such as trial-and-error changes—rely on iterative adjustments and hence are both time-consuming and computationally demanding. IML provides a quick replacement that eliminates the need for constant hand-offering by directly predicting ideal design parameters straight from historical data and ML models. This greatly simplifies the design and optimization process, hence faster and more efficient solutions follow [91]. IML is particularly important for metamaterials and advanced structural systems where the interactions between geometry, material composition, and mechanical performance are highly nonlinear and complicated. Its application to designing buildings ideal for impact absorption, improved buckling resistance, and shape recovery [92] shows its flexibility over many engineering problems. By converting structural design into a data-driven process, IML speeds innovation in demanding structural systems, reduces computing costs, and increases accuracy.

IML Application in Steel Structures

Several studies demonstrating the possibilities of IML in steel construction analysis, fabrication, and performance prediction offer more efficient and statistically driven solutions. IML has been used to maximize steel buildings by improving geometry, mechanical aspects, and material composition.

Teimouri et al. [92] devised an IML framework for designing plate-lattice structures (PLSs) idealized for compression strength and specific recovery force (SFR) in large-scale steel constructions. Their method consisted of three validation sub-frameworks before PLS optimization, 3D-printed using a new shape memory polymer (SMP).

By considering manufacture defect and imperfections, IML has also been used to increase the mechanical robustness of steel-based systems. Glaesener et al. [93] investigated how mechanical response of periodic trusses changed with geometric flaws. Using IML-based defect prediction models, their study examined common defect types over several steel truss topologies to evaluate their impact on effective stiffness. This method actively found and reduced geometric inconsistencies, thus improving the dependability of steel trusses.

Particularly in high-strength and ultra-high-strength steel alloys, IML methods have also been heavily applied to maximize steel materials. Using deep neural networks to improve hohlraum designs for radiation temperature control, McClarren et al. [94] showed an IML-driven optimization framework in a high-performance design environment. Similar uses in steel alloy optimization have been inspired by the approach they used whereby a forward ML model forecasts results and an inverse model finds optimal parameters. Building on this, Lee et al. [95] suggested an inverse design framework for high-strength steels, thus attaining an ultimate tensile strength (UTS) above 2 GPa with improved ductility. Experimentally validated designs with improved mechanical performance resulted from a GA optimized using the Shannon diversity index investigating a high-dimensional search space of chemical compositions and austenitizing conditions. Analogous development of two ML-based materials genome integration systems for inverse microstructure analysis of steels was accomplished by Wang and Adachi [96]. These models allowed inverse design of steel microstructures and direct prediction of material properties to reach desired stress–strain behavior, tensile strength, and elongation. This method helped to create ideal steel compositions with better mechanical properties. Using microstructure image recognition to maximize the composition of martensitic and ferritic steels, Pei et al. [97] introduced an IML-based framework for inverse alloy design, thus improving steel alloy performance. Their method greatly reduced trial-and-error efforts in steel development by using neural networks to identify intricate microstructural patterns and predict alloy properties. Dealing with microstructure-property optimization, Lertkiatpeeti et al. [98] investigated the impact of martensite phase alignment in dual-phase steels by means of a Markov Chain Monte Carlo-based inverse analysis coupled with ML models. Using representative volume element (RVE) simulations incorporated new microstructure descriptors—such as Moran's index, martensite band index, and martensite orientation—and SVR and ANN models trained on them quantify spatial effects in martensitic phase distributions. Their method effectively tuned steel microstructures for intended mechanical characteristics. Adachi et al. [99] examined IML applications in steel material design more broadly and discussed ML techniques integrating process-property-microstructure relations. The study underlined the growing relevance of IML in improving steel material development by means of automation. Important design factors are the lifetime and fatigue resistance of steel buildings; IML has been applied to maximize these features. By means of both direct and inverse analyses, He et al. [100] devised an ML-based framework to estimate fatigue life and fatigue limits of steels. A Bayesian optimization-based inverse analysis was also conducted to ascertain fatigue limits. Optimizing double-layer protective coatings used in spacecraft friction units, Kolesnikov et al. [101] developed an IML-assisted framework. While the Extra Trees method was used for feature importance analysis, their work combined FEM with adaptive sampling techniques to build a training set. By allowing coatings based on target hardness values to be designed, the inverse ML model helped to achieve predictive accuracies above $R^2 = 0.96$, offering an effective method of designing high-performance protective layers.

Relevant for hybrid steel-composite structures, IML has helped to maximize the Automated Fiber Placement (AFP) process in thermoplastic composite manufacture in manufacturing processes. To increase process control, Islam et al. [102] presented a hybrid ML framework combining ANNs, virtual sample generation, physics-based numerical simulations, and experimental data. Using an inverse property estimation framework, Masurkar et al. [88] estimated elastic constants of orthotropic laminates used in steel-composite hybrid structures. Their work structured the problem as an inverse task, whereby a deep neural network model was developed on time-series simulation data to effectively infer material properties.

In recent years, IML has been applied to achieve tangible performance gains in real-world structural components. For example, Challapalli et al. [89] developed an IML-GAN framework that optimized the internal geometry of lightweight metamaterials, leading to a 40–120% improvement in load-bearing capacity compared to traditional lattice designs. Similarly, Shen et al. [103] proposed an inverse design architecture for high-performance gradient honeycomb steel structures, where IML integration resulted in designs that offered improved stress distribution and reduced material usage under high-impact loads. Another study by Challapalli et al. [104] demonstrated how IML can be used to design thin-walled cellular structures made from shape memory polymers, achieving record-high specific recovery stress and a 200% improvement in normalized mechanical performance compared to conventional unit cells. While focused on SMPs, the underlying framework is directly applicable to lightweight steel-composite systems where strength and flexibility must be balanced. IML has also been used in dual-phase and ultra-high-strength steels to refine microstructure design, leading to higher tensile strength and optimized material utilization [105]. These examples show that IML is no longer limited to conceptual or experimental research; it is actively shaping how structural efficiency, geometry optimization, and weight minimization are achieved in modern steel engineering. Using IML on specific structural engineering problems helps to clearly notice its usefulness. In load-bearing capacity estimate, IML enables engineers to define the target resistance and quickly access the design parameters—such as cross-section dimensions or material strength—that satisfy that need. In material selection, IML considers cost, weight, or environmental impact in addition to guiding the identification of perfect material types or grades meeting performance criteria. Without involving many forward simulations, IML guides suitable forms, profiles, or arrangement patterns, thus enabling geometric optimization. These projects demonstrate how much IML can reduce demand for repeated analyses and hand-made iterations by offering a more intelligent and simplified design workflow.

These applications collectively offer not only theoretical frameworks but also clear numerical evidence for IML's effectiveness in structural engineering. For example, IML-driven metamaterial design achieved up to 120% increases in load capacity [89], inverse alloy optimization exceeded 2 GPa tensile strength with validated experiments [95], and fatigue life estimation frameworks reached predictive accuracy above $R^2 = 0.96$ [101]. These results confirm that IML is not merely an abstract modeling concept but a practically validated tool for performance-driven structural optimization.

5. Enhancing Explainability in ML: Concepts, Methods, and Applications in Structural Engineering

Explainability in ML is the ability of models to provide open insights into their decision-making process so ensuring human comprehension of their predictions. Unlike conventional black-box models in which the internal computations are difficult to understand [106], explainable ML techniques aim to reveal the link between input features and output preferences. Transparency of model behavior supports explainability to foster responsibility, trust, and usability in AI applications [11]. One major advantage of explainable models is that they help to ensure that relevant and major events drive forecasts, reducing the risk of erroneous or biased influences [8].

Moreover, explainability facilitates the identification of prejudices in training data, thus advancing fairness and enhancing the quality of decisions made. Open models also contribute to build resilience and reliability by identifying discrepancies or anomalies that might compromise predictive accuracy [107,108]. Beyond only technical improvements, explainability supports ethical considerations since it helps users to critically evaluate model decisions and change them to fit policy, fairness, and operational goals [109–111]. While interpretability stresses how easily a human can grasp the reasoning behind a given prediction, explainability stresses on understanding the internal mechanisms and interactions inside a model. Although the terms explainability and interpretability are sometimes used synonymously. Both components are quite crucial in bridging the gap between advanced ML algorithms and human knowledge so that AI-driven solutions remain transparent, trustworthy, and practically relevant in many different fields [112–115]. XML aims to generate transparent, interpretable, trustworthy ML models guaranteeing their ethical and useful application [116]. Especially in high-stakes decision-making processes, one of its main objectives is to build user confidence by demonstrating that models behave consistently and regularly. While explainability is only one aspect influencing trust in AI, giving users tools to assess the reliability of a model will help to raise confidence in its predictions [117,118]. Apart from this, XAI aims to provide explanations that allow the identification of possible causal relationships inside data, thus transcending observations based on correlation [119]. While conventional ML models focus on pattern recognition, explainable methods help to identify underlying cause-and-effect dynamics, which can then be validated using specialized inference techniques. Simplifying challenging models into understandable insights is also fundamental for XAI so that users may grasp how different features interact and support predictions. By allowing AI-generated decisions to be more readily available, these justifications help to enable informed human supervision and improved integration into useful applications [120]. XAI also emphasizes the transferability of knowledge, that is, the capacity of knowledge to be applied to other problems once the decision-making process of a model is well known. On the other hand, transferring models across tasks depends on a clear awareness of their limitations and presumptions so that predictions remain relevant in new contexts [121,122]. Yet another crucial emphasis of explainability is keeping confidence and consistency in predictions. Users have to be able to assess their certainty, especially in situations when model predictions directly influence outputs produced by AI [123,124]. XAI transparency open communication of constraints and uncertainty, thus improving the dependability of automated decision-making systems. The main goal of XAI is to close the gap between technical complexity and human-centered usability so ensuring that AI solutions remain responsible, interpretable, and compliant with ethical and operational guidelines.

5.1. SHapley Additive exPlanations (SHAP)

Different approaches have developed from enhancing the interpretability of ML models; these can be usually categorized as either model-specific or model-agnostic [125]. Designed to investigate internal dynamics of particular algorithms, model-specific methods offer insights dependent on the structure of the model. By contrast, model-agnostic approaches provide a more flexible approach appropriate for any ML model independent of its underlying design.

One of the most often used model-agnostic interpretability methods [9] is SHAP, which measures the contribution of particular features to model predictions using ideas from game theory. This method views the predictive model as a cooperative game based on Shapley values, in which case features play as players and their influence on the final prediction is assessed methodically either including or excluding them from the model. The Shapley value is the average contribution of a feature across all possible combinations of input variables, guaranteeing a mathematically sound evaluation of feature importance [9].

SHAP is globally, as well as locally, interpretable. Globally, it identifies which traits have the most influence over a whole dataset, providing a general picture of the model's decision-making pattern [9,126]. SHAP shows how specific input values produced a given result, clarifying local level individual predictions. Using binary representations, SHAP produces linear approximations clarifying their influence on predictions by indicating the presence or absence of features.

Beyond only increasing model transparency, SHAP is quite crucial for the ethical development of AI since it supports responsibility and fairness. By precisely defining the function of every feature in decision-making, SHAP helps to detect possible biases and guarantees that models remain interpretable for end users. Even as ML advances, SHAP remains a simple tool for enhancing trust and openness in complex predictive systems [127].

SHAP's surrogate model links complex ML models with human interpretability [11] using both local and global explanations. SHAP quantifies its individual influence on a given prediction since it generates Shapley values for each local input feature. This clarifies for consumers which truly influence an outcome and which have either little or no impact on it. Globally, SHAP aggregates these local explanations to provide a more complete view of feature significance over the entire dataset [9,107]. By means of methodically evaluating the contribution of every feature, SHAP guarantees more transparency in model behavior, separating between variables influencing decision-making from those less important.

5.2. XML Application in Steel Structures

To address the complexity of several instabilities in channel sections under axial compression or bending, Mojtabaei et al. [128] created an XML-based model to forecast the buckling behavior of thin-walled structural elements. High accuracy was obtained by the trained ANN models; SHAP analysis revealed feature contributions. With regard to material properties and geometry, Hu et al. [129] then developed XML models for the probabilistic prediction of buckling stress in steel shear panel dampers. While core plate thickness (t) had little effect, SHAP and feature importance analysis found yield strength (fy), height-to-width ratio (α), width-to-thickness ratio (β), and initial imperfection (δ) as critical factors affecting buckling stress. Building on these developments, Hou et al. [130] developed an interpretable ML model to predict the probabilistic axial buckling strength of steel circular hollow section members, addressing the impact of geometric, material, and initial imperfections not explicitly accounted for in current design codes. Parameter impact was assessed using global sensitivity analysis and SHAP. While greatly lowering computational time, the ML-based probabilistic model showed that it closely matches numerical and design code predictions and provides a more detailed and effective method for axial buckling strength evaluation. Extending XML applications to more general structural systems, Samadian et al. [131] developed meta databases and a surrogate modeling framework for steel moment-resisting frames, providing a computationally efficient alternative to nonlinear time history analysis for disaster risk management. Feature importance was evaluated using random forest and SHAP analysis pointing up important elements in SMRF design. XML studies have also concentrated much on fire resistance. Liu et al. [132] created an XML-based framework to maximize cold-formed steel wall design for the economy and forecast their fire resistance time. Although SHAP analysis found important input parameters, the XGBoost model—optimized with Bayesian tuning—showcased great predictive accuracy. Further advancing ML applications in fire-resistant design, Tang et al. [133] developed an XML framework to predict the axial capacity of cold-formed steel channel sections at elevated temperatures, offering an effective alternative to conventional techniques. Features were interpreted using SHAP analysis. The results show ML's potential to provide faster and more consistent axial capacity predictions as well as to enhance fire-resistant structural design. Apart from fire resistance, XML has helped to enhance the design and reliability of perforated and cellular steel sections. Addressing constraints in traditional design provisions, Degtyarev et al. [134] presented a probabilistic XML model based on Natural Gradient Boosting for predicting the resistance of laterally restrained cellular steel beams. Reliability studies identified the resistance reduction factors needed for compliance with European and US design frameworks; SHAP analysis was then used to improve interpretability. Beyond static loading situations, Widanage et al. [135] investigated the use of XML models for predicting blast loads on rigid structures, thus offering a time-efficient alternative to experimental and numerical methods. Model transparency was guaranteed using XML methods, thus confirming that predictions follow fundamental blast physics rules. Anand et al. [136] similarly used XML models to forecast engineering demand parameters in buckling-restrained braced frames under seismic load, thus reducing reliance on computationally costly simulations. XML methods emphasized important seismic parameters affecting BRBF behavior, thus enhancing model transparency. Fan et al. [137] suggested another creative use for an XML-based framework to forecast the axial compressive capacity of Σ-shaped cold-formed steel with web opening, thus addressing the inefficiencies of conventional finite element and experimental techniques. SHAP analysis revealed important design parameters affecting axial capacity and their interactions, thus offering interpretability. Extending on connections and fastener-related issues, Sarfarazi et al. [138] used XML techniques to forecast the shear strength of stainless-steel column web panels, addressing constraints in current design standards that do not entirely reflect the strain hardening behavior of stainless steel. Twelve machine learning models were tested; Extra Trees Regression had the best predictive accuracy. SHAP analysis identified bolt diameter and the column's second moment of inertia as the most influential factors in shear strength. Predicting axial capacity and failure modes, Aloko et al. [139] investigated further the use of XML models in cold-formed steel built-up columns. Their work showed that, surpassing conventional strength prediction approaches, ML techniques could efficiently record buckling interactions and load-bearing behavior. The interpretability of the models supports even more the relevance of ML in design validation and optimization. Still, it is difficult to guarantee generalizability over different section geometries and loading conditions. Furthermore, helping seismic performance optimization are XML techniques. Gharagoz et al. [140] presented an XML framework including XGBoost for structural engineering optimizing seismic retrofitting techniques. Considered for seismic energy dissipation and self-centering mechanisms, the framework incorporates the Spring-rotational Friction Damper system. Transparency of models was improved using interpretable models, guaranteeing interpretability for use in decision-making. For the quick evaluation of seismic resistance in steel frames, Su et al. [141] presented an XML-based method including active learning techniques. The dataset comprised 1056 seismic response records from 250 steel frames with varying geometric characteristics and steel grades. Evaluated were seven ML models (DT, RF, SVM, KNN, ADA, XG, CB); Extreme Gradient Boosting (XG) obtained the best accuracy (97.79%). SHAP analysis provided interpretability for seismic response predictions. Applying XML models, Gatheeshgar et al. [142] projected web-crippling strength in cold-formed steel beams with staggered sloped perforations. Four ML models—KNN, RF, SVR, ANN—were assessed. Slope length and bearing plate length were found by SHAP analysis to be main determinants of web-crippling strength. Sarfarazi et al. [143] presented a hybrid XML framework for examining the mechanical response of stainless-steel beam-to-column connections.

Beyond their academic importance, these interpretability techniques now serve practical roles in engineering decision-making. For example, when SHAP values identify specific parameters (such as flange width, thickness ratio, or bolt diameter) as dominant influencers in capacity prediction, engineers can confidently prioritize these variables during section design or material selection. Similarly, in XML-based fire resistance studies, the clear identification of wall thickness or insulation parameters helps design safer cold-formed steel walls without extensive physical testing. In seismic applications, SHAP analysis pinpoints key damping and stiffness properties, allowing for informed retrofitting or configuration changes. These tools enable engineers to move from 'black-box' models to traceable and code-aligned decisions, strengthening both model trust and practical design outcomes.

6. Addressing Challenges, Limitations, and Future Directions

Although ML models show very good accuracy in estimating structural responses, several elements limit their usability in professional engineering practice including data availability, model reliability, interpretability, integration with conventional design methods, and regulatory acceptance. This section methodically addresses the shortcomings of current approaches, investigates the main barriers preventing the full-scale implementation of ML in structural engineering, and proposes future directions of research to close the distance between ML-driven discoveries and useful applications.

6.1. Rethinking Model Training: Advancing from Accuracy to Engineering Reliability

6.1.1. Generalization and Overfitting in Structural ML Models

The main issue with current ML applications in steel construction is the excessive emphasis of model accuracy without considering engineering reliability. Although many studies reveal low error rates and strong R^2 values, these measures do not always translate into safe and reliable structural designs. Overfitting is a main problem whereby a model performs remarkably on training data but fails in real-world or unseen conditions. Unlike conventional engineering models including physical laws (such as equilibrium equations and constitutive models), many ML techniques depend just on statistical learning. This can generate physically contradicting predictions whereby results seem statistically valid but violate fundamental engineering criteria.

6.1.2. Addressing the Gap in Physics-Based Learning

Unlike in computational mechanics and fluid dynamics, where Physics-Informed Neural Networks (PINNs) and hybrid solvers have gained considerable ground—most ML applications in structural engineering remain entirely data-driven [144]. Although on well-curated datasets these black-box models can show remarkable accuracy, they usually lack embedded physical constraints including equilibrium, compatibility, or constitutive relations. As a result, such models risk producing outputs that, although numerically plausible, may contradict fundamental engineering principles, especially under extrapolated or safety-critical conditions [145].

The structural engineering community is starting to investigate Physics-Informed Machine Learning (PIML) as a more reliable substitute to mitigate this shortcoming. These models directly include domain-specific physical laws including elasticity, buckling equations, or dynamic balance into the ML training process. For example, Raissi et al. [145] introduced the foundational framework of PINNs, in which governing partial differential equations (PDEs) are embedded in the neural network's loss function. Haghighat et al. [146] later extended this method to solid mechanics problems, enabling the network to produce stress or displacement fields that respect the governing equations of structural behavior.

In parallel, hybrid ML–FEM approaches have emerged that blend neural network-based surrogates with traditional finite element formulations. These models seek to preserve mechanical integrity of the results while lowering computational effort—especially in repeated simulations. Recent models, for instance, let ML models act as fast approximators for particular FEM sub-processes including stress field estimate or crack growth under variable boundary conditions [145,147].

A comprehensive review by Habib et al. [144] cataloged over 90 PIML applications tailored to structural PDEs, spanning use cases like stiffness identification, crack propagation, and damage localization. When training data are limited or difficult to obtain—such as post-buckling behavior, fatigue evolution, or high-strain-rate responses—these techniques are especially helpful. Physics-guided constraints in such environments not only stabilize learning but also strengthen engineering confidence in model outputs.

Physics-based learning must clearly be central in bridging the gap between academic experimentation and practical adoption as structural ML develops. Including mechanical rules into data-driven models increases generalizability, transparency, and offers a necessary basis for future possible regulatory approval.

6.1.3. Future Directions for Reliability-Oriented Model Training

- Creating PIML Models
- Hybrid Engineering-ML Models
- Uncertainty Quantification (UQ)

6.2. The Overlooked Role of Feature Engineering and Domain Expertise

6.2.1. The Difficulties Using Raw Data Without Engineering Context

In structural engineering, feature engineering—the choice and modification of input variables—is among the most crucial but underappreciated elements of ML modeling. Many studies direct raw numerical data—such as cross-sectional dimensions, material strength, and loading conditions—into ML models without regard for the deeper physical interactions guiding structural behavior. For buckling failure, for instance, just entering the height, width, and thickness of a steel column might not be sufficient. Rather, engineering equation-derived features—such as slenderness ratio, moment of inertia, and effective length factor—may offer more important predictors.

6.2.2. Why Structural Analysis Driven by ML Requires Domain Knowledge

While fields like image recognition [148] or natural language processing [149] where raw data directly reflects the problem space, structural engineering calls for a complete awareness of mechanics and materials. Ignorance of domain knowledge might lead to models that memorize patterns instead of learning fundamental structural behavior. However, an often overlooked yet critical issue in ML-based structural analysis is the limited transferability of models across different steel structural typologies. A model trained on I-section beams under flexure may not perform well when applied to box-section columns or perforated trusses under axial or torsional loading. This lack of generalizability stems from fundamental differences in governing mechanics, geometric characteristics, failure modes, and load interactions. To address this, several approaches have been proposed. One promising direction is transfer learning, where a pre-trained model is fine-tuned using smaller datasets from a new typology, preserving learned features while adapting to new contexts [150]. Another emerging method is meta-learning, which aims to train models capable of quickly adapting to new structural scenarios with limited data [151]. Domain adaptation techniques are also under exploration to help models account for the shift in data distributions when moving from one structural typology to another [152]. While these techniques offer potential, their use in structural engineering remains limited and largely exploratory. Therefore, future research must systematically investigate cross-typology learning to develop truly generalizable and robust AI models for structural applications.

6.2.3. Future Directions for Feature Engineering and Domain-Driven Learning

- Creating feature selection techniques grounded in engineering
- Incorporating Expert Knowledge into Feature Engineering
- Automated Feature Extraction Techniques

6.3. From Deterministic to Probabilistic Modeling

6.3.1. The Importance of Uncertainty Quantification in Structural Predictions

ML models treat structural properties as fixed parameters rather than stochastic variables, thus providing single-value predictions. Structural materials do, however, exhibit natural variation, and operational and environmental uncertainty results in loads never exactly known. Traditional engineering methods allow for these uncertainties using safety factors; but, ML models typically ignore uncertainty. For example, a neural network forecasting the shear strength of a steel joint could produce a value of 500 kN but does not show a confidence range (e.g., 480–520 kN). The lack of probabilistic insight reduces the reliability of ML-based predictions in applications related to safety.

6.3.2. The Role of Bayesian ML and Monte Carlo Simulations

Two fundamental components of safety-critical structural design, uncertainty quantification and probabilistic risk assessment, can be supported by ML beyond point predictions. Conventional engineering frameworks like Load and Resistance Factor Design (LRFD) embed probabilistic safety margins, yet most ML models remain deterministic. This disconnect limits their practical reliability in engineering settings.

To bridge this gap, Probabilistic ML techniques such Bayesian Neural Networks (BNNs) [153] and Monte Carlo [154] dropout have been adopted to estimate predictive distributions rather than single values. These techniques help quantify epistemic uncertainty—arising from limited training data—and allow engineers to evaluate how much trust can be placed in a model's prediction [155]. For example, a BNN can estimate that a predicted shear capacity of 480 kN has a 95% confidence range of 450–510 kN, helping decision-makers incorporate statistical safety considerations directly into their workflow.

In parallel, Gaussian Process Regression (GPR) is increasingly used to develop probabilistic surrogate models that approximate finite element responses under uncertain loads or material properties [156]. These surrogates dramatically reduce computational cost when thousands of simulations are needed, for instance, in fragility curve generation or reliability-based design optimization (RBDO). When integrated with Monte Carlo simulations or Latin Hypercube Sampling, such models enable full probabilistic risk evaluations without sacrificing computational efficiency [157].

Embedding explainable AI (XAI) tools—such as SHAP or feature sensitivity analysis—within these frameworks further enhances their credibility. Engineers can not only assess the probability of structural failure but also understand which input uncertainties most influence those risks, improving transparency and accountability [9].

These approaches are now beginning to reshape performance-based structural design, particularly under seismic and fatigue loading scenarios. While regulatory approval is still pending, this convergence of AI and probabilistic engineering offers a promising path toward data-informed, risk-conscious structural decision-making.

6.3.3. Future Directions for Probabilistic and Uncertainty-Aware Modeling

- Adopting Bayesian ML techniques
- Integrating ML with probabilistic design codes
- Developing uncertainty-aware ML models

6.4. Bridging the Gap Between Research and Practice

6.4.1. Why ML Still Is Not Often Applied in Industry?

Although many studies indicate that ML is better than conventional methods, very few ML models have been included into commercial engineering tools. Some primary factors include:

- ML models lack a universal benchmarking system unlike FEA, which follows accepted verification processes.
- Some deep learning architectures demand significant computational resources, thus they are useless for real-time analysis.
- Regulatory and Code Compliance Problems; a highly regulated discipline, structural engineering lacks formal approval of engineering codes using ML-based design methods.

Another key factor limiting the industrial uptake of AI-driven structural optimization is the associated computational cost. The training phase—particularly for deep learning models, evolutionary algorithms, or high-dimensional inverse design frameworks—often requires access to high-performance computing (HPC) environments, including GPUs or multi-core CPU clusters [158]. This is especially true when models are coupled with finite element simulations or need to explore large design spaces through surrogate modeling or generative methods.

However, it is important to note that the computational burden is mostly concentrated during the training or model development phase. Once trained, most ML and IML models are highly efficient at inference and can generate design predictions or optimize solutions in real time on standard computing hardware [159]. Additionally, the emergence of cloud-based AI platforms and lightweight optimization strategies (e.g., pruning, quantization, and knowledge distillation) now offers engineers access to powerful tools without owning dedicated HPC resources [160]. Nonetheless, to train advanced models or explore multi-objective optimizations at scale, HPC remains an enabler rather than a barrier, especially in research or enterprise settings.

6.4.2. Future Directions for Bridging the Gap Between Research and Practice

- Creating Industry-Standard Validation Benchmarks
- Developing ML-Integrated Engineering Software
- Regulatory Frameworks for AI in Structural Engineering

6.5. Expanding the Role of IML in Structural Design

6.5.1. IML and Generative AI as the Next Frontier in Structural Design Automation

While most structural ML applications remain focused on forward prediction, IML and generative AI are reshaping the early stages of structural design by allowing engineers to define performance targets and automatically generate optimized design configurations. This shift holds major implications for steel construction, where structural complexity, load-bearing demands, and geometric constraints often lead to highly iterative and labor-intensive design cycles.

In generative design, AI models such as Generative Adversarial Networks (GANs) and Variational Autoencoders (VAEs) are trained to synthesize novel structural topologies based on desired performance criteria. These models do not merely search for optimum values in a fixed design space; instead, they learn the underlying structural logic and generate entirely new configurations that satisfy both architectural intent and engineering performance. For example, Challapalli et al. [104] demonstrated an inverse design framework using GANs to generate cellular structures with record-breaking recovery stress, revealing how generative models can produce designs that outperform conventional configurations.

In the context of steel frameworks, ML-guided topology optimization—which couples finite element analysis with reinforcement learning or surrogate modeling—has emerged as a powerful technique for generating lightweight, efficient structural forms. AI algorithms explore design variants that would be computationally prohibitive to evaluate through brute-force methods. This has enabled engineers to optimize member layout, cross-sectional geometry, and bracing patterns in ways that reduce weight and material usage without sacrificing stiffness or stability.

Despite these advances, practical challenges remain. Generative models often produce non-code-compliant geometries, requiring post-processing to ensure structural safety and constructability. Additionally, real-world datasets for training such models are scarce, and integration with BIM and FEM workflows is still underdeveloped. Nonetheless, the growing synergy between IML, generative design, and performance-based engineering is paving the way for a new paradigm in steel structure design, where AI supports both creativity and compliance.

6.5.2. Future Directions for Inverse and Generative AI in Structural Design

- Integration of GANs and VAEs into Structural Topology Optimization
- Coupling Generative Design with FEM and BIM Systems
- Data-Driven Evolutionary Design Platforms for Custom Steel Systems

6.6. Regulatory Acceptance and Validation of AI Models in Structural Engineering

6.6.1. Current Regulatory Limitations and Engineering Concerns

Although AI and ML are rapidly advancing structural engineering, they are not formally approved in safety-critical design by any major code authority including the Eurocodes, AISC, or ASCE yet. Current building codes still rely on traditional analytical or empirical formulas that have been tested and validated over decades. These codes have not yet adapted to the use of complex, data-driven models like AI, especially in their "black-box" form, where internal logic may not be transparent or physically interpretable [161]. For now, any AI-based model applied in engineering has to be thoroughly tested and demonstrated to satisfy the same safety criteria established in the codes. This typically means comparing AI predictions with experimental data or ensuring that they align with the safety factors prescribed by standards. Particularly in predicting structural resistance when trained on large, high-quality datasets, several studies have shown that AI can achieve either similar or even better accuracy than conventional codes [162]. Notwithstanding this technical promise, there is still no official path for AI models to be certified or approved as direct substitutes for code-based design equations. To address this issue, researchers have proposed the idea of "engineering equivalence". This method develops an ML model to replicate the output of a known code equation, thus guaranteeing consistent and logical behavior with accepted engineering principles [163]. If this equivalence is achieved, the AI model might be accepted as a substitute for the original equation. Although this idea is still at an early stage, it shows promise for future updates to codes. At the same time, international standards organizations such as ISO/IEC JTC1 SC42 are working on high-level AI quality and trust guidelines that could later be adapted to civil and structural engineering needs [164,165]. Even so, there are currently no specific clauses in design codes that allow general approval of AI models. In practice, if AI is used, it must typically be justified under "alternative design methods" and backed by strong validation. Explainability is one of the key causes of this uncertainty. Engineering authorities are naturally wary of adopting models that perform well but lack any evident justification for their projections. Given the generalizability of AI models [166], this becomes even more critical.

The lack of comprehensive training data is another issue. Especially catastrophic structural failures are rare and costly to test. Training datasets thus are often small, biased, or lacking important failure scenarios [167]. There are also concerns about accountability: if an AI-guided decision results in a structural failure, it is unclear who bears the legal or ethical responsibility—the software developer, the design engineer, or the authority that accepted the design [161].

These unresolved problems keep most AI applications in structural engineering in a supporting role. Although they help with design exploration, optimization, and monitoring, they are not yet accepted as replacements for code-approved methods in final decision-making. To move forward, researchers are concentrating on creating explainable and hybrid models, combining AI with physics-based rules, and suggesting more precise validation and certification guidelines [159,164]. AI will remain a complimentary but not central player in the structural engineering code environment until those efforts develop.

6.6.2. Role of Explainable AI (XAI) in Meeting Regulatory Demands

Explainable AI (XAI) models—those that integrate techniques like SHAP, and global sensitivity analysis—are especially promising for regulatory consideration. Their transparency and interpretability let engineers to trace model predictions to specific design inputs, satisfying key principles of engineering traceability. This transparency can support formal validation by offering clear reasoning, which is essential for approval in safety-critical applications. While no design code formally certifies XAI models for structural design, their fit with rising trustworthiness criteria makes them attractive candidates for future code inclusion. XAI tools remain useful as interpretive aids until codified paths develop, bridging the gap between high-performance AI models and the regulatory demand for transparent, verifiable decision-making.

Moving forward, the structural engineering community must work toward developing formal AI regulatory frameworks tailored for design approval, certification, and standardization. These frameworks should define quality control protocols, minimum interpretability thresholds, validation benchmarks against code-based methods, and pathways for certifying hybrid models that combine AI with physical principles. Establishing such standards will require coordinated input from engineers, AI researchers, code authorities (e.g., Eurocodes, AISC), and software developers. Without these regulatory foundations, the full industrial adoption of AI in structural design will remain restricted to non-critical applications or limited to use under alternative design justifications.

6.6.3. Future Directions for Regulatory Approval and Model Certification

- Establishing formal AI regulatory frameworks tailored for civil/structural engineering
- Defining interpretability thresholds and validation protocols
- Creating certification pathways for hybrid models combining AI and physical principles

6.7. Practical Barriers to Industry Adoption of AI in Structural Engineering

6.7.1. Data Scarcity, Trust, and the Need for Validation

Although many scholarly studies show the strength of AI models in structural engineering, their actual application in practice is still rather restricted. Data availability presents one of the key difficulties. Many engineering companies find that high-quality structural datasets are either difficult to gather or nonexistent. Many datasets are kept privately by businesses or organizations, and experimental results are often few. Most AI models stay limited to particular cases or academic use [162,164] without access to large and well-structured data. Lack of interpretability is another main problem. Most engineers are taught to operate using transparent models based on exact physical principles. Engineers often find AI models unreliable when they make predictions without displaying how or why, particularly in safety-critical contexts. This lack of trust becomes a serious concern when the model output could affect structural performance, risk evaluation, or code compliance [168]. The need for engineering-level validation forms a third practical obstacle. In academic research, an AI model may be sufficient to exhibit great test data accuracy. In industry, however, models are expected to satisfy code-based safety margins, pass experimental comparisons, and offer repeatable results under several design environments.

These expectations are not only technical but also linked to professional liability and public safety. As long as validation procedures for AI remain unclear or unsupported by formal standards, adoption will remain slow [161,165,167].

6.7.2. Future Directions for Industry Adoption of AI in Structural Engineering

- Enhancing access to shared structural datasets
- Promoting model transparency in safety-critical use cases
- Aligning industrial validation with regulatory expectations

6.8. Advancing Seismic Modeling Through Multi-Physics AI

6.8.1. Limitations of Current Seismic AI Models

Most structural AI applications today treat seismic performance as a simplified dynamic problem—often ignoring the coupling of multi-physical factors such as temperature variation, soil-structure interaction, and real-time degradation during earthquakes. This reductionist approach limits prediction accuracy in real-world scenarios. As structural systems become more complex, and the demand for resilient infrastructure increases, integrating AI with multi-physics modeling becomes important for seismic design and real-time monitoring.

6.8.2. Emerging Research on Multi-Physics AI Integration

Recent studies have explored the fusion of AI with multi-physics finite element methods, including soil-fluid–structure interaction, thermal-mechanical behavior under seismic excitation, and even coupled damage evolution. These models allow AI to learn from both physical laws and real-time data, creating better generalization under rare or extreme events [168,169].

Recent advancements in seismic performance prediction for steel structures highlight the effectiveness of physics-informed machine learning (PiML). By embedding fundamental physical laws into neural networks, particularly using techniques like LSTM, these models reliably predict nonlinear seismic behaviors such as interstory drift, plastic hinge formation, and collapse mechanisms. This integration enhances prediction accuracy and interpretability compared to traditional methods. However, computational demand and data calibration challenges still limit their widespread adoption in design offices.

Despite promising progress, multi-physics AI models require improved data infrastructure, standardized validation protocols, and transparent uncertainty handling to gain broader trust and usability in seismic safety-critical decisions.

6.8.3. Feature Direction for Advancing Seismic Modeling Through Multi-Physics AI

- Coupling PINNs with seismic time-history simulations
- Creating benchmark datasets for earthquake-driven ML training
- Developing real-time digital twins for seismic assessment
- Standardizing validation protocols in Eurocode 8, ASCE 41, etc.

Together, ML, IML, and XAI provide a complementary framework for reshaping structural engineering processes. Their integration has the potential to modernize design strategies while upholding safety, transparency, and regulatory integrity. To summarize their respective strengths and challenges, Table 3 provides a comparative overview of these three approaches in terms of their structural applications, benefits, and limitations. The future trajectory of AI in structural engineering depends on bridging the gap between academic advances and industry practices through explainable methods, hybrid physics-guided models, and reliable validation procedures. Furthermore, AI will increasingly contribute to sustainable, climate-adaptive design and life-cycle optimization. Inverse ML will redefine how engineers generate solutions from desired outcomes, opening new frontiers in structural optimization. Achieving this vision requires cross-disciplinary collaboration, industrial validation, and stronger AI literacy in engineering education.

Table 3. Summary of ML, IML, and XAI.

Technique	Strengths	Limitations	Key Structural Applications
Machine Learning (ML)	- Rapid prediction of structural responses once trained. - Capable of handling large, nonlinear, high-dimensional datasets. - Useful for surrogate modeling, failure classification, and load capacity estimation.	- Often acts as a "black box" with low interpretability. - Requires large and high-quality labeled datasets. - Limited generalization outside trained domains.	- Performance prediction under complex loads. - Data-driven structural health monitoring. - Load-carrying capacity and failure mode prediction.
Inverse Machine Learning (IML)	- Directly maps performance goals to optimal design parameters. - Reduces manual iteration in parametric design. - Efficient for optimization in multi-variable, constrained problems.	- Inverse problems can be ill-posed and unstable. - Often requires regularization or surrogate models to ensure convergence. - Experimental validation still limited in structural contexts.	- Automated design of cross-sections and steel profiles. - Topology optimization. - Material and microstructure tuning in steel alloy design.
Explainable AI (XAI)	- Automated design of cross-sections and steel profiles. - Topology optimization. - Material and microstructure tuning in steel alloy design.	- Still emerging in regulatory practice. - Trade-off between complexity and explainability. - Interpretations can be misused if not domain-verified.	- Code validation and transparency for AI-driven designs. - SHAP/LIME interpretation of failure risk. - Engineering decision support in safety-critical systems.

7. Conclusions

Including AI into structural engineering transforms a paradigm change in the design, analysis, and optimization of steel constructions. Over the past three decades, fast development of ML techniques has given engineers great tools to increase predictive accuracy, automate challenging tasks, and derive data-driven insights. But rather than only replacing traditional engineering knowledge, the evolution of AI in this field is about raising its capacity to negotiate the complexity of contemporary building challenges. Using AI, engineers can adopt adaptive, real-time solutions that account for uncertainty, variability, and evolving design constraints, moving beyond traditional simulation-based approaches.

Particularly in fields where more conventional methods have limits, one of the most transforming consequences of AI is its ability to enable more exact decision-making. From load-bearing studies and failure predictions to material selection and seismic performance evaluations, AI models have shown their capacity to provide engineers more complete knowledge of structural behavior. Particularly with IML, engineers can let AI select the best design solutions and directly specify target performance criteria, thus adding a new dimension in optimization. This shift represents a fundamental move away from trial-and-error methods, streamlining the design process and reducing dependence on repetitive simulations

Notwithstanding these advances, the path toward widespread acceptance of AI in structural engineering remains a long one. Practical implementation of AI-driven models still requires greater transparency, standardized practices, and alignment with regulatory codes to ensure their reliability and safety in real-world applications. Moreover, the demand for high-quality, diverse datasets is still a main challenge, since AI models are only as successful as the data they are trained on. Overlooking issues such as data limitations, overfitting, and model interpretability will confine the role of AI in structural engineering to academic research, rather than enabling its practical application

Looking ahead, AI will influence developing fields including digital twin technology, sustainability, and automated building methods beyond its traditional applications, impacting steel structure engineering. Real-time monitoring systems allow AI-powered design frameworks to interact and apply proactive structural assessments and predictive maintenance programs. Moreover, as engineering education evolves, AI literacy will become an essential skill for future professionals, ensuring that engineers not only know how to use AI tools but also critically assess their outputs within the context of engineering principles.

In structural engineering, AI is fundamentally reshaping the approach to design and analysis, going beyond mere technological advancement. Though challenges still exist, the continuous development of AI-driven technologies offers interesting paths for acquiring more intelligent, resilient, and efficient steel buildings. By supporting cooperation between academia, industry, and regulatory authorities, the engineering community can maximize AI's possibilities and so guarantee that its application stays rooted in safety, reliability, and engineering intuition.

Author Contributions: S.S.: Writing—original draft, Writing—review and editing, Data curation, Software. I.M.: Supervision, Visualization. M.M.: Supervision. F.G.: Supervision. All authors have read and agreed to the published version of the manuscript.

Funding: This research was funded by Italian Ministry of Education, University and Research (MIUR), Departments of Excellence (grant number L.232/2016), Italian Ministry of Education, University and Research (MIUR), Next-Generation EU-Prin (PNRR) (grant number 2022P7PF8J) and Italian Ministry of Education, University and Research (MIUR), Next-Generation EU-Prin (PNRR) (grant number P2022Y9ZJ2).

Data Availability Statement: The data presented in this study are available on request from the corresponding author.

Conflicts of Interest: The authors declare no conflicts of interest.

References

1. Thai, H.-T. Machine learning for structural engineering: A state-of-the-art review. *Structures* **2022**, *38*, 448–491. [CrossRef]
2. Sun, H.; Burton, H.V.; Huang, H. Machine learning applications for building structural design and performance assessment: State-of-the-art review. *J. Build. Eng.* **2021**, *33*, 101816. [CrossRef]
3. Salehi, H.; Burgueño, R. Emerging artificial intelligence methods in structural engineering. *Eng. Struct.* **2018**, *171*, 170–189. [CrossRef]
4. Málaga-Chuquitaype, C. Machine learning in structural design: An opinionated review. *Front. Built Environ.* **2022**, *8*, 815717. [CrossRef]
5. Haber, E.; Ascher, U.M.; Oldenburg, D. On optimization techniques for solving nonlinear inverse problems. *Inverse Probl.* **2000**, *16*, 1263. [CrossRef]
6. Gallet, A.; Rigby, S.; Tallman, T.N.; Kong, X.; Hajirasouliha, I.; Liew, A.; Liu, D.; Chen, L.; Hauptmann, A.; Smyl, D. Structural engineering from an inverse problems perspective. *Proc. R. Soc. A Math. Phys. Eng. Sci.* **2022**, *478*, 20210526. [CrossRef]
7. Barredo Arrieta, A.; Díaz-Rodríguez, N.; Del Ser, J.; Bennetot, A.; Tabik, S.; Barbado, A.; Garcia, S.; Gil-Lopez, S.; Molina, D.; Benjamins, R.; et al. Explainable artificial intelligence (XAI): Concepts, taxonomies, opportunities, and challenges toward responsible AI. *Inf. Fusion* **2020**, *58*, 82–115. [CrossRef]
8. Angelov, P.P.; Soares, E.A.; Jiang, R.; Arnold, N.I.; Atkinson, P.M. Explainable artificial intelligence: An analytical review. Wiley Interdiscip. *Rev. Data Min. Knowl. Discov.* **2021**, *11*, e1424. [CrossRef]
9. Lundberg, S.M.; Lee, S.I. A unified approach to interpreting model predictions. *Adv. Neural Inf. Process. Syst.* **2017**, *30*, 2493.
10. Minh, D.; Wang, H.X.; Li, Y.F.; Nguyen, T.N. Explainable artificial intelligence: A comprehensive review. *Artif. Intell. Rev.* **2022**, *55*, 3503–3568. [CrossRef]
11. Jiang, T.; Gradus, J.L.; Rosellini, A.J. Supervised machine learning: A brief primer. *Behav. Ther.* **2020**, *51*, 675–687. [CrossRef] [PubMed]
12. Alloghani, M.; Al-Jumeily, D.; Mustafina, J.; Hussain, A.; Aljaaf, A.J. A systematic review on supervised and unsupervised machine learning algorithms for data science. In *Supervised and Unsupervised Learning for Data Science*; Springer: Cham, Switzerland, 2020; pp. 3–21.
13. Nian, R.; Liu, J.; Huang, B. A review on reinforcement learning: Introduction and applications in industrial process control. *Comput. Chem. Eng.* **2020**, *139*, 106886. [CrossRef]
14. Adeli, H. Neural networks in civil engineering: 1989–2000. *Comput.-Aided Civ. Infrastruct. Eng.* **2001**, *16*, 126–142. [CrossRef]
15. Song, B.; Zhou, R.; Ahmed, F. Multi-modal machine learning in engineering design: A review and future directions. *J. Comput. Inf. Sci. Eng.* **2024**, *24*, 010801. [CrossRef]
16. Daneshfar, R.; Esmaeili, M.; Mohammadi-Khanaposhtani, M.; Baghban, A.; Habibzadeh, S.; Eslamian, S. Advanced machine learning techniques: Multivariate regression. In *Handbook of Hydroinformatics*; Elsevier: Amsterdam, The Netherlands, 2023; pp. 1–38. [CrossRef]
17. Mienye, I.D.; Jere, N. A survey of decision trees: Concepts, algorithms, and applications. *IEEE Access* **2024**, *12*, 86716–86727. [CrossRef]
18. Schonlau, M.; Zou, R.Y. The random forest algorithm for statistical learning. *Stata J.* **2020**, *20*, 3–29. [CrossRef]
19. Tanveer, M.; Rajani, T.; Rastogi, R.; Shao, Y.H.; Ganaie, M.A. Comprehensive review on twin support vector machines. *Ann. Oper. Res.* **2024**, *339*, 1223–1268. [CrossRef]
20. Abd-elaziem, A.H.; Soliman, T.H. A multi-layer perceptron (MLP) neural network for stellar classification: A review of methods and results. *Int. J. Adv. Appl. Comput. Intell.* **2023**, *3*, 54216.
21. Abiodun, O.I.; Jantan, A.; Omolara, A.E.; Dada, K.V.; Mohamed, N.A.; Arshad, H. State-of-the-art in artificial neural network applications: A survey. *Heliyon* **2018**, *4*, e00938. [CrossRef]
22. Halder, R.K.; Uddin, M.N.; Uddin, M.A.; Aryal, S.; Khraisat, A. Enhancing K-nearest neighbor algorithm: A comprehensive review and performance analysis of modifications. *J. Big Data* **2024**, *11*, 113. [CrossRef]
23. Freund, Y.; Schapire, R.E. Experiments with a new boosting algorithm. *ICML* **1996**, *96*, 148–156.
24. Friedman, J.H. Greedy function approximation: A gradient boosting machine. *Ann. Stat.* **2001**, *29*, 1189–1232. [CrossRef]
25. Chen, T.; Guestrin, C. XGBoost: A scalable tree boosting system. In Proceedings of the 22nd ACM SIGKDD International Conference on Knowledge Discovery and Data Mining, San Francisco, CA, USA, 13–17 August 2016; pp. 785–794. [CrossRef]
26. Ke, G.; Meng, Q.; Finley, T.; Wang, T.; Chen, W.; Ma, W.; Liu, T.Y. LightGBM: A highly efficient gradient boosting decision tree. *Adv. Neural Inf. Process. Syst.* **2018**, *30*, 1998–2007. [CrossRef]
27. Prokhorenkova, L.; Gusev, G.; Vorobev, A.; Dorogush, A.V.; Gulin, A. CatBoost: Unbiased boosting with categorical features. In *Advances in Neural Information Processing Systems*; Bengio, S., Wallach, H., Larochelle, H., Grauman, K., Cesa-Bianchi, N.,

Garnett, R., Eds.; Curran Associates, Inc.: Red Hook, NY, USA, 2018; Volume 31, pp. 6638–6648. Available online: https://proceedings.neurips.cc/paper/2018/file/14491b756b3a51daac41c24863285549-Paper.pdf (accessed on 1 April 2025).
28. Mascolo, I.; Guarracino, F.; Sarfarazi, S.; Della Corte, G. A proposal for a simple characterization of stainless steel connections through an equivalent yield strength. *Structures* **2024**, *68*, 107043. [CrossRef]
29. Sarfarazi, S.; Shamass, R.; Mascolo, I.; Della Corte, G.; Guarracino, F. Some considerations on the behaviour of bolted stainless-steel beam-to-column connections: A simplified analytical approach. *Metals* **2023**, *13*, 753. [CrossRef]
30. Sarfarazi, S.; Shamass, R.; Della Corte, G.; Guarracino, F. Assessment of design approaches for stainless-steel joints through an equivalent FE modelling technique. *ce/papers* **2022**, *5*, 271–281. [CrossRef]
31. Sarfarazi, S.; Saffari, H.; Fakhraddini, A. Shear behavior of panel zone considering axial force for flanged cruciform columns. *Civ. Eng. Infrastruct. J.* **2020**, *53*, 359–377. [CrossRef]
32. Sarfarazi, S.; Fakhraddini, A.; Modaresahmadi, K. Evaluation of panel zone shear strength in cruciform columns, box-columns and double web-columns. *Int. J. Struct. Civ. Eng.* **2016**, *5*, 52–56.
33. Saffari, H.; Sarfarazi, S.; Fakhraddini, A. A mathematical steel panel zone model for flanged cruciform columns. *Steel Compos. Struct.* **2016**, *20*, 851–867. [CrossRef]
34. Paral, A.; Roy, D.K.S.; Samanta, A.K. A deep learning-based approach for condition assessment of semi-rigid joint of steel frame. *J. Build. Eng.* **2021**, *34*, 101946. [CrossRef]
35. Kueh, A.B.H. Artificial neural network and regressed beam-column connection explicit mathematical moment-rotation expressions. *J. Build. Eng.* **2021**, *43*, 103195. [CrossRef]
36. Tran, V.L. Investigating the behavior of steel flush endplate connections at elevated temperatures using FEM and ANN. *Int. J. Steel Struct.* **2022**, *22*, 1433–1451. [CrossRef]
37. Sarothi, S.Z.; Ahmed, K.S.; Khan, N.I.; Ahmed, A.; Nehdi, M.L. Predicting bearing capacity of double shear bolted connections using machine learning. *Eng. Struct.* **2022**, *251*, 113497. [CrossRef]
38. Jiang, K.; Liang, Y.; Zhao, O. Machine-learning-based design of high-strength steel bolted connections. *Thin-Walled Struct.* **2022**, *179*, 109575. [CrossRef]
39. Rabbani, A.; Ghiami Azad, A.R.; Rahami, H. Enhancing prediction of the moment-rotation behavior in flush end plate connections using Multi-Gene Genetic Programming (MGGP). *Struct. Eng. Mech.* **2024**, *91*, 643–656. [CrossRef]
40. Ferreira, F.P.V.; Shamass, R.; Limbachiya, V.; Tsavdaridis, K.D.; Martins, C.H. Lateral–torsional buckling resistance prediction model for steel cellular beams generated by artificial neural networks (ANN). *Thin-Walled Struct.* **2022**, *170*, 108592. [CrossRef]
41. Shamass, R.; Ferreira, F.P.V.; Limbachiya, V.; Santos, L.F.P.; Tsavdaridis, K.D. Web-post buckling prediction resistance of steel beams with elliptically-based web openings using artificial neural networks (ANN). *Thin-Walled Struct.* **2022**, *180*, 109959. [CrossRef]
42. Silva de Carvalho, A.; Hosseinpour, M.; Rossi, A.; Martins, C.H.; Sharifi, Y. New formulas for predicting the lateral–torsional buckling strength of steel I-beams with sinusoidal web openings. *Thin-Walled Struct.* **2022**, *181*, 110067. [CrossRef]
43. Xing, Z.; Wu, K.; Su, A.; Wang, Y.; Zhou, G. Intelligent local buckling design of stainless steel I-sections in fire via Artificial Neural Network. *Structures* **2023**, *58*, 105356. [CrossRef]
44. Rossi, A.; Hosseinpour, M.; Silva de Carvalho, A.; Martins, C.H.; Sharifi, Y. Lateral torsional capacity of steel beams in different loading conditions by neural network. *Proc. Inst. Civ. Eng. Struct. Build.* **2023**, *177*, 892–910. [CrossRef]
45. Cheng, J.; Li, X.; Jiang, K.; Li, S.; Su, A.; Zhao, O. Machine-learning-assisted design of high-strength steel I-section columns. *Eng. Struct.* **2024**, *308*, 118018. [CrossRef]
46. Xu, Y.; Zhang, M.; Zheng, B. Design of cold-formed stainless steel circular hollow section columns using machine learning methods. *Structures* **2021**, *33*, 2755–2770. [CrossRef]
47. Nguyen, T.H.; Tran, N.L.; Nguyen, D.D. Prediction of axial compression capacity of cold-formed steel oval hollow section columns using ANN and ANFIS models. *Int. J. Steel Struct.* **2022**, *22*, 1–26. [CrossRef]
48. Fang, Z.; Roy, K.; Padiyara, S.; Chen, B.; Raftery, G.M.; Lim, J.B.P. Web crippling design of cold-formed stainless steel channels under interior-two-flange loading condition using deep belief network. *Structures* **2023**, *47*, 1967–1990. [CrossRef]
49. Lu, Y.; Wu, B.; Li, W.; Zhou, T.; Li, Y. Regression-classification ensemble machine learning model for loading capacity and buckling mode prediction of cold-formed steel built-up I-section columns. *Thin-Walled Struct.* **2024**, *205*, 112427. [CrossRef]
50. Shaheen, M.A.; Presswood, R.; Afshan, S. Application of machine learning to predict the mechanical properties of high-strength steel at elevated temperatures based on the chemical composition. *Structures* **2023**, *52*, 17–29. [CrossRef]
51. Shahin, R.I.; Ahmed, M.; Liang, Q.Q.; Yehia, S.A. Predicting the web crippling capacity of cold-formed steel lipped channels using hybrid machine learning techniques. *Eng. Struct.* **2024**, *309*, 118061. [CrossRef]
52. Yılmaz, Y.; Demir, S.; Öztürk, F. Predicting the load-bearing capacity of lipped channel section cold-formed steel profiles under combined effects using machine learning. *Structures* **2024**, *66*, 106898. [CrossRef]

53. Kim, B.; Yuvaraj, N.; Park, H.W.; Sri Preethaa, K.R.; Pandian, R.A.; Lee, D.-E. Investigation of steel frame damage based on computer vision and deep learning. *Autom. Constr.* **2021**, *132*, 103941. [CrossRef]
54. Truong, V.-H.; Pham, H.-A.; Van, T.H.; Tangaramvong, S. Evaluation of machine learning models for load-carrying capacity assessment of semi-rigid steel structures. *Eng. Struct.* **2022**, *273*, 115001. [CrossRef]
55. Jahjouh, M. An experience-based artificial neural network in the design optimization of steel frames. *Eng. Res. Express* **2022**, *4*, 045031. [CrossRef]
56. Shan, W.; Liu, J.; Zhou, J. Integrated method for intelligent structural design of steel frames based on optimization and machine learning algorithm. *Eng. Struct.* **2023**, *284*, 115980. [CrossRef]
57. Pal, J.; Sikdar, S.; Banerjee, S. A deep-learning approach for health monitoring of a steel frame structure with bolted connections. *Struct. Control Health Monit.* **2022**, *29*, e2873. [CrossRef]
58. Naresh, M.; Kumar, V.; Pal, J. A machine learning approach for health monitoring of a steel frame structure using statistical features of vibration data. *Asian J. Civ. Eng.* **2024**, *25*, 39–49. [CrossRef]
59. Vu, V.T.; Thom, D.V.; Tran, T.D. Identification of damage in steel beam by natural frequency using machine learning algorithms. *Proc. Inst. Mech. Eng. Part C J. Mech. Eng. Sci.* **2024**, *238*, 9644–9659. [CrossRef]
60. Li, H.; Yin, X.; Sha, L.; Yang, D.; Hu, T. Data-driven prediction model for high-strength bolts in composite beams. *Buildings* **2023**, *13*, 2769. [CrossRef]
61. Dissanayake, M.; Nguyen, H.; Poologanathan, K.; Perampalam, G.; Upasiri, I.; Rajanayagam, H.; Suntharalingam, T. Prediction of shear capacity of steel channel sections using machine learning algorithms. *Thin-Walled Struct.* **2022**, *175*, 109152. [CrossRef]
62. Dai, Y.; Roy, K.; Fang, Z.; Chen, B.; Raftery, G.M.; Lim, J.B.P. A novel machine learning model to predict the moment capacity of cold-formed steel channel beams with edge-stiffened and un-stiffened web holes. *J. Build. Eng.* **2022**, *53*, 104592. [CrossRef]
63. Liu, J.-Z.; Li, S.; Guo, J.; Xue, S.; Chen, S.; Wang, L.; Zhou, Y.; Luo, T.X. Machine learning (ML)-based models for predicting the ultimate bending moment resistance of high-strength steel welded I-section beam under bending. *Thin-Walled Struct.* **2023**, *191*, 111051. [CrossRef]
64. Su, A.; Cheng, J.; Li, X.; Zhong, Y.; Li, S.; Zhao, O.; Jiang, K. Unified machine-learning-based design method for normal and high-strength steel I-section beam–columns. *Thin-Walled Struct.* **2024**, *199*, 111835. [CrossRef]
65. Tusnin, A.R.; Alekseytsev, A.V.; Tusnina, O.A. Load identification in steel structural systems using machine learning elements: Uniform length loads and point forces. *Buildings* **2024**, *14*, 1711. [CrossRef]
66. Sediek, O.A.; Wu, T.-Y.; McCormick, J.; El-Tawil, S. Prediction of seismic collapse behavior of deep steel columns using machine learning. *Structures* **2022**, *40*, 163–175. [CrossRef]
67. Imam, M.H.; Mohiuddin, M.; Shuman, N.M.; Oyshi, T.I.; Debnath, B.; Liham, M.I.M.H. Prediction of seismic performance of steel frame structures: A machine learning approach. *Structures* **2024**, *69*, 107547. [CrossRef]
68. Shin, D.-H.; Kim, H.-J. Machine learning-based prediction of hysteretic behaviors of two-side clamped steel shear walls. *Structures* **2024**, *60*, 105875. [CrossRef]
69. Cho, E.; Han, S.W. A numerical model simulating cyclic behavior of high-strength steel. *Adv. Struct. Eng.* **2024**, *27*, 1490–1508. [CrossRef]
70. Hu, S.; Zhu, S.; Alam, M.S.; Wang, W. Machine learning-aided peak and residual displacement-based design method for enhancing seismic performance of steel moment-resisting frames by installing self-centering braces. *Eng. Struct.* **2022**, *271*, 114935. [CrossRef]
71. Samadian, D.; Muhit, I.B.; Occhipinti, A.; Dawood, N. Surrogate models for seismic and pushover response prediction of steel special moment resisting frames. *Eng. Struct.* **2024**, *314*, 118307. [CrossRef]
72. Salama, A.H.E.S. Optimization seismic resilience: A machine learning approach for vertical irregular buildings. *Asian J. Civ. Eng.* **2024**, *25*, 6233–6248. [CrossRef]
73. Huang, M.; Zhang, J.; Li, J.; Deng, Z.; Luo, J. Damage identification of steel bridge based on data augmentation and adaptive optimization neural network. *Struct. Health Monit.* 2024; in press. [CrossRef]
74. Iacussi, L.; Chiariotti, P.; Cigada, A. AI-enhanced IoT system for assessing bridge deflection in drive-by conditions. *Sensors* **2025**, *25*, 158. [CrossRef]
75. Wang, X.; Yue, Q.; Liu, X. SBDNet: A deep learning-based method for the segmentation and quantification of fatigue cracks in steel bridges. *Adv. Eng. Inform.* **2025**, *65*, 103186. [CrossRef]
76. Svendsen, B.T.; Øiseth, O.; Frøseth, G.T.; Rønnquist, A. A hybrid structural health monitoring approach for damage detection in steel bridges under simulated environmental conditions using numerical and experimental data. *Struct. Health Monit.* **2022**, *22*, 540–561. [CrossRef]
77. Zhou, K.; Duan, M.-G.; Wu, Z.-L.; Zhi, L.-H.; Hu, F. Dynamic behavior monitoring of twin supertall buildings during Super Typhoon Soksuri using social sensing data. *J. Build. Eng.* **2024**, *95*, 110119. [CrossRef]

78. Ghaffari, A.; Shahbazi, Y.; Mokhtari Kashavar, M.; Fotouhi, M.; Pedrammehr, S. Advanced predictive structural health monitoring in high-rise buildings using recurrent neural networks. *Buildings* **2024**, *14*, 3261. [CrossRef]
79. Wang, M.; Incecik, A.; Tian, Z.; Zhang, M.; Kujala, P.; Gupta, M.; Krolczyk, G.; Li, Z. Structural health monitoring on offshore jacket platforms using a novel ensemble deep learning model. *Ocean Eng.* **2024**, *301*, 117510. [CrossRef]
80. Martzikos, N.; Ruzzo, C.; Malara, G.; Fiamma, V.; Arena, F. Applying neural networks to predict offshore platform dynamics. *J. Mar. Sci. Eng.* **2024**, *12*, 2001. [CrossRef]
81. Kouchaki, M.; Salkhordeh, M.; Mashayekhi, M.; Mirtaheri, M.; Amanollah, H. Damage detection in power transmission towers using machine learning algorithms. *Structures* **2023**, *56*, 104980. [CrossRef]
82. Kiyoki, S.; Yoshida, S.; Rushdi, M.A. Machine learning-based prediction of 2 MW wind turbine tower loads during power production based on nacelle behavior. *Energies* **2025**, *18*, 216. [CrossRef]
83. Vlasenko, T.; Hutsol, T.; Vlasovets, V.; Glowacki, S.; Nurek, T.; Horetska, I.; Kukharets, S.; Firman, Y.; Bilovod, O. Ensemble learning based sustainable approach to rebuilding metal structures prediction. *Sci. Rep.* **2025**, *15*, 1210. [CrossRef]
84. Shang, Z.; Qin, X.; Zhang, Z.; Jiang, H. Bolt loosening and preload loss detection technology based on machine vision. *Buildings* **2024**, *14*, 3897. [CrossRef]
85. Huang, X.; Duan, Z.; Hao, S.; Hou, J.; Chen, W.; Cai, L. A deep learning framework for corrosion assessment of steel structures using Inception v3 model. *Buildings* **2025**, *15*, 512. [CrossRef]
86. Badini, S.; Regondi, S.; Pugliese, R. Enhancing mechanical and bioinspired materials through generative AI approaches. *Next Mater.* **2025**, *6*, 100275. [CrossRef]
87. Arridge, S.; Maass, P.; Öktem, O.; Schönlieb, C.-B. Solving inverse problems using data-driven models. *Acta Numer.* **2019**, *28*, 1–174. [CrossRef]
88. Masurkar, F.; Aggarwal, S.; Tham, Z.W.; Zhang, L.; Yang, F.; Cui, F. Estimating the elastic constants of orthotropic composites using guided waves and an inverse problem of property estimation. *Appl. Acoust.* **2024**, *216*, 109750. [CrossRef]
89. Challapalli, A.; Patel, D.; Li, G. Inverse machine learning framework for optimizing lightweight metamaterials. *Mater. Des.* **2021**, *208*, 109937. [CrossRef]
90. Liao, W.; Lu, X.; Huang, Y.; Zheng, Z.; Lin, Y. Automated structural design of shear wall residential buildings using generative adversarial networks. *Autom. Constr.* **2021**, *132*, 103931. [CrossRef]
91. Danhaive, R.; Mueller, C.T. Design subspace learning: Structural design space exploration using performance-conditioned generative modeling. *Autom. Constr.* **2021**, *127*, 103664. [CrossRef]
92. Teimouri, A.; Challapalli, A.; Konlan, J.; Li, G. Machine learning assisted design and optimization of plate-lattice structures with superior specific recovery force. *Giant* **2024**, *18*, 100282. [CrossRef]
93. Glaesener, R.N.; Kumar, S.; Lestringant, C.; Butruille, T.; Portela, C.M.; Kochmann, D.M. Predicting the influence of geometric imperfections on the mechanical response of 2D and 3D periodic trusses. *Acta Mater.* **2023**, *254*, 118918. [CrossRef]
94. McClarren, R.G.; Tregillis, I.L.; Urbatsch, T.J.; Dodd, E.S. High-energy density hohlraum design using forward and inverse deep neural networks. *Phys. Lett. A* **2021**, *396*, 127243. [CrossRef]
95. Lee, J.-Y.; Kim, S.-H.; Jeong, H.-B.; Lee, K.; Cho, K.; Lee, Y.-K. Inverse design of high-strength medium-Mn steel using a machine learning-aided genetic algorithm approach. *J. Mater. Res. Technol.* **2024**, *33*, 2672–2682. [CrossRef]
96. Wang, Z.-L.; Adachi, Y. Property prediction and properties-to-microstructure inverse analysis of steels by a machine-learning approach. *Mater. Sci. Eng. A* **2019**, *744*, 661–670. [CrossRef]
97. Pei, Z.; Rozman, K.A.; Doğan, Ö.N.; Wen, Y.; Gao, N.; Holm, E.A.; Hawk, J.A.; Alman, D.E.; Gao, M.C. Machine-learning microstructure for inverse material design. *Adv. Sci.* **2021**, *8*, 2101207. [CrossRef]
98. Lertkiatpeeti, K.; Janya-Anurak, C.; Uthaisangsuk, V. Effects of spatial microstructure characteristics on mechanical properties of dual-phase steel by inverse analysis and machine learning approach. *Comput. Mater. Sci.* **2024**, *245*, 113311. [CrossRef]
99. Adachi, Y.; Chen, T.-T.; Sun, F. A review on inverse analysis models in steel material design. *MGE Adv.* **2024**, *2*, e71. [CrossRef]
100. He, L.; Wang, Z.; Akebono, H.; Sugeta, A. Machine learning-based predictions of fatigue life and fatigue limit for steels. *J. Mater. Sci. Technol.* **2021**, *90*, 9–19. [CrossRef]
101. Kolesnikov, I.; Pashkov, D.M.; Belyak, O.A.; Guda, A.A.; Danilchenko, S.A.; Manturov, D.S.; Novikov, E.S.; Kudryakov, O.V.; Guda, S.A.; Soldatov, A.V.; et al. Design of double-layer protective coatings: Finite element modeling and machine learning approximations. *Acta Astronaut.* **2023**, *204*, 869–877. [CrossRef]
102. Islam, F.; Wanigasekara, C.; Rajan, G.; Swain, A.; Prusty, B.G. An approach for process optimization of the Automated Fibre Placement (AFP) based thermoplastic composites manufacturing using machine learning, photonic sensing and thermo-mechanics modelling. *Manuf. Lett.* **2022**, *32*, 10–14. [CrossRef]
103. Shen, X.; Yan, K.; Zhu, D.; Hu, Q.; Wu, H.; Qi, S.; Yuan, M.; Qian, X. Inverse machine learning framework for optimizing gradient honeycomb structure under impact loading. *Eng. Struct.* **2024**, *309*, 118079. [CrossRef]

104. Challapalli, A.; Konlan, J.; Li, G. Inverse machine learning discovered metamaterials with record high recovery stress. *Int. J. Mech. Sci.* **2023**, *244*, 108029. [CrossRef]
105. Kusampudi, N.; Diehl, M. Inverse design of dual-phase steel microstructures using generative machine learning model and Bayesian optimization. *Int. J. Plast.* **2023**, *171*, 103776. [CrossRef]
106. Confalonieri, R.; Coba, L.; Wagner, B.; Besold, T.R. A historical perspective of explainable artificial intelligence. WIREs Data Min. *Knowl. Discov.* **2021**, *11*, e1391. [CrossRef]
107. Taffese, W.Z.; Zhu, Y.; Chen, G. Explainable AI based slip prediction of steel-UHPC interface connected by shear studs. *Expert Syst. Appl.* **2025**, *259*, 125293. [CrossRef]
108. Wang, Z.; Liu, T.; Long, Z.; Wang, J.; Zhang, J. Predicting the drift capacity of precast concrete columns using an explainable machine learning approach. *Eng. Struct.* **2023**, *282*, 115771. [CrossRef]
109. Lai, D.; Demartino, C.; Xiao, Y. Interpretable machine-learning models for maximum displacements of RC beams under impact loading predictions. *Eng. Struct.* **2023**, *281*, 115723. [CrossRef]
110. Karathanasopoulos, N.; Singh, A.; Hadjidoukas, P. Machine learning-based modelling, feature importance and Shapley additive explanations analysis of variable-stiffness composite beam structures. *Structures* **2024**, *62*, 106206. [CrossRef]
111. Wang, S.; Liu, J.; Wang, Q.; Dai, R.; Chen, K. Prediction of non-uniform shrinkage of steel-concrete composite slabs based on explainable ensemble machine learning model. *J. Build. Eng.* **2024**, *88*, 109002. [CrossRef]
112. Le, H.-A.; Le, D.-A.; Le, T.-T.; Le, H.-P.; Le, T.-H.; Hoang, H.-G.T.; Nguyen, T.-A. An extreme gradient boosting approach to estimate the shear strength of FRP reinforced concrete beams. *Structures* **2022**, *45*, 1307–1321. [CrossRef]
113. Zhang, S.; Lei, H.; Zhou, Z.; Wang, G.; Qiu, B. Fatigue life analysis of high-strength bolts based on machine learning method and SHapley Additive exPlanations (SHAP) approach. *Structures* **2023**, *51*, 275–287. [CrossRef]
114. Junda, E.; Málaga-Chuquitaype, C.; Chawgien, K. Interpretable machine learning models for the estimation of seismic drifts in CLT buildings. *J. Build. Eng.* **2023**, *70*, 106365. [CrossRef]
115. Parvizi, M.; Nasserasadi, K.; Tafakori, E. Development of fragility functions of low-rise steel moment frame by artificial neural networks and identifying effective parameters using SHAP theory. *Structures* **2023**, *58*, 105315. [CrossRef]
116. Shahnazaryan, D.; O'Reilly, G.J. Next-generation non-linear and collapse prediction models for short- to long-period systems via machine learning methods. *Eng. Struct.* **2024**, *306*, 117801. [CrossRef]
117. Liu, T.; Cakiroglu, C.; Islam, K.; Wang, Z.; Nehdi, M.L. Explainable machine learning model for predicting punching shear strength of FRC flat slabs. *Eng. Struct.* **2024**, *301*, 117276. [CrossRef]
118. Ribeiro, M.T.; Singh, S.; Guestrin, C. Why should I trust you? Explaining the predictions of any classifier. *arXiv* **2016**, arXiv:1602.04938.
119. Kim, B.; Glassman, E.; Johnson, B.; Shah, J. iBCM: Interactive Bayesian Case Model Empowering Humans via Intuitive Interaction. *MIT-CSAIL Tech. Rep.* **2015**, *30*, 03z.
120. Rani, P.; Liu, C.; Sarkar, N.; Vanman, E. An empirical study of machine learning techniques for affect recognition in human–robot interaction. *Pattern Anal. Appl.* **2006**, *9*, 58–69. [CrossRef]
121. Huysmans, J.; Dejaeger, K.; Mues, C.; Vanthienen, J.; Baesens, B. An empirical evaluation of the comprehensibility of decision table, tree and rule-based predictive models. *Decis. Support Syst.* **2011**, *51*, 141–154. [CrossRef]
122. Szegedy, C.; Zaremba, W.; Sutskever, I.; Bruna, J.; Erhan, D.; Goodfellow, I.; Fergus, R. Intriguing properties of neural networks. *arXiv* **2014**, arXiv:1312.6199.
123. James, G.; Witten, D.; Hastie, T.; Tibshirani, R. *An Introduction to Statistical Learning*; Springer: New York, NY, USA, 2013. [CrossRef]
124. Yu, B.; Chen, X.; Gupta, A.; Ribeiro, C. Stability. *Bernoulli* **2013**, *19*, 1484–1500. [CrossRef]
125. Samek, W.; Müller, K.R. Towards explainable artificial intelligence. In *Explainable AI: Interpreting, Explaining and Visualizing Deep Learning*; Samek, W., Montavon, G., Vedaldi, A., Hansen, L., Müller, K.R., Eds.; Springer: Cham, Switzerland, 2019; Volume 11700, pp. 5–22. [CrossRef]
126. Došilović, F.K.; Brčić, M.; Hlupić, N. Explainable artificial intelligence: A survey. In Proceedings of the 41st International Convention on Information and Communication Technology, Electronics and Microelectronics (MIPRO), Opatija, Croatia, 21–25 May 2018; pp. 210–215. [CrossRef]
127. Ahmed, G.; Jeon, G.; Piccialli, F. From artificial intelligence to explainable artificial intelligence in Industry 4.0: A survey on what, how, and where. *IEEE Trans. Ind. Inform.* **2022**, *18*, 5031–5042. [CrossRef]
128. Mojtabaei, S.M.; Becque, J.; Hajirasouliha, I.; Khandan, R. Predicting the buckling behaviour of thin-walled structural elements using machine learning methods. *Thin-Walled Struct.* **2023**, *184*, 110518. [CrossRef]
129. Hu, S.; Wang, W.; Lu, Y. Explainable machine learning models for probabilistic buckling stress prediction of steel shear panel dampers. *Eng. Struct.* **2023**, *288*, 116235. [CrossRef]

130. Hou, Z.; Hu, S.; Wang, W. Interpretable machine learning models for predicting probabilistic axial buckling strength of steel circular hollow section members considering discreteness of geometries and material. *Adv. Struct. Eng.* **2024**, *28*, 828–844. [CrossRef]

131. Samadian, D.; Muhit, I.B.; Occhipinti, A.; Dawood, N. Meta databases of steel frame buildings for surrogate modelling and machine learning-based feature importance analysis. *Resilient Cities Struct.* **2024**, *3*, 20–43. [CrossRef]

132. Liu, K.; Yu, M.; Liu, Y.; Chen, W.; Fang, Z.; Lim, J.B.P. Fire resistance time prediction and optimization of cold-formed steel walls based on machine learning. *Thin-Walled Struct.* **2024**, *203*, 112207. [CrossRef]

133. Tang, P.; Dai, Y.; Lu, C.; Hu, S. A machine learning framework for predicting the axial capacity of cold-formed steel face-to-face built-up channel sections at elevated temperatures. *Structures* **2024**, *68*, 107144. [CrossRef]

134. Degtyarev, V.V.; Hicks, S.J.; Ferreira, F.P.V.; Tsavdaridis, K.D. Probabilistic resistance predictions of laterally restrained cellular steel beams by natural gradient boosting. *Thin-Walled Struct.* **2024**, *205*, 112367. [CrossRef]

135. Widanage, C.; Mohotti, D.; Lee, C.K.; Wijesooriya, K.; Meddage, D.P.P. Use of explainable machine learning models in blast load prediction. *Eng. Struct.* **2024**, *312*, 118271. [CrossRef]

136. Anand, T.P.; Pandikkadavath, M.S.; Mangalathu, S.; Sahoo, D.R. Machine learning models for seismic analysis of buckling-restrained braced frames. *J. Build. Eng.* **2024**, *98*, 111398. [CrossRef]

137. Fan, X.; Yang, L.; Zhao, X.; Yan, G.; Yang, Y.; Zhang, H.; Chen, S. Prediction of axial compressive capacity and interpretability analysis of web perforated Σ-shaped cold-formed steel. *Structures* **2024**, *70*, 107880. [CrossRef]

138. Sarfarazi, S.; Shamass, R.; Guarracino, F.; Mascolo, I.; Modano, M. Advanced predictive modeling of shear strength in stainless-steel column web panels using explainable AI insights. *Results Eng.* **2024**, *24*, 103454. [CrossRef]

139. Aloko, M.N.; De Risi, R.; De Luca, F. Capacity prediction and failure mode classification of cold-formed steel built-up columns using machine learning methods. *Thin-Walled Struct.* **2025**, *210*, 112873. [CrossRef]

140. Gharagoz, M.M.; Noureldin, M.; Kim, J. Explainable machine learning (XML) framework for seismic assessment of structures using Extreme Gradient Boosting (XGBoost). *Eng. Struct.* **2025**, *327*, 119621. [CrossRef]

141. Su, A.; Cheng, J.; Wang, Y.; Pan, Y. Machine learning-based processes with active learning strategies for the automatic rapid assessment of seismic resistance of steel frames. *Structures* **2025**, *72*, 108227. [CrossRef]

142. Gatheeshgar, P.; Ranasinghe, R.S.S.; Simwanda, L.; Meddage, D.P.P.; Mohotti, D. Machine learning prediction of web-crippling strength in cold-formed steel beams with staggered slotted perforations. *Structures* **2025**, *71*, 108079. [CrossRef]

143. Sarfarazi, S.; Shamass, R.; Guarracino, F.; Modano, M. Exploring the stainless-steel beam-to-column connections response: A hybrid explainable machine learning framework for characterization. *Front. Struct. Civ. Eng.* **2025**, *19*, 34–59. [CrossRef]

144. Habib, A.; Houri, A.A.; Junaid, M.T.; Barakat, S. A systematic and bibliometric review on physics-based neural networks applications as a solution for structural engineering partial differential equations. *Structures* **2024**, *69*, 107361. [CrossRef]

145. Raissi, M.; Perdikaris, P.; Karniadakis, G.E. Physics-informed neural networks: A deep learning framework for solving forward and inverse problems involving nonlinear partial differential equations. *J. Comput. Phys.* **2019**, *378*, 686–707. [CrossRef]

146. Haghighat, E.; Raissi, M.; Moure, A.; Gomez, H.; Juanes, R. A physics-informed deep learning framework for inversion and surrogate modeling in solid mechanics. *Comput. Methods Appl. Mech. Eng.* **2021**, *379*, 113741. [CrossRef]

147. Lu, L.; Meng, X.; Mao, Z.; Karniadakis, G.E. DeepXDE: A deep learning library for solving differential equations. *SIAM Rev.* **2021**, *63*, 208–228. [CrossRef]

148. Pak, M.; Kim, S. A review of deep learning in image recognition. In Proceedings of the 4th International Conference on Computer Applications and Information Processing Technology (CAIPT), Kuta Bali, Indonesia, 8–10 August 2017; pp. 1–3. [CrossRef]

149. Olsson, F. A Literature Survey of Active Machine Learning in the Context of Natural Language Processing. *Swed. Inst. Comput. Sci. Tech. Rep.* **2009**, *1*, 59.

150. Weiss, K.; Khoshgoftaar, T.M.; Wang, D. A survey of transfer learning. *J. Big Data* **2016**, *3*, 9. [CrossRef]

151. Hospedales, T.; Antoniou, A.; Micaelli, P.; Storkey, A. Meta-learning in neural networks: A survey. *IEEE Trans. Pattern Anal. Mach. Intell.* **2021**, *44*, 5149–5169.

152. Farahani, A.; Voghoei, S.; Rasheed, K.; Arabnia, H.R. A brief review of domain adaptation. In *Advances in Data Science and Information Engineering: Proceedings from ICDATA 2020 and IKE 2020*; Springer: Cham, Switzerland, 2021; pp. 877–894.

153. Khodabakhshian, A.; Puolitaival, T.; Kestle, L. Deterministic and probabilistic risk management approaches in construction projects: A systematic literature review and comparative analysis. *Buildings* **2023**, *13*, 1312. [CrossRef]

154. Harrison, R.L. Introduction to Monte Carlo simulation. *AIP Conf. Proc.* **2010**, *1204*, 17. [CrossRef]

155. Gal, Y.; Ghahramani, Z. Dropout as a Bayesian approximation: Representing model uncertainty in deep learning. In Proceedings of the 33rd International Conference on Machine Learning (ICML), New York, NY, USA, 20–22 June 2016; Volume 48, pp. 1050–1059. Available online: https://proceedings.mlr.press/v48/gal16.html (accessed on 23 March 2025).

156. Marrel, A.; Iooss, B. Probabilistic surrogate modeling by Gaussian process: A review on recent insights in estimation and validation. *Reliab. Eng. Syst. Saf.* **2024**, *247*, 110094.
157. Sudret, B.; Marelli, S.; Wiart, J. Surrogate models for uncertainty quantification: An overview. In Proceedings of the 11th European Conference on Antennas and Propagation (EUCAP), Paris, France, 19–24 March 2017; IEEE: Piscataway, NJ, USA, 2017; pp. 793–797.
158. García-Risueño, P.; Ibáñez, P.E. A review of high performance computing foundations for scientists. *Int. J. Mod. Phys. C* **2012**, *23*, 1230001.
159. Parekh, R.; Mitchell, O. Progress and obstacles in the use of artificial intelligence in civil engineering: An in-depth review. *Int. J. Sci. Res. Arch.* **2024**, *13*, 1059–1080.
160. Lundberg Patel, D.; Raut, G.; Cheetirala, S.N.; Nadkarni, G.N.; Freeman, R.; Glicksberg, B.S.; Timsina, P.; Klang, E. Cloud platforms for developing generative AI solutions: A scoping review of tools and services. *arXiv* **2024**, arXiv:2412.06044.
161. American Society of Civil Engineers (ASCE). *Policy Statement 573—Artificial Intelligence and Engineering Responsibility*; American Society of Civil Engineers: Reston, VA, USA, 2024. Available online: https://www.asce.org/advocacy/policy-statements/ps573---artificial-intelligence-and-engineering-responsibility (accessed on 25 March 2025).
162. American Society of Civil Engineers (ASCE). *AI and Civil Engineering*; American Society of Civil Engineers: Reston, VA, USA, 2024. Available online: https://www.asce.org/topics/ai-and-civil-engineering (accessed on 25 March 2025).
163. Regona, M.; Yigitcanlar, T.; Xia, B.; Li, R.Y.M. Opportunities and adoption challenges of AI in the construction industry: A PRISMA review. *J. Open Innov. Technol. Mark. Complex.* **2022**, *8*, 45. [CrossRef]
164. International Organization for Standardization/International Electrotechnical Commission (ISO/IEC) Joint Technical Committee 1, Subcommittee 42 (JTC 1/SC 42). *Artificial Intelligence—Overview of Trustworthiness*; ISO: Geneva, Switzerland, 2023. Available online: https://www.iso.org/committee/6794475.html (accessed on 25 March 2025).
165. WorkOrb. Barriers to Adopting AI in AEC Firms. WorkOrb Blog. 2024. Available online: https://www.workorb.com/blog/barriers-to-adopting-ai-in-aec-firms (accessed on 25 March 2025).
166. RICS. *To AI or Not to AI: Five Trends in the Adoption of AI in Construction*; Royal Institution of Chartered Surveyors: London, UK, 2024. Available online: https://www.rics.org/news-insights/wbef/to-ai-or-not-to-ai-five-trends-in-the-adoption-of-ai-in-construction (accessed on 25 March 2025).
167. American Society of Civil Engineers (ASCE). What Do Civil Engineers Need to Know About Artificial Intelligence? *Civ. Eng. Mag.* 2024, 94, pp. 46–53. Available online: https://www.asce.org/publications-and-news/civil-engineering-source/civil-engineering-magazine/issues/magazine-issue/article/2024/11/what-do-civil-engineers-need-to-know-about-artificial-intelligence (accessed on 25 March 2025).
168. Habib, A.; Yildirim, U. Developing a physics-informed and physics-penalized neural network model for preliminary design of multi-stage friction pendulum bearings. *Eng. Appl. Artif. Intell.* **2022**, *113*, 104953. [CrossRef]
169. Yao, H.; Gao, Y.; Liu, Y. FEA-Net: A physics-guided data-driven model for efficient mechanical response prediction. *Comput. Methods Appl. Mech. Eng.* **2020**, *363*, 112892. [CrossRef]

metals MDPI

Review

Progress in Simulation Modeling Based on the Finite Element Method for Electrical Discharge Machining

Liwei Li [1], Shuo Sun [1], Wenbo Xing [1], Yuyan Zhang [1], Yonglei Wu [1], Yingjie Xu [1,2], Hongyan Wang [2], Guojun Zhang [2,3,*] and Guofu Luo [4,*]

1 Henan Key Laboratory of Intelligent Manufacturing of Mechanical Equipment, Zhengzhou University of Light Industry, Zhengzhou 450002, China; liliwei@zzuli.edu.cn (L.L.); sunshuo202304@163.com (S.S.); zzuwenbo@gmail.com (W.X.); 2020022@zzuli.edu.cn (Y.Z.); wylei156@gmail.com (Y.W.); xyj15203873137@163.com (Y.X.)
2 Guangdong Provincial Key Laboratory of Digital Manufacturing Equipment, Guangdong HUST Industrial Technology Research Institute, Dongguan 523808, China; 332304040404@email.zzuli.edu.cn
3 School of Mechanical Science & Engineering, Huazhong University of Science and Technology, Wuhan 430074, China
4 Faculty of Engineering, Huanghe Science and Technology University, Zhengzhou 450005, China
* Correspondence: guojun_zhang_stu@163.com (G.Z.); luoguofu@zzuli.edu.cn (G.L.)

Abstract: Electrical Discharge Machining (EDM) is a machining method commonly used to produce complex shapes and deep holes by eroding hard metals with an electric arc. There is a growing demand for process simulation using finite element models in order to improve the quality and efficiency of EDM, to reduce costs, to improve resource efficiency, and to facilitate its application in critical areas such as aerospace and mechanical engineering. Finite element models have greatly improved the prediction accuracy of EDM processes, simulated complex hybrid machining processes, and provided important guidance for the optimization of EDM processes. This paper systematically reviews the research progress of finite element modeling for EDM. Finite element method modeling is evaluated mainly in terms of four indicators: material removal rate, surface roughness, tool wear ratio, and recast layer thickness. Firstly, the importance and application of EDM are described, and the EDM finite element method modeling and its advantages are summarized. Then, the single-spark simulation model and the multi-spark simulation model of EDM are compared and discussed. Among the mainstream finite element models, the prediction error of the material removal rate for single-spark simulation ranges from 8.2% to 14.75%, while the prediction error of the recast layer thickness for multi-spark simulation can be as low as 1.98%. Finally, the applications of finite element modeling in EDM hybrid machining processes' performance prediction and new material machining are summarized, and future research directions and trends in EDM finite element modeling are predicted.

Keywords: single-spark simulations; multi-spark simulations; material removal rate; surface roughness; tool wear ratio; recast layer thickness; hybrid machining simulations

1. Introduction

EDM has a long history, with Benjamin Chew Tilghman of Philadelphia receiving a U.S. Patent (No. 416873) in 1889 for an invention titled the "Cutting of Metals by Electricity" [1], and Karl Strobel filing a patent application for an electric arc cutter in 1932 (patented in 1936) [2]. In the 1940s, Soviet physicists pioneered the application of electroerosion principles to manufacturing, leading to the emergence of electrical discharge machining (EDM) technology. This technology exhibited distinct advantages in machining high-hardness materials and complex workpieces. EDM technology finds extensive use in machining aerospace alloys and manufacturing molds. For aerospace system production, computerized numerical control EDM has emerged as a highly suitable machining method [3,4].

Studies have demonstrated that EDM offers significant advantages in machining tough and structurally complex materials, which are often challenging to process using conventional mechanical techniques [5,6]. In precision machining, EDM has found wide-ranging applications, and researchers have developed precision manufacturing methods based on EDM to meet the performance needs of high-frequency electrical devices [7–9]. According to Research and Markets, the global market for die-sinking and wire-cut EDM was valued at USD 2.347 billion in 2022 and is forecasted to reach USD 6.67 billion by 2030, with a compound annual growth rate of 6.5% from 2023 to 2030 [10]. The indispensable importance of EDM is readily apparent [11].

In EDM, the electrode materials of both the tool and workpiece rapidly evaporate and melt within a matter of microseconds due to the elevated temperatures within the confined space filled with dielectric fluid [12]. Hence, it is imperative to determine the appropriate process parameters to achieve satisfactory machining results [13]. One effective approach is to employ suitable modeling and optimization techniques to establish the relationship between process performance and its key input parameters. Nevertheless, the complexity of electrical discharges during EDM renders it challenging to experimentally observe the discharge process and quantify the material removal mechanisms [14,15]. EDM has found extensive applications, as illustrated in Figure 1, which delineates its characteristics, and various applications in the aerospace, mold, nuclear, defense, electronics, and medical sectors [16]. In the aerospace industry, wire EDM is employed to machine the commonly used titanium alloy, Ti_6Al_4V [17]. In the mold industry, EDM is commonly used to manufacture molds with complex shapes [18]. Furthermore, EDM is extensively employed for machining nickel-based superalloys in nuclear applications [19]. Hence, the development of precise EDM process models, which involve a comprehensive understanding of machining mechanisms and the identification of appropriate process parameters, is of utmost importance for reliable EDM process predictions.

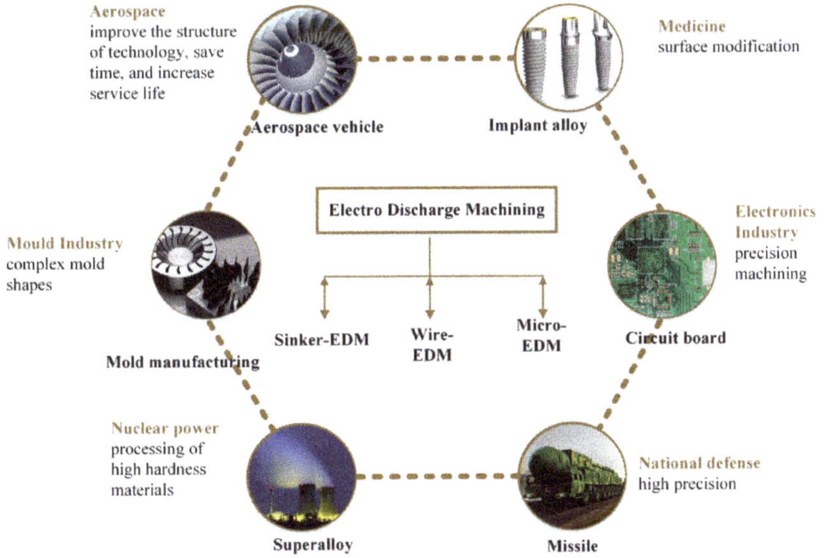

Figure 1. Characteristics and applications of EDM.

EDM's performance depends on multiple process parameters, and combining the results of multiple discharges is not as simple as linearly superimposing individual discharge outcomes. Consequently, simulating EDM presents challenges. Nevertheless, scientists have made significant progress in recent years in the development and optimization of various EDM simulation models, encompassing finite element method (FEM) models [20], molecular dynamics models [21], and computational fluid dynamics models [22]. Of these, FEM models have gained widespread usage due to their straightforward structure, high prediction accuracy, and direct experimental validation. Scientists frequently use FEM models to predict machining performance [23]. Scholars have employed the FEM to investigate the dynamic process of generating three-dimensional surface topography when continuous electrical discharge machining is applied to the workpiece surface. FEM models have also been used to explore topics such as energy distribution ratios during machining [24,25], discharge channel radii [26], and plasma motion [27]. Moreover, FEM models have been developed for ultrasonic-assisted EDM (US-EDM), magnetic field-assisted EDM (MF-EDM), EDM employing non-conductive ceramic electrodes, and the EDM of novel materials [28–31].

This paper's primary focus is on finite element modeling, specifically emphasizing the FEM models of EDM. The FEM model of EDM is a mathematical representation created using the finite element analysis method to describe the electrical discharge phenomenon and simulate machining results. This model is usually rooted in the theories of continuous media and electromagnetic fields. It deconstructs the complex electrical discharge process into a discrete system model composed of three-dimensional unit meshes to address specific concerns [24]. The main goal of this paper is to offer a comprehensive overview of the latest advancements in FEM models for EDM. The article thoroughly explores the theoretical modeling and application domains of FEM models. In Section 2, the review examines single-spark FEM models, used to evaluate essential performance indicators in EDM, such as the material removal rate (MRR), surface roughness (SR), tool wear ratio (TWR), and recast layer thickness (RLT). Section 3 discusses multi-spark FEM models, which are employed to simulate the machining process and conduct thorough investigations of four performance metrics: the MRR, Ra, TWR, and RLT. In Section 4, the review focuses on the practical application areas of electrical discharge machining, while Section 5 provides a summary of this investigation's content and offers insights into future research directions.

2. Single-Spark Simulations

EDM is widely utilized but has inherent limitations. The assessment of EDM's performance and the exploration of avenues for its enhancement are of utmost significance. The principal performance metrics for EDM assessment comprise the MRR, SR, TWR, and RLT. Numerous researchers strive to predict machining performance using single-spark FEM models. Single-spark discharge simulation entails the finite element modeling of an individual discharge event within the EDM process. It is favored for predicting individual discharge events because of its modeling and analysis simplicity. Single-spark FEM models excel in predicting workpiece surface crater profiles and can be extended to multi-spark scenarios to anticipate the surface structures achieved in EDM [32]. The EDM process faces several challenges, including its low machining efficiency, high SR, and difficulties in meeting high-precision requirements, which constrain its broad applicability [33]. Therefore, this section offers a comprehensive review of FEM models for predicting four critical performance indicators when evaluating EDM's performance: the MRR, SR, TWR, and RLT. Additionally, we reevaluate the accuracy of various heat source models in performance prediction, providing valuable insights for future research. In the EDM process, the choice of thermal input in the model is a crucial consideration. Hence, this section categorizes models, based on the type of thermal source, into non-Gaussian heat sources and Gaussian heat sources to explore their influence on performance prediction.

2.1. Material Removal Rate

The MRR quantifies the volume of material removed per minute, calculated as the ratio of the removed material's volume to machining time [34]. In the early stages of finite element modeling of the heat transfer process, various assumptions were employed, including cylindrical spark plasma, a uniform disc heat source, a point heat source, a constant heat source or expanding heat flux radius, and constant thermophysical properties [13]. Tariq Jilani and Pandey [35] developed a finite element model incorporating a disc-shaped heat source to study heat transfer and its impact on metal removal and machining in EDM. They observed that with an increase in the heat source's diameter, the pulse power density decreased, leading to a reduction in MRR. Later, Tariq Jilani and Pandey [36] further introduced an analytical model to compute metal removal rates and wear rates, specifically, of non-rectangular pulse current conditions in EDM. They presented an analytical calculation model for the single-spark erosion of metal in EDM. They investigated the influence of plasma channel growth on metal removal and found a strong correlation between theoretical and experimental results. Dibitonto et al. [37] introduced a cathode erosion model to predict the temperature distribution influenced by a single pulse discharge, taking into account a point heat source, constant material properties, fixed values of f_c (the fraction of power entering the cathode) = 0.18, and a plasma flushing efficiency (PFE) = 100%. In their experimental work, they found that for low discharge energies (<100 mJ), the predicted SR (R_{max}) closely matched the experimental value, while for high discharge energies ($\geq$100 mJ), the predicted MRR approached the experimental value. Additionally, their research group [38] introduced an anode erosion model for predicting the temperature distribution in spark discharges. This model had the capability to re-solidify melted discharge craters. In the case of the short pulse-on times associated with wire electrical discharge machining (WEDM), the model could provide a qualitative prediction of the maximum of the anode erosion curve. Salonitis et al. [39] created a straightforward thermal FEM model that projected that the MRR increased with a rising discharge current and arc voltage. Additionally, they noted that the MRR increased with a longer spark duration, with arc voltage exerting the most substantial influence on the MRR. The model exhibited an average deviation of approximately 8.2% between the experimental average MRR and the theoretically predicted values.

To better represent the heat source distribution in the EDM process, Eubank et al. [40] introduced Gaussian heat source models for the anode, cathode, and cylindrical plasma. Their comparative analysis revealed that Gaussian heat sources could more accurately simulate the thermal distribution during machining than disc heat sources. In FEM models for EDM, there has been a progressive shift from disc or point heat sources to Gaussian heat sources, aiming to better reflect real-world conditions and enhance simulation accuracy. Figure 2a illustrates a schematic diagram of a Gaussian distribution. According to Dibitonto et al.'s research [30], the average prediction error for the MRR exceeded 50%. In addition, the prediction error of the model proposed by Joshi and Pande [41] was lower than that of the model by Dibitonto et al. [37]. This was primarily attributed to the utilization of a Gaussian heat flux distribution in the FEM, a phenomenon also corroborated in the research conducted by Ming et al. [42]. The pulse-on time significantly influenced the prediction error in the FEM models. In the context of micro-electrical discharge machining (micro-EDM), the errors in predicting the crater radius and depth were minimal [43]. To mitigate model prediction errors, Joshi and Pande [14] devised an FEM combining a Gaussian heat source with temperature-dependent material properties and a spark radius contingent on the discharge current and duration. The results demonstrated that the average prediction error for the MRR was less than 10%. Additionally, Joshi and Pande [41] formulated a thermophysical model for the EDM process using FEM and conducted a numerical analysis of single-spark operations in EDM. The simulation results of the single-spark simulation model are depicted in Figure 2c,d. In the experiment, the workpiece material is AISI P20 mold steel and the tool material is copper, which produces several single-spark pits on the workpiece specimen. These results indicated that the trends in the simulation analysis closely aligned with experimental results, affirming that the developed thermophysical model could facilitate comprehensive parameter studies to comprehend EDM's process performance without necessitating physical experiments.

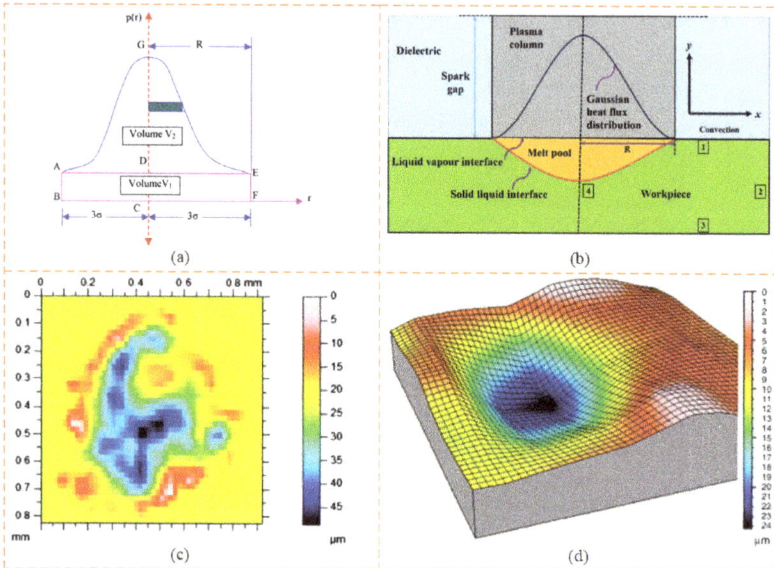

Figure 2. Prediction and analysis of MRR in single-spark simulations. (**a**) Gaussian distribution, reproduced from [44], permission with Elsevier, 2008. (**b**) Schematic representation of the 2D domain used in numerical simulation, reproduced from [45], permission with Elsevier, 2020. (**c**) Scanned two-dimensional view of discharge pits, reproduced from [41], permission with Elsevier, 2010. (**d**) Three-dimensional zoomed-in view of the crater, reproduced from [41], permission with Elsevier, 2010.

Singh [25] employed heat transfer equations to experimentally determine the distribution of input discharge energy in electrical discharge machining. Across various machining parameters, the MRR exhibited remarkable consistency, and the MRR was contingent on the fraction of energy transferred to the workpiece (F_{cw}). Moreover, this fraction was determined by the pulse duration, current, electrode polarity, and the specific combination of the workpiece and tool electrodes. It ranged from 6.1% to 26.82%, varying with changes in current and pulse duration, while numerous thermal models for predicting MRR used a constant value, resulting in less accurate predictions. Wang et al. [46] developed an FEM incorporating a Gaussian heat source to forecast the temperature distribution at the cathode discharge point in a single EDM discharge. The simulation results of the heat-affected zone revealed the dynamics of the temperature distribution, offering a more comprehensive insight into the material removal mechanism in EDM. Furthermore, Ming et al. [42] incorporated variable a fc and PFE dependent on the discharge energy in their proposed single-spark simulation model. Validation experiments demonstrated that the proposed hybrid intelligent process model exhibited strong accuracy in predicting the MRR, with an average error of 14.75% for the MRR. Kuriachen et al. [47] developed a transient three-dimensional thermophysical model employing the finite volume method to calculate the geometries of craters in single-spark micro-EDM processing. The model incorporated a Gaussian heat flux distribution, a constant spark radius, and a percentage distribution of supplied energy to estimate the temperature distribution along three directions on the workpiece. They conducted a thorough thermal analysis of a single spark, comparing the model's outcomes with previously published results and their experimental findings. The results demonstrated good agreement between the model calculations and the experimental outcomes. Singh et al. [45] conducted a numerical simulation of the micro-EDM process, employing a thermal-fluid coupling model. They validated the numerically simulated craters against individually generated experimental craters. Reasonable predictions were made for the diameters and depths of the discharge cavities, with errors of 12% and 13%, respectively. Figure 2b depicts a schematic diagram of the two-dimensional model utilized in the single-spark numerical simulation of the micro-EDM process. In summary, the establishment of EDM FEM models using suitable heat source distributions and parameters allows them to relatively accurately predict the MRR, offering a foundation for further optimization.

2.2. Surface Roughness

SR is employed to characterize the smoothness of a workpiece's surface, with Ra commonly serving as the standard metric. It delineates the geometric texture and quality of the workpiece's surface, encompassing parameters such as the SR and profile. Two-dimensional roughness parameters continue to be the most commonly used metrics for characterizing roughness. The arithmetic mean deviation of the assessed profile, frequently denoted as Ra, is the most widely employed roughness parameter in industry for quantifying surface roughness [48]. SR holds critical significance for the functionality and longevity of workpieces, with measurements obtained from instruments such as roughness measuring machines commonly employed to assess surface quality. Marafona et al. [49] created a two-dimensional finite element model based on a heat source. They compared the results of the finite element analysis with experimental data from AGIE SIT tables, which have been utilized by other researchers. As illustrated in Figure 3a, the SR trend at the cathode, as predicted by the finite element analysis model, was closely aligned with that of AGIE SIT. Salonitis et al. [39] formulated a straightforward two-dimensional finite element model based on a heat source. In this model, the average deviation between experimentally obtained average SRs and theoretically predicted values was approximately 6.1%. The deviation also increased with longer spark discharge durations. The study concluded that arc voltage exerted the most significant influence on the average SR, whereas pulse-off time had a minimal discernible impact. This trend aligned with the findings from other published experimental studies. Figure 3b shows the SR of any machined surface, depending on the size of the pits on the surface.

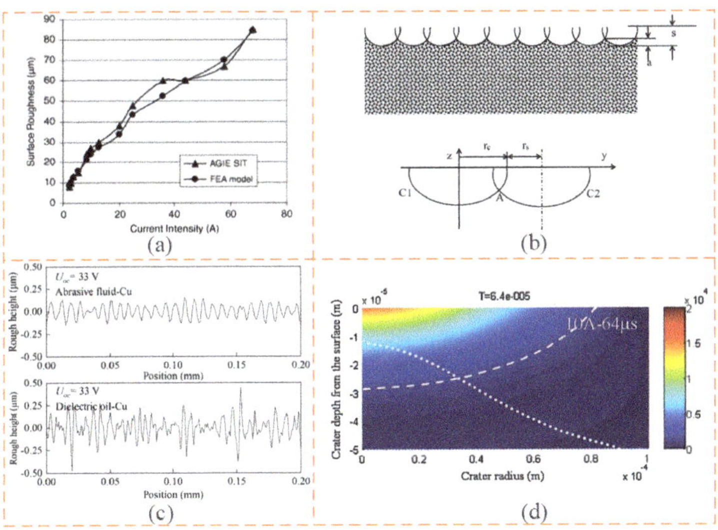

Figure 3. Prediction and analysis of SR in single-spark simulations. (**a**) Surface roughness R_{max} trends of AGIE SIT and FEA, reproduced from [49], permission with Elsevier, 2006. (**b**) Surface roughness model, reproduced from [50], permission with Elsevier, 2022. (**c**) Surface roughness profile of craters with pulsed discharges, reproduced from [51], permission with Elsevier, 2023. (**d**) Workpiece temperature distribution results from the single-spark simulation, reproduced from [52], permission with Elsevier, 2018.

Likewise, models that incorporate Gaussian heat sources can provide more accurate predictions of the SR. Ming et al. [42] integrated the variables fc and PFE, dependent on discharge energy, into their proposed single-spark simulation model. Validation experiments confirmed that the proposed hybrid intelligent process model exhibited good accuracy in predicting Ra, with an average error of 20.74%. Assarzadeh and Ghoreishi [53] introduced a novel approach for accurately predicting the roughness parameter of the electrical EDM process surface. Their method considered the impact of continuous discharges and PFE. They developed an iterative statistical method that considered a normal distribution of surface heights and the influence of each crater cavity formed due to repeated discharges on the surface profile. Reasonable experimental validation was conducted, and the results closely matched the model's predictions. Philip et al. [54] developed a mathematical model for SR on electrically discharge-textured Ti_6Al_4V using Gaussian heat sources. The empirical examinations used to substantiate the formulated modeling approach were carried out on a SPARKONIX Die Sinker EDM apparatus (Model: ENC 20/35/50) and Electronica (SMART CNC/XPERT). The observed error between the predicted Ra values and actual Ra values ranged from approximately 8% to 25%. Figure 3c illustrates the surface texture's Ra of the indentation with intermittent discharge; following the impact discharge in an abrasive liquid slurry, the surface roughness pattern exhibited reduced oscillation in contrast to the discharge in insulating oil. George et al. [55] constructed a finite element model using Gaussian heat sources. Their model predicted the three-dimensional SR of face-milled Inconel 825 specimens. Validation results for the SR closely aligned with the proposed model, with an average absolute error of 10.05%. Ming et al. [13], when comparing prediction errors in the machining performance of prominent single-spark FEM simulations, observed that at low discharge energies (<100 mJ) the predicted R_{max} error was lower, while at high discharge energies (≥100 mJ) the predicted R_{max} error was higher. The results indicated that Ra prediction errors were more favorable at low pulse-on times (<200 μs) compared to high pulse-on times (≥200 μs). Figure 3d illustrates a MATLAB simulation portraying the heat flux distribution in the workpiece at the conclusion of the "pulse-on" stage. The dotted white line represents the radial temperature distribution on the top surface of the workpiece from the center of the single discharge, which determines the size of the crater. The dashed white line represents the consequent boundary of the molten material with the bulk substrate. It also shows the temperature distributions across the entire top surface (indicated by a white dashed line) and the interface between the melting material and the substrate (also marked by a white dashed line) [52]. Li et al. [56] developed a thermal-fluid coupling model for the single pulse discharge process, utilizing Gaussian heat sources. Their findings indicated that the continuous discharge EDM material removal model exhibited a SR prediction error of 8.26%.

2.3. *Tool Wear Ratio*

The TWR is a critical metric for assessing productivity and efficiency in EDM. It is directly linked to machining parameters and is influenced by the machining power and the energy of single pulse discharges. Marafona et al. [49] developed a two-dimensional axisymmetric thermal–electric model to analyze spark generation in the liquid dielectric. They compared their finite element analysis results with experimental data from the AGIE SIT tables utilized by other researchers. The TWR results closely matched the values reported by the researchers who used the table. Figure 4a illustrates the temperature distribution at the cathode and anode. Joshi and Pande [14] created an intelligent EDM model that combined FEM models and artificial neural networks. The ideal BPNN (backpropagation neural network) structure is acquired by adjusting the quantity of the concealed layers, and the quantity of neurons within these covert layers for this instruction is 4–8–12–4. The initial "4" denotes four inputs, and the ultimate "4" indicates four outputs. Eight and 12 signify the count of neurons in the covert layer. They discovered that the optimal network architecture of 4–8–12–4 achieved highly accurate predictions for TWR, with an accuracy of 17.34%. Additionally, Joudivand Sarand et al. [57] developed a comprehensive numerical model for tool wear using finite element modeling and inverse heat conduction techniques. The experiments were conducted using a TA-EDM 204H concave die EDM machine, and high-strength AISI H13 steel was selected as the workpiece material. Comparisons between the numerical and experimental data demonstrated that the developed finite element model and inverse heat conduction technology could estimate tool wear with an average error of 4.83%. Papazoglou et al. [58] conducted experimental and simulation studies on the EDM of grade 2 titanium. They measured and calculated machining performance metrics such as the TWR under various applied currents and pulse numbers. Figure 4b displays a comparison between the actual and predicted TWR values. The results indicated that the TWR was primarily influenced by pulse-on time, with longer pulse durations leading to a reduced TWR. The TWR increased with an increasing number of pulse-on cycles.

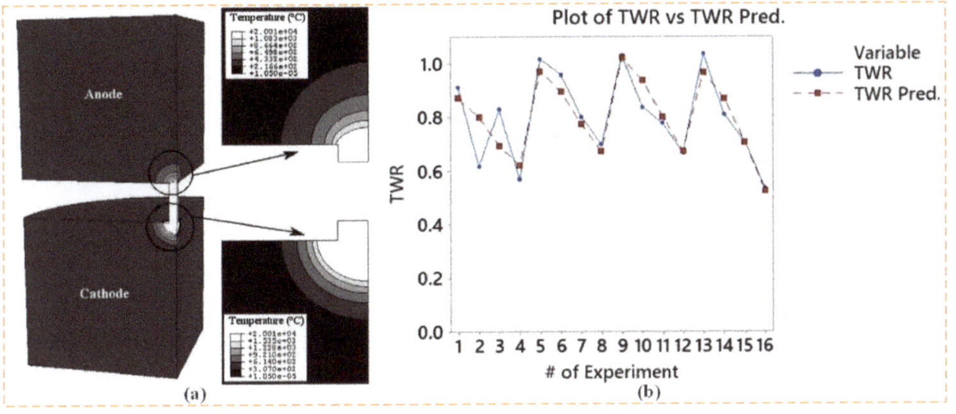

Figure 4. Prediction and analysis of TWR in single-spark simulations. (**a**) Temperature distribution at both cathode and anode, reproduced from [49], permission with Elsevier, 2006. (**b**) Plot of TWR vs. TWR, reproduced from [58], permission with Springer Nature, 2021.

2.4. Recast Layer Thickness

In the process of EDM, electrical discharges melt the workpiece surface. Some of the melted material is expelled by the electrode, while the remaining material resolidifies in place. This resolidified layered structure is referred to as the recast layer. The RLT represents the depth or height of the layer of material that is melted and re-deposited onto the workpiece surface during EDM discharges, typically measured in microns. The RLT significantly affects part performance because thicker recast layers tend to reduce the wear resistance and fatigue life of the part's surface. Vignesh and Ramanujam [59] developed a physical model to predict the MRR and validated it through experiments. Research results demonstrated that at discharge pulse durations of 106, 109, and 112 μs, the percentage error between their experimental and predicted values was 41.6%, 21.2%, and 15%, respectively. This discrepancy was due to the re-deposition of the removed material onto the workpiece surface. Therefore, it is necessary to investigate the RLT in the context of EDM. Pandey and Jilani [60] developed a two-dimensional disc heat source model that accurately calculated the heat-affected zone's thickness caused by a single spark. They achieved a strong correlation between predicted and experimental values. The model could also effectively estimate the extent of the thermal damage to the electrode materials during the EDM process. Shabgard et al. [61] conducted simulations of single discharges in magnetic-assisted EDM using FEM models to obtain the temperature distribution and crater dimensions on the workpiece surface. Their numerical and experimental results for the RLT showed good agreement, with a maximum error of 8.8%.

Figure 5 depicts the prediction and analysis of the RLT using FEM models.

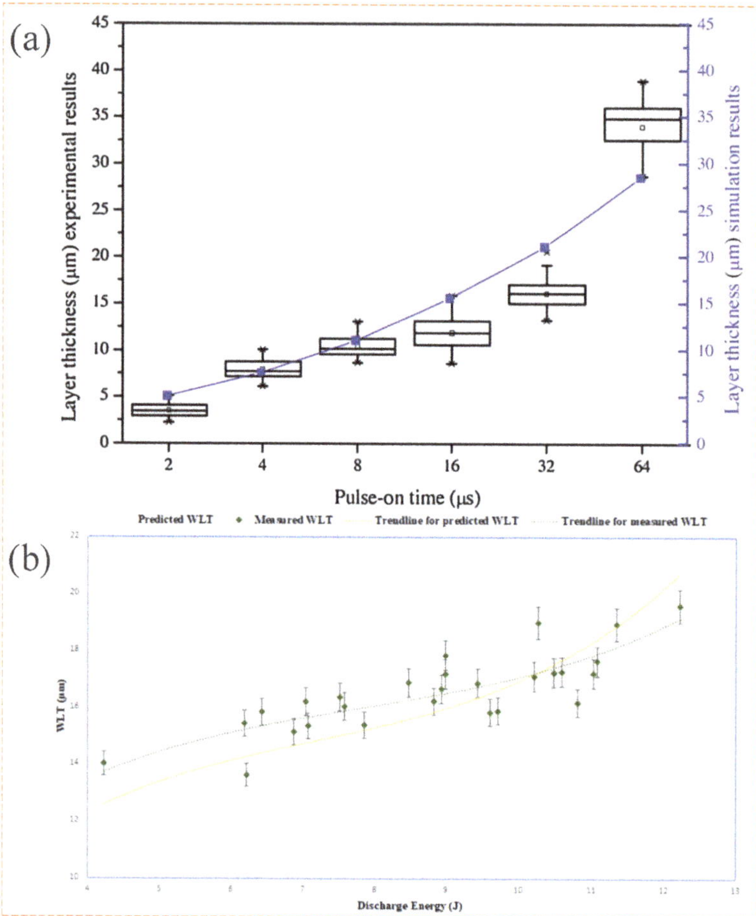

Figure 5. Prediction and analysis of RLT in single-spark simulations. (**a**) Comparison of simulated crater depth from a single spark event with experimentally determined electrical discharge layer thicknesses, reproduced from [52], permission with Elsevier, 2018. (**b**) Variation of predicted and measured WLT concerning discharge energy, reproduced from [62], permission with Elsevier, 2022.

Joudivand Sarand et al. [57] considered the individual and combined effects of thermophysical and electrical parameters in their modeling, achieving an average error of 3.96% in estimating the RLT. Zhang et al. [63] introduced a mathematical thermophysical model that incorporated the wire vibration during wire-cutting. They found that their relative error in estimating the RLT was less than 20% compared to experimental results. Figure 5a illustrates a noteworthy rise in the depth of the indentation, attributable to a rise in the duration of pulsation, aligning with an escalation in the thickness of the layer. Gholipoor et al. [64] conducted a finite element simulation of the single-spark discharge process in EDM. They utilized a numerical model to investigate the effects of pulse current and duration as input parameters on the RLT of molten material in machined surface craters. The results demonstrated that an increasing pulse current improved the PFE, whereas a longer pulse duration decreased the PFE and increased the RLT. George et al. [62] developed an analytical model of the RLT during wire-cut EDM. The model exhibited strong performance in predicting the RLT, with an average absolute error of 5.12%. Figure 5b illustrates the variation between the predicted and measured wire layer thickness (WLT) with respect to discharge energy. In single-spark discharge machining, the RLT and SR are two interrelated but distinct concepts. Achieving excellent surface quality involves eliminating defects on the parent surface, minimizing the RLT, and controlling the SR.

2.5. Summary

This section primarily reviews the research advancements in modeling and simulating key technical parameters in EDM, including the MRR, SR, TWR, and RLT. The researchers discovered that refining the forecast precision of the model involved incorporating a Gaussian distribution of the thermal flux and considering the fluctuation in the radius of the EDM with respect to the discharge current and the duration of the discharge. This adjustment aimed to bring the model into closer alignment with the real conditions of the process. Early models employed basic disk and point heat sources, with a gradual shift towards adopting Gaussian heat source distributions for enhanced accuracy. The research indicates that Gaussian heat sources offer superior accuracy in predicting the thermal distribution and machining outcomes in EDM. Additionally, this section provides a summary of the modeling accuracy for each key indicator, with the average MRR error typically limited to within approximately 10%, the average SR error to within 20%, and the average TWR error to within 5%. Table 1 presents a comparison of the prominent FEM models simulating the single-spark process in EDM. Research findings illustrate that the utilization of suitable models and parameters can proficiently forecast EDM performance metrics, offering theoretical guidance for further optimizing the EDM process.

Table 1. Comparison of FEM models using different heat sources to predict machining performance.

Heat Source type	Purposes	Authors, Year	Findings	Remarks
Non-Gaussian heat source	MRR	Tariq Jilan and Pandey [35], 1982	As the diameter of the heat source increased, the pulse power density decreased and the MRR decreased.	The removal of metal in EDM is due not only to melting but also to evaporation, thus greatly reducing the amount of heat available for conduction.
		Tariq Jilan and Pandey [36], 1982	From the point of view of relative electrode wear, the optimum pulse shape for pulse durations of 50 and 100 μs was an initial current of about 50% of its value at the end of the pulse duration.	The effects of evaporation and re-solidification can be incorporated into the EDM model.
		Dibitonto et al. [37], 1989	In the confirmed experiments, the predicted R_{max} matched the experimental values at low discharge energies (<100 mJ), while the predicted MRR, at high discharge energies (≥100 mJ), was close to the experimental values.	Finite element modeling at high discharge energies should be further investigated.
		Salonitis et al. [39], 2009	The average deviation between the experimental average MRR and the theoretical predictions of the model was about 8.2%.	These deviations are due to neglecting the formation of the recast layer and assuming that the idling time is insignificant compared to the discharge time.

Table 1. *Cont.*

Heat Source type	Purposes	Authors, Year	Findings	Remarks
	Ra	Marafona et al. [49], 2006	The predicted SR agrees well with the experimental results.	Two-dimensional axisymmetric finite elements have simpler formulations than three-dimensional finite elements.
		Salonitis et al. [39], 2009	The average deviation between the experimental average SR and the theoretical prediction of the model was about 6.1%.	The error may arise because the formation of the recast layer is enhanced when dealing with larger peak currents.
	RLT	Pandey and Jilani. [60], 1986	The heat source model can calculate the thickness of the heat-affected zone caused by a single spark more accurately, and there was a good correlation between the predicted and experimental values.	This heat source model can more accurately calculate the thickness of the heat-affected zone caused by single sparks.
		Shabgard et al. [61], 2019	The maximum error between the RLT obtained by the numerical and experimental methods was 8.8%.	The application of an external magnetic field during EDM has a positive effect on increasing the PFE and reducing the RLT.
Gaussian heat source	MRR	Eubank et al. [40], 1993	An anode model, a cathode model, and a cylindrical plasma model were developed using a Gaussian heat source.	Gaussian heat sources can more accurately simulate the heat distribution during machining than disc heat sources.
		Joshi et al. [14], 2009	The optimal network architecture 4–8–12–4 has a prediction error of 1.53% for the MRR.	The ANN-based process model can be used to select optimum process conditions to improve EDM's process productivity and finishing capability.
		Joshi and Pande [41], 2010	The results show that the trends of the simulation analysis are consistent with the experimental results.	This thermophysical model can be used to perform a wide range of parametric studies.
		Singh, H. [25], 2012	For different machining parameters the results of the MRR match very well.	It is possible to study the effect of different material properties and machining parameters on the proportion of energy transferred to the workpiece.
		Wang et al. [46], 2012	The temperature distribution at the cathode discharge point during a single discharge in the EDM process was predicted.	The mechanism of material removal during EDM is better explained.

Table 1. *Cont.*

Heat Source type	Purposes	Authors, Year	Findings	Remarks
		Ming et al. [42], 2014	The validation experiments show that the mean error of the MRR is 14.75%.	Through the GPR model, it is found that the responses of the EDM process can be accurately predicted for the chosen process conditions.
		Kuriachen et al. [47], 2015	The model calculations were in good agreement with the experimental results.	Combining different spark radii with the effects of capacitance and voltage, plasma pressure, and the enthalpy of the material results in a better simulation model.
		Singh et al. [45], 2020	The prediction errors for the diameter and depth of the discharge crater were 12% and 13%, respectively.	Material removal is mainly evaporation and melt discharge caused by recoil pressure, which determines the size of the discharge crater.
	Ra	Ming et al. [42], 2014	The mean error of Ra was 20.74%.	This model can be used to select optimal process conditions to improve EDM performance.
		Philip et al. [54], 2020	The model employed a Gaussian heat source, and the observed minimum-to-maximum error between the predicted Ra and the actual roughness Ra values ranges from about 8 to 25%.	The formation of radially enlarged craters is the main cause of highly rough surfaces.
		George et al. [55], 2020	The average absolute error of the forecast was 10.05%.	This study allows for the online monitoring of the surface quality of difficult-to-machine materials.
		Ming et al. [13], 2022	The predicted R_{max} error was found to be lower at low discharge energies (<100 mJ) and higher at high discharge energies ($\geq$100 mJ).	The accuracy of the predictions needs to be further improved, especially for EDM processes.
		Li et al. [56], 2022	The error in the prediction of the SR was 8.26% for the continuous discharge material removal model of wire-cutting EDM.	This study contributes to a better understanding of the principles of the wire-cutting process and the formation of single-spark discharge craters.
	TWR	Marafona et al. [49], 2006	The TWR results are fairly consistent with the values the researchers found for the AGIE SIT table itself.	Two-dimensional axisymmetric finite elements have simpler formulations than three-dimensional finite elements and can reduce CPU time with very similar results.

Table 1. *Cont.*

Heat Source type	Purposes	Authors, Year	Findings	Remarks
		Joshi and Pande [14], 2009	The prediction accuracy of this model artificial neural network's optimal network architecture, 4–8–12–4, for the TWR was 17.34%.	Artificial neural network-based process modeling can be used to select optimal process conditions to improve the productivity and finishing capabilities of EDM machining.
		Joudivand Sarand et al. [57], 2017	The average error in predicting tool wear was 4.83%.	This study considers the direct and interactive effects of thermal, electrical, and physical variables on the EDM process, with accurate predictions.
		Papazoglou et al. [58], 2021	The TWR was mainly dependent on the pulse-on time, and an increase in pulse duration led to a decrease in that ratio.	Optimal processing schedules can be achieved on a case-by-case basis using recommended MRR, TWR, and average white layer thickness correlations.
	RLT	Joudivand Sarand et al. [57], 2017	The average error in the estimation of the RLT was 3.96%.	Pulse current, pulse on-time, and the thermal diffusion coefficient have a significant effect on the thickness of the recast layer.
		Zhang et al. [63], 2019	Upon comparison with the experimental results, the relative error of the RLT was found to be less than 20%.	Controlling wire vibration helps improve EDM's wire-cutting performance.
		Gholipoo-r et al. [64], 2020	The increase in pulse duration decreased the PFE and increased the thickness of the recast layer.	Above a certain value of pulse current, the increase in the PFE is counterbalanced by the depth of thermal penetration and the volume of molten material, and the RLT decreases.
		George et al. [62], 2022	The model predicted the RLT better, with an average absolute error of 5.12%.	The RLT tends to increase with increasing discharge energy.

3. Multi-Spark Simulations

In finite element modeling and the simulation of EDM, two categories exist: single-spark simulations and multi-spark simulations. Shahane and Pande [65] introduced an innovative approach for simulating the occurrence of multiple sparks with overlapping craters by modifying the transient heat transfer model originally designed for single sparks. In multi-spark simulations, EDM is conducted through a sequence of single-spark events. The initiation of sparks is inherently random, often occurring in clusters while maintaining a minimum distance between the tool and the workpiece [66]. Research has shown that, in contrast to single-spark models, multi-spark models provide MRR predictions that align more closely with experimental findings. Multi-spark simulations offer several advantages over single-spark simulations, including enhanced realism, accuracy, the capability to simulate intricate machining processes, improved simulation precision, and broader applicability. However, due to variations in the position and pressure of dielectric media, such as bubbles and solid particles within the gap, which change with time and location, it is not possible to obtain the result of multiple discharges through a simple linear superposition of individual discharge outcomes [67,68]. Therefore, the study of multi-spark simulations is crucial.

3.1. Material Removal Rate

In FEM models of EDM, simulations of multi-spark EDM predominantly employ Gaussian heat sources. Izquierdo et al. [69] developed a numerical model for the EDM process that accounted for the impacts of multiple discharges. This model computed the internal temperature distribution of the workpiece using finite difference methods and utilized Gaussian heat sources to simulate the EDM surface. EDM trials were conducted on an ONA TECHNO H300 SEDM apparatus, employing a copper electrode with a square cross-sectional area measuring 30×30 mm^2, and external rinsing facilitated by a flushing nozzle. With optimal input parameters, the MRR prediction error remained below 3%. Figure 6d illustrates the evolution of the material removal volume per discharge during the EDM process. The modeling of Philip et al. [70] focused on the propagation of multiple sparks. Shahane and Pande [65] introduced and implemented an innovative strategy to create a multi-spark model with overlapping craters. This model considered factors such as spark aggregation, Gaussian heat flux, temperature-dependent material properties, and the influence of the latent heat of fusion. The predictions of the multi-spark model closely matched experimental values, with a deviation of approximately 50%. Figure 6a depicts the experimental MRR data, alongside the ratios of the MRR from both single-spark and multi-spark models to the experimental MRR. H. Saikiran et al. [71] conducted MRR modeling of the Orvar Supreme H-13 EDM process using an axisymmetric model. They extended this single-spark model into multi-spark machining by calculating the MRR based on the number of pulses. The experiments were carried out on a CNC die-sinking EDM machine and compared with the analytical results, and a high degree of agreement was observed. Almacinha et al. [72] observed that the presence of multiple discharges had a significant impact on the MRR during the machining process. Incorporating this effect into the volume removed by a single discharge led to predicted MRR values that closely matched the experimental data. Razeghiyadaki et al. [73] employed a simplified two-dimensional numerical heat conduction equation along with supplementary assumptions. They demonstrated that their model exhibited good agreement with experimental results, thereby validating their numerical approach. In comparison to the standard finite element model, the finite element model considering instantaneous evaporation enhanced its concurrence with the MRR experimental values by 1.5%. Figure 6a illustrates the comparison of the MRR between the numerical model and the experimental values obtained by Almacinha [72]. Yakup Yildiz [74] employed a three-dimensional finite element model to forecast white layer thickness and the MRR. The numerical solutions from the model were compared with experimental data, yielding an average MRR prediction error of 3.34%. Figure 6b illustrates the MRR as a function of the discharge current and the modeling approach. In comparison to the theoretical thermal model, the results from the finite element model closely approximated the experimental findings.

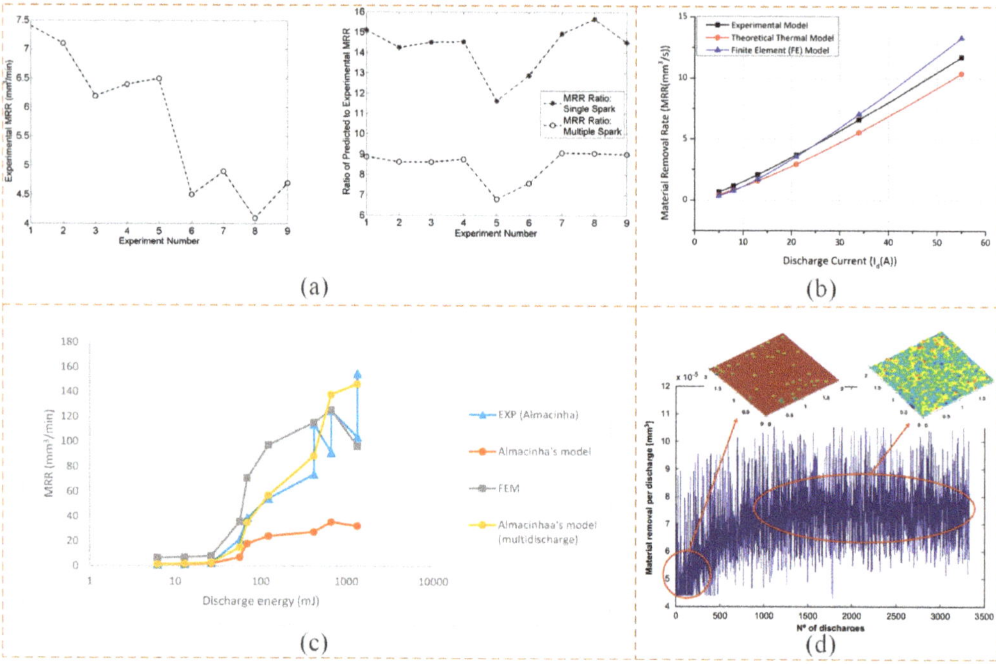

Figure 6. Prediction and analysis of MRR in multi-spark simulations. (**a**) Validation of single- and multi-spark models, reproduced from [65], permission with Elsevier, 2016. (**b**) MRR results of the models, reproduced from [74], permission with Elsevier, 2016. (**c**) Comparison of MRR obtained from the numerical model with experimental values from Almacinha (discharge voltage: 20 V), reproduced from [73], permission with MDPI, 2019. (**d**) Variation of the volume of material removed per discharge as the operation progresses, reproduced from [69], permission with Elsevier, 2009.

3.2. Surface Roughness

Figure 7 presents the prediction and analysis of SR using multi-spark simulation FEM models. Kurnia et al. [75] introduced a micro-EDM SR model based on estimating crater geometry. The model incorporated surface factors, accounting for effects, such as overlapping craters, scratches, microcracks, and re-attached debris, from the real machining process. In this initial study, the theoretical results from the proposed model were compared to the corresponding experimental results, revealing an average Ra error of 6.5%. Ti_6Al_4V was employed as a validation material to assess the applicability of the SR prediction model to different materials. Izquierdo et al. [69] developed a numerical model of the EDM process that considered the effects of multiple discharges. The model calculated the internal temperature field of the workpiece using finite difference methods and considered the impact of continuous discharges on the EDM surface. Under optimal input conditions, the prediction error for the surface finish was below 6%. Razeghiyadaki et al. [73] similarly predicted the SR, observing an increase in the predicted SR with a higher discharge energy. Figure 7a shows that the gray curve "FEM" is the predicted data from this model. The SR predicted by this model is very close to the experimental data, compared to the Salonitis model [39]. One possible explanation for the variance between Salonitis' model and this model is that Salonitis' model relied on numerous simplified assumptions, including a uniform heat source. This model and Salonitis' model exhibited deviations of 6.4% and 7.5%, respectively, from their experimental values.

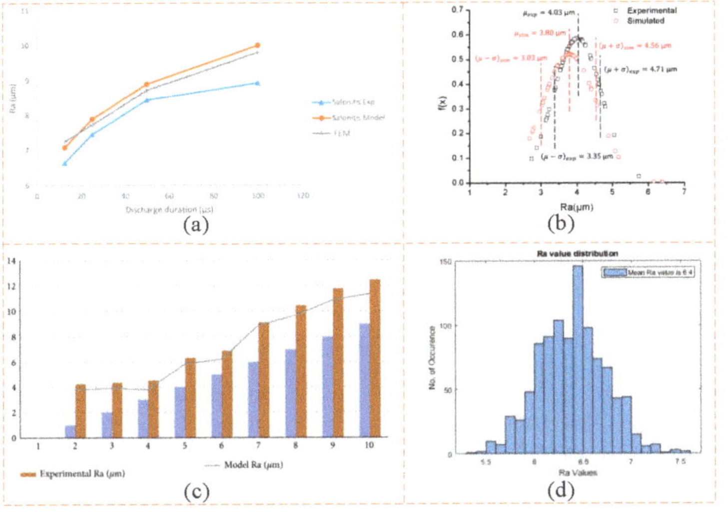

Figure 7. Prediction and analysis of SR in multi-spark simulations. (**a**) Comparison of SR obtained from a numerical model and experimental values from Salonitis, reproduced from [73], permission with MDPI, 2019. (**b**) Ra distribution of the experimental and simulated surfaces of Ti_6Al_4V under a parameter combination of 10 A, 20 μs, and 50 V, reproduced from [76], permission with Elsevier, 2020. (**c**) Comparison of Ra values of the simulation and experiment, reproduced from [77], permission with Hindawi, 2022. (**d**) Ra distribution for 100 V–0.5 A trial, reproduced from [78], permission with Elsevier, 2022.

Multi-spark simulation is continuously advancing due to its accurate predictions. Jithin et al. [76] employed numerical simulation methods to replicate the gradual development of surface topography resulting from the continuous application of spark discharges to the surface texture formed by previous discharges. The multi-spark model adopted random spark distributions in position, energy level, and time sequence. For surface generation, the validation of the SR prediction values for different parameter combinations, using Ti_6Al_4V steel and M300 steel as workpiece materials, revealed a strong agreement between the predicted and measured values, with prediction errors ranging from 6% to 17.5%. Figure 7b presents another comparison of the surface topographies by measuring Ra along 50 cross-sections of both surfaces and comparing their Ra distributions. Jithin et al. [79] developed a multi-spark discharge model to predict average Ra, taking into account the random distribution of crater profiles. The multi-spark model was used to simulate the surface profile, while the single-spark model assessed crater profiles. With the adoption of more realistic assumptions, the model reduced the average Ra prediction error to 11.5%. The average relative error in predicting surface roughness was 8.26% for the EDM continuous discharge removal model developed by Abed et al. [77]. Figure 7c shows that the predicted values of the model are in general agreement with the experimental measurements. Jamunkar and Sundaram [78] introduced a multi-spark model that yielded the normal distribution of SR values for various EDM process parameters. The simulated EDM surface was created through finite element analysis and online EDM monitoring techniques. The measured EDM SR values fell within the predicted range. The simulated surface included multiple overlapping, small impact craters produced by individual sparks. For the validation experiments, SS304 and AISI 304 workpieces were machined, with a 4 mm hollow copper tool, using the EDM process. The maximum deviation between the predicted and measured values was 0.76 μm. Figure 7d illustrates the normal distribution of the SR values.

3.3. Tool Wear Ratio

The TWR is a crucial indicator of machining quality and economic efficiency. Predicting and controlling the TWR is of paramount importance for optimizing the electrical discharge machining process. Figure 8a illustrates the estimated response surface of electrode wear while varying the parameters of intensity and pulse time. Observably, as the intensity factor decreases to its minimum value, there is a declining trend in wear value, which reverses and begins to increase after reaching the minimum. Puertas et al. [80] developed a finite element model using design experimentation techniques to determine the optimal machining conditions for the finishing stage. In the context of electrode wear, it is evident that the intensity factor exerts the most significant influence, followed by its pure quadratic effect and interaction effect with pulse time. To achieve reduced electrode wear values within the defined working range, it is advisable to use an intensity factor value near the tool's central value (i.e., I = 4) or slightly higher, in conjunction with low pulse time values. Kim et al. [81] conducted both experimental and numerical investigations on the micro-EDM of molybdenum. Their findings indicated a significant increase in the TWR when the pulse duration exceeded 2 μs. The combination of high voltage and extended pulse duration led to severe tool wear. In multi-spark micro-EDM experiments, debris composed primarily of a mixture of molybdenum and tungsten has been observed around the generated holes. Barenji et al. [82] applied a response surface methodology to predict and optimize the TWR in the EDM process of AISI D6 tool steel. They also developed a numerical model for predicting the TWR. For the validation experiments, AISI D6 tool steel of 20 mm diameter and 20 mm thickness was used for the workpiece. In addition, round electrolytic copper of 18 mm diameter was used as the electrode and commercial kerosene was used as the dielectric. Their findings revealed that minimizing the TWR occurred when the pulse-on time, pulse current, and input voltage were set at 40 μs, 14 A, and 150 V, respectively. As depicted in Figure 8b, within the spectrum of EDM process parameters, the predicted response values closely aligned with the actual response values, as indicated by the even distribution of data points along the 45° line. In micro-EDM drilling, it is common practice to employ regression models for the prediction of the MRR and TWR based on the applied machining parameters. To enhance prediction accuracy, Bellotti et al. [83] investigated the effectiveness of data-driven regression models in predicting tool wear and material removal. By incorporating process monitoring data as inputs into the regression models, they achieved a significant reduction in errors when predicting the TWR, approximately 85%.

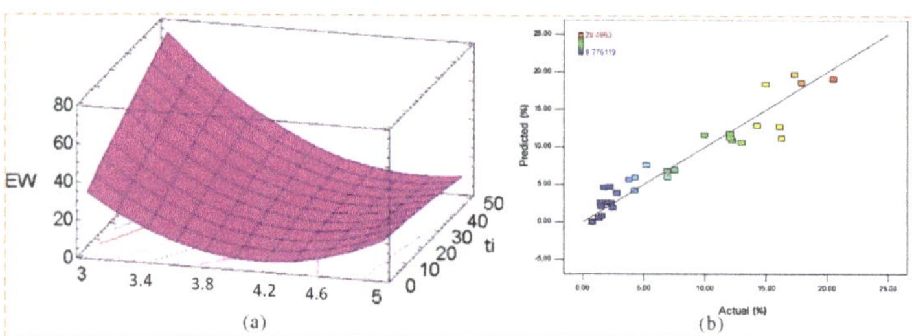

Figure 8. Prediction and analysis of TWR in multi-spark simulations. (**a**) Estimated response surface of EW vs. I and ti, reproduced from [80], permission with Elsevier, 2004. (**b**) Predicted versus actual response plot for TWR, reproduced from [82], permission with Elsevier, 2016.

3.4. Recast Layer Thickness

Research has indicated that the surface erosion caused by sparks on workpieces results in the formation of a hard recast layer, commonly referred to as a white layer. During the EDM process, sparks generated by the discharge melt and vaporize a small region on the electrode's surface. When the pulse-on time concludes, a small quantity of molten material is expelled from the surface, and the residual liquid solidifies. The resulting recast layer typically exhibits fine grains and high hardness. It can also form alloys with carbon derived from the cracked dielectric or materials from the tool [84]. The RLT generated in the EDM process serves as a critical process performance indicator. It offers insights into the degree of surface crack propagation and the thickness of the alloyed functional layer on the machined surface. Manufacturers typically employ finishing machining operations such as grinding to eliminate the recast layer. The white recast layer can be effectively removed to achieve a polished surface finish by utilizing silicon carbide grinding with the aid of a lubricant. Polishing entails the removal of material from the EDM surface, but excessive material removal can result in a loss of tolerance. It is of the utmost importance to determine the average RLT to establish the appropriate conditions for polishing and grinding [84].

Tan and Yeo [85] introduced a numerical model that relies on multiple discharges to predict the RLT generated in micro-EDM. In this approach, the recast layer was regarded as a composite formed by individual crater recast zones resulting from a series of multiple discharges. The model featured a peak discharge current of 1.45 A, pulse-on times spanning from 166 ns to 606 ns, and projected RLT values within the range of 1.0 μm to 1.82 μm. For pulse times of 166 ns, 362 ns, and 606 ns, the corresponding average RLT values were 1.10 μm, 1.37 μm, and 1.75 μm, respectively. Shabgard et al. [86] developed a three-dimensional axisymmetric model employing FEM models to simulate the temperature distribution in EDM and estimate the surface integrity characteristics of AISI H13 tool steel workpieces. The examined surface integrity characteristics included the white layer thickness, heat-affected zone depth, and average thickness. The study investigated the effects of pulse time and pulse current. Both numerical and experimental findings indicated that increasing the pulse time resulted in greater white layer thickness, heat-affected zone depth, and surface porosity. The finite element model successfully predicted workpiece white layer thickness (WT) with an average deviation of 9.65%. Yakup Yildiz [74] employed a three-dimensional finite element model to predict white layer thickness, and his results demonstrated a strong agreement between his finite element analysis and experimental findings. The average errors for predicting white layer thickness using the FEM were 1.98%. In Figure 9a, it is evident that at lower discharge currents, the outcomes from the thermal model significantly deviated from those of the experimental and FEM models. Conversely, the experimental results closely matched the results obtained from the finite element model. Yadav and Pradhan [87] determined the thicknesses of the recast layer and martensite layer during the wire-cut EDM of D2 steel by employing numerical and quantitative expression methods. Recast zones were identified through multiple discharge sequences. In prior experiments, Rajurkar and Pandit [88] achieved a RLT of 8.0 μm in D2 steel using a 20 A current. Conversely, their current numerical simulation model predicts an RLT value of 10.2 μm, demonstrating strong agreement. Figure 9b visually presents the alignment between the numerically estimated RLT and the published experimental values. The simulation results are depicted in blue, while the experimental results are shown in red.

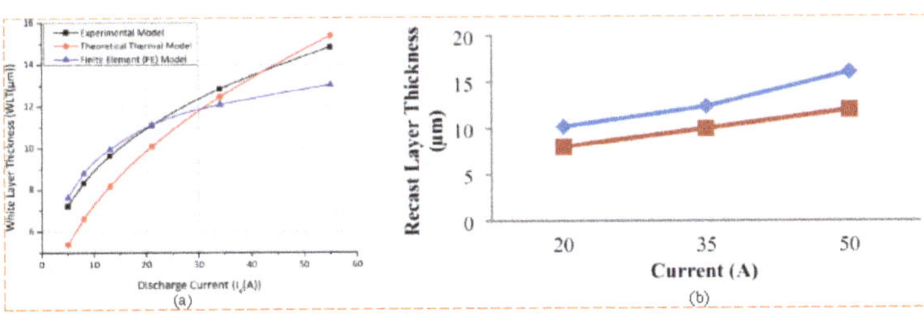

Figure 9. Prediction and analysis of RLT in multi-spark simulations. (**a**) WLT results of the models, reproduced from [74], permission with Elsevier, 2016. (**b**) Comparison of RLTs of D2 steel obtained from ANSYS, reproduced from [87], permission with Elsevier, 2021.

3.5. Summary

This section primarily presents the advancements in multi-spark simulation within the context of finite element modeling for EDM. In contrast to single-spark simulation, multi-spark simulation takes into account the randomness and overlapping effects of sparks. Researchers have found that the discharge locations of the multi-spark simulations were characterized as follows: the discharge locations were random, taking into account the effect of bubbles or debris on the discharge locations; and the discharge locations were located at the maximum point of the electric field. This approach enables a more realistic and accurate simulation of complex machining processes, producing results that closely align with experimental data. This section provides a systematic summary of the modeling efforts in multi-spark simulations to predict MRRs, SR, TWRs, and RLTs. Research findings indicate that multi-spark models can effectively manage the prediction errors for these critical parameters, maintaining them at approximately 3%, 6.5%, 5%, and 2%, respectively. Table 2 provides a comparison of the existing methods for multi-spark finite element simulation in EDM processes. In summary, multi-spark simulation represents the evolving direction of finite element modeling for EDM, offering improved accuracy in predicting machining effects. However, it is crucial to consider the interactions between different sparks.

Table 2. Comparison of existing major multi-spark simulations by FEM models for the process of EDM.

Purposes	Authors, Year	Findings	Remarks
MRR	Izquierdo et al. [69], 2009	The finite difference method was used and the MRR prediction error was less than 3%.	The inverse discrimination method can be used to estimate the discharge characteristics for a particular operation.
	Yakup Yildiz [74], 2016	The average error in predicting the MRR using the FEM was 3.34%.	The effect of additional parameters on the measurement of the properties of different types of materials can be studied in the same way.
	Shahane and Pande [65], 2016	The results predicted by the multi-spark model were closer to the experimental values by about 50%.	The multi-spark model can be further used for extensive parametric studies, which can be used to predict the MRR by varying the input process parameters.
	H.SAIKIRAN et al. [71], 2016	An axisymmetric model was used for the MRR modeling of the Orvar Supreme H-13 EDM process, and the MRR was calculated by counting the number of pulses in multi-discharge machining. The experimental results were in good agreement with the analytical results.	This model can be used to obtain the residual stress distribution, and to enhance the thermal stress distribution of the particle rupture phenomenon.
	Razeghiyadaki et al. [73], 2019	Compared to the standard finite element model, the finite element model considering transient evaporation showed a 1.5% better agreement with the experimental values.	Consideration can be given to the study of EDM in non-conductive ceramic materials.
Ra	Kurnia et al. [75], 2009	The theoretical results produced by the model had an average error of 6.5% in Ra when compared to the respective experimental results.	This surface roughness prediction model can be applied to other workpiece materials.
	Izquierdo et al. [69], 2009	A numerical model of the EDM process considering the effect of multiple discharges was developed with an SR prediction error of less than 6%.	It is possible to generate surfaces comparable to those obtained by the actual EDM process and, after adjusting the model parameters, the predicted results are in good agreement with the measured results.
	Jithin et al. [76], 2020	The results show that the predicted values are in good agreement with the measured values and the prediction error is between 6~17.5%.	This model takes into account the expansion of the discharge column at the critical time and the dependence of the spark radius on the discharge power and discharge duration.
	Jithin et al. [79], 2020	A multi-spark discharge model was developed to reduce the average prediction error of Ra to 11.5%.	In each simulation of the multi-spark model, the position, size, and shape of the individual peaks and valleys of the sparks vary due to their random distribution.
	Jamunkar and Sundaram [78], 2022	The model gives a normal distribution of SR values for any given EDM process parameter. The maximum deviation between the predicted and measured values is 0.76 μm.	Finite element analysis and online EDM monitoring techniques were used to create a simulated EDM surface.
TWR	Puertas et al. [80], 2004	To be able to obtain low electrode wear values, an intensity factor value close to the electrode's center value (i.e., I = 4), or slightly higher, should be used, along with a low pulse duration value, for the operating interval under consideration.	The relationship between the TWR, SR, MRR, and design factors was investigated.

Table 2. *Cont.*

Purposes	Authors, Year	Findings	Remarks
	Kim et al. [81], 2010	The relative TWR increased significantly when the pulse duration was greater than 2 μs, and the combination of large voltages and large pulse durations produced significant tool wear.	Tool wear is the main cause of chip formation.
	Reza Vatankhah Barenji et al. [82], 2016	The TWR was minimized when the pulse-on time, pulse current, and input voltage were 40 μs, 14 A, and 150 V, respectively.	Increasing the pulse current increases the TWR. The higher the input voltage, the lower the TWR.
	Bellotti et al. [83], 2020	By using process monitoring data as inputs to the regression model, the errors in predicting the TWR were reduced by approximately 85%.	Expanding the dataset with higher aspect ratio holes could be considered in the future to investigate the applicability of data-driven regression modeling to a wider range of industrial applications of the MEMS process.
RLT	Tan and Yeo [85], 2010	The model has a peak discharge current of 1.45 A, a pulse on-time of 166 ns to 606 ns, and a predicted thickness of the recast layer ranging between 1.0 μm and 1.82 μm.	The model developed using the multiple discharge method is suitable for the estimation of the RLT in microfabricated EDM.
	Shabgard et al. [86], 2011	An axisymmetric 3D finite element model was developed and the average deviation between its predicted and actual values was 9.65%.	The increase in pulse current resulted in a slight decrease in the thickness of the recast layer and the depth of the heat-affected zone, but with a rougher surface.
	Yakup Yildiz [74], 2016	The average error in predicting the thickness of the recast layer via the FEM was 1.98%.	This study conveniently relates the discharge current to white layer thickness and the MRR in EDM machining.
	Yadav and Pradhan [87], 2021	Numerical simulations predicted a value of 10.2 μm for the thickness of the recast layer, which is in good agreement with previous experimental results.	This method can be used for the RLT prediction for different materials and boundary conditions.

4. Hybrid Machining Simulations

Within EDM, the residual material fragments near the workpiece and the tool electrode tend to amass in the machining gap. This accumulation impedes the ionization process and leads to the generation of secondary discharge pulses [89]. These abnormal discharges, primarily short circuits and arcing, are the principal causes of this impediment. The excessive accumulation of debris can result in a low MRR, a high electrode wear rate, compromised surface integrity, pronounced microcracks, thick recast layers, the substantial dissipation of discharge energy, and the emission of harmful byproducts [90,91]. The development of effective strategies for purging or eliminating debris from the spark gap has emerged as a critical challenge in EDM technology. In the context of EDM, materials must exhibit a degree of electrical conductivity. While traditional EDM methods can effectively machine advanced conductive ceramics, the EDM of semiconductors and advanced non-conductive ceramics presents substantial challenges [92]. To tackle these challenges, researchers have devised hybrid process models. In the case of non-conductive materials, scientists construct an auxiliary conductive electrode on the surface of such materials to facilitate spark discharges. When it comes to machining novel materials, EDM has demonstrated noteworthy advantages [93]. Researchers have consistently explored the use of EDM with new materials [94], and the applications of EDM are on the rise.

4.1. Material Removal Rate

Given the complex effects of magnetic fields and ultrasonic vibrations on EDM, which add complexity to simulation modeling, many of the existing FEM models have resorted to simplifications. Magnetic-assisted EDM FEM models integrate the influences of magnetic fields into EDM models primarily by adjusting pertinent parameters within the electrical and thermal conduction equations to accommodate the physical impacts of the magnetic field. In Figure 10b, a schematic illustrates the debris propelled by magnetic forces within the machining gap. This magnetic barrier streamlines the process of removing machining debris emitted from the gap, making it more efficient. Bonny et al. investigated the influences of secondary electroconductive phases (comprising 40 vol% WC, TiCN, or TiN) on the EDM of ZrO_2 ceramic composites [95]. As depicted in Figure 10c, with a discharge current of 350 A, a pulse-on time of 2.4 μs, and a pulse cycle of 20 μs, the maximum MRR for ZrO_2-WC reached 33.54 mm^2/min. Figure 10d illustrates the trade-off of achieving the desired surface roughness at the expense of a reduced MRR, a pattern similar to that observed in the EDM of metallic materials [96]. Gupta and Joshi [97] formulated a mathematical model that accounts for the influence of plasma confinement under a magnetic field, the reduction in the mean free path, and the effects of magnetic pulsations. Within dry EDM, the magnetic field affects the dimensions of both single and multiple discharge craters. According to the model's predictions, under multi-spark conditions, the hole diameter decreased with the increasing magnetic field strength. However, at higher magnetic field values, the hole diameter increased due to the prevailing electrostatic repulsion force. In the context of single-spark discharge machining, the model's forecasts for crater diameters and depths closely approximated the corresponding experimental values, irrespective of the presence of a magnetic field. At moderate energy levels, the model exhibited an average error of 10% in predicting crater diameters and a 9% error in predicting depths for single spark discharges. Beravala and Pandey [98] formulated a mathematical model to predict the MRR in the context of EDM, considering the combined influences of air and magnetic fields. The developed mathematical model primarily accounted for three physical phenomena: the decrease in energy density resulting from plasma expansion, the increase in the MRR due to assistance from a liquid–gas medium, and the reduction in the electron mean free path due to the presence of a magnetic field. The MRR expression derived from the model was validated under all experimental conditions except those utilized to derive constants. The experimental results demonstrated that the prediction error of the model for MRRs was less than 10%.

The US-EDM finite element model builds upon the conventional EDM model by incorporating the physical effects of ultrasonic vibration. As illustrated in Figure 10a, a direct ultrasonic vibration-assisted method was employed, along with abrasives, to polish the conductive layer and craters of a workpiece, resulting in the attainment of a high-quality SR. In a study by Shabgard et al. [99], both experimental investigations and mathematical modeling were conducted to analyze the characteristics of EDM and US-EDM on AISI H13. Their findings revealed that the application of ultrasonic vibration to the workpiece significantly enhanced the MRR in the finish machining mode, tripling its value. Zhang et al. [100] introduced a hybrid machining technique that combines ultrasonic vibration and magnetic field assistance with electric spark wire-cutting. This approach aims to enhance machine tool performance. The study examined the effects of critical process parameters on the MRR and surface quality, including SR and surface crack density, in the context of Ti_6Al_4V machining. A comprehensive comparison revealed that the hybrid technique, combining ultrasonic vibration and magnetic field assistance with electric spark wire-cutting, continued to offer clear advantages in striking a balance between machining efficiency and surface quality.

In addition to the well-known MF-EDM and US-EDM, it is worth highlighting other noteworthy hybrid processes. H.K. Kansal et al. [44] developed a two-dimensional axisymmetric model for powder-mixed electrical discharge machining (PMEDM) using FEM models. The model incorporated various critical elements, such as temperature-dependent material properties and heat source characteristics (with a Gaussian heat distribution), to forecast the thermal dynamics and material removal mechanisms in the PMEDM process. A strong correlation (correlation coefficient = 0.91) was observed between the calculated MRR and the predicted MRR. The validation experiments were conducted using AISI D2 mold steel with a workpiece size of 100 mm × 50 mm × 10 mm. After adding graphite powder with an average particle size of 30 μm, the workpiece was processed in commercially available kerosene. Both in theoretical predictions and experimental validations, the MRR achieved through PMEDM consistently surpassed that of traditional EDM. This superiority was attributed to the presence of powder suspensions in the EDM dielectric, which promoted a uniform dispersion of thermal energy in all directions within the plasma channel.

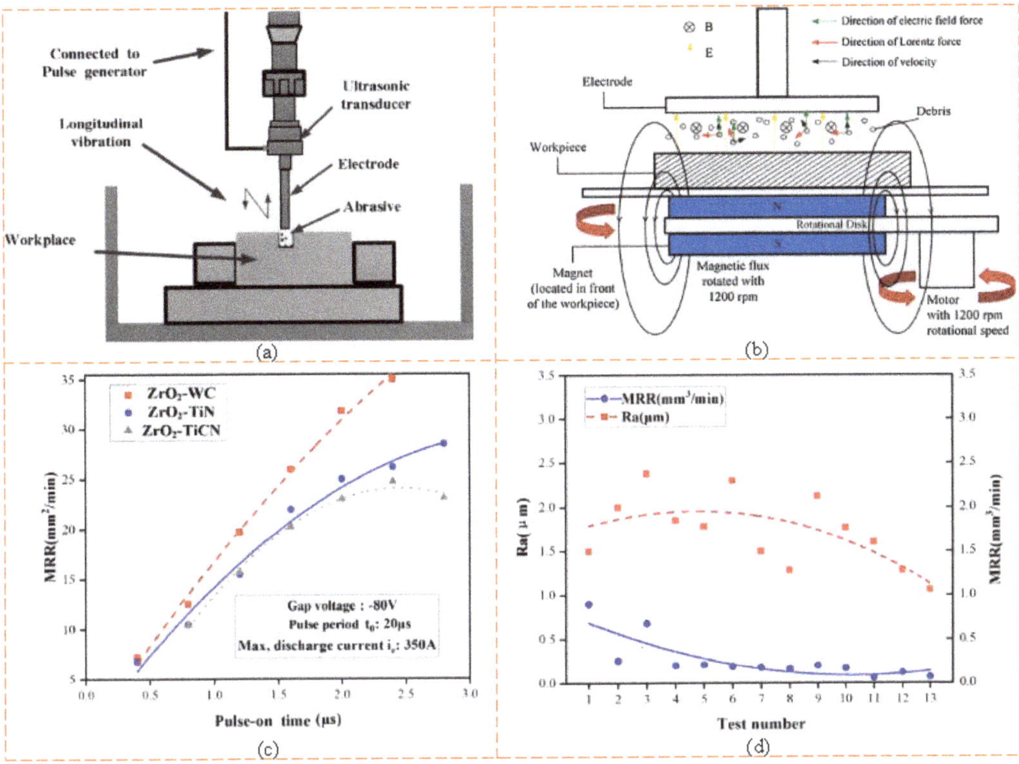

Figure 10. Prediction and analysis of MRR in hybrid machining simulations. (**a**) Schematic diagram of ultrasonic vibration-assisted EDM with abrasives, reproduced from [101], permission with J-STAGE, 2010. (**b**) Schematic diagram of the debris driven by the magnetic force in the machining gap, reproduced from [102], permission with Elsevier, 2008. (**c**) Relationship between MRR and pulse-on time (pulse duration) when machining advanced ZrO_2 ceramic composites with different secondary electroconductive phases, reproduced from [95], permission with Elsevier, 2008. (**d**) The relationship between MRR and Ra in each experiment on processing Si-SiC, reproduced from [96], permission with Elsevier, 2010.

4.2. Surface Roughness

The hybrid EDM process with intelligent optimization algorithms [103,104] offers a significant enhancement in surface quality, which holds critical importance for improving workpiece performance, reducing the need for secondary machining, increasing manufacturing efficiency, broadening application domains, and reducing the process's environmental impact [105]. Ming et al. [31] conducted a comprehensive study that amalgamated theoretical and experimental research to introduce an innovative thermophysical model for EDM. The model computes the energy flow and temperature distribution within workpieces, tool electrodes, and discharge channels for both single-spark and continuous-spark scenarios. During the experiment, a modifiable magnetic apparatus producing a magnetic force was positioned around the SKD11 workpiece. The machining temperatures at the upper and lower sections of the workpiece were tracked using thermocouples of the Pt100 type. Both simulation and experimental investigations demonstrate the superiority of magnetic field-assisted electrical discharge machining over conventional EDM, particularly in terms of surface integrity and discharge stability. Furthermore, the magnetic field effect enhances its energy efficiency by 15.2%. To achieve a balance between energy efficiency and Ra, the optimal combination of cutting parameters is as follows: a magnetic field strength of 0.2 T, a discharge current ranging from 7 to 9 A, and a pulse duration of 300 µs. As depicted in Figure 11a, with a pulse duration of 300 µs, the Ra achieved by hybrid EDM consistently surpasses that of conventional EDM. Ming et al. [106] carried out a comparative investigation of MF-EDM on both magnetic and non-magnetic materials. Initially, a series of MF-EDM experiments were conducted on specific magnetic materials, such as SKD11, and non-magnetic materials, like Ti_6Al_4V, to assess their surface quality. Subsequently, a comprehensive analysis was carried out to examine the impact of the magnetic field and servo voltage on the performance of MF-EDM machines, considering both SKD11 and Ti_6Al_4V materials. In the case of the SKD11 material, a significant increase in the servo voltage results in a greater surface roughness. Conversely, for the Ti_6Al_4V material, a servo voltage of 71V produces the desired SR.

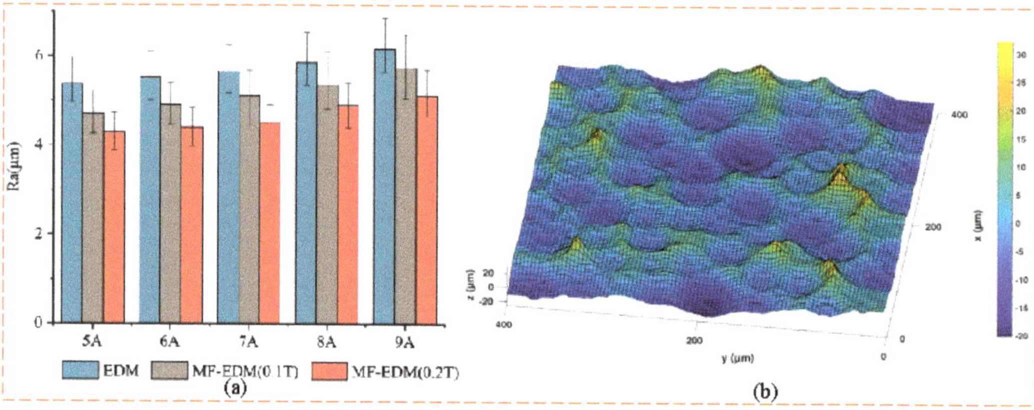

Figure 11. Prediction and analysis of SR in hybrid machining simulations. (**a**) Comparison of Ra between EDM and MF-EDM, reproduced from [31], permission with Elsevier, 2019. (**b**) Simulated surface topography of SS 304 at a 10 A, 50 µs, and 50 V parameter combination, reproduced from [76], permission with Elsevier, 2020.

Ultrasonic vibration enhances the surface quality of EDM. It can reduce the SR, micro-cracks, bubbles, and other defects, thereby enhancing the smoothness and uniformity of the machined surface [107]. Ghiculescu et al. [28] conducted a study on the modeling the material removal mechanisms at a microscopic geometrical level using computer-aided finite element analysis for ultrasonic-aided electrical discharge machining. The results of the finite element analysis revealed that accumulated micro-jets nearly eliminated the molten material caused by discharging, resulting in a 500% increase in the machining rate. The SR improved by up to 50%. Pupaza et al. [108] investigated the phenomena that occurring during the US-EDM of conductive metal matrix composites (EC-MMCs-Al/Al_2O_{3p}). Using finite element techniques, the study described the thermal shock phenomenon and its influence on the material removal process. The efficiency of US-EDM was lower during rough machining compared to finish machining. It was observed that by using shorter pulse durations and longer pause intervals to maximize the number of discharges within a single oscillation cycle, the use of ultrasonic vibration increased the machining rate and reduced the pulse energy to achieve a lower SR. In their study, Wang et al. [107] examined the mechanisms involved in the surface generation in ultrasonic vibration-assisted electrical EDM. They developed a microscopic material removal volume model for single-spark EDM based on the characteristics of the EDM removal mechanism and the influence of ultrasonic vibration energy. The study revealed that ultrasonic vibration led to an increased overlap of adjacent molten craters, resulting in a reduction in the carbon area and a smoother surface. The introduction of ultrasonics reduced the SR by 15.8% to 29.1%. Furthermore, Tang et al. [109] conducted a thermally and electrically coupled simulation of powder-mixed EDM on functionally graded SiC/Al materials, validating the results against experimental data. In comparison to conventional EDM, PMEDM improved the machining efficiency by 16.34% and reduced the SR by 29.42%. Figure 11b shows the surface topography simulation of the EDT performed on SS 304 at the 20 A, 50 μs, and 50 V parameter combination; it matches the measured surface morphology quite well.

4.3. Tool Wear Ratio

Tool wear is a prevalent and unavoidable phenomenon in EDM, with significant implications for machining precision, surface quality, and efficiency [110,111]. The monitoring and analysis of tool wear facilitate timely interventions to preserve machining quality and prolong the tool's lifespan. Ming et al. [31] combined theoretical and experimental research to introduce an innovative thermo-physical model for MF-EDM. The model computes the energy flow and temperature distribution within workpieces, tool electrodes, and discharge channels under both single-spark and continuous-spark conditions. Zhang et al. [112] introduced MF-EDM for the processing of high volume fraction SiCp/Al materials to improve its sustainable machining performance, including productivity and the most detrimental aerosol emissions to operators. The study investigated its effects on the TWR. The results showed that for a SiCp/Al material with a volume fraction of 65%, the average optimal solution reduced the TWR and aerosol emissions by around 5.7% and 10%, respectively, while the MRR increased by 5.5%. Figure 11c depicts the optimal Pareto surface for a 65% volume fraction. Namboodiri et al. [113] discovered that near-dry EDM holds promise as a technique for enhancing the TWR and machining performance, offering advantages over dry and wet EDM. Joudivand Sarand et al. [57] developed an extensive numerical model for tool wear using finite element modeling and reverse heat conduction techniques. This approach accounts for the independent and interactive effects of thermal and electrical parameters, modeling the contributing factors. Both experimental and numerical results demonstrated the significant impacts of parameters like the thermal diffusion coefficient of the tool electrode, the pulse current, and the pulse duration on tool wear. Nadda and Nirala [114] developed an analytical model for micro-EDM based on electrothermal theory and conducted finite element simulations of the Gaussian heat flux distribution during a single discharge in micro-EDM. They introduced a tool wear compensation technique based on the thermal model. When combined with the current pulse discrimination system, this approach may produce improved results. Figure 12a presents finite element simulation results for a capacitance and voltage of 27 nF and 90 V. Allen and Chen [115] utilized a thermal numerical model to simulate material removal in fine micro-EDM on molybdenum during a single-spark discharge. The model predicted that increasing the pulse duration would result in a decreased TWR, with molybdenum demonstrating significantly higher TWRs under identical machining conditions compared to steel. Figure 12b illustrates the temperature distribution after the heating phase.

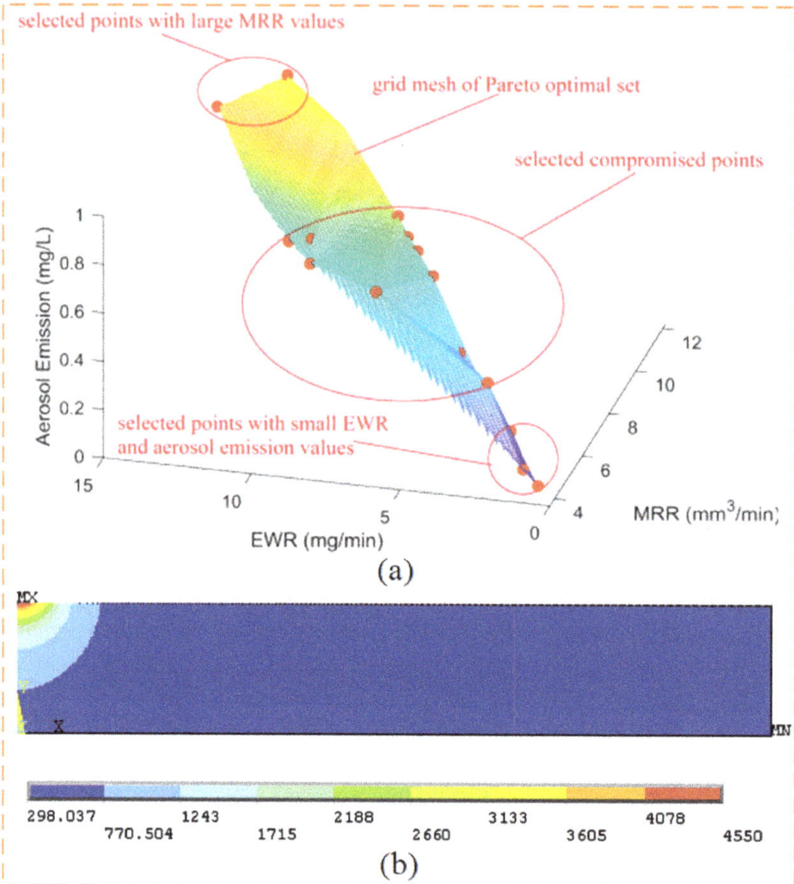

Figure 12. Prediction and analysis of TWR in hybrid machining simulations. (**a**) Optimal Pareto surface for 65% volume fraction SiCp/Al, reproduced from [112], permission with Elsevier, 2021. (**b**) Temperature (K) distribution after a heat flux (power = 30 W, pulse duration = 2 s), reproduced from [115], permission with Elsevier, 2007.

4.4. Recast Layer Thickness

The selection of the RLT in EDM requires careful consideration, taking into account specific machining requirements and material characteristics. In practical applications, various factors, including machining efficiency, surface quality, precision, electrode wear, and thermal effects, must be considered to achieve optimal machining results. Shabgard et al. [116] created an innovative mathematical model for the plasma channel radius and incorporated it into finite element modeling. They examined the impact of ultrasonic vibration on the RLT of the tool electrode under varying pulse currents and durations. A CHARMILLES ROBOFORM 200 EDM with an iso-pulse generator was used for the experiments and, in order to apply ultrasonic vibrations to the tool electrodes, a HD 2200 Bandelin ultrasonic vibration generator was used. The results showed that the numerical simulation of the RLT had a maximum error of 6.1% when compared to experimental data. The application of ultrasonic vibration reduced the RLT produced on the machined surface for all pulse currents and durations. Additionally, Shabgard et al. [61] introduced a finite element model to investigate the influence of applied magnetic fields on the pit size and the RLT during the EDM process. The results demonstrated good agreement between the numerical

RLT and experimental data, with a maximum error of 8.8%. Tan and Yeo [117] developed a numerical model to describe the process of powder-mixed fine micro-EDM. They employed a finite element analysis along with a multi-discharge-based approach. Verification revealed that the simulated RLT exhibited a consistent deviation from the measured values, although the overall trend was similar. Shabgard et al. [86] utilized a three-dimensional axisymmetric finite element model to estimate the temperature distribution during the EDM process for AISI H13 tool steel. Both numerical and experimental results revealed that an increase in the pulse duration resulted in a greater white layer thickness, increased heat-affected zone depth, and enhanced surface permeability. Conversely, an increase in pulse current marginally reduced the white layer thickness and heat-affected zone depth but led to coarser surface permeability. Liu and Guo [118] introduced an FEM for predicting and analyzing the mechanisms of the recast layer and heat-affected zone formation in the EDM of ASP 2023 tool steel, as well as the associated phase transformations. The predicted transformations of the white layer (WL) and heat-affected zone (HAZ) closely matched the experimental data, showing good accuracy. Figure 13a shows their predicted subsurface WL and HAZ. Both the solid white layers and heat-affected zones displayed non-uniform structures formed by multiple random discharges. Zhang et al. [119] elaborated on the complex thermal deformation phenomena and their underlying causes during the wire-cutting of thin-walled parts made of Inconel 718. They developed a thermophysical model and, by examining the surface characteristics of the deformed samples, discovered that the alterations in the RLT were fundamentally in line with the variations in thermal deformation. Figure 13b illustrates the relationship between the RLT and thermal deformation under these processing parameters.

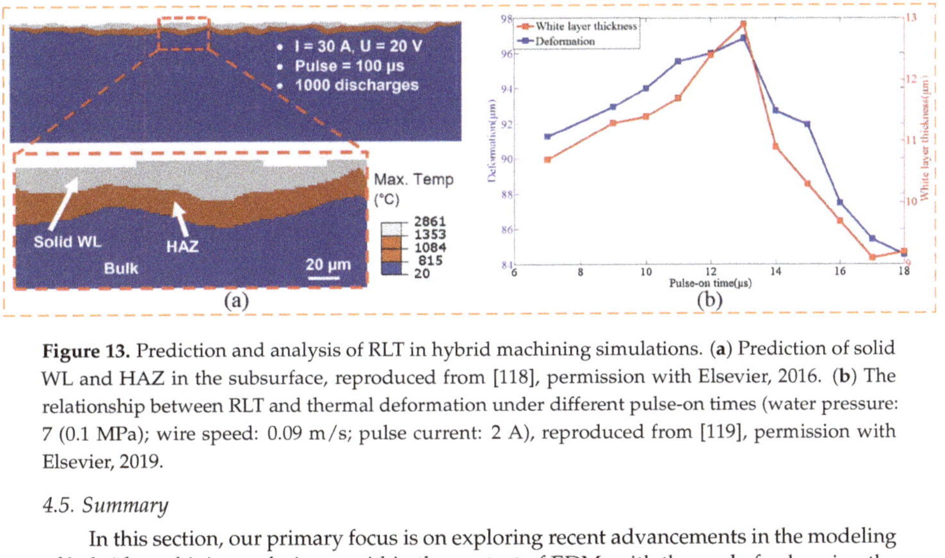

Figure 13. Prediction and analysis of RLT in hybrid machining simulations. (**a**) Prediction of solid WL and HAZ in the subsurface, reproduced from [118], permission with Elsevier, 2016. (**b**) The relationship between RLT and thermal deformation under different pulse-on times (water pressure: 7 (0.1 MPa); wire speed: 0.09 m/s; pulse current: 2 A), reproduced from [119], permission with Elsevier, 2019.

4.5. Summary

In this section, our primary focus is on exploring recent advancements in the modeling of hybrid machining techniques within the context of EDM, with the goal of enhancing the MRR, SR, TWR, and RLT. Researchers have found that hybrid EDM significantly improves machining performance, however, the machining principles are complex and difficult to model, so the models are mostly simplified. Challenges related to debris removal and machining non-conductive materials have persisted throughout the EDM process. To address these challenges, we have proposed various FEM models for hybrid machining techniques, encompassing models that incorporate magnetic field assistance, ultrasonic vibration assistance, and powder mixing. In summary, these hybrid machining technologies hold the potential to substantially enhance the processing efficiency and surface quality of EDM. Additionally, we present relevant simulation results, including the potential for hybrid machining to more than double the MRR and reduce Ra by over 20%. Table 3 offers a comparative overview of the major existing FEM models for the hybrid machining in EDM. Presently, the integration of hybrid machining techniques with traditional EDM shows significant potential for enhancing machining performance. However, modeling in this context still encounters certain challenges and necessitates further development, such as combining it with deep learning [120].

Table 3. Comparison of existing major hybrid machining simulations using FEM models for the process of EDM.

Purposes	Authors, Year	Findings	Remarks
MRR	H.K. Kansal et al. [44], 2008	In theory and experiment, the MRR values obtained by PMEDM were always higher than those of conventional EDM.	This model can be used to predict the evolution of temperatures, stresses, strains, and cracks that may occur on the surface of a PMEDM-machined workpiece.
	Shabgard et al. [99], 2011	The application of ultrasonic vibration to the workpiece had a significant effect on the MRR of the finishing method, increasing it three-fold.	The ultrasonic vibration of the workpiece changes the pressure in the gap, resulting in better flushing and improved machining stability.
	Zhang et al. [100], 2016	The hybrid technology of USV-MF-assisted EDM wire-cutting still offers significant advantages in balancing machining efficiency and surface quality.	The hybrid EDM-WEDM process has enormous advantages and application potential in the field of practical machining and manufacturing.
	Gupta and Joshi [97],	At moderate energy levels, the error in the predicted value of the mean diameter of the single-spark discharge crater was 10%, while the error in the mean depth was 9%.	This model neglects the effect of the self-magnetic field.
	Beravala and Pandey [98], 2020	The model developed predicts an MRR error of less than 10%.	The prediction error of this model is mainly caused by the assumed conditions and uncontrollable experimental errors.
Ra	Ming et al. [31], 2009	Simulation and experimental studies showed that the machining performance of MF-EDM was significantly better than that of EDM, especially in terms of surface integrity and discharge stability.	MF-EDM offers greater potential and advantages than conventional methods in improving machinability and surface quality.
	Ming et al. [106], 2020	When machining the SKD11 material, the SR increased when the servo voltage was increased excessively. However, when machining the Ti_6Al_4V material, when the servo voltage was 71V, the SR was satisfactory.	The energy efficiency of magnetic materials was significantly improved compared to that of non-magnetic materials.
	Ghiculescu et al. [28], 2009	Cumulative micro-jet flow almost eliminated the material melted by discharge, resulting in a 500% increase in the machining rate. SR was improved by up to 50%.	Future research will focus on synchronizing EDM and US generators and improving the flexibility of EDM+US technology.
	Pupaza et al. [108], 2015	A lower SR can be obtained by using lower values of pulse time and longer values of pause time.	Low pulse time values and high pause time values can be used to increase the machining rate.
	Wang et al. [107], 2020	Ultrasonic vibration caused more adjacent melting pits to overlap, resulting in a smaller area of carbon buildup and a smoother surface, and the introduction of ultrasound reduced the SR by 15.8–29.1%.	Crater depth and surface residual height decrease with increasing ultrasonic amplitude.
	Tang et al. [109], 2019	Compared with conventional EDM, PMEDM's machining efficiency increased by 16.34% and SR decreased by 29.42%.	This research provides a new way to improve efficiency in high-performance material processing and related fields.

Table 3. *Cont.*

Purposes	Authors, Year	Findings	Remarks
TWR	Ming et al. [31], 2009	Used to calculate the energy flow and temperature distribution of the workpiece, tool electrode, and discharge channel under single/continuous-spark conditions with good predictions.	The application of magnetic field-assisted technology in EDM also reduces the negative impact on the health of the operator.
	Zhang et al. [112], 2021	For 65% SiCp/Al, the average best solution for EWR and aerosol emissions was reduced by about 5.7% and 10%, respectively, while MRR increased by 5.5%.	MF-EDM outperforms conventional EDM in terms of energy efficiency and environmental impact.
	Joudivand Sarand et al. [57], 2017	Parameters such as the heat diffusion coefficient of the tool electrode, the pulse current, and the pulse duration had a significant effect on the degree of tool wear.	The strong dependence of this model improvement factor on the thermo-physical variables of the EDM process dramatically alters the numerical methods for accurately predicting the amount of tool wear.
	Nadda and Nirala [114], 2020	A finite element simulation of the Gaussian heat flow density distribution of a single discharge micro-EDM was performed, and a thermal modeling-based tool wear compensation technique was proposed.	This research can be used to establish a real-time tool wear monitoring and compensation system based on thermal modeling.
	Allen and Chen [115], 2007	The model estimated that the TWR decreased with increased pulse duration and was much higher for molybdenum than for steel under the same machining conditions.	Tensile residual stresses remaining on the workpiece may superimpose after several spark cycles, leading to surface damage such as microcracks.
RLT	Shabgard et al. [116], 2018	At all pulse currents and durations, ultrasonic vibration reduces the thickness of the recast layers produced on machined surfaces.	This study simulates the US-EDM process from an electrothermal perspective.
	Shabgard et al. [61], 2019	The numerical results of the RLT were in good agreement with the experimental results, with a maximum error of 8.8%.	Applying a magnetic field near the EDM gap increases the PFE while reducing the height of the RLT generated in all pulse currents and pulse-on time.
	Tan and Yeo [117], 2013	To validate the model, the simulated R max and RLT values were compared with their corresponding measured values. The trends of the simulated and measured results are approximately the same.	The process model takes into account the effect of powder particles on the plasma channel size and the proportion of heat flux distributed to the workpiece.
	Shabgard et al. [86], 2011	The increase in pulse current resulted in a slight decrease in the white layer thickness and heat-affected zone depth.	Each increase in discharge duration results in more heat being distributed into the workpiece, so more of the underlying material is exposed to high temperatures.
	Liu and Guo [118], 2016	The predicted WL and HAZ transformations of martensite match the experimental data with reasonable accuracy.	This provides us with the ability to predict macro-scale cumulative thermal damage during EDM processes.
	Zhang et al. [119], 2019	A thermophysical model was established, and by studying the surface characteristics of the deformed specimens, the changes in the thickness of the recast layer were consistent with the changes in their thermal deformation.	The change in the thickness of the recast layer almost corresponds to the change in its thermal deformation.

5. Outlooks

(1) In single-spark finite element modeling, a distinction is made between two categories: those that utilize Gaussian heat sources and those that do not. This differentiation

aims to offer a comprehensive analysis of how single-spark FEM models predict the four major machining performance indicators: the MRR, SR, TWR, and RLT. FEM models that employ Gaussian heat sources show more precise predictions of machining performance metrics. Future FEM models are anticipated to incorporate higher-precision Gaussian heat sources. Currently, single-spark models frequently assume the electric discharge to be in a plasma state, thereby overlooking the impact of the non-plasma state. Moreover, these models often neglect the influence of the microstructure on the electrode and workpiece surfaces during the EDM process. Future modeling efforts will explore the effects of the non-plasma state on the EDM process in greater depth and enable the models to consider the influence of the microstructure on the electrode and workpiece surfaces. Furthermore, considering that real EDM processes often involve continuous sequences of multiple pulse discharges, future research should concentrate on developing multi-spark models based on FEM models. In the current modeling framework, the four performance indicators—MRR, SR, TWR, and RLT—may display conflicting behaviors. Future modeling should consider the following aspects to simultaneously predict these four performance indicators and provide optimization strategies that take into account their interactions, thus saving both time and resources.

(2) Multi-spark FEM models in EDM, when compared to single-spark models, provide enhanced precision in capturing the stochastic nature of the EDM process and the effects of overlapping multiple discharges. Existing research suggests that multi-spark models demonstrate considerably smaller errors in predicting critical machining parameters compared to single-spark models, aligning more closely with real-world machining results. With the continuous advancement of computational technology, the growing adoption of Gaussian heat source distribution enables multi-spark models to effectively simulate intricate three-dimensional surface morphologies. In the future, these multi-spark models can be integrated with hot technologies like deep learning and artificial intelligence to develop intelligent models capable of accurately predicting the machining process. Furthermore, it is essential to expand the application of multi-spark models to novel materials such as ceramics and composite materials. Future directions for multi-spark models include dynamic adaptive simulations and the modeling of the process–performance relationship. Combining multi-spark models with optimization algorithms, online monitoring, and other technologies enables the closed-loop control and optimization of the EDM process. Additionally, the development of multi-spark models for machining intricate shapes and their integration into digital twin platforms can result in fully digitized and intelligent processes. Multi-spark models represent a critical developmental trend in numerical simulations of EDM.

(3) Various finite element models have been developed in research on hybrid EDM processes, encompassing models for ultrasonic vibrations, magnetic fields, auxiliary electrodes, and powder integration. Typically, these models are simplified for hybrid machining simulations. One future research direction is to develop hybrid models that integrate finite element methods and flow field simulations for a more precise simulation of hybrid EDM processes. Furthermore, a thorough investigation of the distinctions between the discharge states of magnetic and non-magnetic materials is necessary. In magnetic field-assisted machining, an effective discharge waveform sampling system is also required to monitor variations in the discharge ratio. Hybrid EDM encompasses multiple techniques, including magnetic fields, ultrasonic vibration, powder mixing, chemical etching, and laser ablation, which are fully utilized to address the limitations of the EDM process. These processes present further opportunities and potential while also promoting energy conservation and enhanced surface integrity. In the future, there is a need for the continued development of finite element modeling for emerging hybrid EDM technologies to facilitate efficient, high-quality, environmentally friendly, safe, and sustainable manufacturing. Given the inherent high hardness and brittleness of advanced ceramics, conventional machining techniques pose a risk of adversely affecting their surface. As a result, advanced ceramics are frequently machined using EDM. As the utilization of innovative materials, includ-

ing advanced ceramics, continues to rise, there is a burgeoning demand for improved finite element modeling studies. This is imperative for effectively addressing the diverse requirements of EDM's various applications.

Author Contributions: Conceptualization, G.Z. and G.L.; validation, W.X. and Y.Z.; writing—original draft preparation L.L. and S.S.; writing—review and editing Y.W., Y.X. and H.W.; supervision, L.L.; funding acquisition, G.Z. and G.L. All authors have read and agreed to the published version of the manuscript.

Funding: This work was supported by the Guangdong Basic and Applied Basic Research Foundation (No. 2022A1515140066) and by the Guangdong Provincial Key Laboratory of Manufacturing Equipment Digitization (No. 2023B1212060012). In addition, this work was also supported by the National Natural Science Foundation of China (No. 52105536).

Conflicts of Interest: The authors declare no conflict of interest.

References

1. Tilghman, B.C. Cutting Metal by Electricity. U.S. Patent 416,873, 10 December 1889.
2. Strobel, K. Electric Arc Cutting Machine. U.S. Patent 2007225, 7 September 1936.
3. Ho, K.H.; Newman, S.T. State of the Art Electrical Discharge Machining (EDM). *Int. J. Mach. Tools Manuf.* **2003**, *43*, 1287–1300. [CrossRef]
4. Das, S.; Klotz, M.; Klocke, F. EDM Simulation: Finite Element-Based Calculation of Deformation, Microstructure and Residual Stresses. *J. Mater. Process. Technol.* **2003**, *142*, 434–451. [CrossRef]
5. Mahajan, A.; Devgan, S.; Kalyanasundaram, D. Electrical Discharge Machining- a Futuristic Technique for Surface Engineering of Biomedical Alloys: A Concise Review. *Surf. Topogr. Metrol. Prop.* **2022**, *10*, 043001. [CrossRef]
6. Lu, Y.; Sun, P.; Yang, X.; Guo, X.; Li, X.; Ming, W. Review of Research Progress in Nontraditional Machining of Ultrahigh-Temperature Ceramic Matrix Composites. *Coatings* **2023**, *13*, 187. [CrossRef]
7. Liu, K.; Lauwers, B.; Reynaerts, D. Process Capabilities of Micro-EDM and Its Applications. *Int. J. Adv. Manuf. Technol.* **2010**, *47*, 11–19. [CrossRef]
8. Kumar, D.; Porwal, R.K. Recent Advances in Machining of Composite Materials by Electrical Discharge Machine. *Key Eng. Mater.* **2023**, *965*, 3–20. [CrossRef]
9. Pandey, A.K.; Anas, M. Sustainability and Recent Trends in Micro-Electric Discharge Machining (μ-EDM): A State-of-the-Art Review. *Mater. Today Proc.* **2022**, *57*, 2049–2055. [CrossRef]
10. Die Sinking and Wire Cutting Edm Market Size, Trends and Growth Opportunity, by Service, by Type, by Hole Size, by Price Range by Application, by Industry Vertical, by Region and Forecast to 2030. Available online: https://www.researchandmarkets.com/reports/5783625/global-die-sinking-wire-cutting-edm-market-size (accessed on 4 May 2023).
11. Ming, W.; Xie, Z.; Ma, J.; Du, J.; Zhang, G.; Cao, C.; Zhang, Y. Critical Review on Sustainable Techniques in Electrical Discharge Machining. *J. Manuf. Process.* **2021**, *72*, 375–399. [CrossRef]
12. Guo, X.; Tan, L.; Xie, Z.; Zhang, L.; Zhang, G.; Ming, W. Simulation and Experimentation of Renewable Dielectric Gap Flow Fields in EDM. *Int. J. Adv. Manuf. Technol.* **2023**, 170. [CrossRef]
13. Ming, W.; Zhang, S.; Zhang, G.; Du, J.; Ma, J.; He, W.; Cao, C.; Liu, K. Progress in Modeling of Electrical Discharge Machining Process. *Int. J. Heat Mass Transf.* **2022**, *187*, 122563. [CrossRef]
14. Joshi, S.N.; Pande, S.S. Development of an Intelligent Process Model for EDM. *Int. J. Adv. Manuf. Technol.* **2009**, *45*, 300–317. [CrossRef]
15. Descoeudres, A.; Hollenstein, C.; Waelder, G.; Demellayer, R.; Perez, R. Time- and Spatially-Resolved Characterization of Electrical Discharge Machining Plasma. *Plasma Sources Sci. Technol.* **2008**, *17*, 024008. [CrossRef]
16. He, W.; Wang, D.; Ming, W.; Ma, J.; Liu, K.; Du, J. Research Progress on Cutting Machining Simulation Technology of Metallic Glasses. *Int. J. Adv. Manuf. Technol.* **2022**, *122*, 1167–1193. [CrossRef]
17. Unses, E.; Cogun, C. Improvement of Electric Discharge Machining (EDM) Performance of Ti_6Al_4V Alloy with Added Graphite Powder to Dielectric. *Stroj. Vestn.-J. Mech. Eng.* **2015**, *61*, 409–418. [CrossRef]
18. Yu, Z.; Li, D.; Yang, J.; Zeng, Z.; Yang, X.; Li, J. Fabrication of Micro Punching Mold for Micro Complex Shape Part by Micro EDM. *Int. J. Adv. Manuf. Technol.* **2019**, *100*, 743–749. [CrossRef]
19. Bindya Devi, M.; Kumar Birru, A.; Kumar Bannaravuri, P. The Recent Trends of EDM Applications and Its Relevance in the Machining of Aluminium MMCs: A Comprehensive Review. *Mater. Today Proc.* **2021**, *47*, 6870–6873. [CrossRef]
20. Chekuri, R.B.R.; Eshwar, D.; Kotteda, T.K.; Srikanth Varma, R.S. Experimental and Thermal Investigation on Die-Sinking EDM Using FEM and Multi-Objective Optimization Using WOA-CS. *Sustain. Energy Technol. Assess.* **2022**, *50*, 101860. [CrossRef]
21. Yue, X.; Yang, X. Molecular Dynamics Simulation of Machining Properties of Polycrystalline Copper in Electrical Discharge Machining. *Proc. Inst. Mech. Eng. Part B J. Eng. Manuf.* **2019**, *233*, 371–380. [CrossRef]
22. Zhang, W. Research on the Influence of the Discharge Breakdown Process on the Crater Formation during EDM. *Multidiscip. Model. Mater. Struct.* **2021**, *17*, 841–860. [CrossRef]

23. Chen, Z.; Hu, S.; Zhang, S.; Zhang, Q.; Zhang, Z.; Ming, W. Simulation and Experimental Study on the Precision Molding of Irregular Vehicle Glass Components. *Micromachines* **2023**, *14*, 1974. [CrossRef]
24. Ming, W.; Zhang, Z.; Wang, S.; Huang, H.; Zhang, Y.; Zhang, Y.; Shen, D. Investigating the Energy Distribution of Workpiece and Optimizing Process Parameters during the EDM of Al6061, Inconel718, and SKD11. *Int. J. Adv. Manuf. Technol.* **2017**, *92*, 4039–4056. [CrossRef]
25. Singh, H. Experimental Study of Distribution of Energy during EDM Process for Utilization in Thermal Models. *Int. J. Heat Mass Transf.* **2012**, *55*, 5053–5064. [CrossRef]
26. Zhang, Y.; Liu, Y.; Shen, Y.; Li, Z.; Ji, R.; Cai, B. A Novel Method of Determining Energy Distribution and Plasma Diameter of EDM. *Int. J. Heat Mass Transf.* **2014**, *75*, 425–432. [CrossRef]
27. Li, Q.; Yang, X. Study on Arc Plasma Movement and Its Effect on Crater Morphology during Single-Pulse Discharge in EDM. *Int. J. Adv. Manuf. Technol.* **2020**, *106*, 5033–5047. [CrossRef]
28. Ghiculescu, D.; Marinescu, N.I.; Nanu, S. Modelling Aspects of Removal Mechanism at Ultrasonic Aided Electrodischarge Machining. *Int. J. Mater. Form.* **2009**, *2*, 685–688. [CrossRef]
29. Khatri, B.C.; Rathod, P.; Valaki, J.B. Ultrasonic Vibration–Assisted Electric Discharge Machining: A Research Review. *Proc. Inst. Mech. Eng. Part B J. Eng. Manuf.* **2016**, *230*, 319–330. [CrossRef]
30. Mansoor, S.M.; Raj, S.O.N.; Pal, V.K.; Roy, S. Finite Element Analysis of Magnetic Field Assisted Wire Electric Discharge Machine. *IOP Conf. Ser. Mater. Sci. Eng.* **2018**, *402*, 012051. [CrossRef]
31. Ming, W.; Zhang, Z.; Wang, S.; Zhang, Y.; Shen, F.; Zhang, G. Comparative Study of Energy Efficiency and Environmental Impact in Magnetic Field Assisted and Conventional Electrical Discharge Machining. *J. Clean. Prod.* **2019**, *214*, 12–28. [CrossRef]
32. Jithin, S.; Raut, A.; Bhandarkar, U.V.; Joshi, S.S. FE Modeling for Single Spark in EDM Considering Plasma Flushing Efficiency. *Procedia Manuf.* **2018**, *26*, 617–628. [CrossRef]
33. Chen, Y.; Hu, S.; Li, A.; Cao, Y.; Zhao, Y.; Ming, W. Parameters Optimization of Electrical Discharge Machining Process Using Swarm Intelligence: A Review. *Metals* **2023**, *13*, 839. [CrossRef]
34. Chao, Z.; Mingbo, Q.; Jing, Y.; Yunxiao, H.; Jiongbo, F. Influence of Electrode Temperature on Electric Erosion Pit Morphology during Cool Electrode EDM. *Int. J. Adv. Manuf. Technol.* **2017**, *93*, 2173–2181. [CrossRef]
35. Tariq Jilani, S.; Pandey, P.C. Analysis and Modelling of Edm Parameters. *Precis. Eng.* **1982**, *4*, 215–221. [CrossRef]
36. Tariq Jilani, S.; Pandey, P.C. An Analysis of Surface Erosion in Electrical Discharge Machining. *Wear* **1983**, *84*, 275–284. [CrossRef]
37. DiBitonto, D.D.; Eubank, P.T.; Patel, M.R.; Barrufet, M.A. Theoretical Models of the Electrical Discharge Machining Process. I. A Simple Cathode Erosion Model. *J. Appl. Phys.* **1989**, *66*, 4095–4103. [CrossRef]
38. Patel, M.R.; Barrufet, M.A.; Eubank, P.T.; DiBitonto, D.D. Theoretical Models of the Electrical Discharge Machining Process. II. The Anode Erosion Model. *J. Appl. Phys.* **1989**, *66*, 4104–4111. [CrossRef]
39. Salonitis, K.; Stournaras, A.; Stavropoulos, P.; Chryssolouris, G. Thermal Modeling of the Material Removal Rate and Surface Roughness for Die-Sinking EDM. *Int. J. Adv. Manuf. Technol.* **2009**, *40*, 316–323. [CrossRef]
40. Eubank, P.T.; Patel, M.R.; Barrufet, M.A.; Bozkurt, B. Theoretical Models of the Electrical Discharge Machining Process. III. The Variable Mass, Cylindrical Plasma Model. *J. Appl. Phys.* **1993**, *73*, 7900–7909. [CrossRef]
41. Joshi, S.N.; Pande, S.S. Thermo-Physical Modeling of Die-Sinking EDM Process. *J. Manuf. Process.* **2010**, *12*, 45–56. [CrossRef]
42. Ming, W.; Zhang, G.; Li, H.; Guo, J.; Zhang, Z.; Huang, Y.; Chen, Z. A Hybrid Process Model for EDM Based on Finite-Element Method and Gaussian Process Regression. *Int. J. Adv. Manuf. Technol.* **2014**, *74*, 1197–1211. [CrossRef]
43. Yeo, S.H.; Kurnia, W.; Tan, P.C. Electro-Thermal Modelling of Anode and Cathode in Micro-EDM. *J. Phys.-Appl. Phys.* **2007**, *40*, 2513–2521. [CrossRef]
44. Kansal, H.K.; Singh, S.; Kumar, P. Numerical Simulation of Powder Mixed Electric Discharge Machining (PMEDM) Using Finite Element Method. *Math. Comput. Model.* **2008**, *47*, 1217–1237. [CrossRef]
45. Singh, M.; Sharma, S.; Ramkumar, J. Numerical Simulation of Melt-Pool Hydrodynamics in μ-EDM Process. *Procedia CIRP* **2020**, *95*, 226–231. [CrossRef]
46. Wang, Z.L.; Xie, B.C.; Wang, Y.K.; Zhao, W.S. Numerical Simulation of Cathode Erosion in EDM Process. *Adv. Mater. Res.* **2012**, *462*, 109–115. [CrossRef]
47. Kuriachen, B.; Varghese, A.; Somashekhar, K.P.; Panda, S.; Mathew, J. Three-Dimensional Numerical Simulation of Microelectric Discharge Machining of Ti_6Al_4V. *Int. J. Adv. Manuf. Technol.* **2015**, *79*, 147–160. [CrossRef]
48. Gadelmawla, E.S.; Koura, M.M.; Maksoud, T.M.A.; Elewa, I.M.; Soliman, H.H. Roughness Parameters. *J. Mater. Process. Technol.* **2002**, *123*, 133–145. [CrossRef]
49. Marafona, J.; Chousal, J.A.G. A Finite Element Model of EDM Based on the Joule Effect. *Int. J. Mach. Tools Manuf.* **2006**, *46*, 595–602. [CrossRef]
50. Wandra, R. Finite Element Analysis of Powder Mixed Electric Discharge Machining on Ti-Alloy. *Mater. Today Proc.* **2022**, *50*, 593–599. [CrossRef]
51. Chen, Z.; Xie, J.; Zhu, X. Experimental Study on Surface Physical Behaviour and Integrity of Polycrystalline Diamond (PCD) in Micron-Scale Abrasive Flow Assisted Impulse-Discharge Machining. *Int. J. Refract. Met. Hard Mater.* **2023**, *116*, 106330. [CrossRef]
52. Algodi, S.J.; Clare, A.T.; Brown, P.D. Modelling of Single Spark Interactions during Electrical Discharge Coating. *J. Mater. Process. Technol.* **2018**, *252*, 760–772. [CrossRef]

53. Assarzadeh, S.; Ghoreishi, M. Prediction of Root Mean Square Surface Roughness in Low Discharge Energy Die-Sinking EDM Process Considering the Effects of Successive Discharges and Plasma Flushing Efficiency. *J. Manuf. Process.* **2017**, *30*, 502–515. [CrossRef]
54. Philip, J.T.; Mathew, J.; Kuriachen, B. Numerical Simulation of the Effect of Crater Morphology for the Prediction of Surface Roughness on Electrical Discharge Textured Ti_6Al_4V. *J. Braz. Soc. Mech. Sci. Eng.* **2020**, *42*, 248. [CrossRef]
55. George, J.; Mathew, J.; Manu, R. Determination of Crater Morphology and 3D Surface Roughness in Wire Electrical Discharge Turning of Inconel 825. *Arab. J. Sci. Eng.* **2020**, *45*, 5109–5127. [CrossRef]
56. Li, Z.; Liu, Y.; Cao, B.; Li, W. Modeling of Material Removal Morphology and Prediction of Surface Roughness Based on WEDM Successive Discharges. *Int. J. Adv. Manuf. Technol.* **2022**, *120*, 2015–2029. [CrossRef]
57. Joudivand Sarand, M.; Shabgard, M.; Jodari Saghaie, M.; Joudivand Sarand, L. Tool Wear Estimating Model Considering Modeling Improvement Factors in Electrical Discharge Machining Process Based on Physical Properties of Tool Electrodes. *Proc. Inst. Mech. Eng. Part B J. Eng. Manuf.* **2017**, *231*, 850–866. [CrossRef]
58. Papazoglou, E.L.; Karmiris-Obratański, P.; Leszczyńska-Madej, B.; Markopoulos, A.P. A Study on Electrical Discharge Machining of Titanium Grade2 with Experimental and Theoretical Analysis. *Sci. Rep.* **2021**, *11*, 8971. [CrossRef] [PubMed]
59. Vignesh, M.; Ramanujam, R. Numerical Modelling and Experimental Validation of Crater Formation in WEDM Hybrid Turning of Ti_6Al_4V Alloy. *Proc. Inst. Mech. Eng. Part E J. Process Mech. Eng.* **2021**, *235*, 392–404. [CrossRef]
60. Pandey, P.C.; Jilani, S.T. Plasma Channel Growth and the Resolidified Layer in Edm. *Precis. Eng.* **1986**, *8*, 104–110. [CrossRef]
61. Shabgard, M.R.; Gholipoor, A.; Mohammadpourfard, M. Investigating the Effects of External Magnetic Field on Machining Characteristics of Electrical Discharge Machining Process, Numerically and Experimentally. *Int. J. Adv. Manuf. Technol.* **2019**, *102*, 55–65. [CrossRef]
62. George, J.; Philip, J.T.; Mathew, J.; Manu, R. Prediction and Analysis of Material Removal Rate and White Layer Thickness during Wire Electrical Discharge Turning (WEDT) Process. *CIRP J. Manuf. Sci. Technol.* **2022**, *39*, 210–222. [CrossRef]
63. Zhang, G.; Huang, H.; Zhang, Z.; Zhang, Y. Study on the Effect of Three Dimensional Wire Vibration on WEDM Based on a Novel Thermophysical Model. *Int. J. Adv. Manuf. Technol.* **2019**, *100*, 2089–2101. [CrossRef]
64. Gholipoor, A.; Shabgard, M.R.; Mohammadpourfard, M. A Novel Approach to Plasma Channel Radius Determination and Numerical Modeling of Electrical Discharge Machining Process. *J. Braz. Soc. Mech. Sci. Eng.* **2020**, *42*, 185. [CrossRef]
65. Shahane, S.; Pande, S.S. Development of a Thermo-Physical Model for Multi-Spark Wire EDM Process. *Procedia Manuf.* **2016**, *5*, 205–219. [CrossRef]
66. Banerjee, S.; Prasad, B. Numerical Evaluation of Transient Thermal Loads on a WEDM Wire Electrode under Spatially Random Multiple Discharge Conditions with and without Clustering of Sparks. *Int. J. Adv. Manuf. Technol.* **2010**, *48*, 571–580. [CrossRef]
67. Erdem, O.; Cogun, C.; Uslan, I.; Erbas, M. Thermo-Fluid Multi-Physics Modeling and Experimental Verification of Volumetric Workpiece Material Removal by a Discharge Pulse in Electric Discharge Machining Process. *J. Phys. D Appl. Phys.* **2020**, *53*, 395501. [CrossRef]
68. Hinduja, S.; Kunieda, M. Modelling of ECM and EDM Processes. *CIRP Ann.* **2013**, *62*, 775–797. [CrossRef]
69. Izquierdo, B.; Sánchez, J.A.; Plaza, S.; Pombo, I.; Ortega, N. A Numerical Model of the EDM Process Considering the Effect of Multiple Discharges. *Int. J. Mach. Tools Manuf.* **2009**, *49*, 220–229. [CrossRef]
70. Philip, J.T.; Kuriachen, B.; Mathew, J. Numerical Modelling and Simulation of Single and Multi-spark Impacts in Electrical Discharge Machining. In *Advances in Simulation, Product Design and Development: Proceedings of AIMTDR 2018*; Springer: Singapore, 2019; pp. 479–488.
71. Saikiran, H.; Kaushal, I.V.; Chandra, M.A. Experimental and numerical analysis of metal removal rate and surface finish on orvar supreme. In Proceedings of the 38th IRF International Conference, Chennai, India, 20 March 2016.
72. Almacinha, J.A.; Lopes, A.M.; Rosa, P.; Marafona, J.D. How Hydrogen Dielectric Strength Forces the Work Voltage in the Electric Discharge Machining. *Micromachines* **2018**, *9*, 240. [CrossRef] [PubMed]
73. Razeghiyadaki, A.; Molardi, C.; Talamona, D.; Perveen, A. Modeling of Material Removal Rate and Surface Roughness Generated during Electro-Discharge Machining. *Machines* **2019**, *7*, 47. [CrossRef]
74. Yildiz, Y. Prediction of White Layer Thickness and Material Removal Rate in Electrical Discharge Machining by Thermal Analyses. *J. Manuf. Process.* **2016**, *23*, 47–53. [CrossRef]
75. Kurnia, W.; Tan, P.C.; Yeo, S.H.; Tan, Q.P. Surface Roughness Model for Micro Electrical Discharge Machining. *Proc. Inst. Mech. Eng. Part B J. Eng. Manuf.* **2009**, *223*, 279–287. [CrossRef]
76. Jithin, S.; Raut, A.; Bhandarkar, U.V.; Joshi, S.S. Finite Element Model for Topography Prediction of Electrical Discharge Textured Surfaces Considering Multi-Discharge Phenomenon. *Int. J. Mech. Sci.* **2020**, *177*, 105604. [CrossRef]
77. Abed, F.N.; Ramesh, V.; Fadhil Jwaid, M.; Agarwal, N.; Koundal, D.; Mohamed Ibrahim, A. Enhancement Modelling Based on Electrical Discharge Machining Successive Discharges. *Adv. Mater. Sci. Eng.* **2022**, *2022*, e8017375. [CrossRef]
78. Jamunkar, T.; Sundaram, M. Prediction of Surface Roughness on Electrical Discharge Machined Surfaces. *Manuf. Lett.* **2022**, *33*, 143–152. [CrossRef]
79. Jithin, S.; Bhandarkar, U.V.; Joshi, S.S. Multi-Spark Model for Predicting Surface Roughness of Electrical Discharge Textured Surfaces. *Int. J. Adv. Manuf. Technol.* **2020**, *106*, 3741–3758. [CrossRef]
80. Puertas, I.; Luis, C.J.; Álvarez, L. Analysis of the Influence of EDM Parameters on Surface Quality, MRR and EW of WC–Co. *J. Mater. Process. Technol.* **2004**, *153–154*, 1026–1032. [CrossRef]

81. Kim, D.; Chen, X.; Allen, P. Micro-Electro Discharge Drilling Characteristics on Molybdenum Aperture Fabrication. *Int. J. Mach. Mach. Mater.* **2010**, *7*, 161–175. [CrossRef]
82. Barenji, R.V.; Pourasl, H.H.; Khojastehnezhad, V.M. Electrical Discharge Machining of the AISI D6 Tool Steel: Prediction and Modeling of the Material Removal Rate and Tool Wear Ratio. *Precis. Eng.* **2016**, *45*, 435–444. [CrossRef]
83. Bellotti, M.; Wu, M.; Qian, J.; Reynaerts, D. Tool Wear and Material Removal Predictions in Micro-EDM Drilling: Advantages of Data-Driven Approaches. *Appl. Sci.* **2020**, *10*, 6357. [CrossRef]
84. Ramasawmy, H.; Blunt, L.; Rajurkar, K.P. Investigation of the Relationship between the White Layer Thickness and 3D Surface Texture Parameters in the Die Sinking EDM Process. *Precis. Eng.* **2005**, *29*, 479–490. [CrossRef]
85. Tan, P.C.; Yeo, S.H. Modeling of Recast Layer in Micro-Electrical Discharge Machining. *J. Manuf. Sci. Eng.* **2010**, *132*, 031001. [CrossRef]
86. Shabgard, M.; Oliaei, S.N.B.; Seyedzavvar, M.; Najadebrahimi, A. Experimental Investigation and 3D Finite Element Prediction of the White Layer Thickness, Heat Affected Zone, and Surface Roughness in EDM Process. *J. Mech. Sci. Technol.* **2011**, *25*, 3173–3183. [CrossRef]
87. Yadav, K.; Pradhan, S.K. Analysis of Recast Layer Thickness and Martensitic Layer Thickness in Wire Electrical Discharge Machining. *Mater. Today Proc.* **2021**, *46*, 10799–10805. [CrossRef]
88. Rajurkar, K.P.; Pandit, S.M. Quantitative Expressions for Some Aspects of Surface Integrity of Electro Discharge Machined Components. *J. Eng. Ind.* **1984**, *106*, 171–177. [CrossRef]
89. Cao, M.R.; Yang, S.Q.; Li, W.H.; Yang, S.C. Chip-Ejection Mechanism and Experimental Study of Water Dispersant Dielectric Fluid on Small-Hole EDM. *Adv. Mater. Res.* **2010**, *97–101*, 4111–4115. [CrossRef]
90. Zhang, Z.; Zhang, Y.; Ming, W.; Zhang, Y.; Cao, C.; Zhang, G. A Review on Magnetic Field Assisted Electrical Discharge Machining. *J. Manuf. Process.* **2021**, *64*, 694–722. [CrossRef]
91. Zhang, Z.; Wei, S.; Wang, P.; Qiu, W.; Zhang, G. Progress in Applications of Laser Induced Cavitation on Surface Processing. *Opt. Laser Technol.* **2024**, *170*, 110212. [CrossRef]
92. Ming, W.; Jia, H.; Zhang, H.; Zhang, Z.; Liu, K.; Du, J.; Shen, F.; Zhang, G. A Comprehensive Review of Electric Discharge Machining of Advanced Ceramics. *Ceram. Int.* **2020**, *46*, 21813–21838. [CrossRef]
93. Jatti, V.S.; Khedkar, N.K.; Jatti, A.V.; Jatti, V.S. Finite Element Simulation of Electric Discharge Machining Process during Machining of NiTi, NiCu and BeCu Alloys. *Mater. Today Proc.* **2023**. [CrossRef]
94. Shen, D.; Ming, W.; Ren, X.; Xie, Z.; Liu, X. Progress in Non-Traditional Processing for Fabricating Superhydrophobic Surfaces. *Micromachines* **2021**, *12*, 1003. [CrossRef]
95. Bonny, K.; De Baets, P.; Vleugels, J.; Salehi, A.; Van der Biest, O.; Lauwers, B.; Liu, W. Influence of Secondary Electro-Conductive Phases on the Electrical Discharge Machinability and Frictional Behavior of ZrO2-Based Ceramic Composites. *J. Mater. Process. Technol.* **2008**, *208*, 423–430. [CrossRef]
96. Clijsters, S.; Liu, K.; Reynaerts, D.; Lauwers, B. EDM Technology and Strategy Development for the Manufacturing of Complex Parts in SiSiC. *J. Mater. Process. Technol.* **2010**, *210*, 631–641. [CrossRef]
97. Gupta, A.; Joshi, S.S. Modelling Effect of Magnetic Field on Material Removal in Dry Electrical Discharge Machining. *Plasma Sci. Technol.* **2017**, *19*, 025505. [CrossRef]
98. Beravala, H.; Pandey, P.M. Modelling of Material Removal Rate in the Magnetic Field and Air-Assisted Electrical Discharge Machining. *Proc. Inst. Mech. Eng. Part C J. Mech. Eng. Sci.* **2020**, *234*, 1286–1297. [CrossRef]
99. Shabgard, M.R.; Sadizadeh, B.; Amini, K.; Pourziaie, H. Comparative Study and Mathematical Modeling of Machining Parameters in Ultrasonic-Assisted EDM of AISI H13 Tool Steel by the Application of Workpiece Vibration. *Adv. Mater. Res.* **2011**, *154–155*, 1604–1613. [CrossRef]
100. Zhang, Z.; Huang, H.; Ming, W.; Xu, Z.; Huang, Y.; Zhang, G. Study on Machining Characteristics of WEDM with Ultrasonic Vibration and Magnetic Field Assisted Techniques. *J. Mater. Process. Technol.* **2016**, *234*, 342–352. [CrossRef]
101. Praneetpongrung, C.; Fukuzawa, Y.; Nagasawa, S.; Yamashita, K. Effects of the Edm Combined Ultrasonic Vibration on the Machining Properties of Si3N4. *Mater. Trans.* **2010**, *51*, 2113–2120. [CrossRef]
102. Lin, Y.-C.; Lee, H.-S. Machining Characteristics of Magnetic Force-Assisted EDM. *Int. J. Mach. Tools Manuf.* **2008**, *48*, 1179–1186. [CrossRef]
103. Ming, W.; Shen, F.; Zhang, G.; Liu, G.; Du, J.; Chen, Z. Green Machining: A Framework for Optimization of Cutting Parameters to Minimize Energy Consumption and Exhaust Emissions during Electrical Discharge Machining of Al 6061 and SKD 11. *J. Clean. Prod.* **2021**, *285*, 124889. [CrossRef]
104. Ma, J.; Yin, C.; Li, X.; Han, X.; Ming, W.; Chen, S.; Cao, Y.; Liu, K. Optimization of EDM Process Parameters Based on Variable-Fidelity Surrogate Model. *Int. J. Adv. Manuf. Technol.* **2022**, *122*, 2031–2041. [CrossRef]
105. Ming, W.; Xie, Z.; Cao, C.; Liu, M.; Zhang, F.; Yang, Y.; Zhang, S.; Sun, P.; Guo, X. Research on EDM Performance of Renewable Dielectrics under Different Electrodes for Machining SKD11. *Crystals* **2022**, *12*, 291. [CrossRef]
106. Ming, W.; Shen, F.; Zhang, Z.; Huang, H.; Du, J.; Wu, J. A Comparative Investigation on Magnetic Field–Assisted EDM of Magnetic and Non-Magnetic Materials. *Int. J. Adv. Manuf. Technol.* **2020**, *109*, 1103–1116. [CrossRef]
107. Wang, Y.; Liu, Z.; Shi, J.; Dong, Y.; Yang, S.; Zhang, X.; Lin, B. Analysis of Material Removal and Surface Generation Mechanism of Ultrasonic Vibration–Assisted EDM. *Int. J. Adv. Manuf. Technol.* **2020**, *110*, 177–189. [CrossRef]

108. Pupaza, C.; Ghiculescu, D.; Opran, C.G. A Model Regarding Material Removal Using FEM at Ultrasonic Aided EDM of Metal Matrix Composites. *Appl. Mech. Mater.* **2015**, *760*, 557–562. [CrossRef]
109. Tang, L.; Ji, Y.; Ren, L.; Zhai, K.G.; Huang, T.Q.; Fan, Q.M.; Zhang, J.J.; Liu, J. Thermo-Electrical Coupling Simulation of Powder Mixed EDM SiC/Al Functionally Graded Materials. *Int. J. Adv. Manuf. Technol.* **2019**, *105*, 2615–2628. [CrossRef]
110. Du, J.; Tian, B.; Duan, L.; Ming, W.; Liu, K.; He, W. Traditional and Non-Traditional Machining Technology of Metallic Glass. *Int. J. Adv. Manuf. Technol.* **2022**, *120*, 2879–2902. [CrossRef]
111. Chen, Y.; Guo, X.; Zhang, G.; Cao, Y.; Shen, D.; Li, X.; Zhang, S.; Ming, W. Development of a Hybrid Intelligent Process Model for Micro-Electro Discharge Machining Using the TTM-MDS and Gaussian Process Regression. *Micromachines* **2022**, *13*, 845. [CrossRef] [PubMed]
112. Zhang, Z.; Zhang, Y.; Lin, L.; Wu, J.; Yu, H.; Pan, X.; Li, G.; Wu, J.; Xue, T. Study on Productivity and Aerosol Emissions of Magnetic Field-Assisted EDM Process of SiCp/Al Composite with High Volume Fractions. *J. Clean. Prod.* **2021**, *292*, 126018. [CrossRef]
113. Namboodiri, V.N.J.; Pazhamannil, R.V.; Govindan, P. Finite Element Based Numerical Modelling of Near-Dry EDM with Glycerin-Air Dielectric Mixture. *IOP Conf. Ser. Mater. Sci. Eng.* **2022**, *1248*, 012019. [CrossRef]
114. Nadda, R.; Nirala, C.K. Thermal Modeling of Single Discharge in Prospect of Tool Wear Compensation in μEDM. *Int. J. Adv. Manuf. Technol.* **2020**, *107*, 4573–4595. [CrossRef]
115. Allen, P.; Chen, X. Process Simulation of Micro Electro-Discharge Machining on Molybdenum. *J. Mater. Process. Technol.* **2007**, *186*, 346–355. [CrossRef]
116. Shabgard, M.R.; Gholipoor, A.; Mohammadpourfard, M. Numerical and Experimental Study of the Effects of Ultrasonic Vibrations of Tool on Machining Characteristics of EDM Process. *Int. J. Adv. Manuf. Technol.* **2018**, *96*, 2657–2669. [CrossRef]
117. Tan, P.C.; Yeo, S.H. Simulation of Surface Integrity for Nanopowder-Mixed Dielectric in Micro Electrical Discharge Machining. *Met. Mater. Trans. B* **2013**, *44*, 711–721. [CrossRef]
118. Liu, J.F.; Guo, Y.B. Modeling of White Layer Formation in Electric Discharge Machining (EDM) by Incorporating Massive Random Discharge Characteristics. *Procedia CIRP* **2016**, *42*, 697–702. [CrossRef]
119. Zhang, Y.; Guo, S.; Zhang, Z.; Huang, H.; Li, W.; Zhang, G.; Huang, Y. Simulation and Experimental Investigations of Complex Thermal Deformation Behavior of Wire Electrical Discharge Machining of the Thin-Walled Component of Inconel 718. *J. Mater. Process. Technol.* **2019**, *270*, 306–322. [CrossRef]
120. Chen, X.; Wu, Y.; He, X.; Ming, W. A Comprehensive Review of Deep Learning-Based PCB Defect Detection. *IEEE Access* **2023**, *11*, 139017–139038. [CrossRef]

metals

MDPI

Review

Parameters Optimization of Electrical Discharge Machining Process Using Swarm Intelligence: A Review

Yanyan Chen [1], Shunchang Hu [2,3], Ansheng Li [2,3,*], Yang Cao [3], Yangjing Zhao [3] and Wuyi Ming [3,4,*]

[1] College of Mechanical Engineering, Yellow River Conservancy Technical Institute, Kaifeng 475000, China; chyy1106@126.com

[2] Hydropower Equipment and Intelligent System Engineering Technology Research Centre of Henan Province, Department of Mechanical and Electronic Engineering, Henan Vocational College of Water Conservancy and Environment, Zhengzhou 450002, China; hushunchang1001@gmail.com

[3] Mechanical and Electrical Engineering Institute, Zhengzhou University of Light Industry, Zhengzhou 450002, China; caoyang8786@zzuli.edu.cn (Y.C.); zhaoyj0529@163.com (Y.Z.)

[4] Guangdong Provincial Key Laboratory of Digital Manufacturing Equipment, Guangdong HUST Industrial Technology Research Institute, Dongguan 523808, China

* Correspondence: liansheng@zzuli.edu.cn (A.L.); mingwuyi@zzuli.edu.cn (W.M.)

Abstract: Electrical discharge machining (EDM) can use soft tool electrodes to process hard workpieces to achieve "soft against hard", because it directly uses electrical energy and thermal energy to remove metal materials. Then, it can generate complex features on harder materials and meet the requirements of excellent surface quality. Since EDM involves many process parameters, including electrical parameters, non-electrical parameters, and materials properties, it is essential to optimize its process parameters to obtain good performance. In this direction, the application of the swarm intelligence (SI) technique has become popular. In this paper, the existing literature is comprehensively reviewed, and the application of the SI technique in the optimization of EDM process parameters is summarized. Sinker-EDM (SEDM), wire-EDM (WEDM), and micro-EDM (MEDM) with various hybrid techniques are among the EDM methods considered in this study because of their broad adoption in industrial sections. The fundamental nature of all review articles will assist engineers/workers in determining the process parameters and processing performance, the SI algorithm, and the optimal technique by which to obtain the desired process parameters. In addition, discussions from the perspectives of the similarity, individuality, and complementarity of various SI algorithms are proposed, and necessary outlooks are predicted, which provides references for the high performance of the EDM processes in the future.

Keywords: EDM process; SI; optimization; parameter; performance

1. Introduction

The tool electrode and the workpiece are submerged in the dielectric during electrical discharge machining (EDM), and they are linked to the two poles of the pulse power source. A large amount of heat energy is concentrated in the tiny discharge channel, the temperature may be as high as 10,000 °C, and the pressure also sharply changes [1–3]. Therefore, at this time, a small amount of metal material on the surface of the workpiece in the discharge area will immediately melt, vaporize, and explode into the working fluid. The material removal in the electric discharge is realized by the electric heating in the discharge process, and the conductivity and thermal properties of the material have a significant impact on its machinability. Therefore, low-rigidity workpieces may be machined and micro-machined particularly well using the EDM technique. In addition, it is particularly suitable for machining a workpiece with a complex surface, because the tool electrode shape can be copied to the workpiece. Since EDM involves many process parameters, including electrical parameters, non-electrical parameters, and material properties, it is essential to optimize its process parameters to obtain good performance [4]. In this direction, the application of the swarm intelligence (SI) technique has become popular.

Ant colonies, bird flocking, bee colonies, eagle hunting, grazing, fish colonies, bacterial development, etc., are all examples of SI in nature. SI is derived from research on the collective behavior of social insects such as ants and bees. A swarm intelligence system is often made up of a collection of simple agents or clusters that interact one with another and with surroundings. Swarm intelligence has been formally proposed as a theory and has steadily caught the attention of a significant number of researchers since Dorigo [5] put forward the Ant Colony Optimization (ACO) hypothesis in 1991, starting a research movement. Kennedy et al. [6] presented the Particle Swarm Optimization (PSO) in 1995. Since then, research on swarm intelligence has rapidly developed, and relevant technologies can be applied in various fields, such as aerospace [7,8], precision medicine [9], and advanced manufacturing [10,11].

This review focuses on the progress in parameters optimization of the EDM process using swarm intelligence. First, the principle of different swarm intelligence techniques is critically reviewed in Section 2, including the ACO, the PSO, the genetic algorithm (GA), the Artificial Bee Colony (ABC), the Glowworm Swarm Optimization (GSO), the Cuckoo Search Algorithm (CSA), the Differential Evolution Algorithm (DEA), and others. Second, the application of the swarm intelligence technique in the optimization of EDM process parameters is summarized, and the sinker-EDM (SEDM), wire-EDM (WEDM), and micro-EDM (MEDM) with various hybrid techniques are reviewed in Sections 3, 4, and 5, respectively. Third, discussions from the perspectives of similarity, individuality, and complementarity of various SI algorithms characteristics are proposed. Finally, necessary outlooks (involving limitations and challenges) are predicted, which provide references for the high performance of the EDM processes in the future.

2. Principal of Swarm Intelligence

2.1. ACO

Ants seek food sources, and there are numerous routes to the food source. As they select a way, they emit a pheromone. This pheromone's concentration decreases as time passes. As a result, the number of ants with shorter paths will grow, as will the pheromone content. The ants will take the path with the most pheromones. With the passage of time, the number of ants using shorter paths will increase. Then, corresponding is the best answer to the problem to be solved [5,12]. The ACO algorithm is a probabilistic technique for determining the best path. It was originally intended to solve the problem of the traveling salesman (TSP). Currently, the ACO algorithm has been continuously improved and gradually built with a mature algorithm framework, becoming one of the most promising algorithms in the field of combinatorial optimization [12,13]. Its application extends to all aspects of optimization problems, such as the assignment problem, the vehicle routing problem, the graph coloring problem, the job-shop scheduling problem, and the network routing problem [12–15].

2.2. PSO

PSO is a group behavior that mimics the predation process of birds and fish, and is a branch of evolutionary computation [6]. A single particle is an individual of a group of birds or fish. It describes the members of the group as individuals without mass or volume, and it is also convenient to describe its speed and acceleration state [6,16]. PSO utilizes three simple behaviors of separation, alignment, and cohesion to describe the population, and moves the individuals in the population to a good region according to their adaptability to the environment, thus guiding the particles to find the global optimal solution. It is worth noting that the GA does not include the "crossover" and "mutation" operations [17]. It seeks the global optimum by following the currently searched optimal value. Compared with other modern optimization methods, the PSO algorithm has an obvious feature in that it requires few parameters to be adjusted and has a fast convergence speed [18]. It has become a key topic in the field of modern optimization methods. The PSO algorithm has now been widely used in optimization problems in various engineering fields [19,20].

2.3. GA

The strong adapt and thrive in nature, whereas the weak typically perish. The GA search optimization algorithm mimics the idea of "survival of the fittest" by basing itself on the mechanism of natural selection [21,22]. The GA is a group-based technique where participants score each other based on how suitable their answers are. The technique uses mathematics and computer simulation to change the problem-solving process into a process such as the crossing, reproduction, and mutation of chromosomal genes in biological evolution [23,24]. In comparison to certain conventional optimization techniques, they may frequently produce better optimization outcomes more rapidly when handling complicated combinatorial optimization issues [23,25,26]. Consequently, the GA has been widely applied in combinatorial optimization, signal processing, artificial life, machine learning, adaptive control, and other fields.

2.4. ABC

A single bee's behavior is quite basic, yet the social group it belongs to exhibits extraordinarily sophisticated behavior. The actual bee population can efficiently gather nectar from flowers (food sources) in settings. They may also adjust to environmental changes at the same time. An optimization technique called the ABC algorithm is put forth by emulating bee behavior [27,28]. The population's overall optimal value is eventually highlighted by the local optimization behavior of individual artificial bees. A novel swarm intelligence algorithm is the ABC algorithm. It merely requires a comparison of the benefits and disadvantages of the problem and has quick convergence; it does not require extensive knowledge of the topic. The algorithm is considered as simple to code as the PSO and the DEA [29,30].

2.5. GSO

The GSO algorithm is a new multi-modal function optimization technology of swarm intelligence proposed by Krishnanad and Ghose [30,31]. The GSO algorithm is memoryless and does not need global and gradient information. It has the characteristics of a fast calculation speed, fewer adjustment parameters, and easy implementation. In the process of population evolution, each iteration consists of five parts: the population initialization stage, the fluorescein update stage, the mobile probability calculation stage, the location update stage, and the neighborhood range update stage. After several years of development, the GSO algorithm has good application prospects in the optimization process of continuous space and some production scheduling.

2.6. CSA

The cuckoo has a unique breeding pattern in nature as a nest parasite bird. Nest parasitism is the practice of using other birds' nests to hatch their own eggs. To increase egg survival, the parasite birds destroy or conceal the host's eggs. If the host discovers the egg, the host will either find a new nest or kill the egg. Because this is a natural selection process, the quality of nest selection has a direct impact on the survival rate of the following generation. The CSA is a method for solving optimization problems by imitating the parasitic brood of a particular species of cuckoo [32–34]. Compared with other methods, the CS algorithm has the advantages of a multi-modal objective function and requires less parameters to be fine-tuned [35–38].

2.7. DEA

The DEA is a population-based algorithm based on the genetic annealing technique. It is comparable to a genetic algorithm due to the employment of similar operators (mutation, crossover, and selection) [39,40]. The primary distinction between the DEA and the GA is the ability to build superior solutions. The DEA is dependent on mutation, whereas the GA is dependent on crossover. In the DEA, regardless of their fitness value, all solutions in the population have the same chance of being chosen as parents. This is the primary operational distinction between the DEA and the GA. Although the DEA offers advantages in terms of improving local search ability and preserving population diversity, its convergence can be gradual and unpredictable [4,41].

2.8. Others

There are also other interesting evolutionary algorithms, such as genetic programming (GP), evolutionary programming (EP), evolutionary strategy (ES), the firefly algorithm (FA), the bat algorithm (BA), and the grey wolf optimizer (GWO) [42–45]. GP is another evolutionary algorithm, involving a process similar to the genetic algorithm [46]. The ES algorithm, using the same method as the GA and DE, is another optimization method. However, it uses an adaptive mutation rate [47]. Initialization, mutation, and evaluation are comparable processes in both the EP and GA. The EP does not employ a crossover operation to produce offspring, which is the primary distinction between it and the GA [45,48]. The FA was influenced by how fireflies interact with one another by flashing their lights. The BA algorithm is another recently introduced optimization technology inspired by the feeding behavior of bats. It is very similar to the PSO and consists of speed and position equations [45]. The GWO achieves the goal of optimization based on the mechanism of wolf group cooperation by simulating the predatory behavior of gray wolf groups [49,50].

3. Sinker-EDM

3.1. Brief Introduction for Sinker-EDM

Sinker EDM is performed in a liquid media, with the machine tool's automated feed adjustment unit accurately controlling the spark to treat conductive materials [51]. Sinker EDM works by eroding conductive materials via a sequence of spatially discrete high-frequency discharges (sparks) between the tool and the workpiece [52,53]. Figure 1 depicts the four stages of EDM. When a high pulse is delivered between these two electrodes (attaining the breakdown voltage of the medium in the gap), the dielectric insulation strength is destroyed at the weakest spot. Because the energy is highly short and concentrated, the instantaneous high temperature in the machining surface partly melts, vaporizes, and evaporates the metal on the two electrodes' surfaces. Under the action of the explosive force, the partially melted and vaporized metal is propelled into the working fluid, cooled into minute metal particles, and is swiftly washed away from the working area, producing a small crater on the surface of the two electrodes. The dielectric insulation strength will recover after the initial discharge and await the next discharge. Therefore, sinker EDM involves many process parameters, including electrical parameters (discharge current, pulse width, pulse frequency, gap voltage, etc.), non-electrical parameters (dielectric fluid, scouring mode, pressure, etc.), and material thermal properties. Therefore, it is essential to optimize its process parameters to obtain the required performance, such as the material removal rate (MRR), Ra, the tool relative wear rate, machining accuracy, etc.

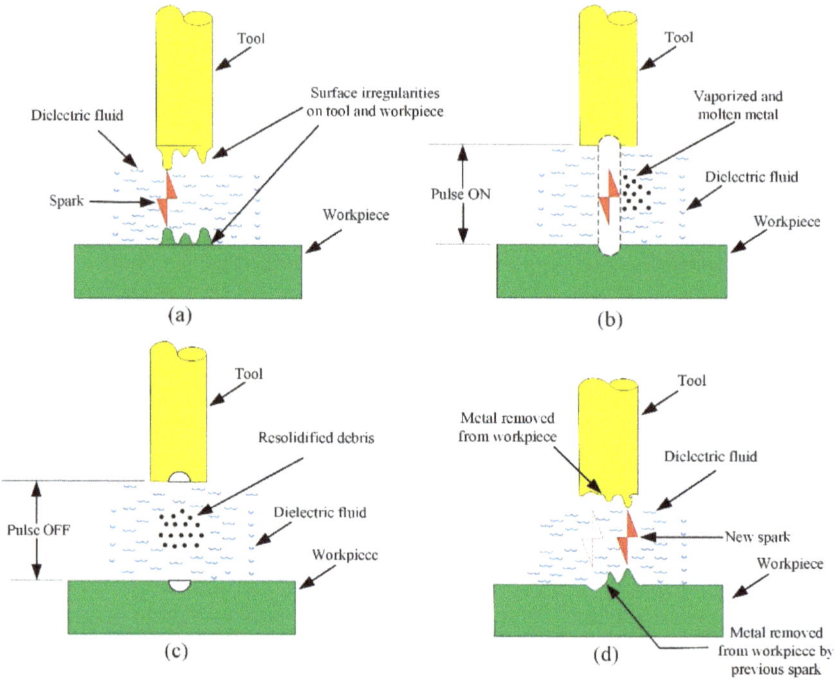

Figure 1. Four stages of the process of EDM; (**a**) spark beginning stage, (**b**) melting and vaporization stage, (**c**) removing debris stage, and (**d**) recovery stage.

3.2. Single-Objective Optimization

ACO is a popular optimization technique that has been used for optimization problem-solving. Mondal et al. [54] investigated the optimization of EDM process parameters using the ACO algorithm while cutting AISI 304 stainless steel. The findings revealed that, after 50 iterations, the highest MRR of 9.232 (mm^3/min) was attained, and at the end of all iterations, the optimal solution was found. The PSO technique is one of the most advanced evolutionary computational intelligence-based optimization methodologies for optimizing real-world multimodal problems. Aich and Banerjee [55] determined the desired process parameters combination of the maximum MRR and the minimum Ra through the PSO algorithm, with an average searching time of 4.334 s. The EDM processing optimization proposed by Saffaran et al. [56] was highly effective, and the error of single-objective and multi-objective optimization was less than 4% and 7%, respectively. Tzeng and Chen [57] optimized the EDM process using the GA approach while machining SKD 11, and discovered that GA had better prediction than RSM. Mahanta et al. [58] carried out GA-based optimization to produce high-quality jobs with low power consumption in the process of EDM. Mukherjee and Chakraborty [59] compared the optimization performance of the EDM process by using various swarm intelligence algorithms, and found that the performance of the ABC algorithm was better that of the ACO or GA algorithms. Zainal et al. [60] employed linear regression and the GSO method to conduct an integrated assessment of surface roughness in the EDM process. Furthermore, the minimal Ra of 2.03 μm could be searched for by the proposed method. Shen et al. [61] proposed an improved beta-distributing cuckoo search (IBCS) algorithm, and utilized this proposed IBCS to obtain the single-objective optimization (Ra) in the EDM process. The results demonstrated that the number of iterations of the IBCS algorithm (eight iterations) was half of that of the BCS algorithm (seventeen iterations). The DEA method was also used in the parameter optimization of the EDM process [62], and the optimization performance could be acceptable. As for the GP algorithm, its effect in terms of EDM parameter optimization was also better than the traditional response surface method (RSM), and the previous study showed that it had better and more accurate estimation [63]. The ES method could be used in the EDM process optimization of machining Inconel 718 alloy [64]. For different materials, such as die steel [65], Al-SiCp [66], FA was applied in the EDM process for optimizing the process parameters to obtain the desired machining performance.

For the hybrid-EDM process, the single-objective optimization can also be performed by various swarm intelligence techniques. A continuous ACO method was used to select the best magnetic-field-assisted-EDM (MF-EDM) process parameters for maximum MRR and specified surface roughness (SR) by Teimouri and Baseri [67]. The optimal input values (process parameters) for different desired SR were obtained, and the variation of SR and MRR were heavily dependent on discharge energy fluctuation. Singh et al. [68] proposed an intelligent hybrid approach, using the PSO algorithm, to obtain a desired prediction of gas-assisted EDM (GA-EDM) performance. The results showed that the improved PSO algorithm was effective for the G-EDM performance estimation. Rouniyar and Shandilya [69] performed the optimization of process parameters in magnetic field assisted powder mixed EDM (MFAPM-EDM) for cutting Al 6061, and compared the optimized performance of GA method with others. Danish et al. [70] used the GWO approach to optimize the hydroxyapatite powder mixed electric discharge machining (HPM-EDM) process in order to improve modified surface characteristics, such as the recast layer thickness (RLT), of 316L stainless steel. The test demonstrated that the gray wolf optimizer's predicted solution sets are very accurate, with less than 10% inaccuracy. As listed below, the comparison of various swarm intelligence techniques for parameters optimization in the sinker-EDM process is depicted in Table 1.

Table 1. Comparison of various swarm intelligence techniques for parameters optimization in the sinker-EDM process.

Techniques	Year, Authors, Process	Parameters	Performance	Findings	Shortcomings or Limitation
ACO	2014, Teimouri and Baseri [67], MF-EDM	Magnetic field intensity (Fi), rotational speed of electrode (Rs), and discharge energy (Ee)	MRR and SR	With the help of the ACO algorithm, the magnetic field assisted rotary EDM process can also be successfully optimized.	It is not compared with other optimization algorithms.
	2022, Mondal et al. [54], EDM	Pulse-on time (Ton), discharge current (Ip), gap voltage (Vg)	MRR and SR	The ACO algorithm had converged after 50 iterations.	The performance of finding the optimal parameters needs to be improved.
PSO	2014, Aich and Banerjee [55], EDM	Ton, pulse-off time (Toff), and Ip	MRR and Ra	The PSO algorithm had converged after 20 iterations.	The search time of the algorithm is longer than 1 h.
	2020, Saffaran et al. [56], EDM	Ton, Toff, Ip, Vg and duty factor (Df)	MRR, SR, and TWR	Optimization error was less than 7%; the performance of PSO was better than SA.	The sample data is too small, which affects the robustness of the model.
	2020, Singh et al. [68], GA-EDM	Ton, Toff, Df, Rs, and gas discharge pressure (Gp)	MRR and SR	Could be an efficient and productive approach.	The performance of the proposed PSO should be mensurated.

Table 1. *Cont.*

Techniques	Year, Authors, Process	Parameters	Performance	Findings	Shortcomings or Limitation
GA	2013, Tzeng and Chen [57], EDM	Ton, Ip, and Vg	MRR, Ra, and TWR	The GA approach had better prediction than the RSM method.	The number of samples needs to be increased to improve robustness.
	2018, Mahanta et al. [58], EDM	Ip, Df, Toff, and Vg	Power consumed and SR	To be an effective tool with minimum effort.	Other important performance is not involved, such as MRR.
	2020, Rouniyar and Shandilya [69], EDM	Ip, Ton, Toff, concentration of powder (Cp), and magnetic field intensity	MRR, and TWR	The GA approach could be applied to solve this process parameters optimization problem.	Performance needs to be improved.
ABC	2011, Mukherjee and Chakraborty [59], EDM	Ip and Ton	MRR, and SR	The number of iterations (<250) of ABC was less than ACO algorithm or GA.	Few process parameters to be optimized.
GSO	2017, Zainal et al. [60], EDM	Ton and Toff, Ip, and servo voltage (Sv)	Ra	The minimal Ra of 2.03 μm could be searched by the proposed method.	Few machining performances to be considered.
CSA	2011, Shen et al. [61], EDM	Ton, Toff, Ip, and Sv	MRR, and Ra	The number of iterations of the IBCS algorithm was eight iterations.	Multi-objective optimization needs to perform for MRR and Ra.
DEA	2020, Kumar et al. [62], EDM	Vg, Ip, Rs, and cycle time	MRR, SR, TWR, overcut and circularity error	The optimization performance could be acceptable.	The accuracy and consistency of the derived optimal solutions needs to be improved.
GP	2020, Ghadai et al. [63], EDM	Ip, Ton and Toff	MRR, and TWR	Both single-objective optimization and multi-objective optimization were investigated.	The experimental data is too small to affect the accuracy of the model.
EP	2020, Jafarian et al. [64], EDM	Vg, Ton, Ip, and Toff	MRR and Ra	The optimal process combination was successfully achieved using the GP algorithm.	The processing performance involved is relatively small.
GWO	2022, Danish et al. [70], HPM-EDM	Ip, Ton, Vg, and hydroxyapatite amount	MRR, Ra, and RLT	With less than 10% inaccuracy.	Three-objective optimization should be performed for MRR, Ra, and RLT.

3.3. Multi-Objective Optimization

Dang et al. [71] employed the PSO method to conduct the restricted multi-objective (MRR, SR, and tool wear rate (TWR)) optimization of process parameters, which was regarded as the intelligent optimization of the EDM machining process. The results demonstrated that effective process parameter selection could assist the EDM operator in reducing cutting time, as well as expenses. Moghaddam and Kolahan [72] utilized the PSO algorithm to optimize several response characteristics (MRR, SR, and TWR). The proposed approach accurately simulated the actual EDM process with less than 1% inaccuracy, according to the results. Zhang et al. [73] adopted the proposed PSO algorithm, combining quantum behavior and Gaussian local attractor, to provide the best processing parameters for the green MF-EDM while machining SiCp/Al, as depicted in Figure 2. The points in Figure 2 are 15 selected optimized results, which contains minimal and maximum MRR, EWR, and aerosol emissions values, as well as some compromised points. Compared with the average experimental data, the average optimal solution for SiCp/Al, TWR (EWR), and aerosol emissions decreased by around 7.8% and 12.5% respectively, while the MRR increased by around 6.05%. Garg and Lam [74] used the GP technique to model the multiple-response features. The multi-response features, namely one manufacturing aspect (TWR) and two environmental aspects (dielectric consumption and heat energy consumption), were considered in their investigation, and the efficacy of the suggested GP models was validated using experimental data.

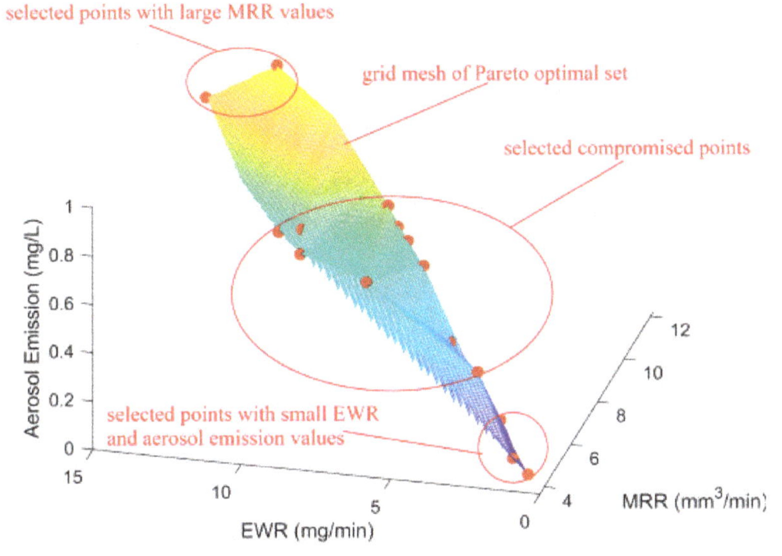

Figure 2. Optimal Pareto for MF-EDM while green machining SiCp/Al, reproduced from [73], permission with Elsevier, 2020.

During the green machining of Al 6061 and SKD 11, Ming et al. [75] recommended utilizing the GSO algorithm to optimize parameters to reduce exhaust emissions and energy consumption. The Pareto frontiers of the multi-objective optimization for Al 6061 and SKD 11 are illustrated in Figure 3, which demonstrate the Pareto frontier of MRR and exhaust emissions characteristics (EEC). The intended EEC in Figure 3a,b was less than 180 $\mu g/m^3$ for integrating cutting productivity with environmental friendliness while milling Al 6061. Similarly, the intended EEC for SKD 11 did not exceed 160 $\mu g/m^3$. As a result, as shown in Figure 3, the left sides below these criteria are colored green. This suggests that the optimized outcomes in the green zone were less harmful to the environment. The targeted energy efficiency per volume (EEV) in Figure 3c,d was less than 0.51 KJ/mm^3 for integrating machining output with environmental friendliness while cutting Al 6061. Moreover, for SKD 11, the targeted EEV does not exceed 5.5 KJ/mm^3. Similarly, the desired outcomes in the green region were energy efficient. Hence, the optimized combination of cutting process parameters for multi-objective optimization could meet the prediction accuracy.

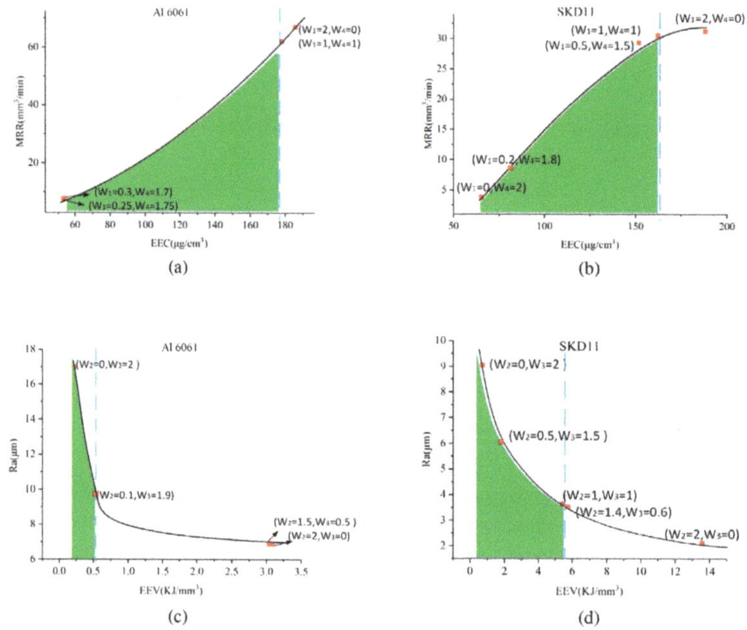

Figure 3. Optimal Pareto for EDM while green machining Al6061 and SKD 11: (**a**) Pareto frontier of MRR and EEC for Al 6061, (**b**) Pareto frontier of MRR and EEC for SKD 11, (**c**) Pareto frontier of Ra and EEV for Al 6061, (**d**) Pareto frontier of Ra and EEV for SKD 11 reproduced from [75], permission with Elsevier, 2021.

3.4. Summary

From the existing optimization of EDM process parameters, most of the research is still focused on single-objective optimization, and the number of research papers on multi-objective optimization is relatively small. This may be related to the characteristics of the swarm intelligence algorithm, which is not suitable for multi-objective optimization. Moreover, in actual production, the demand for single targets may be more urgent under specific circumstances. In addition, since most of the studies are traditional sinker-EDMs, the amount of hybrid sinker-EDMs is also relatively small. Therefore, there have not been many swarm intelligence algorithms involved. Furthermore, it is found that the PSO and GA are the most widely used in the optimization, and the number of studies is also significantly more than other swarm intelligence algorithms. One of the most important reasons for this may be that the PSO and GA algorithms are easy to implement, and their robustness is relatively good, which can meet the needs of engineering applications. Optimization algorithms provide the results of EDM improvements, and the optimized processing parameters can significantly guide the EDM process in an economical and environmentally friendly manner and meet sustainable manufacturing needs.

4. Wire-EDM

4.1. Brief Introduction for Wire-EDM

According to wire speed, wire-EDM (WEDM) can be divided into two categories: high speed wire-EDM (HS-WEDM) and low speed wire-WEDM (LS-WEDM) [76–78]. The wire feed speed of the HS-WEDM ranges from 6 to 12 m/s, and it is commonly employed in low/medium quality mold manufacture and special component processing, and the equipment of the process is depicted in Figure 4a [77]. The HS-WEDM tool electrode is a molybdenum wire, which is fed in a cyclic manner and can be reused. Unfortunately, the electrode wire jitter is high due to the difficulty in applying consistent tension control to the electrode wire, and the wire is easily broken during processing. Furthermore, because the electric wire is reciprocally employed, it will result in electrode wire loss, decreased processing accuracy, and surface quality deterioration. The LS-WEDM electrode wire employs copper wire as the tool electrode and normally travels in a single direction at less than 0.2 m/s, the equipment of the process is depicted in Figure 4b. A pulse voltage of 60–300 V is delivered between the copper wire and the treated material, such as copper, steel, or a superhard alloy, while maintaining a spacing of 550 μm. The gap is filled by deionized water (like distilled water) and other insulating media, resulting in a spark discharge between the electrode and the treated material, which consume and corrode each other [78]. The discharge phenomena are uniform due to digital program control, monitoring and control, and servo mechanism execution, allowing the treated item to be processed, and producing a product with the requisite size and shape precision [79,80]. The HS-WEDM affects the machining accuracy as the electrode wire is recycled, and as the electrode wire continuously wears out. Generally, the accuracy of products machined by the HS-WEDM is ±0.015~0.02 mm. The LS-WEDM electrode wire is not recycled, which greatly improves the machining accuracy, and the machining accuracy of the machine can reach ±0.002 mm.

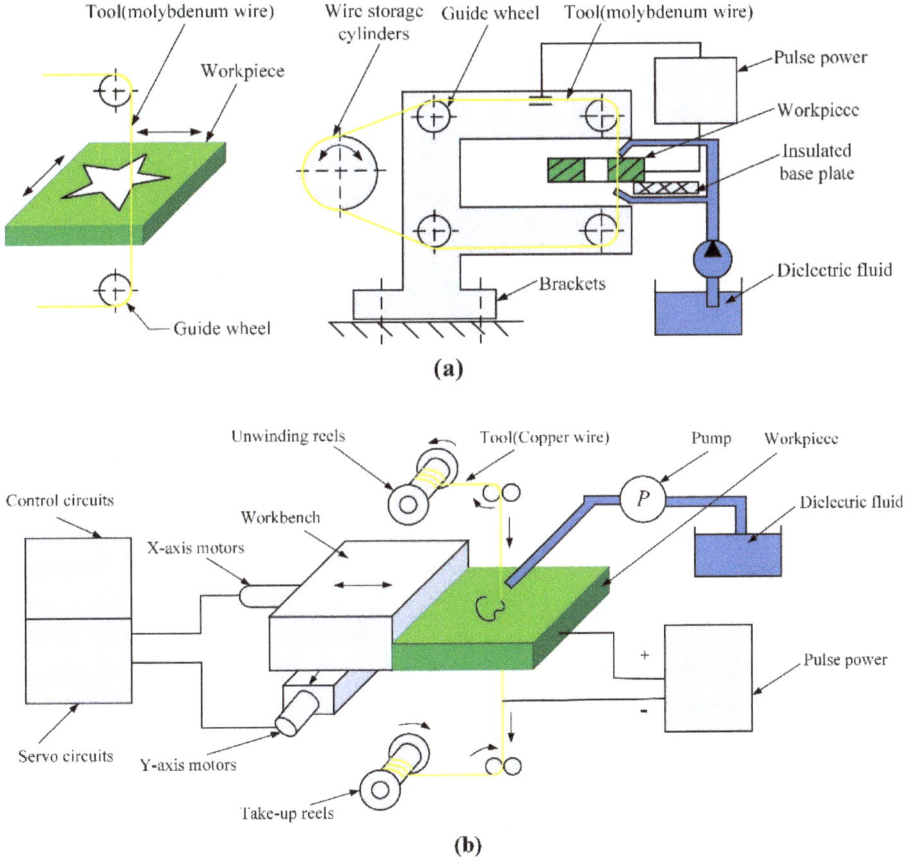

Figure 4. Equipment of the process of WEDM; (**a**) HS-WEDM, (**b**) LS-WEDM.

4.2. Single-Objective Optimization

In single-objective optimization, the optimization of one desired response at a time is carried out. Sharma et al. [81] adopted the PSO method, integrating evaluation based on distance from the average solution (EDAS), to optimize the wire-EDM parameters when cutting biomedical materials. Figure 5 depicts the process flow of the PSO method for optimizing the wire-EDM parameters. The best configuration for reducing the surface imperfections when milling titanium alloy on LS-WEDM was suggested as Ton: 8 μs; Toff: 13 μs; Vs: 45 V; and wire tense: 8 N. Moreover, the forthcoming solution at the optimal parametric values had the following values: an Ra of 3.163 μm, an Rz of 22.99 μm, a wire loss of 0.0182 g, and a dimensional accuracy of 95% [81]. To find the best parameter settings for the LS-WEDM process, Tzeng et al. [82] suggested a hybrid approach combining the GA, a back-propagation neural network (BPNN), and the RSM. The outcomes showed that the GA approach's suggested algorithm provided superior prediction and confirmation results compared to the RSM technique. This means that the GA optimization techniques have a lot of promise for challenging applications in industry. Furthermore, the similar integrating method was also used in the process of the HS-WEDM by our team [83]. For fabricating microchannels in industrial applications, Singh et al. [84] used the GA approach to accomplish the multi-response optimization of the LS-WEDM process parameters for the fabrication of a brass microchannel. They found that the single-response optimization through the GA approach for a maximum MRR = 7.10 mm^3/min and a minimum Ra = 3.36 μm could be obtained. The validation outcomes demonstrated the effectiveness of the proposed optimization models. Kuruvilla and Ravindra [85] applied the Taguchi and GA methods to determine the influence of process parameters on the oil-hardened non-shrinkage steel (OHNS) material with a thickness of 40 mm by LS-WEDM. The variation of the performance parameters with machining parameters was mathematically modeled using the Regression analysis method. The results showed that, in the processing parameters, it was better to use a smaller pulse closing time to achieve a good performance overall.

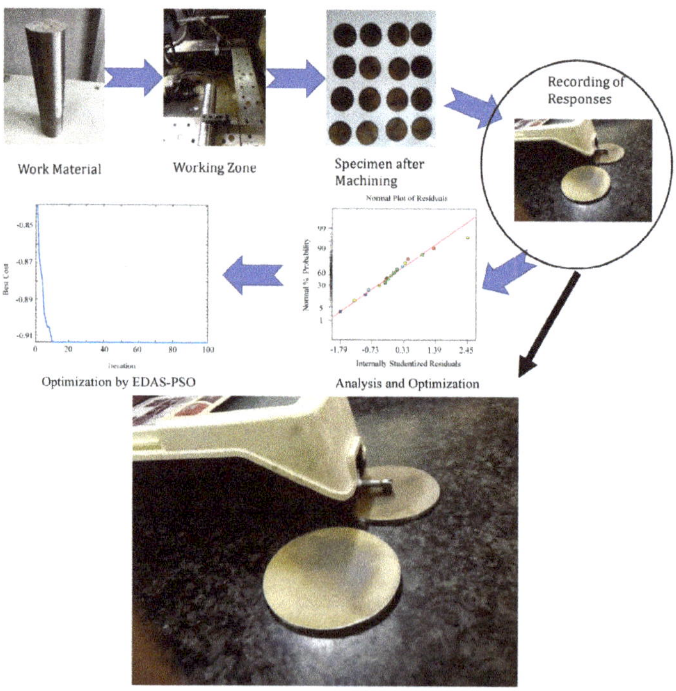

Figure 5. Process flow of the PSO method for optimizing the wire-EDM parameters, reproduced from [81].

Rao and Pawar [86] modeled and optimized the process parameters of LS-WEDM using the RSM and ABC methods. Through the proposed methods, a maximum MRR of 7.10 mm^3/min and a minimum Ra of 3.36 μm were obtained. This demonstrated that the

ABC method, an advanced algorithm of swarm intelligence, could solve the engineering problems. In order to optimize the multiple performance of the dry LS-WEDM process of Al-SiC metal matrix composite, Fard et al. [87] proposed an intelligent model that integrated the ABC algorithm. In this model, the impact of the discharge current, pulse-on time, pulse-off time, wire tension, gap voltage, and wire feed on cutting velocity and Ra was examined. The results showed that brass wire and oxygen gas ensured superior cutting velocity. Additionally, it was discovered that the ABC algorithm could predict the ideal set of process parameters with accuracy. Rao and Venkaiah [88] proposed a modified CSA to optimize the LS-WEDM process when machining Inconel-690 to significantly enhance the performance of the cuckoo search. In comparison to the GA, PSO, and the standard cuckoo search, the proposed method was determined to be accurate and quick. The industry will also benefit from the machining data produced by this endeavor. Similar results could also be confirmed in our previous study in another engineering field [38]. Saravanan et al. [89] investigated the parameters (pulse-on time, pulse-off time, cutting speed, discharge current, wire tension, wire feed, servo voltage, and servo feed) optimization of LS-WEDM through CSA. The optimum machining performance for different conditions could be obtained in no more than 100 iterations, as depicted in Figure 6. Figure 6a represents the variation of the obtained objective function values, where it is observed that the molybdenum wire outperforms the other two wires, namely plain brass and galvanized, during the iterative process. Figure 6b shows that the optimum value of the input parameter cutting speed is 75%, and the speed of molybdenum wire is relatively higher than that of galvanized and ordinary brass wire.

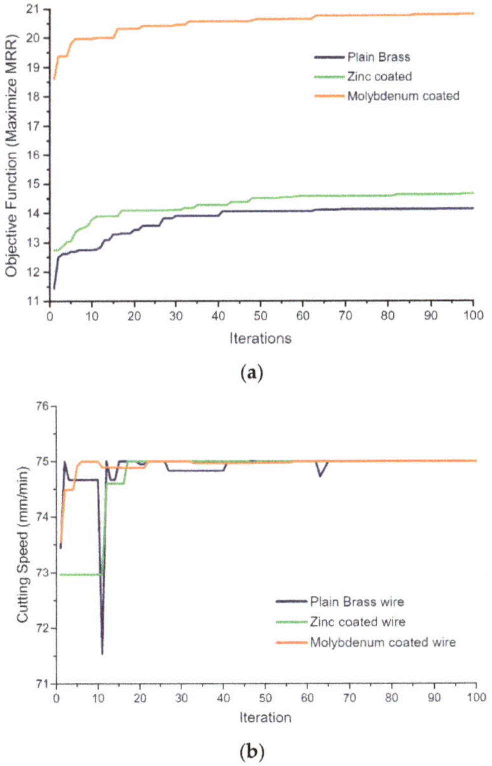

Figure 6. Optimum machining performance for different conditions, reproduced from [89], with permission from Elsevier, 2020; (**a**) MRR, (**b**) cutting speed.

Kulkarni et al. [90] analyzed the machinability and optimized the LS-WEDM of NiTiNOL memory alloy using the modified DEA method. The results confirmed that the convergence curve of the modified DEA was much better than that of the conventional DEA, and the convergence speed was twice as fast as before, thus saving calculation time. Nayak et al. [91] established a genetic programming (GP) model to study six process parameters to optimize angle error and minimize surface roughness. The results demonstrated that GP was a cost-effective and time-saving method compared with typical prediction methods. Xu et al. [92] optimized the process parameters, as depicted in Figure 7, of wire cutting nickel-titanium shape memory alloy (NiTi-SMA) with the help of the BA algorithm and multiple regression (MLR)/BPNN to obtain the maximum processing speed and ideal kerf width. The experimental results showed that the prediction error of the proposed optimization method could be within ±2%, which was valuable for engineering application. As listed below, the comparison of various swarm intelligence techniques for parameters optimization in the wire-EDM process is depicted in Table 2.

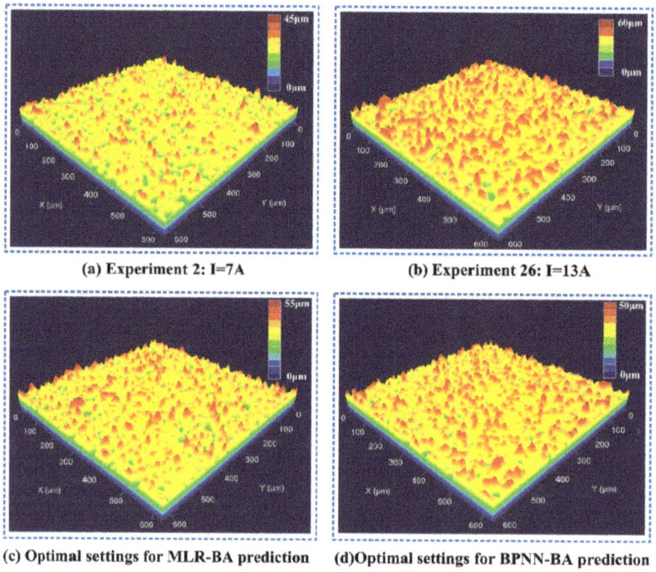

Figure 7. Optimal settings results of 3D CLSM images of the sample surface: (**a**) low discharge energy, (**b**) high discharge energy, (**c**) optimal settings for MLR-BA prediction, (**d**) optimal settings for BPNN-BA prediction, reproduced from [92], with permission from Elsevier, 2022.

Table 2. Comparison of various swarm intelligence techniques for parameters optimization in the wire-EDM process.

Techniques	Year, Authors, Process	Parameters	Performance	Findings	Shortcomings or Limitation
PSO	2023, Sharma et al. [81], LS-WEDM	Ton, Toff, vs. and wire tension (Wt)	Surface roughness (Ra and Rz), wire loss, and dimensional accuracy	The PSO's efficiency in accurately forecasting the outcomes of WEDM-processed machining of pure titanium (Grade 2) was validated.	Multi-objective optimization does not perform.

Table 2. *Cont.*

Techniques	Year, Authors, Process	Parameters	Performance	Findings	Shortcomings or Limitation
GA	2011, Tzeng et al. [82], LS-WEDM	Ton, Toff, arc-off time (Tarc), Vs, wire feed rate (Wf), Wt and Wp	MRR and Ra	The GA optimization techniques have a lot of promise for challenging applications.	The number of samples needs to be increased to improve robustness.
	2013, Zhang et al. [83], HS-WEDM	Ton, Toff, Ip, Wf, and tracking coefficient	MRR and Ra	The prediction error of the model was no more than 10%.	Multi-objective optimization is not considered.
	2020, Singh et al. [84], LS-WEDM	Ip, Ton, and Toff	MRR, and Ra	The maximum MRR = 7.10 mm^3/min and minimum Ra = 3.36 μm could be obtained.	The engineering application is worthy of further study.
ABC	2009, Rao and Pawar [86], LS-WEDM	Ton, Toff, Ip, and servo feed setting	Cutting velocity (CV), and SR	The desired value of Ra = 2.1 μm and CV = 1.106 mm/min could be achieved.	There are few process parameters to be optimized.
	2012, Fard et al. [87], dry LS-WEDM	Ip, Ton, Toff, Wt, Vg, and Wf	CV and SR	The ideal set of process parameters could be predicted.	It is not compared with other optimization algorithms.
CSA	2017, Rao and Venkaiah et al. [88], LS-WEDM	Ton and Toff, Ip, and Vs	MRR, SR, and kerf width	Both single-objective and multi-objective optimization were investigated.	The engineering application should be investigated.
MDE	2020, Kulkarni et al. [90], LS-WEDM	Ton, Toff, vs. and Wf	MRR, SR, and TWR	The convergence speed of modified DEA was twice as fast as conventional DEA.	Multi-objective optimization of MRR, SR, and TWR should be involved.
GP	2022, Nayak et al. [91], LS-WEDM	Ip, Tom, Wt, wire speed, workpiece thickness, and taper angle	Angular error, and SR	GP model was an effective tool for solving this problem.	Other processing properties are not involved.
BA	2022, Xu et al. [92], LS-WEDM	Ip, Wp, Wt, Wf, and discharge frequency	CV and kerf width	The prediction error was within ±2%.	Multi-objective optimization of CV and kerf width should be performed.

4.3. Multi-Objective Optimization

Multi-objective optimization is the process of optimizing two or more conflicting objectives. In machining operations, there are often situations where the desired responses are inherently conflicting. An improvement in one response can worsen another. Ranjan et al. [93] performed the multi-objective optimization of an abrasive powder, including SiC and Al_2O_3, mixed WEDM (APM-EDM) of Inconel 718 using the PSO method. Using a hybrid artificial neural network (ANN) linked GA technique, Vaidyaa et al. [94] sought to maximize the multi-objective optimization of LS-WEDM on AlSi10Mg. In their investigation, the following optimal process parameters were established: vs. of 42 V, Ip of 12 A, and Ton of 12 μs result in a maximum micro-hardness of 478 VHN and a minimum SR of 4.3 μm. Zhang et al. [95] used BPNN combined with the GA (BPNN-GA) and the non-dominated sorting genetic algorithm-II (NSGA-II) to optimize the process parameters on surface integrity, such as white layer thickness (WLT), surface crack density (SCD), and SR, in the LS-WEDM process of the tungsten tool YG15. The best optimum solutions for WEDM machining, which balanced the performance of SR, SCD, and WL, as well as the

Pareto-optimal front of three-objective optimization, were discovered by their proposed methods. Similarly, the gray relational analysis technique (GRA) and the backpropagation neural network-genetic algorithm (BPNN-GA) were used by Soepangkat et al. [96] to accomplish multi-objective optimization in the LS-WEDM process of SKD 61 tool steel. Setting Ton, Toff, and vs. to 3 ms, 10 ms, and 38 V, respectively, would yield the lowest RLT, SR, and SCD.

When performing multi-objective optimization for LS-WEDM on Ti6Al4V alloy, Jain and Parashar [97] compared a priori and a posteriori approaches. A multi-objective ABC with GRA was selected as an a priori approach, and a multi-objective grasshopper optimization algorithm (MO-GOA) was selected as an a posteriori approach. According to the amount of calculating time, it was discovered that the a priori approach to multi-objective optimization was superior to the a posteriori approach. Similarly, the Pareto front was found by extending the modified CSA to simultaneously optimize the MRR, SR, and kerf [88]. Majumder et al. [98] studied the multi-objective optimization of the LS-WEDM process when cutting RAFM steel manufactured in India. They compared the performance of giving optimal results of the FA, the PSO, and the DEA, and found that the FA was better than the others (DEA and PSO). Kondayya et al. [99] described an approach of an evolutionary strategy for the optimization of a LS-WEDM process. The result of Pareto optimal solutions, as depicted in Figure 8, confirmed that the all-encompassing evolutionary strategy was a solution for the process optimization. In the beginning, it was possible to machine components with the best surface quality, but with the lowest MRR. Finally, the highest material removal rate could be achieved, but with the worst surface quality.

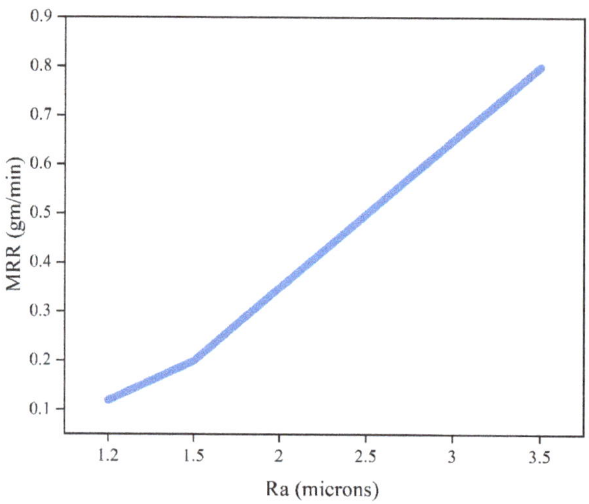

Figure 8. Pareto front of LS-WEDM for the selection of optimum values.

4.4. Summary

One of the most researched areas in machining is process modeling and optimization, as this can reduce production costs and improve product quality. However, experimental optimization of any machining process is expensive and time consuming due to the complexity, coupling, and nonlinear nature of the input and output variables of the process. There are various optimization methods for process parameter optimization. Based on the existing research literature, most of the process parameter optimizations of wire-EDM have been aimed at LS-WEDM, with very few involving HS-WEDM. This is because the machining performance of HS-WEDM equipment is poor, and its processing dimensional accuracy and surface quality are not as good as LS-WEDM. As a result, its application scope is also limited, and the number of relevant research results is very scarce. Most of the machining performances studied in the literature have mainly involved the MRR and the SR, while others such as kerf width, TWR, and angular error are relatively few. Similarly, it has been found that PSO and GA are the most widely applied in the optimization of wire-EDM process parameters, and the number of research articles is also significantly higher than that of other swarm intelligence algorithms, such as the ABC, GP, or BA methods. Furthermore, it is concluded that the attention of researchers has focused on single-objective optimization, and the involved methods of wire-EDM are traditional wire-EDM.

5. Micro-EDM

5.1. Brief Introduction for Micro-EDM

With the increasing miniaturization and precision of products worldwide, microfabrication and small-hole EDM, one of the non-contact microfabrication methods, has become an important part of micromechanics and is widely used in the manufacturing industry because of its superfine and high-precision machining characteristics. Although micro-EDM and sinker-EDM are based on the same physical principle of spark erosion, there are still obvious differences. This is because, during its processing, the discharge energy is reduced to the order of 10^{-6}~10^{-7} J to minimize the amount of material removal per unit time and realize the micro-level EDM process. This means that, compared with sinker-EDM, micro-EDM has more difficulties in terms of power control, manufacturing methods of micro-tools, real-time processing monitoring, and other aspects [100–103]. Micro-EDM technology can be categorized into five different types: micro sinker-EDM, micro wire-EDM, micro drilling-EDM, micro milling-EDM, and micro wire electro-discharge grinding [102,104]. Figure 9 demonstrates the principle of the process of micro drilling-EDM and micro milling-EDM.

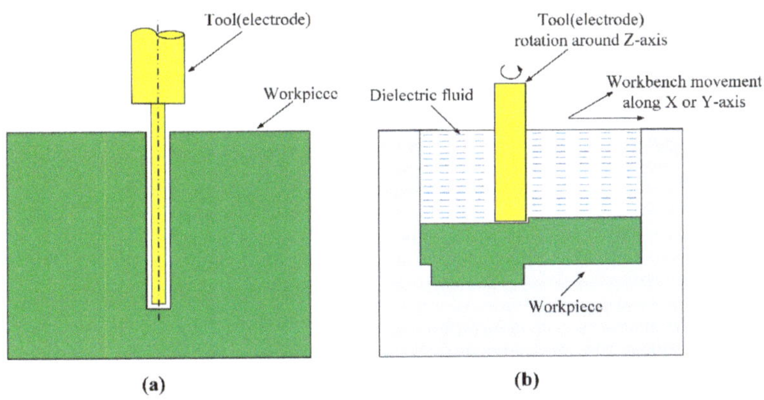

Figure 9. Principle of the process of micro-EDM; (**a**) micro drilling-EDM, (**b**) micro milling-EDM.

5.2. Parameters Optimization

With the use of the PSO optimization technique, Chen et al. [105] suggested a hybrid optimization strategy that increased MRR by a maximum of 49.1% and decreased Ra by a maximum of 37.8% in the micro-WEDM process while machining 65 vol.% SiCp/Al composite. Moreover, the Pareto optimum solution sets, as depicted in Figure 10, had great accuracy and dependability. Similarly, Quarto et al. [106] completed a micro-EDM optimization with the PSO approach. The results showed that the suggested optimization approach may adapt to the automated environment in both scenarios of externally imposed material or processing performance. For optimizing the multi-objective performance of the circularity at the entrance, the hole overcut, and the circularity at the exit of drilled micro holes, a hybrid approach of the PSO integrating gray relational analysis was used in the process of micro hole drilling in Mg alloy [107]. Regarding the prediction of micro-hole quality (radial over-cutting, recast layer thickness, and MRR) on Inconel 718 superalloy treated by micro-EDM, Rao [108] suggested a prediction model based on the bionic intelligent hybrid algorithm. The prediction approach was based on the combination of the adaptive neuro-fuzzy inference system (ANFIS) and the PSO/GA algorithms (ANFIS-PSO/ANFIS-GA). Comparing the ANFIS-PSO against the ANFIS-GA, ANFIS, and ANN models, it was discovered that ANFIS-PSO was superior.

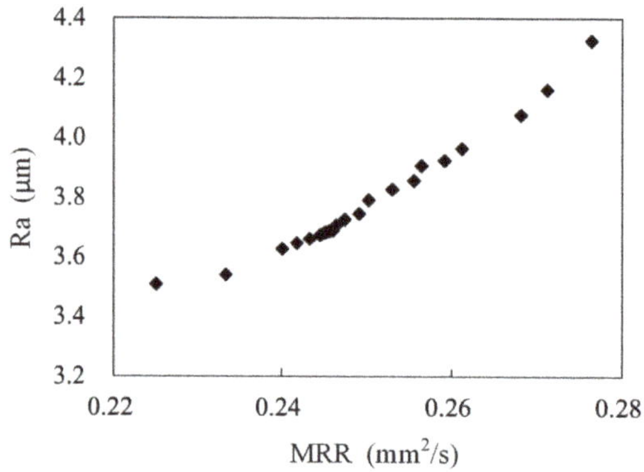

Figure 10. The Pareto optimal solution of MRR and Ra in the micro-WEDM process while machining 65 vol.% SiCp/Al composite, reproduced from [105], with permission from Elsevier, 2021.

To predict the performance (machining time, MRR, and dimensional deviation) of micro drilling-EDM, Quarto et al. [109] compared a finite element model (FEM) simulation and an integrated ANN and PSO technique. In order to meet various industrial objectives, the combined ANN-PSO approach included a twofold direction functionality. It provided a way to anticipate process performance while simultaneously optimizing the process parameters (discharge current, voltage, frequency, tool diameter, workpiece, and tool materials) in relation to the needed performance levels. Figure 11 depicts the bar chart of the prediction error between the experimental results and models simulation, while drilling micro-EDM. The findings demonstrated that the combined ANN-PSO technique provided performance predictions that were more accurate. The ANN-PSO approach was also quicker and simpler to use, but it needed a lot of historical data to train the ANN. In contrast, setting up the FEM was more difficult because it required several physical and thermal properties of the materials, and single simulation took a long time.

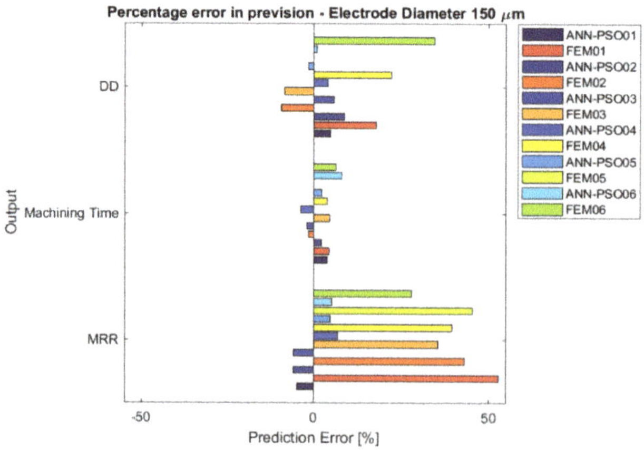

(a)

Figure 11. *Cont.*

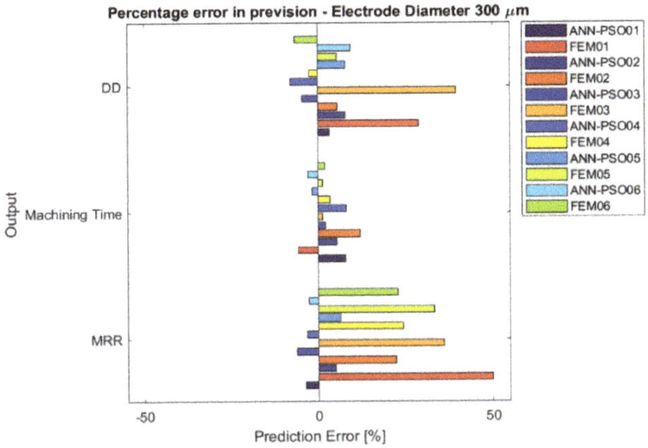

(b)

Figure 11. Bar chart of prediction error between experimental results and models simulation while drilling micro-EDM with a tool diameter of (**a**) 150 μm, and (**b**) 300 μm, reproduced from [109].

In micro-EDM, processing parameters greatly affect processing efficiency and stability. Based on Taguchi experiments with support vector machines, Zhang et al. [110] developed a process model for micro-EDM and suggested a novel multi-objective optimization GA to optimize process parameters and decrease machining time and TWR. The GA is an evolutionary algorithm that uses genetic operators to obtain optimal solutions without any assumptions about the search space. GA works with a population of feasible solutions and, therefore, it can be used in multi-objective optimization problems to simultaneously capture several solutions. The suggested multi-objective GA was accurate and efficient in producing Pareto optimum solutions for parameter settings, according to the experimental data. The optimal parameter combination could significantly decrease machining time while maintaining a manageable TWR. Hence, the suggested technology could increase the stability and efficiency of machining while also being appropriate for the parameter adjustment of micro-EDM. Dilip et al. [111] carried out the multi-objective optimization of process parameters on Inconel 718 using weighted objective summation and GA techniques to enhance the processing quality of drilling cooling holes for turbine blades. According to the experimental data, the inner wall's surface roughness could be reduced to 13,587 μm under the ideal cutting circumstances. Furthermore, the confirmed experiments confirm that the maximum relative error obtained was less than 10%.

5.3. Summary

Compared to sinker-EDM and wire-EDM, the number of publications on process parameter optimization for micro-EDM is significantly lower. The reason is that compared to sinker-EDM and wire-EDM, micro-EDM is not widely used, especially in industrial applications. Most of the existing literature on micro-EDM remains at the laboratory stage, mainly focusing on the research of processing mechanisms. Indeed, micro-ED is a branch of EDM, and the machining mechanism of electrical discharge machining, like other special machining, is very complex [112–114]. The existing research has not fully understood its working mechanism [3,115–117]. In addition, the swarm intelligence algorithms for the process parameter optimization of micro-EDM have mainly focused on PSO and GA methods. This is primarily because PSO and GA methods are relatively simple and robust compared to other methods. In micro-EDM, the machining parameters have a great influence on the machining efficiency and stability. In order to correctly set the process parameters of micro-EDM, the process is generally modeled and several new multi-objective optimization GA algorithms are proposed. At the same time, properly selected process parameters can obtain a good clearance state, ensure the stability of the process, and result in low electrode wear, thus greatly improving the machining efficiency.

6. Discussion

This review focuses on the progress in the application of the SI technique in the optimization of EDM process parameters. In this study, the parameter optimizations of sinker-EDM (SEDM), wire-EDM (WEDM), and micro-EDM (MEDM) with various hybrid techniques using SI technique are comprehensively reviewed. This section discusses the three aspects of similarity, individuality, and complementarity of these different SI techniques and machining methods.

6.1. Similarity

The PSO method, which is frequently employed in the optimization of these three categories of EDM process parameters, is the primary representative algorithm of swarm intelligence. The fundamental idea, which looks for the best solution through individual cooperation, comes from research on the behavior of bird flocks moving in groups. First designed to graphically represent the ethereal and erratic movements of flocks of birds in two dimensions, it was later extended to the multidimensional space and used to the solution of optimization problems. The ACO method, which uses the collective optimization capability of biological ant colonies that can find the shortest path from the ant nest to food by straightforward information flow between individuals, is another example of a representative algorithm of swarm intelligence.

The similar characteristics of swarm intelligence can be summarized as follows: At first, because control is decentralized and not centralized, the system is better equipped to adapt to the network and is less likely to be adversely affected by the failure of one or more individual. Second, through indirect communication, each member of a group has the power to influence the environment, spread knowledge, and work together. Third, because each member of a group has very basic skills or behavioral norms, group intelligence is simpler and easier to use. Lastly, because of their ability to self-organize, groups exhibit complex actions that are the result of intelligence that develops via the interaction of individuals [118].

6.2. Individuality

Of course, different swarm intelligence algorithms also have certain differences. The swarm intelligence optimization algorithm mainly simulates the swarm behavior of insects, herds of animals, flocks of birds, and schools of fish. Due to the differences between species, there are certain differences in the implementation methods of different algorithms. For example, the GA has a strong global search ability but a weak local search ability, often only obtaining suboptimal solutions rather than optimal solutions. It can achieve over 90% of the optimal solution with extremely fast speed, but it takes a long time to achieve the true optimal solution, which means that the local search ability is insufficient [88]. The ACO parameter settings are complex, and if the parameters are not properly set, it is easy to deviate from the high-quality solution. Therefore, for different EDM methods, one or more swarm intelligence algorithms can be selected according to the actual situation to optimize their process parameters. In addition, it is worth noting that although sinker-EDM, wire-EDM, and micro-EDM are all branches of electrical discharge machining, their machining mechanisms are basically similar. However, due to significant differences in processing objects, control methods, and processing performance, there are still certain differences in the optimization of process parameters.

Single-objective optimization evaluates only one objective and requires only the optimal value to be found according to the specific satisfaction function conditions. Multi-objective optimization has multiple evaluation functions in place and the solutions, using different evaluation functions, are different. Single-objective optimization problems are simple to implement and have well-established algorithms for solving them, with the disadvantage that they can only be solved for a single objective. Multi-objective optimization is more comprehensive and detailed for objective solving. The disadvantages of multi-objective optimization are that the units are not consistent among objectives of different nature, which are not easy to compare, and the assignment of weighted values for each objective is more subjective. The main algorithms for single objective optimization are the ACO, the PSO, the GA, ABC, the GSO, etc. The main types of multi-objective optimization are weighted methods, constrained methods, and hybrid methods combining weighted and constrained methods and multi-objective genetic algorithms. From the existing research on the optimization of EDM process parameters, most of the research is still focused on single-objective optimization, with relatively few research papers on multi-objective optimization. This may be related to the characteristics of group intelligence algorithms, which are not suitable for multi-objective optimization.

6.3. Complementarity

At present, the mathematical and theoretical foundation of swarm intelligence algorithms is relatively weak. For example, there is no exact theoretical basis for the parameter settings of related algorithms, which significantly depends on specific issues and application environments. In order to better solve practical engineering problems, such as the optimization of EDM process parameters, various types of swarm intelligence algorithms can be used for optimization, thereby increasing the probability of obtaining a global optimal solution. In addition, the existing comparative research on the optimization of EDM process parameters is insufficient, and there is a lack of standard test sets for performance evaluation. This means that there is no absolute credibility or application risk. Therefore, it is necessary to establish a diverse and unified evaluation platform to facilitate the complementary application of swarm intelligence algorithms and improve the engineering application's ability to solve practical problems.

7. Outlooks

Based on the literature review in this study, the future development directions of parameters optimization of EDM using swarm intelligence are as follows.

(1) As one of the five major intelligent forms focused on the development of the new generation of artificial intelligence, swarm intelligence has important application prospects in both civil and military fields. At present, swarm intelligence is still in its infancy in basic theory and mechanism innovation and key technology applications, and various algorithms still need to be continuously studied, improved, and expanded in scope of application. Especially in the field of electrical discharge process parameter optimization, swarm intelligence still has broad application and development space [3,26]. Integrating different swarm intelligence algorithms for optimizing electrical discharge process parameters and better searching for global optimal solutions may be a future development direction.

(2) The existing optimization of EDM process parameters is mainly oriented towards machining performance, such as MRR, SR, TWR, machining accuracy, etc. With the increasing attention paid to sustainable manufacturing, green EDM will become a key feature in the future. The response of processing output not only involves processing performance, but also involves environmental impacts, such as toxic gas emissions, processing noise, green dielectric, and so on [75,116,119,120]. Therefore, there will be more goals to optimize and the difficulty will further increase.

(3) Swarm Intelligent is a heuristic search algorithm based on the behavior of a population to find optimization for a given goal, and is centered on the ability of a population of simple individuals to achieve a more complex function through simple cooperation between them. There are many existing swarm intelligence algorithms, such as ACO, ABC, GSO, etc. These algorithms will be improved as they are applied, and it is believed that future artificial intelligence will also produce more new algorithms that

the optimization algorithm will apply to the EDM, such as selfish herds optimization, bald eagle search, etc.

(4) With the fast advancement of technology, machine learning (ML) has found widespread use in a variety of industries, including industrial testing [121–123], medical diagnostics [124,125], life sciences [126,127], and renewable energy [128–130]. AlphaFold2, for instance, created a protein structure prediction model using ML, which can predict the properties of proteins based on gene sequences and achieve 98.5% of the structure of human proteins [126]. With preliminary artificial intelligence, combining ML techniques with swarm intelligence algorithms to achieve autonomous parameter setting, the dynamic adjustment of search directions, etc., may become a research focus in the future.

Author Contributions: Y.C. (Yanyan Chen), A.L. and W.M. carried out the concept, methodology, writing the first draft. Y.C. (Yanyan Chen), S.H. and Y.C. (Yang Cao) carried out formal analysis and data analysis. Y.C. (Yanyan Chen), A.L., Y.C. (Yang Cao), S.H., Y.Z. and W.M. mainly did the investigation of literature review. W.M. undertook fund acquisition. All authors have read and agreed to the published version of the manuscript.

Funding: This work was supported by the Local Innovative and Research Teams Project of Guangdong Pearl River Talents Program (2017BT01G167), Guangdong Basic and Applied Basic Research Foundation (2022A1515140066) and partially support by Science and Technology Project of Henan Province (222102220011).

Institutional Review Board Statement: Not applicable.

Informed Consent Statement: Not applicable.

Data Availability Statement: Not applicable.

Conflicts of Interest: The authors declare no conflict of interest.

References

1. Ho, K.H.; Newman, S.T. State of the art electrical discharge machining (EDM). *Int. J. Mach. Tools Manuf.* **2003**, *43*, 1287–1300. [CrossRef]
2. Singh, V.; Sharma, A.K.; Goyal, A.; Saxena, K.K.; Negi, P.; Rao, P.C.S. Electric discharge machining performance measures and optimisation: A review. *Adv. Mater. Process. Technol.* **2023**, 1–14. [CrossRef]
3. Ming, W.Y.; Zhang, S.F.; Zhang, G.J.; Du, J.G.; Ma, J.; He, W.B.; Cao, C.; Liu, K. Progress in modeling of electrical discharge machining process. *Int. J. Heat Mass Transf.* **2022**, *187*, 122563. [CrossRef]
4. Slowik, A.; Kwasnicka, H. Evolutionary algorithms and their applications to engineering problems. *Neural Comput. Appl.* **2020**, *32*, 12363–12379. [CrossRef]
5. Dorigo, M. Optimization, Learning and Natural Algorithms. Ph.D. Thesis, Politecnico di Milano, Milan, Italy, 1992.
6. Kennedy, J.; Eberhart, R. Particle swarm optimization. In Proceedings of the ICNN'95-International Conference on Neural Networks, Perth, WA, Australia, 27 November–1 December 1995; Volume 4, pp. 1942–1948.
7. Qiu, H.X.; Zhou, Q.R.; Sun, C.H.; Wang, X.C.; Feng, Y.T. Distributed Satellite Formation Based on Swarm Intelligence of Pigeon Flocks. In Proceedings of the 2020 Chinese Automation Congress (CAC), Shanghai, China, 6–8 November 2020; pp. 4525–4530. [CrossRef]
8. Escamilla-Salazar, I.G.; Torres-Treviño, L.M.; González-Ortíz, B.; Zambrano, P.C. Machining optimization using swarm intelligence in titanium (6Al 4V) alloy. *Int. J. Adv. Manuf. Technol.* **2013**, *67*, 535–544. [CrossRef]
9. Pinheiro, C.A.D.P.; Nedjah, N.; Mourelle, L.D.M. Detection and classification of pulmonary nodules using deep learning and swarm intelligence. *Multimed. Tools Appl.* **2020**, *79*, 15437–15465. [CrossRef]
10. Ebrahimzadeh, A.; Ranaee, V. Recognition of control chart patterns using swarm intelligence and neural networks based on the statistical and shape features. *IEEJ Trans. Electr. Electron. Eng.* **2011**, *6*, 80–85. [CrossRef]
11. Kale, I.R.; Pachpande, M.A.; Naikwadi, S.P.; Narkhede, M.N. Optimization of advanced manufacturing processes using socio inspired cohort intelligence algorithm. *Int. J. Simul. Multidiscip. Des. Optim.* **2022**, *13*, 6. [CrossRef]
12. Wei, X.M. Hybrid Behavior Ant Colony Algorithm. *Adv. Mater. Res.* **2012**, *433–440*, 4496–4499. [CrossRef]
13. Tang, J.; Liu, G.; Pan, Q.T. A Review on Representative Swarm Intelligence Algorithms for Solving Optimization Problems: Applications and Trends. *IEEE/CAA J. Autom. Sin.* **2021**, *8*, 1627–1643. [CrossRef]
14. Blum, C. Ant colony optimization: Introduction and recent trends. *Phys. Life Rev.* **2005**, *2*, 353–373. [CrossRef]
15. Mohan, B.C.; Baskaran, R. A survey: Ant Colony Optimization based recent research and implementation on several engineering domain. *Expert Syst. Appl.* **2012**, *39*, 4618–4627. [CrossRef]

16. Del Valle, Y.; Venayagamoorthy, G.K.; Mohagheghi, S.; Hernandez, J.-C.; Harley, R.G. Particle Swarm Optimization: Basic Concepts, Variants and Applications in Power Systems. *IEEE Trans. Evol. Comput.* **2008**, *12*, 171–195. [CrossRef]
17. Bonyadi, M.R.; Michalewicz, Z. Particle Swarm Optimization for Single Objective Continuous Space Problems: A Review. *Evol Comput.* **2017**, *25*, 1–54. [CrossRef]
18. Zhang, Y.D.; Wang, S.H.; Ji, G.L. A Comprehensive Survey on Particle Swarm Optimization Algorithm and Its Applications. *Math. Probl. Eng.* **2015**, *2015*, 931256. [CrossRef]
19. Zhang, Z.; Ming, W.Y.; Zhang, Y.; Yin, L.; Xue, T.; Yu, H.S.; Chen, Z.J.; Liao, D.M.; Zhang, G.J. Analyzing sustainable performance on high-precision molding process of 3D ultra-thin glass for smart phone. *J. Clean. Prod.* **2020**, *255*, 120196. [CrossRef]
20. Zhang, G.J.; Liu, M.; Li, J.; Ming, W.Y.; Shao, X.Y.; Huang, Y. Multi-objective optimization for surface grinding process using a hybrid particle swarm optimization algorithm. *Int. J. Adv. Manuf. Technol.* **2014**, *71*, 1861–1872. [CrossRef]
21. Goldberg, D.E. *Genetic Algorithms in Optimization, Search and Machine Learning*; Addison Wesley: Boston, MA, USA, 1987; p. 432.
22. Holland, J.H. Genetic Algorithms. *Sci. Am.* **1992**, *267*, 66–73. [CrossRef]
23. Katoch, S.; Chauhan, S.S.; Kumar, V. A review on genetic algorithm: Past, present, and future. *Multimed. Tools Appl.* **2021**, *80*, 8091–8126. [CrossRef]
24. Gao, K.Z.; Cao, Z.G.; Zhang, L.; Chen, Z.H.; Han, Y.Y.; Pan, Q.K. A Review on Swarm Intelligence and Evolutionary Algorithms for Solving Flexible Job Shop Scheduling Problems. *IEEE/CAA J. Autom. Sin.* **2019**, *6*, 904–916. [CrossRef]
25. He, W.B.; Chen, Z.J.; Ming, W.Y.; Du, J.G.; Cao, Y.; Ma, J.; Wei, A.Y. Multi-objective optimization of glass multi-station bending machining for smartphone curved screen. *J. Braz. Soc. Mech. Sci. Eng.* **2019**, *41*, 476. [CrossRef]
26. Ming, W.Y.; Ma, J.; Zhang, Z.; Huang, H.; Shen, D.L.; Zhang, G.J.; Huang, Y. Soft computing models and intelligent optimization system in electro-discharge machining of SiC/Al composites. *Int. J. Adv. Manuf. Technol.* **2016**, *87*, 201–217. [CrossRef]
27. Karaboga, D. *An Idea Based on Honeybee Swarm for Numerica Optimization*; Technical Report TR06; Erciyes University: Kayseri, Turkey, 2005.
28. Karaboga, D.; Basturk, B. A powerful and efficient algorithm for numerical function optimization: Artificial bee colony (ABC) algorithm. *J. Glob. Optim.* **2007**, *39*, 459–471. [CrossRef]
29. Li, G.Q.; Niu, P.F.; Xiao, X.J. Development and investigation of efficient artificial bee colony algorithm for numerical function optimization. *Appl. Soft Comput.* **2012**, *12*, 320–332. [CrossRef]
30. Karaboga, D.; Gorkemli, B.; Ozturk, C.; Karaboga, N. A comprehensive survey: Artificial bee colony (ABC) algorithm and applications. *Artif. Intell. Rev.* **2014**, *42*, 21–57. [CrossRef]
31. Krishnanand, K.N.; Ghose, D. Glowworm Swarm Optimization for Simultaneous Capture of Multiple Local Optima of Multimodal Functions. *Swarm Intell.* **2009**, *3*, 87–124. [CrossRef]
32. Krihnanand, K.N.; Ghose, D. Glowworm Swarm Optimization: A new method for optimizing multi-modal function. *J. Comput. Intell. Stud.* **2009**, *1*, 93–119. [CrossRef]
33. Yang, X.S.; Deb, S. Cuckoo Search via Levy Flights. In Proceedings of the 2009 World Congress on Nature & Biologically Inspired Computing (NaBIC), Coimbatore, India, 9–11 December 2009; pp. 210–214.
34. Yang, X.S.; Deb, S. Engineering optimization by cuckoo search. *Int. J. Math. Model.* **2009**, *1*, 330–343.
35. Yang, X.S.; Deb, S. Multi-objective cuckoo search for design optimization. *Comput. Oper. Res.* **2013**, *40*, 1616–1624. [CrossRef]
36. Abdel-Basset, M.; Hessin, A.-N.; Abdel-Fatah, L. A comprehensive study of cuckoo-inspired algorithms. *Neural Comput. Appl.* **2018**, *29*, 345–361. [CrossRef]
37. Chiroma, H.; Herawan, T.; Fister, J.I.; Fister, I.; Abdulkareem, S.; Shuib, L.; Hamza, M.F.; Saadi, Y.; Abubakar, A. Bio-inspired computation: Recent development on the modifications of the cuckoo search algorithm. *Appl. Soft Comput.* **2017**, *61*, 149–173. [CrossRef]
38. Ming, W.Y.; Zhang, Y.; Li, X.K.; Shen, D.L.; He, W.B.; Ma, J.; Shen, F. Multi-objective optimization based IBCS for surface roughness and textural feature in MCVE piston machining. *Int. J. Adv. Manuf. Technol.* **2018**, *97*, 1285–1304. [CrossRef]
39. Storn, R.; Price, K. Differential Evolution—A Simple and Efficient Heuristic for Global Optimization over Continuous Spaces. *J. Glob. Optim.* **1997**, *11*, 341–359. [CrossRef]
40. Price, K.; Storn, R.; Lampinen, J.A. *Differential Evolution: A Practical Approach to Global Optimization*; Springer: Berlin/Heidelberg, Germany, 2006; Volume 104, p. 542.
41. Wu, Y.C.; Lee, W.P.; Chien, C.W. Modified the Performance of Differential Evolution Algorithm with Dual Evolution Strategy. In Proceedings of the 2009 International Conference on Machine Learning and Computing, Singapore, 30 August–3 September 2011; Volume 3, pp. 57–63.
42. Bonabeau, E.; Dorigo, M.; Theraulaz, G.; Theraulaz, G. *Swarm Intelligence: From Natural to Artificial Systems*; Oxford University Press: Oxford, UK, 1999.
43. Peng, X.Y.; Yu, P.; Dai, Y.F. Swarm Intelligence Theory and Applications. *Acta Electron. Sin.* **2003**, *31*, 1982.
44. Yang, X.S.; Cui, Z.H.; Xiao, R.B.; Gandomi, A.H.; Karamanoglu, M. *Swarm Intelligence and Bio-Inspired Computation: Theory and Applications*; Elsevier: Amsterdam, The Netherlands, 2013.
45. Available online: https://zhuanlan.zhihu.com/p/467844674 (accessed on 31 March 2023).
46. Gandomi, A.H.; Alavi, A.H. Multi-stage genetic programming: A new strategy to nonlinear system modeling. *Inf. Sci.* **2011**, *181*, 5227–5239. [CrossRef]

47. Sinha, N.; Chakrabarti, R.; Chattopadhyay, P.K. Evolutionary programming techniques for economic load dispatch. *IEEE Trans. Evol. Comput.* **2003**, *7*, 83–94. [CrossRef]
48. Yang, X.S. Firefly Algorithm, Stochastic Test Functions and Design Optimisation. *Int. J. Bio-Inspir. Comput.* **2010**, *2*, 78–84. [CrossRef]
49. Mirjalili, S.; Saremi, S.; Mirjalili, S.M.; dos S. Coelho, L. Multi-objective grey wolf optimizer: A novel algorithm for multi-criterion optimization. *Expert Syst. Appl.* **2015**, *47*, 106–119. [CrossRef]
50. Ma, J.; Ming, W.Y.; Du, J.G.; Huang, H.; He, W.B.; Cao, Y.; Li, X.K. Integrated optimization model in wire electric discharge machining using Gaussian process regression and wolf pack algorithm approach while machining SiCp/Al composite. *Adv. Mech. Eng.* **2018**, *10*, 168781401878740. [CrossRef]
51. Jameson, E.C. Description and Development of Electrical Discharge Machining (EDM). In *Electrical Discharge Machining*; Society of Manufacturing Engineers: Dearbern, MI, USA, 2001; p. 12.
52. Llanes, L.; Idanez, E.; Martínez, E.; Casas, B.; Esteve, J. Influence of electrical discharge machining on the sliding contact response of cemented carbides. *Int. J. Refract. Met. Hard Mater.* **2001**, *19*, 35–40. [CrossRef]
53. Ming, W.Y.; Zhang, G.J.; Li, H.; Guo, J.W.; Zhang, Z.; Huang, Y.; Chen, Z. A hybrid process model for EDM based on finite-element method and Gaussian process regression. *Int. J. Adv. Manuf. Technol.* **2014**, *74*, 1197–1211.
54. Mondal, N.; Nishant; Mandal, M.C.; Das, S.; Banerjee, T. Comparative study on EDM process parameters optimization using BBO and ACO algorithms. *Mater. Today Proc.* **2022**, *62*, 6601–6605. [CrossRef]
55. Aich, U.; Banerjee, S. Modeling of EDM responses by support vector machine regression with parameters selected by particle swarm optimization. *Appl. Math. Model.* **2014**, *38*, 2800–2818. [CrossRef]
56. Saffaran, A.; Moghaddam, M.A.; Kolahan, F. Optimization of backpropagation neural network-based models in EDM process using particle swarm optimization and simulated annealing algorithms. *J. Braz. Soc. Mech. Sci. Eng.* **2020**, *42*, 73. [CrossRef]
57. Tzeng, C.-J.; Chen, R.-Y. Optimization of Electric Discharge Machining Process Using the Response Surface Methodology and Genetic Algorithm Approach. *Int. J. Precis. Eng. Manuf.* **2013**, *14*, 709–717. [CrossRef]
58. Mahanta, S.; Chandrasekaran, M.; Samanta, S. GA based Optimization for the Production of Quality Jobs with Minimum Power Consumption in EDM of Hybrid MMCs. *Mater. Today Proc.* **2018**, *5*, 7788–7796.
59. Mukherjee, R.; Chakraborty, S. Selection of EDM Process Parameters Using Biogeography-Based Optimization Algorithm. *Mater. Manuf. Process.* **2012**, *27*, 954–962. [CrossRef]
60. Zainal, N.; Zain, A.M.; Sharif, S.; Hamed, H.N.A.; Yusuf, S.M. An integrated study of surface roughness in EDM process using regression analysis and GSO algorithm. *J. Phys. Conf. Ser.* **2017**, *892*, 012002. [CrossRef]
61. Shen, D.L.; Ming, W.Y.; Ren, X.G.; Xie, Z.B.; Zhang, Y.; Liu, X.W. A Cuckoo Search Algorithm Using Improved Beta Distributing and Its Application in the Process of EDM. *Crystals* **2021**, *11*, 916. [CrossRef]
62. Kumar, V.; Diyaley, S.; Chakraborty, S. Teaching-Learning-Based Parametric Optimization Of An Electrical Discharge Machining Process. *Facta Univ. Ser. Mech. Eng.* **2020**, *18*, 281–300. [CrossRef]
63. Ghadai, R.K.; Kalita, K.; Gao, X.-Z. Symbolic Regression Metamodel Based Multi-Response Optimization of EDM Process. *FME Trans.* **2020**, *48*, 404–410. [CrossRef]
64. Jafarian, F. Electro discharge machining of Inconel 718 alloy and process optimization. *Mater. Manuf. Process.* **2020**, *35*, 95–103. [CrossRef]
65. Raja, S.B.; Pramod, C.V.S.; Krishna, K.V.; Ragunathan, A.; Vinesh, S. Optimization of electrical discharge machining parameters on hardened die steel using Firefly Algorithm. *Eng. Comput.* **2015**, *31*, 1–9. [CrossRef]
66. UdhayaKumar, A.R.; Kumar, S.S. Multiobjective optimization of electric discharge machining of an Al-SiCp composite using the Taguchi-PCA method as well as the firefly and cuckoo search algorithms. *Trans. Can. Soc. Mech. Eng.* **2022**, *46*, 503–523. [CrossRef]
67. Teimouri, R.; Baseri, H. Optimization of magnetic field assisted EDM using the continuous ACO algorithm. *Appl. Soft Comput.* **2014**, *14*, 381–389. [CrossRef]
68. Singh, N.K.; Upadhyay, R.K.; Singh, Y.; Sharma, A. Intelligent hybrid approaches for ensuring better prediction of gas-assisted EDM responses. *SN Appl. Sci.* **2020**, *2*, 914. [CrossRef]
69. Rouniyar, A.K.; Shandilya, P. Optimization of process parameters in magnetic field assisted powder mixed EDM of aluminium 6061 alloy. *Proc. Inst. Mech. Eng. Part C J. Mech. Eng. Sci.* **2021**, *235*, 2998–3014. [CrossRef]
70. Danish, M.; Amin, A.; Abdul-Rani, A.M.; Rubaiee, S.; Ahmed, A.; Zohura, F.T.; Ahmed, R.; Yildirim, M.B. Optimization of hydroxyapatite powder mixed electric discharge machining process to improve modified surface features of 316L stainless steel. *Proc. Inst. Mech. Eng. Part E J. Process. Mech. Eng.* **2022**. [CrossRef]
71. Dang, X.-P. Constrained multi-objective optimization of EDM process parameters using kriging model and particle swarm algorithm. *Mater. Manuf. Process.* **2018**, *33*, 397–404. [CrossRef]
72. Moghaddam, M.A.; Kolahan, F. Modeling and Optimization of the Electrical Discharge Machining Process Based on a Combined Artificial Neural Network and Particle Swarm Optimization Algorithm. *Sci. Iran.* **2020**, *27*, 1206–1217. [CrossRef]
73. Zhang, Z.; Zhang, Y.; Lin, L.Q.; Wu, J.H.; Yu, H.S.; Pan, X.; Li, G.L.; Wu, J.; Xue, T. Study on productivity and aerosol emissions of magnetic field-assisted EDM process of SiCp/Al composite with high volume fractions. *J. Clean. Prod.* **2021**, *292*, 126018. [CrossRef]

74. Garg, A.; Lam, J.S.L. Modeling multiple-response environmental and manufacturing characteristics of EDM process. *J. Clean. Prod.* **2016**, *137*, 1588–1601. [CrossRef]
75. Ming, W.Y.; Shen, F.; Zhang, G.J.; Liu, G.D.; Du, J.G.; Chen, Z.J. Green machining: A framework for optimization of cutting parameters to minimize energy consumption and exhaust emissions during electrical discharge machining of Al 6061 and SKD 11. *J. Clean. Prod.* **2021**, *285*, 124889. [CrossRef]
76. Liao, Y.S.; Huang, J.T.; Chen, Y.H. A study to achieve a fine surface finish in Wire-EDM. *J. Mater. Process. Technol.* **2004**, *149*, 165–171. [CrossRef]
77. Zhang, G.J.; Zhang, Z.; Ming, W.Y.; Guo, J.W.; Huang, Y.; Shao, X.Y. The multi-objective optimization of medium-speed WEDM process parameters for machining SKD11 steel by the hybrid method of RSM and NSGA-II. *Int. J. Adv. Manuf. Technol.* **2014**, *70*, 2097–2109. [CrossRef]
78. Huang, Y.; Ming, W.Y.; Guo, J.W.; Zhang, Z.; Liu, G.D.; Li, M.Z.; Zhang, G.J. Optimization of cutting conditions of YG15 on rough and finish cutting in WEDM based on statistical analyses. *Int. J. Adv. Manuf. Technol.* **2013**, *69*, 993–1008. [CrossRef]
79. Zhang, Z.; Ming, W.Y.; Zhang, G.J.; Huang, Y.; Wen, X.Y.; Huang, H. A new method for on-line monitoring discharge pulse in WEDM-MS process. *Int. J. Adv. Manuf. Technol.* **2015**, *81*, 1403–1418. [CrossRef]
80. Ming, W.Y.; Cao, C.; Shen, F.; Zhang, Z.; Liu, K.; Du, J.G.; Jia, H.J. Numerical and experimental study on WEDM of BN-AlN-TiB_2 composite ceramics based on a fusion FEM model. *J. Manuf. Process.* **2022**, *76*, 138–154. [CrossRef]
81. Sharma, V.S.; Sharma, N.; Singh, G.; Gupta, M.K.; Singh, G. Optimization of WEDM Parameters While Machining Biomedical Materials Using EDAS-PSO. *Materials* **2023**, *16*, 114. [CrossRef]
82. Tzeng, C.-J.; Yang, Y.-K.; Hsieh, M.-H.; Jeng, M.-C. Optimization of wire electrical discharge machining of pure tungsten using neural network and response surface methodology. *Proc. Inst. Mech. Eng. Part B J. Eng. Manuf.* **2011**, *225*, 841–852. [CrossRef]
83. Zhang, G.J.; Zhang, Z.; Guo, J.W.; Ming, W.Y.; Li, M.Z.; Huang, Y. Modeling and Optimization of Medium-Speed WEDM Process Parameters for Machining SKD11. *Mater. Manuf. Process.* **2013**, *28*, 1124–1132. [CrossRef]
84. Singh, H.; Kumar, V.; Kapoor, J. Multi-response optimization of WEDM process parameters during the fabrication of microchannels for industrial applications. *Mater. Today Proc.* **2021**, *46*, 81–88. [CrossRef]
85. Kuruvila, N.; Ravindra, H.V. Parametric Influence And Optimization Of Wire Edm On Oil Hardened Non Shrinking Steel. In Proceedings of the ASME 2011 International Mechanical Engineering Congress and Exposition, Denver, CO, USA, 11–17 November 2011; Volume 54891, pp. 891–899.
86. Rao, R.V.; Pawar, P.J. Modelling and optimization of process parameters of wire electrical discharge machining. *Proc. Inst. Mech. Eng. Part B J. Eng. Manuf.* **2009**, *223*, 1431–1440. [CrossRef]
87. Fard, R.K.; Afza, R.A.; Teimouri, R. Experimental investigation, intelligent modeling and multi-characteristics optimization of dry WEDM process of Al–SiC metal matrix composite. *J. Manuf. Process.* **2013**, *15*, 483–494. [CrossRef]
88. Rao, M.S.; Venkaiah, N. A modified cuckoo search algorithm to optimize Wire-EDM process while machining Inconel-690. *J. Braz. Soc. Mech. Sci. Eng.* **2017**, *39*, 1647–1661. [CrossRef]
89. Saravanan, M.; Thiyagarajan, C.; Somasundaram, S. Parametric optimization of wirecut-electrical discharge machining through cuckoo search algorithm. *Mater. Today Proc.* **2020**, *22*, 681–687. [CrossRef]
90. Kulkarni, V.N.; Gaitonde, V.N.; Karnik, S.R.; Manjaiah, M.; Davim, J.P. Machinability Analysis and Optimization in Wire EDM of Medical Grade NiTiNOL Memory Alloy. *Materials* **2020**, *13*, 2184. [CrossRef]
91. Nayak, B.B.; Kumar, N.; Singh, A.; Sahu, S.; Das, S.S.; Behera, D.D.; Sahu, R.K. Genetic programming approach to predict the performance characteristics of WEDM taper cutting process. *Mater. Today Proc.* **2022**, *62*, 4504–4508. [CrossRef]
92. Xu, J.K.; Li, M.Y.; Zhong, J.Z.; Hou, Y.G.; Xia, S.J.; Yu, P. Process parameter modeling and multi-response optimization of wire electrical discharge machining NiTi shape memory alloy. *Mater. Today Commun.* **2022**, *33*, 104252. [CrossRef]
93. Ranjan, A.; Chakraborty, S.; Kumar, D.; Bose, D. Multi-objective optimization of a hybrid machining process abrasive powder mixed WEDM of Inconel 718 using particle swarm optimization technique. *Mater. Today Proc.* **2019**, *18*, 3606–3616. [CrossRef]
94. Vaidyaa, P.; John, J.J.; Puviyarasan, M.; Prabhu, T.R.; Prasad, N.E. Wire EDM Parameter Optimization of AlSi10Mg Alloy. *Trans. Indian Inst. Met.* **2021**, *74*, 2869–2885. [CrossRef]
95. Zhang, Z.; Ming, W.Y.; Huang, H.; Chen, Z.; Xu, Z.; Huang, Y.; Zhang, G.J. Optimization of process parameters on surface integrity in wire electrical discharge machining of tungsten tool YG15. *Int. J. Adv. Manuf. Technol.* **2015**, *81*, 1303–1317. [CrossRef]
96. Soepangkat, B.O.P.; Norcahyo, R.; Rupajati, P.; Effendi, M.K.; Agustin, H.C.K. Multi-objective optimization in wire-EDM process using grey relational analysis method (GRA) and backpropagation neural network–genetic algorithm (BPNN–GA) methods. *Multidiscip. Model. Mater. Struct.* **2019**, *15*, 1016–1034. [CrossRef]
97. Jain, S.; Parashar, V. Comparison of priori and posteriori approach of multi-objective optimization for WEDM on Ti6Al4V alloy. *Mater. Res. Express* **2022**, *9*, 076504. [CrossRef]
98. Majumder, A.; Das, A.; Das, P.K. A standard deviation based firefly algorithm for multi-objective optimization of WEDM process during machining of Indian RAFM steel. *Neural Comput. Appl.* **2018**, *29*, 665–677. [CrossRef]
99. Kondayya, D.; Krishna, A.G. An integrated evolutionary approach for modelling and optimization of wire electrical discharge machining. *Proc. Inst. Mech. Eng. Part B J. Eng. Manuf.* **2011**, *225*, 549–567. [CrossRef]
100. Rahman, M.; Asad, A.B.M.A.; Masaki, T.; Saleh, T.; Wong, Y.S.; Kumar, A.S. A multi-process machine tool for compound micromachining. *Int. J. Mach. Tool Manuf.* **2010**, *50*, 344–356. [CrossRef]

101. Wong, Y.S.; Rahman, M.; Lim, H.S.; Han, H.; Ravi, N. Investigation of micro-EDM material removal characteristics using single RC-pulse discharges. *J. Mater. Process. Technol.* **2003**, *140*, 303–307. [CrossRef]
102. Jahan, M.P.; Rahman, M.; Wong, Y.S. A review on the conventional and micro-electro discharge machining of tungsten carbide. *Int. J. Mach. Tool Manuf.* **2011**, *51*, 837–858. [CrossRef]
103. Chen, Y.Y.; Guo, X.D.; Zhang, G.J.; Cao, Y.; Shen, D.L.; Li, X.K.; Zhang, S.F.; Ming, W.Y. Development of a Hybrid Intelligent Process Model for Micro-Electro Discharge Machining Using the TTM-MDS and Gaussian Process Regression. *Micromachines* **2022**, *13*, 845. [CrossRef]
104. Pham, D.T.; Dimov, S.S.; Bigot, S.; Ivanov, A.; Popov, K. Micro-EDM—Recent developments and research issues. *J. Mater. Process. Tech.* **2004**, *149*, 50–57. [CrossRef]
105. Chen, Z.; Zhou, H.B.; Yan, Z.J.; Han, F.L.; Yan, H.Z. Machining characteristics of 65 vol.% SiCp/Al composite in micro-WEDM. *Ceram. Int.* **2021**, *47*, 13533–13543. [CrossRef]
106. Quarto, M.; D'Urso, G.; Giardini, C. Micro-EDM optimization through particle swarm algorithm and artificial neural network. *Precis. Eng.* **2021**, *73*, 63–70. [CrossRef]
107. Ahuja, N.; Batra, U.; Kumar, K. Multicharacteristics optimization of electrical discharge micro hole drilling in Mg alloy using hybrid approach of GRA-regression-PSO. *Grey Syst. Theory Appl.* **2021**, *11*, 136–151. [CrossRef]
108. Rao, T.B. Prediction of EDMed micro-hole quality characteristics using hybrid bio-inspired machine learning-based predictive approaches. *Int. J. Interact. Des. Manuf.* **2022**, *17*, 747–764. [CrossRef]
109. Quarto, M.; D'Urso, G.; Giardini, C.; Maccarini, G.; Carminati, M. A Comparison between Finite Element Model (FEM) Simulation and an Integrated Artificial Neural Network (ANN)-Particle Swarm Optimization (PSO) Approach to Forecast Performances of Micro Electro Discharge Machining (Micro-EDM) Drilling. *Micromachines* **2021**, *12*, 667. [CrossRef]
110. Zhang, L.X.; Jia, Z.Y.; Wang, F.J.; Liu, W. A hybrid model using supporting vector machine and multi-objective genetic algorithm for processing parameters optimization in micro-EDM. *Int. J. Adv. Manuf. Technol.* **2010**, *51*, 575–586. [CrossRef]
111. Dilip, D.G.; Panda, S.; Mathew, J. Characterization and Parametric Optimization of Micro-hole Surfaces in Micro-EDM Drilling on Inconel 718 Superalloy Using Genetic Algorithm. *Arab. J. Sci. Eng.* **2020**, *45*, 5057–5074. [CrossRef]
112. Cao, Y.; Zhang, Y.C.; Ming, W.Y.; He, W.B.; Ma, J. Review: The Metal Additive-Manufacturing Technology of the Ultrasonic-Assisted Wire-and-Arc Additive-Manufacturing Process. *Metals* **2023**, *13*, 398. [CrossRef]
113. Du, J.G.; Tian, B.; Duan, L.Y.; Ming, W.Y.; Liu, K.; He, W.B. Traditional and non-traditional machining technology of metallic glass. *Int. J. Adv. Manuf. Technol.* **2022**, *120*, 2879–2902. [CrossRef]
114. Zhang, Z.; Qiu, W.Z.; Zhang, G.J.; Liu, D.H.; Wang, P. Progress in applications of shockwave induced by short pulsed laser on surface processing. *Opt. Laser Technol.* **2023**, *157*, 108760. [CrossRef]
115. Ming, W.Y.; Xie, Z.B.; Ma, J.; Du, J.G.; Zhang, G.J.; Cao, C.; Zhang, Y. Critical review on sustainable techniques in electrical discharge machining. *J. Manuf. Process.* **2021**, *72*, 375–399. [CrossRef]
116. Ming, W.; Cao, C.; Xie, Z.; Liu, X.; Xu, Y.; Jiang, Z.; Li, X.; Liu, K.; Guo, X.; Yuan, J.; et al. Green manufacturing: A comparative study of renewable dielectrics in the EDM process. *J. Braz. Soc. Mech. Sci. Eng.* **2022**, *44*, 580. [CrossRef]
117. Zhang, Z.; Zhang, Y.; Ming, W.; Zhang, Y.; Cao, C.; Zhang, G. A review on magnetic field assisted electrical discharge machining. *J. Manuf. Process.* **2021**, *64*, 694–722. [CrossRef]
118. Available online: https://zhuanlan.zhihu.com/p/428166363 (accessed on 31 March 2023).
119. Ma, J.; Yuan, J.; Ming, W.Y.; He, W.B.; Zhang, G.J.; Zhang, H.M.; Cao, Y.; Jiang, Z.W. Non-traditional processing of carbon nanotubes: A review. *Alex. Eng. J.* **2022**, *61*, 597–617. [CrossRef]
120. Ming, W.Y.; Zhang, Z.; Wang, S.Y.; Zhang, Y.M.; Shen, F.; Zhang, G.J. Comparative study of energy efficiency and environmental impact in magnetic field assisted and conventional electrical discharge machining. *J. Clean. Prod.* **2019**, *214*, 12–28. [CrossRef]
121. Ming, W.Y.; Zhang, S.F.; Liu, X.W.; Liu, K.; Yuan, J.; Xie, Z.B.; Sun, P.Y.; Guo, X.D. Survey of Mura Defect Detection in Liquid Crystal Displays Based on Machine Vision. *Crystals* **2021**, *11*, 1444. [CrossRef]
122. He, W.B.; Jiang, Z.W.; Ming, W.Y.; Zhang, G.J.; Yuan, J.; Yin, L. A critical review for machining positioning based on computer vision. *Measurement* **2021**, *184*, 109973. [CrossRef]
123. Ming, W.Y.; Cao, C.; Zhang, G.J.; Zhang, H.M.; Zhang, F.; Jiang, Z.W.; Yuan, J. Review: Application of Convolutional Neural Network in Defect Detection of 3C Products. *IEEE Access* **2021**, *9*, 135657–135674. [CrossRef]
124. Kandubothula, V.; Uppada, R.; Nandan, D. A review on detection of breast cancer cells by using various techniques. *Soft Comput. Theor. Appl.* **2020**, *1154*, 813–823. [CrossRef]
125. He, W.; Liu, T.; Han, Y.; Ming, W.; Du, J.; Liu, Y.; Yang, Y.; Wang, L.; Jiang, Z.; Wang, Y.; et al. A review: The de-tection of cancer cells in histopathology based on machine vision. *Comput. Biol. Med.* **2022**, *146*, 105636. [CrossRef]
126. Jumper, J.; Evans, R.; Pritzel, A.; Green, T.; Figurnov, M.; Ronneberger, O.; Tunyasuvunakool, K.; Bates, R.; Žídek, A.; Potapenko, A.; et al. Highly accurate protein structure prediction with AlphaFold. *Nature* **2021**, *596*, 583–589. [CrossRef]
127. Sherman, M.A.; Yaari, A.U.; Priebe, O.; Dietlein, F.; Loh, P.-R.; Berger, B. Genome-wide mapping of somatic mutation rates uncovers drivers of cancer. *Nat. Biotechnol.* **2022**, *40*, 1634–1643. [CrossRef]
128. Ming, W.; Sun, P.; Zhang, Z.; Qiu, W.; Du, J.; Li, X.; Zhang, Y.; Zhang, G.; Liu, K.; Wang, Y.; et al. A systematic review of machine learning methods applied to fuel cells in performance evaluation, durability prediction, and application monitoring. *Int. J. Hydrogen Energy* **2022**, *48*, 5197–5228. [CrossRef]

129. Chen, Y.Y.; Liu, Y.K.; Xu, Y.J.; Guo, X.D.; Cao, Y.; Ming, W.Y. Review: Modeling and Simulation of Membrane Electrode Material Structure for Proton Exchange Membrane Fuel Cells. *Coatings* **2022**, *12*, 1145. [CrossRef]
130. Zhang, Z.; Liu, D.; Zhang, Y.; Xue, T.; Huang, Y.; Zhang, G. Fabrication and droplet impact performance of superhydrophobic Ti6Al4V surface by laser induced plasma micro-machining. *Appl. Surf. Sci.* **2022**, *605*, 154661. [CrossRef]

metals

MDPI

Review

Review: The Metal Additive-Manufacturing Technology of the Ultrasonic-Assisted Wire-and-Arc Additive-Manufacturing Process

Yang Cao *, Yanchao Zhang, Wuyi Ming, Wenbin He and Jun Ma

Mechanical and Electrical Engineering Institute, Zhengzhou University of Light Industry, Zhengzhou 450002, China
* Correspondence: 2015063@zzuli.edu.cn

Abstract: Ultrasonic-assisted wire–arc additive manufacturing (WAAM) can refine microstructures, enhancing performance and improving stress concentration and anisotropy. It has important application prospects in aerospace, weaponry, energy, transportation, and other frontier fields. However, the process parameters of ultrasonic treatment as an auxiliary technology in the WAAM process still have an important impact on product performance indicators, such as the amplitude of the ultrasonic tool, the distance between the points of action of the product, and the scanning speed. The number of ultrasonic impacts influences the performance indexes. Therefore, these parameters must be optimized. This paper describes the advantages and the defects of WAAM components, as well as the principle and development status of ultrasonic treatment technology. Subsequently, this paper also briefly describes how ultrasonic-assisted technology can refine the crystal and improve the mechanical properties of WAAM components. Finally, we review the influence of process parameters (such as ultrasonic amplitude, application direction, and impact times) on the product materials. In this paper, a comprehensive optimization method for ultrasonic parameters is proposed to improve the mechanical properties of WAAM components.

Keywords: metal additive manufacture; wire-and-arc additive manufacture; ultrasonic assistance; mechanical properties; anisotropy; process parameters

1. Introduction

With the rapid development of aerospace, shipbuilding, automobile, and rapid manufacturing of molds, modern equipment manufacturing has developed a lightweight, complex structure, few parts, a poor service environment, and high-performance requirements, which result in higher requirements for the accuracy, cost, and manufacturing cycle of such parts. However, it is difficult for traditional casting, forging, and machining methods to meet the manufacturing needs. Thus, there is an urgent need to adopt advanced production technologies to overcome the limitations of traditional molding and manufacturing methods [1–3]. With a high degree of freedom and flexible manufacturing, additive manufacturing (AM) is suitable for the low-cost, small-batch production and manufacture of complex structural and functional parts, especially in the direct forming of metal parts [4]. AM technology involves computer-aided design (CAD). A digital model of the component is discretized into points, lines, or surfaces. It then accumulates the material layer-by-layer to form the 3D model, so that the forming process is highly flexible, and the entire manufacture of mechanical parts or models with a complex structure can be realized without any tools or molds. This layer-by-layer manufacturing feature increases the free degree of design and manufacturing flexibility and can realize complex structure customization and rapid manufacturing and shorten the manufacturing cycle. Compared with traditional manufacturing methods, AM is a "bottom-up" material accumulation manufacturing technology [5–9]. It has been widely recognized and adopted by advanced manufacturing powers worldwide and has gradually become a research hotspot worldwide [10]. Per ISO/ASTM standards, AM divides the techniques used to create the layers into seven categories, of which the first four on the list are suitable for metals.

1. Binder jetting;
2. Directed energy deposition;
3. Powder bed fusion;
4. Sheet lamination;
5. Material extrusion;
6. Material jetting;
7. Vat photo polymerization.

The directed energy deposition (DED) process is defined as the "additive manufacturing process in which focused thermal energy is used to fuse materials by melting as they are being deposited" per the ISO/ASTM 52900 standard [11,12]. According to the type of heat source, DED technology can be divided into three categories: laser, electron beam, and arc [13–17]. Over the past 20 years, the research has mainly centered on powder-based metal-additive manufacturing that uses laser and electron beams as a source of heat, in which complex structural parts are constantly prepared layer-by-layer through continuous melting or sintering of metal powder. This process has been used in key components of aerospace, defense, military, energy, and other high-tech areas. The additive-manufacturing process of metal includes a complex physical metallurgical process, and the material melting, solidification, and cooling experienced during part forming are all carried out quickly. The solidification process has the common characteristics of a high-temperature gradient, high cooling rate, and local remelting of deposited materials induced by a complex thermal cycle. However, because of the characteristics of its raw materials and heat sources, metal-powder-based laser and electron-beam additive-manufacturing technology is limited to specific structures or components and cannot be formed. Even if it could be formed, the cost of raw materials and time would also be high, and the technology would have many shortcomings, such as: (1) a slow forming rate for the laser heat source and the low laser absorption rate of the aluminum alloy, (2) the vacuum furnace size for the electron-beam heat source being limited to the component volume, and (3) the high cost of the integral forming equipment of the laser heat source and electron-beam heat source [18–22].

As shown in Figure 1, DED technologies include processes with a wire feed and processes with a powder feed. The ability to use yarn as a raw material instead of powder reduces the price per kilogram, increases the efficiency of material utilization, reduces the need for powder recycling systems, creates an environmentally friendly process, and adds the possibility of easy material handling by the operator without health and safety concerns [23]. The most typical example is WAAM. WAAM uses a low-cost arc as the heat source and inert gas as protection; it can be processed in an open environment, and the size of the component is not limited, so the arc-increasing technology has a low manufacturing cost and high processing efficiency. The size of the processing components can be large, so it is especially suitable for metal parts with large-scale complex sizes [24,25].

However, when examining the possibilities and limitations of these methods, it was concluded that additive processes for metal fabrication have many challenges to overcome. In terms of forming morphology, the arc fuse component has the characteristics of large surface roughness and low dimensional accuracy. In the microstructure, the arc fuse component has the characteristics of a rough structure, directed growth, and separation of the composition. In terms of mechanical properties, the arc fuse component has the characteristics of significant anisotropy and heterogeneity. There are many process parameters of WAAM that greatly influence the above characteristics of the product [26–28]. To address these problems, many universities and research institutions have initiated relevant research and proposed various interlaminar strengthening processes and processing methods, such as laser impact peening, mechanical hammering, interlayer rolling, ultrasonic impact, ultrasonic vibration, and micro-forging strengthening technologies [29]. These technologies can effectively reduce the residual stress, thermal deformation, porosity, and surface roughness of the deposited parts. At the same time, the directed growth of the microstructure caused by the temperature gradient of the metal parts is improved, and the coarse column crystalline structure transforms into a fine equiaxed crystal structure. Then, the grain refinement and homogenization of the microstructure distribution are obtained, which can improve the microhardness and tensile properties of the parts and reduce the degree of anisotropy [30–34]. The ultrasonic vibration interlayer-strengthening technology is based on controlling the solidification process of the molten pool to refine the microstructure and achieve the purpose of strengthening. The means of laser impact, mechanical hammering, inter-laminar rolling, and micro-forging interlayer strengthening mainly rely on the plastic deformation on the surface of the deposition layer, and the dislocation wall is formed after a large number of dislocations are multiplied and then transformed into sub-grains to realize grain refinement. In addition, plastic deformation greatly increases the deformation energy storage in the treatment area. Promoting recrystallization under the comprehensive thermal effect of the subsequent deposition layer can effectively change the internal structure of the parts, weaken or even eliminate the anisotropy, and improve the overall properties. Among these post-processing technologies, ultrasonic-assisted manufacturing is the mainstream method for products at present. Ultrasonic-assisted manufacturing includes the technology of ultrasonic impact and ultrasonic vibration. The treatment process and equipment are relatively simple but can greatly improve the properties of the products, which has broad research and application prospects in AM [35–42]. Wang et al. [43] reduced several problems (e.g., heterogeneous microstructures, internal cavities, and large porosity) that exist in laser-engineered net-shaping (LENS)-fabricated IN718 parts. Ultrasonic vibration-assisted (UV-A) LENS was developed based on the ultrasonic vibration's actions in material melting and solidification. This process is an effective manufacturing method in the fabrication of IN718 parts.

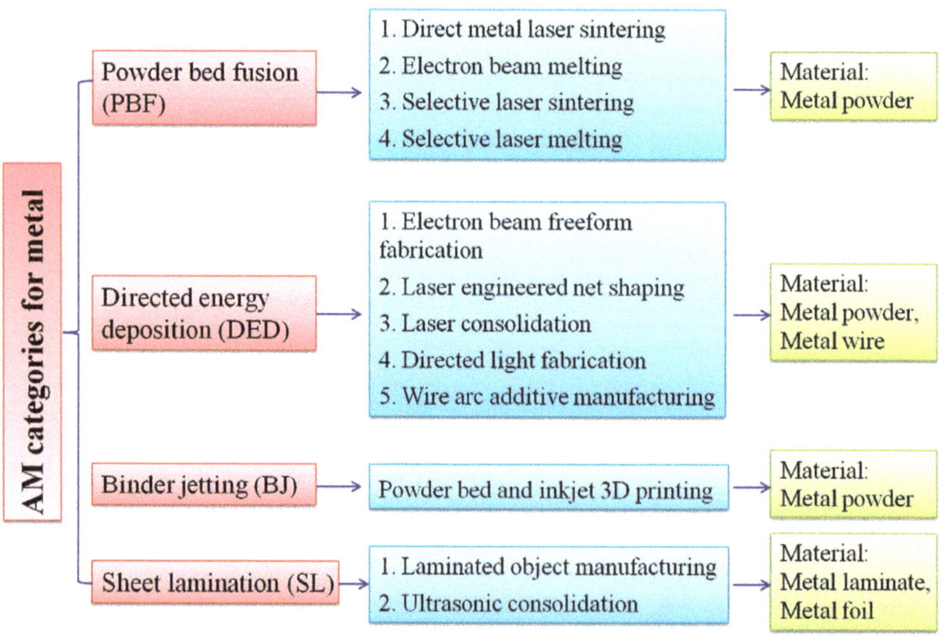

Figure 1. Classification of metal additive manufacturing, reproduced or adapted from [13], with permission from Elsevier, 2021.

At present, the research on ultrasonic-assisted additive manufacturing generally investigates the effect of ultrasonic vibration on the microstructure and mechanical properties of additively manufactured parts under the conditions of a single parameter, and there are a few studies on the effect of ultrasonic multi-parameter optimization. Based on this situation, this paper briefly introduces the advantages of WAAM and the unavoidable shortcomings of the WAAM workpiece, summarizes the development status and existing problems of ultrasonic-assisted technology, and discusses the characteristics and advantages of ultrasonic-assisted WAAM. Finally, the ultrasonic-assisted process parameters are analyzed in detail. A comprehensive optimization method for ultrasonic parameters is proposed to improve the mechanical properties of WAAM components, which can lay the foundation for the application of wire–arc manufacturing components in the aerospace field and forecast the development trend.

2. Introduction of Wire–Arc Additive Manufacturing

WAAM is an advanced material-increasing manufacturing technology [44] that uses the arc as the heat source and the wire as the raw material. The melted wire is deposited layer-by-layer under the control of the program. According to the established three-dimensional digital model, the metal parts are gradually formed by line to the surface and then to the body. Because it uses fewer materials for the manufacturing of components than other AM techniques, WAAM has proven useful in terms of consumption, and the equipment cost is comparatively low. In general terms, a comparison of the main characteristics of AM technologies is shown in Table 1. A schematic diagram of WAAM is shown in Figure 2.

Table 1. Comparison of main metal additive-manufacturing processes [45].

Factors	PBF	LMD	EBAM	WAAM
Accuracy	High: ±0.05–0.2 mm	Mean: ±0.2 mm	Requires final machining	Requires final machining
Structural integrity	High: vacuum chamber/protected atmosphere	High: protected atmosphere	High: empty chamber	High: protected Atmosphere
Productivity	Low: ~0.1–0.2 kg/h	Average: ~0.5–1 kg/h	High: 3–11 kg/h	High: ~10 kg/h
Part size	Limited by working space (max. 800 × 400 × 500 mm)	Large, limited by machine range	Large, limited by machine range	Large, limited by the range of the machine
Geometric complexity	High	Media	Low–medium	Low–medium
Industrial application	Direct manufacture of complex parts	Repair of parts, coatings, direct	Repair of parts, coatings, direct	Repair of parts, coatings, direct
Price of equipment	High	High	High	Under
Raw material cost	Very High	High	Under	Under

2.1. Characteristics of Wire–Arc Additive Manufacturing

WAAM can be used to manufacture large components because of its high deposition efficiency, low deposition cost, and high material utilization. The high flexibility of WAAM equipment can realize the deposition of metal parts with different shapes [46]. Furthermore, more traditional processes can be combined with the technology to solve the problems of coarse microstructure and the low mechanical properties caused by the large heat input to further improve the comprehensive mechanical properties of deposited materials.

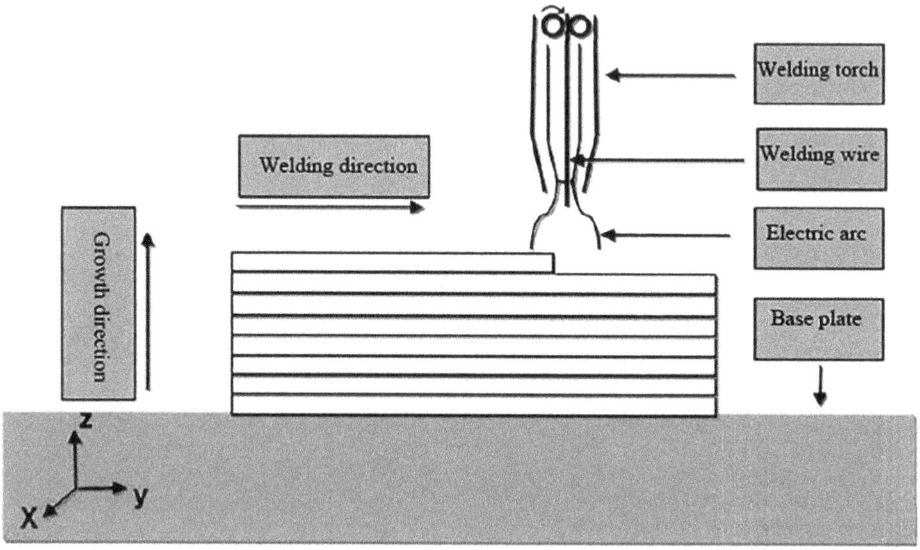

Figure 2. Schematic diagram of wire–arc additive-manufacturing principle.

(1) The material utilization and manufacturing efficiency

The high material utilization rate of WAAM is one of its most important benefits for the aerospace industry. Martina et al. [47] formed a Ti-6Al-4V thin-walled wall by WAAM and determined its process window. They found that the stacking width of the thin-walled wall can reach 17.4 mm, the material utilization rate can reach 93%, and the deposition rate can reach 1.8 kg/h.

(2) The size of the product

In the laser- and electron-beam additive-manufacturing technology, the forming size of the product is limited by the forming chamber, powder bed, and electron-beam vacuum chamber, which is a bottleneck in its development. WAAM does not require strict atmosphere protection, and the size of the product and forming equipment is almost unlimited, which effectively solves the forming size problems of laser and electron-beam additive manufacturing [48].

(3) The composition and high compactness of the product

There are many defects in the forming process of laser-forming metal powder technology, such as porosity and unfused components. WAAM has a long high-temperature arc time and a large molten pool. Therefore, the components have uniform composition and high compactness. Compared with ingot metallurgy, the mechanical properties and overall quality of the component are better and have the characteristics of high toughness, high heat treatment stability, and high thermal fatigue cracking resistance [49–51].

2.2. *Defects in the WAAM*

After years of development, the performance of the WAAM process has been significantly improved. However, in the preparation process, the materials easily produce defects such as micro-pores, micro-cracks, and deformation. The reasons may include improper setting of welding parameters, low precision of the welding machine, unstable wire transportation, welding heat accumulation, and the manufacturing environment. As shown in Table 2 [52], there are many forms of processing: Gas Metal Arc Welding (GMAW), Gas Tungsten Arc Welding (GTAW), Tungsten Inert-Gas (TIG), Cold Metal Transfer (CMT), Direct Current Electrode Positive-GMAW(DCEP-GMAW), Double-electrode gas metal arc welding (DE-GMAW), Variable-polarity gas metal arc welding (VP-GMAW), CMT pulse advanced (CMT-PADV), Pulsed plasma arc deposition (PPAD), Pulse metal inert gas (PMIG).

Table 2. Defects in WAAM [52].

Material	Welding Process	Defect Type					
		Porosity	Crack	Deformation	Anisotropy	Substrate Adherence	Surface Quality
Ti-6AL-4V	TIG	-	-	-	Serious	Good	Good
	Plasma	-	-	-	Light	Good	Good
	CMT	-	-	-	Serious	Good	Good
	DCEP-GMAW	-	-	-	Light	Medium	Poor
HO8Mn2Si steel	DE-GMAW	Low	-	-	Light	Good	Medium
Copper-coated steel	GMAW	-	-	-	Light	Good	Medium
ER4043 aluminum alloy	CMT	High	-	-	Light	Good	Good
	VP-GMAW	-	-	-	Light	Good	Medium
AA2319 aluminum alloy	CMT	High	-	-	Light	Good	Good
	CMT-PADV	-	-	-	Light	Good	Good
5356 aluminum alloy	VP-GMAW	-	√	-	Light	Good	Good
Inconel-625	PPAD	High	√	-	Light	Good	Good
	GTAW	-	-	-	Light	Good	Good
Inconel-718	GMAW	High	√	√	Light	Good	Good
AZ31 magnesium alloy	PMIG	-	-	-	Serious	Medium	Medium
Nickel–aluminum–copper	CMT	-	-	-	Light	Good	Good
Steel–bronze bimetal	GMAW	-	-	-	Light	Good	Good

Researchers have analyzed the microstructure and properties of WAAM products and found that the mechanical properties were reduced. Taking pores and residual stress as examples, the disadvantages of the materials caused by the defects in the WAAM process are briefly described.

(1) Porosity

As one of the main defects in metal additive manufacturing, porosity defects seriously affect the quality of manufacturing workpieces. Effective control and elimination of stomata is the key technology for improving the quality of the product in AM. Kobayashi et al. studied the cast alloy Al-Si-Mg [53], and they found that the porosity seriously affects the degree of the error. Because of the reduction in the tensile bearing capacity of the pore zone, the pore zone yields first, resulting in strain concentration near the pore and thus premature fracture. The study result suggested that the elimination of pores in aluminum is crucial to improving its mechanical properties. Yi et al. [54] welded 6061 aluminum alloy T-joints using a double pulse MIG welding process. Then, the welded T-joints were post-weld heat-treated. The microstructure of the weld with different aging temperatures and times was studied by transmission electron microscopy and scanning electron microscopy. The mechanical properties were tested by a hardness test and tensile test. They found that the aluminum alloy easily produces pores and voids in the WAAM process because of the high thermal expansion coefficient, high solidification shrinkage, and other welding characteristics of aluminum alloy. The high melting point dense oxide film (alumina) is easily formed on the alloy surface, and the molten metal is blocked in the molten pool, which prevents the metal from being deposited.

In order to control pores, most scholars considered the cleanliness of raw materials themselves, thereby reducing porosity at the source. If there is a degree of contamination, such as water, grease, or other hydrocarbons, on the surface of the wire or substrate, it is difficult to remove, and these pollutants can be easily absorbed by the molten pool and then produce pores after solidification.

(2) Residual stress and deformation

Because of the large heat input of arc fuse technology and the complex temperature-field distribution of the workpiece, the deformation and residual stress produced by this process are inevitable. In the WAAM of titanium alloys, Zhang et al. [55] found that the existence of welding residual stress at the interface causes microscopic deformation of the parts, including longitudinal and transverse thermal shrinkage, bending deformation, shear deformation, and torsion deformation. Chi et al. [56] studied the microstructure of metals manufactured by WAAM technology and found that there were many columnar grains that created harmful residual tensile stress in the product. After heat treatment and laser impact strengthening, serious plastic deformation occurred on the surface of the product. The plastic deformation transformed the residual tensile stress into compressive stress, which was beneficial to the product. Malcolm Dinovitzer et al. [57] used 304 stainless-steel plates as the substrate for WAAM technology. They used the Taguchi method and analysis of variance (ANOVA) to determine the travel speed, wire feed rate, current, and argon flow rate. They found that increasing travel speed or decreasing current caused a decrease in melt-through depth and an increase in roughness. Fernando Veiga et al. [58] applied WAAM to Invar. They utilized and unified research material previously investigated in WAAM technology and took a new approach based on the study of symmetrical phenomena that guarantee the quality of the process. On the other hand, the monitoring of the symmetry of the melting pool utilized thermography techniques. The geometry of the zero beads was determined through the analysis of the coefficient of symmetry in the longitudinal and transverse planes. The results indicated that residual stress and deformation are generated in the process of WAAM.

3. The Current Status of Ultrasound-Assisted Development

Ultrasonic-assisted additive manufacturing mainly includes ultrasonic vibration technology and ultrasonic impact treatment (UIT). The two methods can be traced back to the 1960s, beginning with the research of Mukhanov et al. [59]. The researchers applied the continuous output ultrasonic vibration produced by the ultrasonic transducer directly on the surface of treated hard materials such as carbide alloy and rhinestone. A thin plastic deformation layer was formed on the surface of the treated material, and the microstructure and the residual stress distribution of the deformation were changed. In the 1970s, Statnikov formally proposed the UIT technology [60]. In recent years, ultrasonic assistance as a post-processing technology has been widely studied.

3.1. Ultrasonic Vibration-Assisted Technology

The ultrasonic vibration system consists of an adjustable height frame, generator, transducer, and tool head. First, the ultrasonic generator converts ordinary alternating current into ultrasonic-frequency alternating current, then the ultrasonic-frequency alternating current is converted into mechanical vibration with the same frequency by the transducer. The mechanical vibration then acts on the tool head through the horn; under the effect of self-weight and external pressure, the tool head applies high-frequency vibration to the corresponding working parts, such as the welding platform, weld, or corresponding experimental parts, to improve the material performance [61].

When the high-power ultrasonic wave propagates in the liquid metal, the mechanical, thermal, cavitation, and sound flow effects are produced by the high-frequency vibration, which can refine the grain and homogenize and purify the structure, for example by degassing, slag removal, and purification [62]. Some scholars have introduced ultrasonic vibration technology into the AM process to improve the microstructure and reduce residual stress. Cong et al. [63] studied a novel ultrasonic vibration-assisted laser-engineered net-shaping (LENS) technology. They found that the introduction of ultrasonic vibration energy into the molten pool in the LENS process can reduce the voids and microcracks and significantly refine the grains of the deposited samples. Chen et al. [64,65] studied the effect of ultrasonic vibration on the laser cladding process. The results indicated that ultrasonic vibration can reduce cracks, homogenize chemical composition, refine the microstructure, and improve the properties of the cladding layer. Qin et al. [66] investigated the process of ultrasound-assisted laser deposition of a titanium alloy. They suggested that the solid crystal network formed by the staggered connection of primary dendrites is broken by the ultrasonic cavitation and mechanical effect, and the grains are refined. At the same time, the melt is supplemented during continuous crystallization, and the tensile stress between dendrites is reduced. Wang [67,68] discussed the microstructure and properties of IN718 deposited by ultrasonic vibration-assisted laser forming. The results indicated that the surface roughness and residual stress of the parts were significantly improved after ultrasonic vibration. The microstructure, tensile, and yield strength were remarkably improved. Yashwan et al. [69] welded a 6061-T6 aluminum alloy to a 6082-T6 aluminum alloy of a different thickness. The joint was prepared by an ultrasonic-assisted cold metal transfer (U-CMT) welding process, and the mechanical properties and microstructure of the joint were improved. The results revealed that the weld-bead dimensions were increased with the aid of ultrasonic vibrations. The tensile strength and microhardness of the welding joint were enhanced. Yuan of Harbin Engineering University [70] proposed a manufacturing method for the metal additive by introducing ultrasonic vibration in the laser-fuse additive process and studied the microstructure evolution of a titanium alloy in this process with a high-power ultrasonic energy field. The ultrasonic vibration device in this study adopted an innovative structural design, and the amplitude transformer and ultrasonic probe were rigidly connected to maintain better wear resistance and toughness of the device, as shown in Figure 3. The ultrasonic wave acts on the molten pool at ultrasonic frequency (20 kHz), resulting in sound pressure and cavitation in the molten pool, which effectively restrains the epitaxial growth trend of the primary β crystal of the titanium alloy, refines the grain structure of the titanium alloy, and promotes the formation of the equiaxed crystal. According to the above analysis, the introduction of ultrasonic vibration energy improved the internal structure of the material, made the microstructure of the deposition layer more uniform and the grain finer, and at the same time effectively reduced the residual stress in the material.

3.2. Ultrasonic Impact Treatment Technology

In the UIT technology [71,72] electrical oscillation signals with high frequency in the 15~40 kHz range are output by the ultrasonic generator and then transformed into a high-frequency mechanical vibration by the transducer. The amplitude of the mechanical vibration is amplified and transmitted to the vibrating tool head by the horn connected to the transducer, which promotes the high-frequency impact movement of the impact needle in the impact gun, and the projectile hits the part surface to strengthen it. Figure 4 contains a schematic diagram of the UIT process.

The equipment used in this technology has the advantages of small size, easy operation, and high energy density input, so it is widely used in the welding industry. The UIT process can eliminate the residual tensile stress of welded joints, introduce beneficial compressive stress, reduce the stress concentration on the surface of welded joints, and improve the fatigue performance of welded joints [73,74]. Roy et al. [75] tested the fatigue resistance of welded transverse stiffeners and cover plate details by UIT and determined the residual stress distribution near the weld toe before and after treatment. The results indicated that UIT can smooth the profile of the weld toe, improve its microstructure, and introduce beneficial compressive residual stress in the weld area. Mordyuk et al. [76] studied the effect of UIT on the forming process of AISI 321 stainless steel and found that the nanocrystalline surface layer was formed after UIT. Yang et al. [77] studied the UIT-assisted WAAM-forming Ti-6Al-4V alloy. The results revealed that most of the residual stress was eliminated, and a refined bamboo-like microstructure was formed in the process; the grain size of this bamboo-like structure was much smaller than the coarse columnar grains. Cheng et al. [78] investigated the influence of ultrasonic impact treatment on austenitic stainless-steel welds. The residual stress test results showed that the residual compressive stress was produced in the ultrasonic impact area at a depth of 1.5~1.7 mm and a width of 15 mm. The maximum compressive stress of the treated materials exceeded the yield strength of the base metal. Rao et al. [79] took Q345 steel as the research object to analyze the effect of ultrasonic impact on weld residual stress. The stress measured by the hole-drilling technique was reduced by 34~55% at the weld depth of 2~4 mm under ultrasonic impact conditions. Mordyuk et al. [80] analyzed the microstructure changes of Zr-1%Nb alloy after ultrasonic vibration treatment for 4 min. The results showed that an ultra-fine grain formed on the surface of the Zr-1%Nb alloy, and the grain size was about 100 nm. In the abovementioned studies, UIT technology was shown to significantly reduce the residual stress in the material and improve the internal structure.

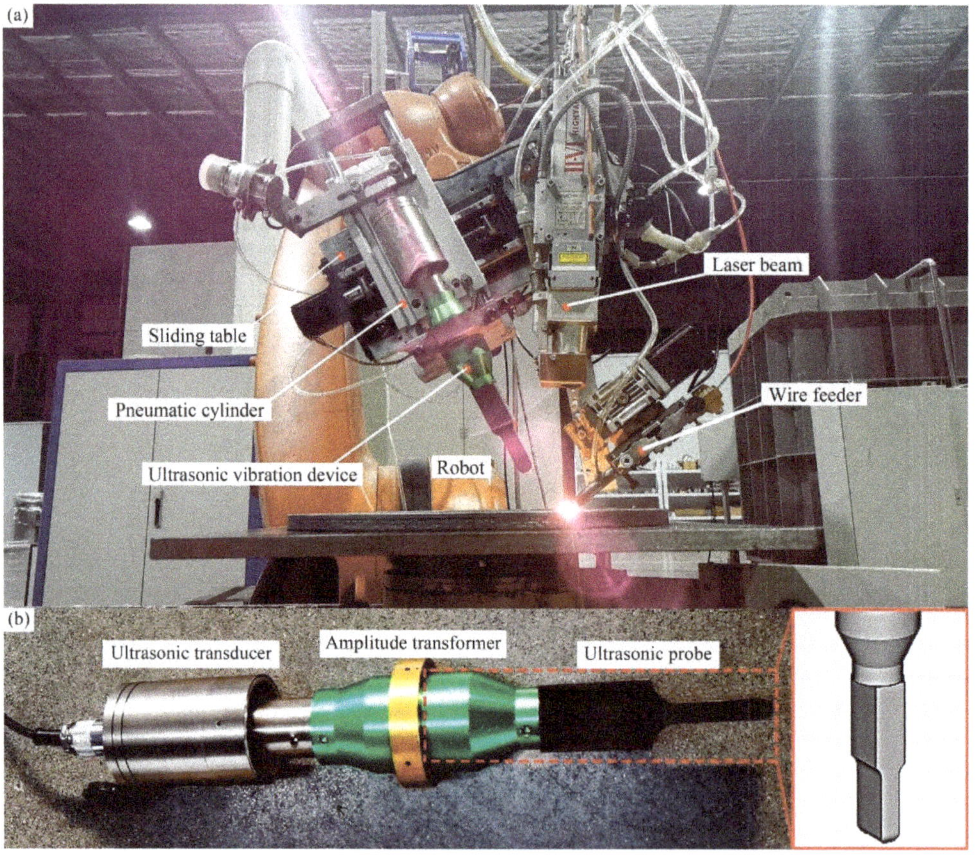

Figure 3. Set-up for LWAM assisted with ultrasonic vibration: (**a**) overall arrangement diagram; (**b**) an ultrasonic vibration device, reproduced or adapted from [70], with permission from Elsevier, 2021.

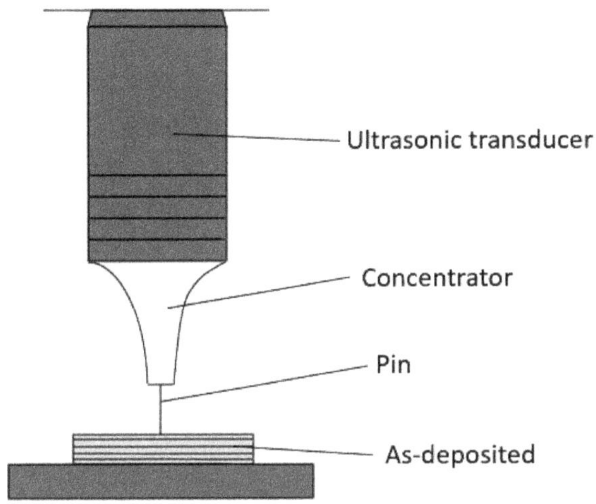

Figure 4. Schematic diagram of the UIT process.

4. UIT-Assisted WAAM

There is a complex physical metallurgical process for the metal parts during the arc fuse forming process. The as-cast microstructure is presented in the materials with the high temperature of the welding heating source, resulting in a dendritic structure, coarse grains, and so on. These problems significantly affect the mechanical properties, such as toughness, strength, and fatigue properties, and cause the anisotropy of the material. To solve these problems, UIT technology was introduced as a post-processing method to improve the synthesized material properties [81].

Figure 5 shows a schematic diagram of UIT-assisted WAAM equipment. The ultrasonic generator is connected to the welding head of the gas metal arc-adding manufacturing (GMA-AM) equipment through the two fixtures. In the process of adding material, the impact thimble is aligned and perpendicular to the deposition layer so that it impacts the deposition surface by self-weight after the movement of the robot [82].

4.1. Ultrasonic-Assisted Elimination of Residual Stress

When the weld toe is subjected to ultrasonic impact treatment, the smooth transition at the weld toe is realized, the stress concentration factor is reduced, and the residual tensile stress is reduced or even eliminated, and it is possible to introduce residual compressive stress. As a result, the mechanical properties such as the tension and fatigue of the weld can be improved [83].

Cheng et al. [78] explored the internal stress of ultrasonic impact and shot peening by neutron diffraction and X-ray diffraction and measured the residual stress under ultrasonic impact conditions for the first time. They found that the compressive stress produced by the two techniques was greater than the yield stress of the matrix material, and the compressive stress caused by the ultrasonic impact was about 1.6 mm, while the depth of the stress layer produced by shot peening was about 0.8 mm. Ye of the Armored Army Engineering College and others studied the effect of ultrasonic impact on the residual stress and fatigue performance of ultra-high-strength steel welded joints and tested the solder joint and weld bead with an X-ray stress analyzer [84]. They found that the fatigue strength of the welded state increased by 33.3% after ultrasonic impact treatment. Amir Abdullah et al. [85] studied the effect of UIT on the residual stress of 304 stainless-steel welded joints. UIT reduced the stress concentration factor at the weld and increased the fatigue life of the components by 29%. Xiao used the method of real-time ultrasonic impact to eliminate the residual stress of the sample and carried out an ANSYS simulation [86]. It was found that the real-time ultrasonic impact on the back of the weld effectively improved the longitudinal residual stress, but the improvement of the transverse residual stress was not obvious. The surfacing on a 10 mm-thick plate produced the stress of 39.36 MPa transversely and 210 MPa longitudinally. After real-time UIT, the longitudinal residual stress was eliminated by about 33.38%. When Gao used the ultrasonic impact to treat the multi-layer welding of high-strength alloy steel [87], it was found that ultrasonic impact could significantly improve the residual stress of multi-layer welding parts. The maximum tensile stress after impact was about 28% lower than that during welding.

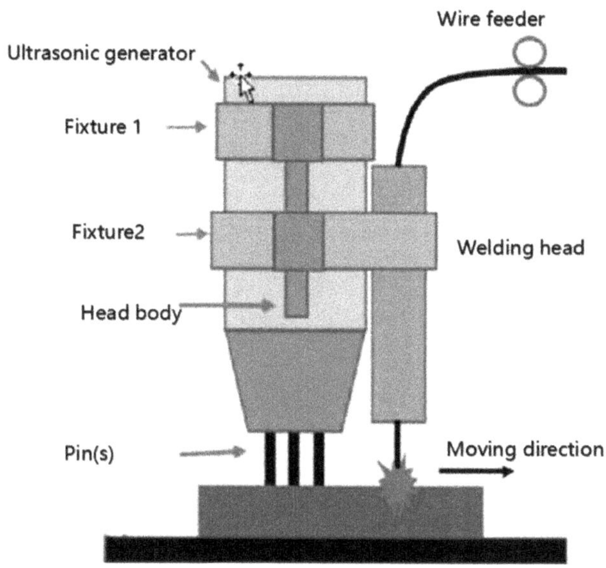

Figure 5. Schematic diagram of joint equipment.

4.2. Ultrasound-Assisted Preparation of Nano-Layer on Metal Surface

Ultrasound-assisted additive manufacturing can help the metal surface to obtain nano-layers, and nanomaterials have excellent mechanical and physical properties because of their unique structure. Preparing a nano-layer on the metal surface can optimize the comprehensive mechanical properties and service environment of the materials.

Shao et al. [88] studied the effect of different levels of ultrasonic power on the grain refinement of the AZ80 alloy. Table 3 shows that ultrasonic impact can refine grains. From 0 to 600 W, the average grain size decreased from 387 to 147 μm. However, when the power increased from 950 to 1400 W, the average grain size increased. The thermal effect of ultrasound may be an important reason for this transformation.

Table 3. Effect of ultrasonic power on the grain size of the AZ80 [88].

Ultrasonic power/W	0	200	400	600	800	1000	1400
Average grain size/μm	387	199	175	147	168	184	207

Wang et al. [89] studied the ultrasonic impact treatment of 4Si steel. Through ultrasonic surface nano-processing, the working head exerts a certain degree of ultrasonic-frequency mechanical vibration on the workpiece along the normal direction of the surface, and under the conditions of a certain static pressure and feed speed, the working head transmits the pressure and ultrasonic impact vibration to the surface of the machined machinery parts in the rotating state, and the material produces elastic–plastic deformation by ultrasonic punching. A nano-layer was obtained on the surface, and the average grain size was about

50 nm. Hu of Qingdao University of Technology (Qingdao, China) and others emitted ultrasonic vibration to the surface of the molten pool through an ultrasonic horn installed on the side axis [90], thus exerting an influence on the laser cladding process. They found that the ultrasonic flow accelerated the flow of the molten pool and suppressed the element segregation. Xu et al. [91] of Jiangsu University introduced ultrasound to assist laser cladding to prepare an iron-based coating by means of substrate vibration. They found that the original columnar crystal in the coating was transformed into the fine equiaxed crystal, and the hardness, elastic modulus, and wear resistance of the coating were improved. Wen of Harbin Engineering University fabricated the AlCrCoFeMn0.5Mo0.1 coating by introducing ultrasound into the substrate vibration to assist laser cladding [92]. It was found that the dilution rate of the coating increased with the increase in laser line energy.

4.3. Ultrasound-Assisted Improvement of Morphology

To improve the morphology of the weld-toe zone through the study of the titanium alloy butt joint and cross joint after UIT, Yang and others [93] found that relative to the transition radius of the weld-toe zone, there is no obvious change in weld height, width, or inclination angle. The stress concentration factor decreases with the increase in the weld-toe transition radius. After UIT, the stress concentration factor of the butt joint and cross joint is reduced by 20% and 22%, respectively. Zhao of Tianjin University (Tianjin, China) and others [94] studied the fatigue properties of TC4 welded joints and calculated the stress concentration factor by ANSYS. They found that the stress concentration factor at the weld toe was 4.1. After ultrasonic impact, the stress concentration factor decreased to 2.79. The geometry of the weld toe was improved, and the stress concentration degree was improved by about 47%.

4.4. Ultrasonic-Assisted Improvement of Fatigue Strength

Per Jahn Haagensen et al. [95] evaluated the ultrasonic impact on high-strength steel and welded AA5083 aluminum plates and found that the ultrasonic impact can greatly improve the fatigue prosperities of samples, in which the fatigue strength of an aluminum alloy and high-strength steel increased by 93% and 121%, respectively. Based on the finite element simulation of ductile damage of GTN, Yang et al. revealed the plastic damage of AISI304 stainless steel caused by ultrasonic impact technology [96]. In their study, they found that the damaged area of the ultrasonic impact process was annular, and the indentation center was not affected. The main pores in the aluminum alloy WAAM process are hydrogen pores, and hydrogen elements mainly come from the surface of the aluminum substrate, welding wire, and protective gas atmosphere. The effects of ultrasonic frequency on the porosity of the fabricated IN718 parts are shown in Figure 6 [43]. Compared with the LENS process without ultrasonic vibration, UV-A LENS produced reduced pore number and average pore size, as shown in Figure 6a–d. Figure 6e shows that the porosity value first decreased after introducing ultrasonic vibration with a frequency of 25 kHz and then increasing the ultrasonic frequency from 25 to 41 kHz.

4.5. Other Methods

He et al. [97] experimented with arc adding and ultrasonic impact to manufacture titanium alloy parts, and the properties of the parts were optimized by ultrasonic impact. The results indicated that the ultrasonic impact broke the microstructure of the sample from columnar to fine equiaxed grains, and the isotropy of the material increased from 12.5% to 1.5%. Xu et al. [98] studied the effect of ultrasonic vibration on the microstructure and properties of the formed parts in the process of CMT addition to TC4 titanium alloy. The results showed that compared with the parts without ultrasonic vibration, the grain size of the parts formed by ultrasonic vibration was reduced, and the mechanical properties were significantly improved. Chen and others [99] studied the effect of ultrasonic vibration on the microstructure and tensile properties of aluminum bronze alloy made by arc addition. The results indicated that under the ultrasonic impact conditions, the microstructure of the sample was refined, the tensile strength increased from 436.5 ± 2.1 to 516.1 ± 0.5 MPa, and the anisotropy was effectively restrained. Ding et al. [70] studied the effect of ultrasound on the grain refinement of the Ti-6Al-4V alloy produced by laser fusing. The results showed that the introduction of ultrasonic impact can effectively restrain the epitaxial growth of the β crystal, weaken the texture strength of the β crystal, and refine the microstructure of the formed parts. Figure 7a [100] shows the inverse pole figure (IPF) orientation map of the austenite (face-centered cubic, FCC) phase along the building direction in the deposited layer without UV. It indicates that the grain structure within a 1.0 mm distance from the substrate has extremely irregular morphology, the columnar dendritic grains grow epitaxially along the building direction, and many fine grains exist between the coarse columnar dendritic grains. In contrast, the grain structure of the deposition layer with UV is more uniform, and there are a few large amounts of coarse columnar dendrite grains in the deposition layer (see Figure 7b).

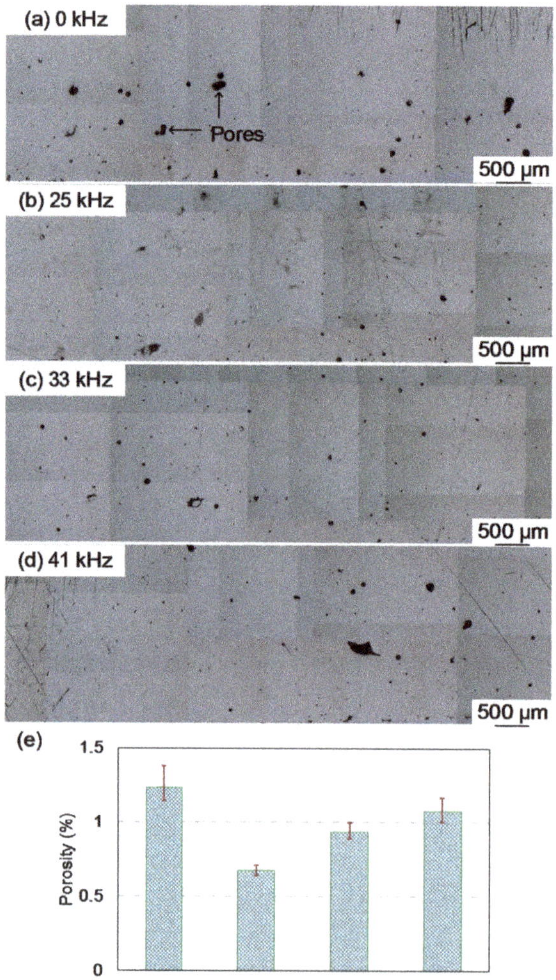

Figure 6. Effects of ultrasonic frequency on porosity: (**a**) 0; (**b**) 25; (**c**) 33; (**d**) 41 kHz; (e) effects of ultrasonic frequency, reproduced or adapted from [43], with permission from Elsevier, 2020.

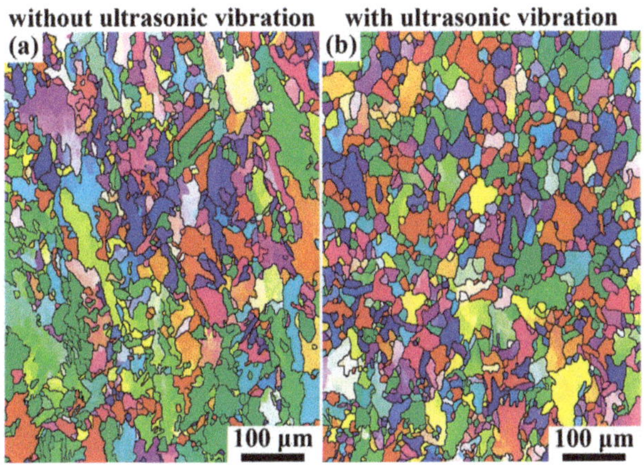

Figure 7. Inverse pole figure (IPF) orientation maps depicting grain structures along the building direction (z) for the austenitic phase in samples: (**a**) no additional ultrasonic vibration; (**b**) additional ultrasonic vibration, reproduced or adapted from [100], with permission from Elsevier, 2021.

Through the comparison of Figure 7a,b, it can be seen that the grain structure of the deposited parts is obviously refined after additional ultrasonic vibration.

5. Ultrasound-Assisted Process Parameters

In the process of ultrasonic-assisted WAAM, the amplitude of the ultrasonic tool, the distance between the action points of the product, the scanning speed, and the number of ultrasonic impacts have an important influence on the performance indexes. This section further analyzes and discusses the mechanism of columnar-crystal and strip-band formation and columnar-crystal transformation into the equiaxial crystal and the influence of ultrasonic-assisted treatment process parameters.

5.1. Ultrasonic Amplitude

Ultrasonic amplitude is one of the main parameters of an ultrasonic device. It determines the maximum amplitude of ultrasound acting on the sedimentary layer. After the action spacing is fixed, the difference in the input ultrasonic amplitude also directly determines the intensity of the ultrasonic energy field acting on the molten pool, which determines the influence of ultrasound on the solidification process of the molten pool.

As shown in Figure 8, the surface of sample 1 without ultrasonic vibration is covered with the metallic luster, its weld width is 5 mm, and reinforcement is 4.2 mm. Compared with sample 1, the surface of sample 2 obviously loses the metallic luster and capillary waves appear (it does not affect the formed appearance), its weld width is 5.7 mm, and the reinforcement is 4 mm. A small number of tiny pores appear on the surface of sample 3, and a larger number of pores gather on the surface at the arc-end point. Its weld width is 5.8 mm, and the reinforcement is 4 mm. The phenomenon of the accumulation of pores at the shallow surface of the weld can be explained as follows: With the promotion of the ultrasonic wave, the tiny bubbles escape upward. When the next layer of cladding is carried out, the cladding metal melts the metal below the weld, and most of the pores are eliminated, leaving only some pores farther away from the weld surface. It can be concluded from samples 1, 2, and 3 that the ultrasonic vibration increases the weld width and lowers the reinforcement.

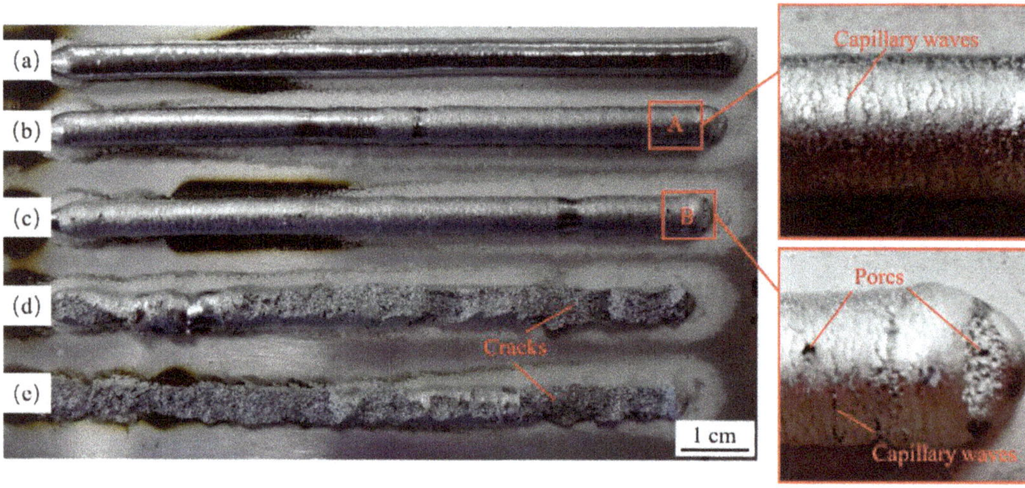

Figure 8. Effect of different ultrasonic amplitudes on weld: (**a**) sample 1; (**b**) sample 2; (**c**) sample 3; (**d**) sample 4; (**e**) sample 5, reproduced or adapted from [81], with permission from MDPI, 2022.

As shown in Figure 8d,e, with the increase in ultrasonic amplitude, it can be seen that the incompletely solidified weld cracks because of the excessive energy of ultrasonic vibration. Thus, when the molten pool is not completely solidified, the weld cracks when the tensile stress generated by ultrasonic vibration during the vibration period exceeds the ultimate tensile strength that can be endured in this state.

In other words, in the process of WAAM assisted by ultrasonic vibration, when the distance between the ultrasonic vibration and the molten pool is fixed, only when the input ultrasonic amplitude reaches a certain threshold can it greatly influence the solidification process of the molten pool [101] to meet the needs of microstructure refinement. Through a comparison of the temperature gradient, cooling rate, and solidification rate with different amplitudes, ultrasonic vibration was found to accelerate the flow in the molten pool with the increase in ultrasonic amplitude, thus homogenizing the temperature field, reducing the temperature gradient in the molten pool, and improving the solidification speed of the molten pool [102].

5.2. *Ultrasonic Frequency*

Ultrasonic frequency is also one of the main parameters in ultrasonic device. The microstructural properties (including grain sizes, phase composition, precipitated phase morphology, bonding interface, and porosity) and mechanical properties (including microhardness, wear rate, and elastic modulus) of parts at different levels of ultrasonic frequencies (0, 25, 33, and 41 kHz) were investigated.

As shown in Figure 9, the grains grew along the building direction with the columnar shape without ultrasonic vibration, while the grains demonstrated uniform distribution and equiaxed shapes with uniform grain boundaries.

However, as shown in Figure 6. The porosity value first decreased after introducing ultrasonic vibration with a frequency of 25 kHz and then increasing the ultrasonic frequency from 25 to 41 kHz. As a result, it is easy to remove the material from the part and then increase wear rate. The elastic modulus was decreased when ultrasonic frequency was increased from 25 to 41 kHz. The change in porosity might play an important role in the change in elastic modulus of the parts. In other words, ultrasonic frequency has a great influence on the microstructural properties, and there is an optimal range of ultrasonic frequency.

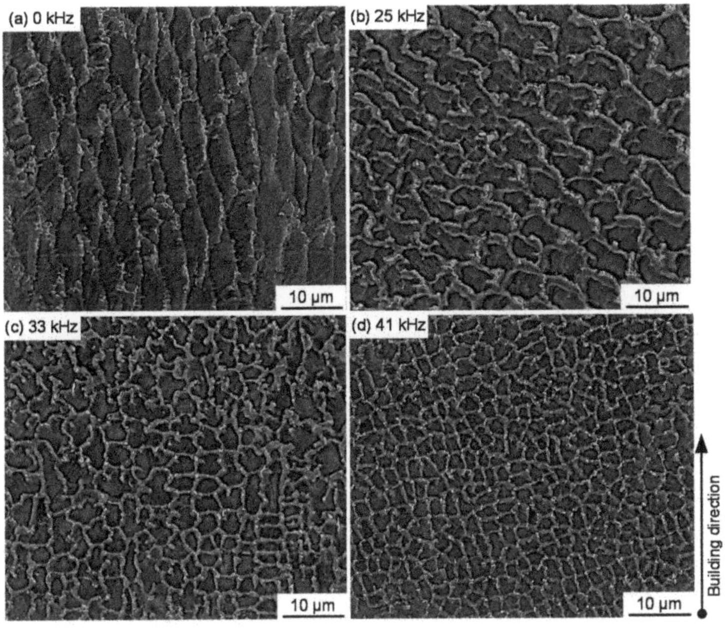

Figure 9. Effect of different ultrasonic frequencies on microstructures and grain sizes: (**a**) 0; (**b**) 25; (**c**) 33; (**d**) 41 kHz, reproduced or adapted from [43], with permission from Elsevier, 2020.

5.3. Ultrasonic Impact Scanning Speed

The top of the sample without ultrasonic impact treatment showed flaky, large, and skeletal columnar crystals, as shown in Figure 10a. By reducing the ultrasonic impact scanning speed and breaking the dendrites, grain refinement was more obvious. When the ultrasonic scanning speed was 6 mm/s, compared with the untreated top arc area, the microstructure mainly developed columnar crystals; some grains were refined, and the columnar crystals were large and not completely broken into sheets, as shown in Figure 10b. When the scanning speed was 4 mm/s, some columnar crystals were interrupted, and the grains were transformed into equiaxed grains, as shown in Figure 10c. When the scanning speed was 2 mm/s, the degree of refinement further increased, as shown in Figure 10d [103].

5.4. Ultrasonic Impact Times

Figure 11 shows the internal pores of a 2219 aluminum alloy manufactured by arc additive manufacturing without ultrasonic impact treatment and with treatment applied once, twice, and three times. The 2219 aluminum alloy without ultrasonic impact treatment contained many micron pores. There were fewer pores in the samples treated by the ultrasonic impact, the pores were extruded into an oval shape, and the tiny pores were closed. The statistical data of porosity number, average diameter, porosity, and porosity roundness (the ratio of the short axis to the long axis) of different UIT samples are shown in Table 4. After one ultrasonic impact, the number of stomata in the sample decreased by 13.2% compared with the non-ultrasonic impact treatment, and the average diameter decreased by 17.1%. After two ultrasonic impacts, the number of pores in the sample decreased by 20.4%, and the average diameter decreased by 13 µm compared with the non-ultrasonic impact treatment. After three ultrasonic impacts, the average pore diameter decreased to less than 20 µm, and the pore roundness was 0.66 [104].

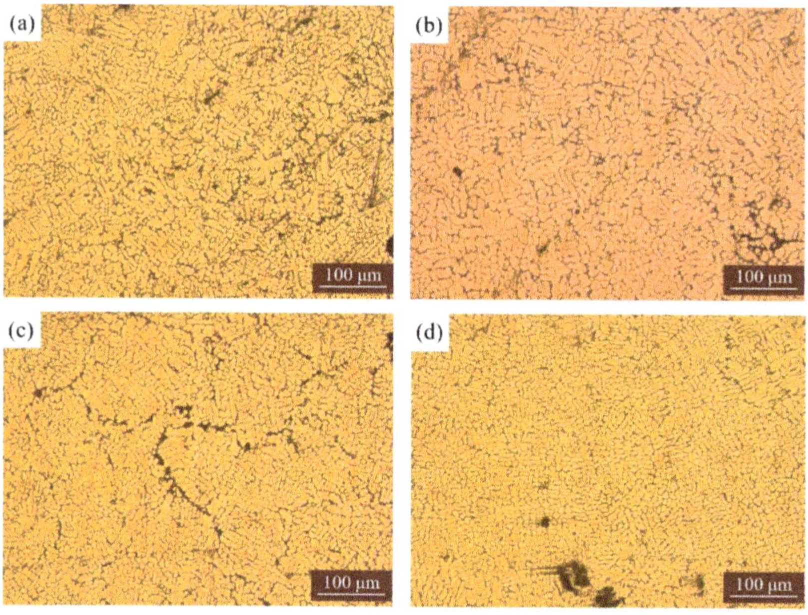

Figure 10. Microstructure of the arc area on the top of the additive sample with the different ultrasonic peening scanning speeds: (**a**) no ultrasonic impact; (**b**) 6 mm/s; (**c**) 4 mm/s; (**d**) 2 mm/s, reproduced or adapted from [81], with permission from MDPI, 2022.

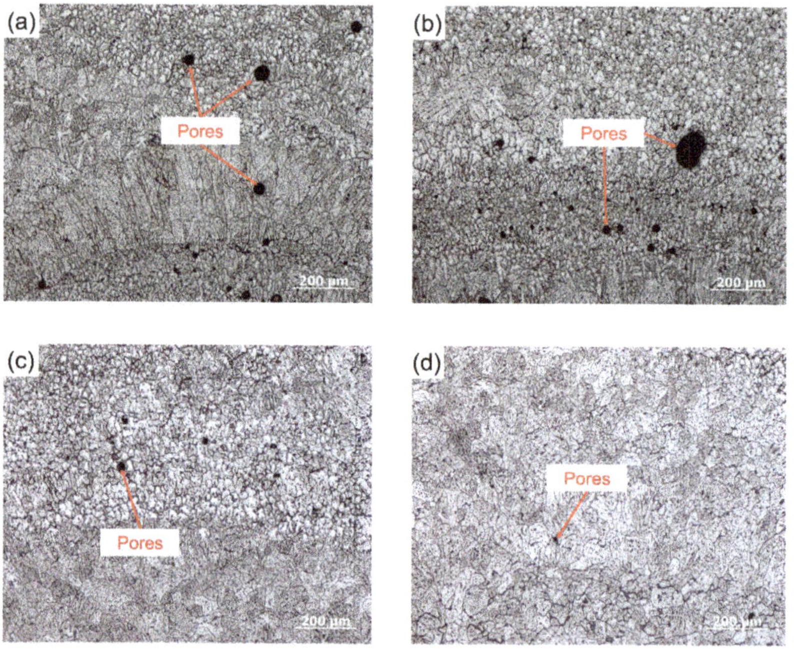

Figure 11. Pores in the samples with the different interlayered ultrasonic peening times: (**a**) no ultrasonic peening; (**b**) ultrasonic peening once; (**c**) ultrasonic peening twice; (**d**) ultrasonic peening three times, reproduced or adapted from [105], with permission from Elsevier, 2022.

Table 4. Pores of samples with the different interlayered ultrasonic peening times [81].

Index/Number of Ultrasonic Peening Treatments	None	One	Two	Three
Number of stomata	986	856	785	679
Average diameter/µm	41	34	28	19
Porosity ratio/%	0.95	0.83	0.78	0.62
Porosity roundness	0.86	0.75	0.71	0.66

5.5. Action Direction of Ultrasonic Vibration

The tensile test results of the single-channel and multi-layer deposited parts are shown in Figure 12. The tensile strength, yield strength, and elongation of the 2319 aluminum alloy deposits assisted by ultrasonic vibration in the vertical direction were strengthened, and the tensile strength, yield strength, and elongation were improved. Among them, the tensile strength increased by about 8.61%, the yield strength increased by 3.44%, and the elongation increased by about 40.51%. This was mainly due to the grain refinement caused by additional ultrasonic vibration. However, when the yield strength of the 2319 aluminum alloy deposits assisted by ultrasonic vibration in the horizontal direction was increased, the elongation decreased significantly. The reason for this phenomenon may be that there are many porosity defects in the horizontal drawing parts with additional ultrasonic vibration, which leads to stress concentration in the process of tension, resulting in brittle fracture of the drawing parts [106].

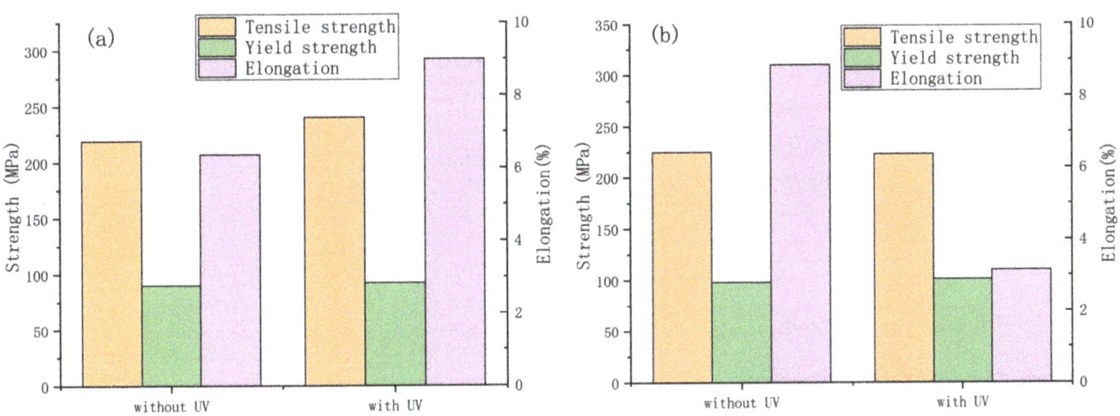

Figure 12. Tensile test results of single-channel multilayer deposition samples: (**a**) horizontal direction; (**b**) vertical direction.

6. Conclusions

The WAAM process has the advantages of high material utilization and deposition efficiency, low production cost, and highly versatile equipment, so it has broad application prospects in the rapid prototyping of large and complex light alloy components in the aerospace field. However, because of the limitations of the high heat input, large residual stress, relatively low forming accuracy, and surface roughness in the manufacturing process, this process is still far from large-scale production and application. Other conclusions drawn from this review are:

- Ultrasonic assistance has been continuously combined with WAAM to consolidate large-size continuous materials and rapid precision manufacturing ability, which can effectively improve the microstructure and mechanical properties.
- Ultrasonic-assisted process parameters (such as ultrasonic amplitude, application direction, and impact times) have an intuitive effect on the product materials. In order

to improve the mechanical properties of WAAM product components, each parameter must be accurately set during the ultrasonic-assisted WAAM process.

7. Overlook

The comprehensive optimization system of ultrasonic-assisted process parameters is not perfect. WAAM components have limitations such as large residual stress and low forming accuracy, which make them unsuitable for large-scale production in the aerospace field. In future research, in order to improve the ultrasonic-assisted process parameter optimization system, it will be necessary to analyze the effects of process parameters such as ultrasonic amplitude and the scanning speed on the average grain size of WAAM components and establish an algorithm to optimize the parameters with process parameters as factors. This is an important research direction to realize the further development of arc-fuse additive-manufacturing accuracy and defect control.

Author Contributions: Conceptualization, methodology, writing—first draft, Y.C. and Y.Z.; literature review, W.M. and J.M.; funding acquisition, Y.C. and W.H. All authors have read and agreed to the published version of the manuscript.

Funding: This work was supported by the Henan Provincial Youth Backbone University Teacher Training Plan (2021GGJS090) and Henan Provincial Key Development Project (221111240200).

Institutional Review Board Statement: Not applicable.

Informed Consent Statement: Not applicable.

Conflicts of Interest: The authors declare no conflict of interest.

References

1. Ding, J.; Colegrove, P.; Mehnen, J.; Ganguly, S.; Almeida, P.M.S.; Wang, F.; Williams, S. Thermo-mechanical analysis of Wire and Arc Additive Layer Manufacturing process on large multi-layer parts. *Comput. Mater. Sci.* **2011**, *50*, 3315–3322. [CrossRef]
2. Zalameda, J.N.; Burke, E.R.; Hafley, R.A.; Taminger, K.M.B.; Domack, C.S.; Brewer, A.; Martin, R.E. Thermal Imaging for Assessment of Electron-Beam Freeform Fabrication (EBF3) Additive Manufacturing Deposits. In Proceedings of the Conference on Thermosense—Thermal Infrared Applications XXXV, Baltimore, MD, USA, 30 April–1 May 2013.
3. King, D.; Tansey, T. Rapid tooling: Selective laser sintering injection tooling. *J. Mater. Process. Technol.* **2003**, *132*, 42–48. [CrossRef]
4. Levy, G.N.; Schindel, R.; Kruth, J.P. Rapid manufacturing and rapid tooling with layer manufacturing (LM) technologies, state of the art and future perspectives. *Cirp Ann. Manuf. Technol.* **2003**, *52*, 589–609. [CrossRef]
5. Lu, B.; Li, D.; Tian, X. Development Trends in Additive Manufacturing and 3D Printing. *Engineering* **2015**, *1*, 85–89. [CrossRef]
6. Wong, K.V.; Hernandez, A. A review of additive manufacturing. *Int. Sch. Res. Not.* **2012**, *2012*, 208760. [CrossRef]
7. Frazier, W.E. Metal additive manufacturing: A review. *J. Mater. Eng. Perform.* **2014**, *23*, 1917–1928. [CrossRef]
8. Vayre, B.; Vignat, F.; Villeneuve, F. Designing for additive manufacturing. *Procedia CIrP* **2012**, *3*, 632–637. [CrossRef]
9. Herzog, D.; Seyda, V.; Wycisk, E.; Emmelmann, C. Additive manufacturing of metals. *Acta Mater.* **2016**, *117*, 371–392. [CrossRef]
10. Milewski, J.O. *Additive Manufacturing of Metals*; Springer International Publishing AG: Cham, Switzerland, 2017.
11. Armstrong, M.; Mehrabi, H.; Naveed, N. An overview of modern metal additive manufacturing technology. *J. Manuf. Process.* **2022**, *84*, 1001–1029. [CrossRef]
12. Ramkumar, P.L.; Rijwani, T. Additive manufacturing of metals and ceramics using hybrid fused filament fabrication. *J. Braz. Soc. Mech. Sci. Eng.* **2022**, *44*, 455. [CrossRef]
13. Li, Y.; Su, C.; Zhu, J.J. Comprehensive review of wire arc additive manufacturing: Hardware system, physical process, monitoring, property characterization, application and future prospects. *Results Eng.* **2022**, *13*, 100330. [CrossRef]
14. Liu, J.; Dong, S.; Sun, C.; Zhang, Z.; Xu, B. Research status of quality control of direct energy deposition forming technology. *Manuf. Technol. Mach. Tools* **2020**, *6*, 44–48.
15. Wohlers, T.; Gornet, T. History of additive manufacturing. *Wohlers Rep.* **2014**, *24*, 118.
16. Gao, W.; Zhang, Y.B.; Ramanujan, D.; Ramani, K.; Chen, Y.; Williams, C.B.; Wang, C.C.L.; Shin, Y.C.; Zhang, S.; Zavattieri, P.D. The status, challenges, and future of additive manufacturing in engineering. *Comput. Aided Des.* **2015**, *69*, 65–89. [CrossRef]
17. Singh, S.; Ramakrishna, S.; Singh, R. Material issues in additive manufacturing: A review. *J. Manuf. Process.* **2017**, *25*, 185–200. [CrossRef]
18. Eskin, G.I. *Ultrasonic Treatment of Light Alloy Melts*; CRC Press: Boca Raton, FL, USA, 1998.
19. Babamiri, B.B.; Indeck, J.; Demeneghi, G.; Cuadra, J.; Hazeli, K. Quantification of porosity and microstructure and their effect on quasi-static and dynamic behavior of additively manufactured Inconel 718. *Addit. Manuf.* **2020**, *34*, 101380. [CrossRef]
20. Li, F.; Qi, B.J.; Zhang, Y.X.; Guo, W.; Peng, P.; Zhang, H.P.; He, G.Z.; Zhu, D.Z.; Yan, J.F. Effects of Heat Treatments on Microstructures and Mechanical Properties of Ti6Al4V Alloy Produced by Laser Solid Forming. *Metals* **2021**, *11*, 346. [CrossRef]

21. Paul, C.; Jinoop, A.; Bindra, K. Metal additive manufacturing using lasers. *Addit. Manuf. Appl. Innov.* **2018**, *57*, 37–94.
22. Wang, F.; Tzanakis, I.; Eskin, D.; Mi, J.; Connolley, T. In situ observation of ultrasonic cavitation-induced fragmentation of the primary crystals formed in Al alloys. *Ultrason. Sonochemistry* **2017**, *39*, 66–76. [CrossRef]
23. Bidare, P.; Jimenez, A.; Hassanin, H.; Essa, K. Porosity, cracks, and mechanical properties of additively manufactured tooling alloys: A review. *Adv. Manuf.* **2022**, *10*, 175–204. [CrossRef]
24. Long, P. *Study on Microstructure and Properties of 316 L Stainless Steel Produced by Arc Fuse*; Huazhong University of Science and Technology: Hangzhou, China, 2020. [CrossRef]
25. Sharma, S.K.; Grewal, H.S.; Saxena, K.K.; Mohammed, K.A.; Prakash, C.; Davim, J.P.; Buddhi, D.; Raju, R.; Mohan, D.G.; Tomkow, J. Advancements in the Additive Manufacturing of Magnesium and Aluminum Alloys through Laser-Based Approach. *Materials* **2022**, *15*, 8122. [CrossRef]
26. Raut, L.P.; Taiwade, R.V. Wire Arc Additive Manufacturing: A Comprehensive Review and Research Directions. *J. Mater. Eng. Perform.* **2021**, *30*, 4768–4791. [CrossRef]
27. Li, Z.W.; Xu, Z.W.; Ma, L.; Wang, S.; Liu, X.S.; Yan, J.C. Cavitation at filler metal/substrate interface during ultrasonic-assisted soldering. Part II: Cavitation erosion effect. *Ultrason. Sonochemistry* **2019**, *50*, 278–288. [CrossRef] [PubMed]
28. Wang, B.; Tan, D.Y.; Lee, T.L.; Khong, J.C.; Wang, F.; Eskin, D.; Connolley, T.; Fezzaa, K.; Mi, J.W. Data and videos for ultrafast synchrotron X-ray imaging studies of metal solidification under ultrasound. *Data Brief* **2018**, *17*, 837–841. [CrossRef]
29. Han, C.J.; Fang, Q.H.; Shi, Y.S.; Tor, S.B.; Chua, C.K.; Zhou, K. Recent Advances on High-Entropy Alloys for 3D Printing. *Adv. Mater.* **2020**, *32*, e1903855. [CrossRef]
30. Gu, D.D.; Ma, C.L.; Dai, D.H.; Yang, J.K.; Lin, K.J.; Zhang, H.M.; Zhang, H. Additively manufacturing-enabled hierarchical NiTi-based shape memory alloys with high strength and toughness. *Virtual Phys. Prototyp.* **2021**, *16*, S19–S38. [CrossRef]
31. Cooke, S.; Ahmadi, K.; Willerth, S.; Herring, R. Metal additive manufacturing: Technology, metallurgy and modelling. *J. Manuf. Process.* **2020**, *57*, 978–1003. [CrossRef]
32. Cunningham, C.R.; Flynn, J.M.; Shokrani, A.; Dhokia, V.; Newman, S.T. Invited review article: Strategies and processes for high quality wire arc additive manufacturing. *Addit. Manuf.* **2018**, *22*, 672–686. [CrossRef]
33. Zhang, D.Z.; Li, Y.Z.; Wang, H.; Cong, W.L. Ultrasonic vibration-assisted laser directed energy deposition in-situ synthesis of NiTi alloys: Effects on microstructure and mechanical properties. *J. Manuf. Process.* **2020**, *60*, 328–339. [CrossRef]
34. Chen, Y.H.; Xu, M.F.; Zhang, T.M.; Xie, J.L.; Wei, K.; Wang, S.L.; Yin, L.M.; He, P. Grain refinement and mechanical properties improvement of Inconel 625 alloy fabricated by ultrasonic-assisted wire and arc additive manufacturing. *J. Alloys Compd.* **2022**, *910*, 164957. [CrossRef]
35. Diao, M.X.; Guo, C.H.; Sun, Q.F.; Jiang, F.C.; Li, L.Y.; Li, J.F.; Xu, D.; Liu, C.M.; Song, H.L. Improving mechanical properties of austenitic stainless steel by the grain refinement in wire and arc additive manufacturing assisted with ultrasonic impact treatment. *Mater. Sci. Eng. A Struct. Mater. Prop. Microstruct. Process.* **2022**, *857*, 144044. [CrossRef]
36. Kumar, S.; Wu, C.S.; Padhy, G.K.; Ding, W. Application of ultrasonic vibrations in welding and metal processing: A status review. *J. Manuf. Process.* **2017**, *26*, 295–322. [CrossRef]
37. Xu, M.; Chen, Y.; Zhang, T.; Xie, J.; Wei, K.; Wang, S.; Yin, L. Effect of post-heat treatment on microstructure and mechanical properties of nickel-based superalloy fabricated by ultrasonic-assisted wire arc additive manufacturing. *Mater. Sci. Eng. A* **2023**, *863*, 144548. [CrossRef]
38. Wu, W.Z.; Li, J.L.; Jiang, J.L.; Liu, Q.P.; Zheng, A.D.; Zhang, Z.; Zhao, J.; Ren, L.Q.; Li, G.W. Influence Mechanism of Ultrasonic Vibration Substrate on Strengthening the Mechanical Properties of Fused Deposition Modeling. *Polymers* **2022**, *14*, 904. [CrossRef] [PubMed]
39. Li, G.X.; Bie, W.B.; Zhao, B.; Chen, F.; Zhao, C.Y.; Zhang, Y.M. Ultrasonic assisted machining of gears with enhanced fatigue resistance: A comprehensive review. *Adv. Mech. Eng.* **2022**, *14*, 57. [CrossRef]
40. Shu, S.; Guo, J.X.; Liu, X.L.; Wang, X.J.; Yin, H.Q.; Luo, D.M. Effects of pore sizes and oxygen-containing functional groups on desulfurization activity of Fe/NAC prepared by ultrasonic-assisted impregnation. *Appl. Surf. Sci.* **2016**, *360*, 684–692. [CrossRef]
41. Yi, Z.Y.; Song, C.C.; Zhang, G.H.; Tong, T.Q.; Ma, G.Y.; Wu, D.J. Microstructure and Wear Property of ZrO2-Added NiCrAlY Prepared by Ultrasonic-Assisted Direct Laser Deposition. *Materials* **2021**, *14*, 5785. [CrossRef]
42. Nový, F.; Petrů, M.; Trško, L.; Jambor, M.; Bokůvka, O.; Lago, J. Fatigue Properties of Welded Strenx 700 MC HSLA Steel after Ultrasonic Impact Treatment Application. *Mater. Today Proc.* **2020**, *32*, 174–178. [CrossRef]
43. Wang, H.; Hu, Y.B.; Ning, F.D.; Cong, W.L. Ultrasonic vibration-assisted laser engineered net shaping of Inconel 718 parts: Effects of ultrasonic frequency on microstructural and mechanical properties. *J. Mater. Process. Technol.* **2020**, *276*, 116395. [CrossRef]
44. Williams, S.W.; Martina, F.; Addison, A.C.; Ding, J.; Pardal, G.; Colegrove, P. Wire plus Arc Additive Manufacturing. *Mater. Sci. Technol.* **2016**, *32*, 641–647. [CrossRef]
45. Uralde, V.; Veiga, F.; Aldalur, E.; Suarez, A.; Ballesteros, T. Symmetry and Its Application in Metal Additive Manufacturing (MAM). *Symmetry* **2022**, *14*, 1810. [CrossRef]
46. Liu, J.N.; Xu, Y.L.; Ge, Y.; Hou, Z.; Chen, S.B. Wire and arc additive manufacturing of metal components: A review of recent research developments. *Int. J. Adv. Manuf. Technol.* **2020**, *111*, 149–198. [CrossRef]
47. Wang, T.T.; Zhang, Y.B.; Xie, Y.L. Research status and prospect of wire arc additive manufacturing technology. *Electr. Weld.* **2017**, *47*, 5.

48. Xiong, J.; Zhang, G.J. Adaptive control of deposited height in GMAW-based layer additive manufacturing. *J. Mater. Process. Technol.* **2014**, *214*, 962–968. [CrossRef]
49. Tamayo, J.A.; Riascos, M.; Vargas, C.A.; Baena, L.M. Additive manufacturing of Ti6Al4V alloy via electron beam melting for the development of implants for the biomedical industry. *Heliyon* **2021**, *7*, e06892. [CrossRef] [PubMed]
50. Mahmoud, D.; Magolon, M.; Boer, J.; Elbestawi, M.A.; Mohammadi, M.G. Applications of Machine Learning in Process Monitoring and Controls of L-PBF Additive Manufacturing: A Review. *Appl. Sci.* **2021**, *11*, 11910. [CrossRef]
51. An, N.Y.; Shuai, S.S.; Hu, T.; Chen, C.Y.; Wang, J.; Ren, Z.M. Application of Synchrotron X-Ray Imaging and Diffraction in Additive Manufacturing: A Review. *Acta Metall. Sin. Engl. Lett.* **2022**, *35*, 25–48. [CrossRef]
52. Wu, B.T.; Pan, Z.X.; Ding, D.H.; Cuiuri, D.; Li, H.J.; Xu, J.; Norrish, J. A review of the wire arc additive manufacturing of metals: Properties, defects and quality improvement. *J. Manuf. Process.* **2018**, *35*, 127–139. [CrossRef]
53. Kobayashi, M.; Dorce, Y.; Toda, H.; Horikawa, H. Effect of local volume fraction of microporosity on tensile properties in Al-Si-Mg cast alloy. *Mater. Sci. Technol.* **2010**, *26*, 962–967. [CrossRef]
54. Yi, J.; Guan, G.; Li, S.K.; Liu, Z.W.; Gong, Y.L. Effect of post-weld heat treatment on microstructure and mechanical properties of welded joints of 6061-T6 aluminum alloy. *Trans. Nonferrous Met. Soc. China* **2019**, *29*, 2035–2046. [CrossRef]
55. Zhang, J.K.; Chen, B.H.; Zhang, X. Residual stress at the interface of titanium alloy produced by arc addition and its effect. *Rare Met. Mater. Eng.* **2018**, *47*, 7.
56. Chi, J.; Cai, Z.; Wan, Z.; Zhang, H.; Chen, Z.; Li, L.; Li, Y.; Peng, P.; Guo, W. Effects of heat treatment combined with laser shock peening on wire and arc additive manufactured Ti17 titanium alloy: Microstructures, residual stress and mechanical properties. *Surf. Coat. Technol.* **2020**, *396*, 125908. [CrossRef]
57. Dinovitzer, M.; Chen, X.H.; Laliberte, J.; Huang, X.; Frei, H. Effect of wire and arc additive manufacturing (WAAM) process parameters on bead geometry and microstructure. *Addit. Manuf.* **2019**, *26*, 138–146. [CrossRef]
58. Veiga, F.; Suarez, A.; Aldalur, E.; Artaza, T. Wire arc additive manufacturing of invar parts: Bead geometry and melt pool monitoring. *Measurement* **2022**, *189*, 110452. [CrossRef]
59. Mukhanov, I.; Golubev, Y.M. Strengthening Steel Components by Ultrasonically Vibrating Ball. *Vestn. Mashinostr.* **1966**, *11*, 52–53.
60. Statnikov, E.S.; Korolkov, O.V.; Vityazev, V.N. Physics and Mechanism of Ultrasonic Impact. *Ultrasonics* **2006**, *44*, e533–e538. [CrossRef]
61. Campbell, J. Effects of vibration during solidification. *Int. Met. Rev.* **1981**, *26*, 71–108. [CrossRef]
62. Jian, X.; Xu, H.; Meek, T.T. Effect of power ultrasound on solidification of aluminum A356 alloy. *Mater. Lett.* **2005**, *59*, 190–193. [CrossRef]
63. Cong, W.; Ning, F. A fundamental investigation on ultrasonic vibration-assisted laser engineered net shaping of stainless steel. *Int. J. Mach. Tools Manuf.* **2017**, *121*, 61–69. [CrossRef]
64. Chen, C.Y.; Deng, Q.L.; Song, J.L. Effect of ultrasonic vibration on laser cladding process. *Electromach. Die* **2005**, *3*, 4.
65. Chen, C.Y.; Deng, Q.L.; Song, J.L. Effect of Ni content and ultrasonic vibration on cracks in laser cladding. *J. Nanjing Univ. Aeronaut. Astronaut.* **2005**, *37*, 5.
66. Qin, L.Y.; Wang, W.; Yang, G. Experimental study on ultrasonic-assisted laser deposition of titanium alloy. *China Laser* **2013**, *40*, 0103001.
67. Wang, T.; Zhang, A.F.; Liang, S.D.; Yan, S.P.; Zhang, L.Z.; Li, D.C. Study on as-deposited microstructure and properties of IN718 by ultrasonic vibration-assisted laser metal forming. *China Laser* **2016**, *43*, 6.
68. Wang, T.; Zhang, A.F.; Zhang, W.L.; Liang, S.D.; Li, S.; Yan, S.P.; Zhang, L.Z. Research progress of ultrasonic vibration-assisted laser metal forming technology. *Appl. Laser* **2015**, *35*, 4.
69. Koli, Y.; Yuvaraj, N.; Aravindan, S. Enhancement of Mechanical Properties of 6061/6082 Dissimilar Aluminium Alloys Through Ultrasonic-Assisted Cold Metal Transfer Welding. *Arab. J. Sci. Eng.* **2021**, *46*, 12089–12104. [CrossRef]
70. Yuan, D.; Shao, S.Q.; Guo, C.H.; Jiang, F.C.; Wang, J.D. Grain refining of Ti-6Al-4V alloy fabricated by laser and wire additive manufacturing assisted with ultrasonic vibration. *Ultrason. Sonochemistry* **2021**, *73*, 105472. [CrossRef] [PubMed]
71. Malaki, M.; Ding, H. A Review of Ultrasonic Peening Treatment. *Mater. Des.* **2015**, *87*, 1072–1086. [CrossRef]
72. Krylov, N.A.; Polischuk, A.M. The use of ultrasonic equipment for metal structure stabilization. *Basic Phys. Ind. Ultrason. Appl. Part* **1970**, *1*, 70.
73. Khurshid, M.; Leitner, M.; Barsoum, Z.; Schneider, C. Residual stress state induced by high frequency mechanical impact treatment in different steel grades—Numerical and experimental study. *Int. J. Mech. Sci.* **2017**, *123*, 34–42. [CrossRef]
74. Ma, Q.S.; Chen, H.Z.; Ren, N.N.; Zhang, Y.Y.; Hu, L.; Meng, W.; Yin, X.H. Effects of Ultrasonic Vibration on Microstructure, Mechanical Properties, and Fracture Mode of Inconel 625 Parts Fabricated by Cold Metal Transfer Arc Additive Manufacturing. *J. Mater. Eng. Perform.* **2021**, *30*, 6808–6820. [CrossRef]
75. Roy, S.; Fisher, J.W.; Yen, B.T. Fatigue resistance of welded details enhanced by ultrasonic impact treatment (UIT). *Int. J. Fatigue* **2003**, *25*, 1239–1247. [CrossRef]
76. Mordyuk, B.N.; Prokopenk, G.I.; Vasylyev, M.A.; Iefimov, M.O. Effect of structure evolution induced by ultrasonic peening on the corrosion behavior of AISI-321 stainless steel. *Mater. Sci. Eng. A Struct. Mater. Prop. Microstruct. Process.* **2007**, *458*, 253–261. [CrossRef]

77. Yang, Y.C.; Jin, X.; Liu, C.M.; Xiao, M.Z.; Lu, J.P.; Fan, H.L.; Ma, S.Y. Residual Stress, Mechanical Properties, and Grain Morphology of Ti-6Al-4V Alloy Produced by Ultrasonic Impact Treatment Assisted Wire and Arc Additive Manufacturing. *Metals* **2018**, *8*, 934. [CrossRef]
78. Cheng, X.H.; Fisher, J.W.; Prask, H.J.; Gnaupel-Herold, T.; Yen, B.T.; Roy, S. Residual stress modification by post-weld treatment and its beneficial effect on fatigue strength of welded structures. *Int. J. Fatigue* **2003**, *25*, 1259–1269. [CrossRef]
79. Rao, D.L.; Chen, L.G.; Ni, C.Z.; Zhu, Z.Q. Effect of ultrasonic impact on residual stress of welded structure. *Weld. J.* **2005**, *26*, 4.
80. Mordyuk, B.N.; Karasevskaya, O.P.; Prokopenko, G.I.; Khripta, N.I. Ultrafine-grained textured surface layer on Zr-1%Nb alloy produced by ultrasonic impact peening for enhanced corrosion resistance. *Surf. Coat. Technol.* **2012**, *210*, 54–61. [CrossRef]
81. Zhang, J.; Xing, Y.F.; Zhang, J.J.; Cao, J.Y.; Yang, F.Y.; Zhang, X.B. Effects of In-Process Ultrasonic Vibration on Weld Formation and Grain Size of Wire and Arc Additive Manufactured Parts. *Materials* **2022**, *15*, 5168. [CrossRef] [PubMed]
82. Yang, D.Q.; Wang, X.W.; Peng, Y.; Zhou, Q.; Wang, K.H. Study on Microstructure and Properties of 316L stainless Steel produced by Ultrasonic impact assisted Melt Arc addition. *Mater. Guide* **2022**, *1*, 122–125.
83. Zhang, Y.-Z.; Wang, Z.-G.; Xie, L.-Y.; Ding, Z.-Y.; Ma, Y.-H.; Ouyang, J.-H.; Liu, Z.-G.; Wang, Y.-J.; Wang, Y.-M. Laser surface nanocrystallization of oxide ceramics with eutectic composition: A comprehensive review. *Heat Treat. Surf. Eng.* **2021**, *3*, 37–54. [CrossRef]
84. Ye, X.L.; You, Z.L.; Kan, W. Effects of Ultrasonic Impact Treatment on the Residual Stress and Fatigue Performance of Ultrahigh Strength Steel Weld Joint. In Proceedings of the International Technology and Innovation Conference 2006, Hangzhou, China, 6–7 November 2006; pp. 150–154. [CrossRef]
85. Abdullah, A.; Malaki, M.; Eskandari, A. Strength enhancement of the welded structures by ultrasonic peening. *Mater. Des.* **2012**, *38*, 7–18. [CrossRef]
86. Xiao, C.H. *Study on the Elimination of Welding Residual Stress and Deformation by Real-Time Ultrasonic Impact*; Harbin Institute of Technology: Harbin, China, 2013.
87. Gao, H.; Dutta, R.K.; Huizenga, R.M.; Amirthalingam, M.; Hermans, M.J.M.; Buslaps, T.; Richardson, I.M. Stress relaxation due to ultrasonic impact treatment on multi-pass welds. *Sci. Technol. Weld. Join.* **2014**, *19*, 505–513. [CrossRef]
88. Shao, Z.W.; Le, Q.C.; Zhang, Z.Q.; Cui, J.Z. Effect of ultrasonic power on grain refinement and purification processing of AZ80 alloy by ultrasonic treatment. *Met. Mater. Int.* **2012**, *18*, 209–215. [CrossRef]
89. Wang, D.P.; Song, N.X.; Wang, T.; Huo, L.X. Nano-treatment of ultrasonic metal surface. *J. Tianjin Univ.* **2007**, *40*, 6.
90. Hu, G.F.; Yang, Y.; Sun, R.; Qi, K.; Lu, X.; Li, J.D. Microstructure and properties of laser cladding NiCrBSi coating assisted by electromagnetic-ultrasonic compound field. *Surf. Coat. Technol.* **2020**, *404*, 126469. [CrossRef]
91. Xu, J.L.; Zhou, J.Z.; Tan, W.S.; Huang, S.; Wang, S.T.; He, W.Y. Ultrasonic vibration on wear property of laser cladding Fe-based coating. *Surf. Eng.* **2020**, *36*, 1261–1269. [CrossRef]
92. Wen, X.; Cui, X.F.; Jin, G.; Zhang, X.R.; Zhang, Y.; Zhang, D.; Fang, Y.C. Design and characterization of FeCrCoAlMn0.5Mo0.1 high-entropy alloy coating by ultrasonic assisted laser cladding. *J. Alloys Compd.* **2020**, *835*, 155449. [CrossRef]
93. Yang, Y.T.; Zhang, Y.Y.; Yu, W. Study on the properties of titanium alloy welded joints treated by ultrasonic impact treatment. *Mater. Dev. Appl.* **2007**, *22*, 5.
94. Zhao, X.H.; Wang, D.P.; Wang, X.B.; Deng, C.Y.; Zu, Z.Q. Loading ultrasonic impact improves the fatigue performance of TC4 titanium alloy welded joints. *J. Weld.* **2010**, *11*, 4.
95. Haagensen, P.J.; Statnikov, E.S.; Lopez-Martinez, L. Introductory fatigue tests on welded joints in high strength steel and aluminium improved by various methods including ultrasonic impact treatment (UIT). *IIW Doc* **1998**, *13*, 1748–1798.
96. Yang, X.; Zhou, J.; Ling, X. Study on plastic damage of AISI 304 stainless steel induced by ultrasonic impact treatment. *Mater. Des. (1980–2015)* **2012**, *36*, 477–481. [CrossRef]
97. He, Z.; Hu, Y.; Qu, H.; Wang, Z.M.; Bu, X.Z. Study on the anisotropy of titanium alloy parts fabricated by ultrasonic impact arc. *Aerosp. Manuf. Technol.* **2016**, *6*, 6.
98. Xu, M.F.; Chen, Y.H.; Deng, H.B.; Ji, D. Study on Microstructure and Mechanical Properties of TC4 Titanium Alloy produced by Ultrasonic assisted CMT Arc addition. *Precis. Form. Eng.* **2019**, *5*, 7.
99. Chen, W.; Chen, Y.; Zhang, T.; Wen, T.; Yin, Z.; Feng, X. Effect of ultrasonic vibration and interpass temperature on microstructure and mechanical properties of Cu-8Al-2Ni-2Fe-2Mn alloy fabricated by wire arc additive manufacturing. *Metals* **2020**, *10*, 215. [CrossRef]
100. Yuan, D.; Sun, X.; Sun, L.; Zhang, Z.; Guo, C.; Wang, J.; Jiang, F. Improvement of the grain structure and mechanical properties of austenitic stainless steel fabricated by laser and wire additive manufacturing assisted with ultrasonic vibration. *Mater. Sci. Eng. A* **2021**, *813*, 141177. [CrossRef]
101. Ming, W.; Guo, X.; Xu, Y.; Zhang, G.; Jiang, Z.; Li, Y.; Li, X. Progress in Non–Traditional Machining of Amor-phous Alloys. *Ceram. Int.* **2023**, *49*, 1585–1604. [CrossRef]
102. Zhang, Z.; Zhang, Y.; Liu, D.; Zhang, Y.; Zhao, J.; Zhang, G. Bubble behavior and its effect on surface integrity in laser-induced plasma micro-machining silicon wafer. *ASME J. Manuf. Sci. Eng.* **2022**, *144*, 091008. [CrossRef]
103. Zhou, C. *Study on Temperature Field and Stress Field of Laser Metal Deposition Assisted by Ultrasonic Impact*; Harbin University of Engineering: Harbin, China, 2021; p. 000087.
104. Wu, Y.; Deng, W.; Liu, K.; Zhang, Z.; Wu, D.; Chen, M.; Bai, J. Effect of interlaminar ultrasonic impact on microstructure and mechanical properties of 2219 aluminum alloy produced by TIG arc addition. *Aviat. Sci. Technol.* **2021**, *32*, 80–86.

105. Wang, C.; Li, Y.; Tian, W.; Hu, J.; Li, B.; Li, P.; Liao, W. Influence of ultrasonic impact treatment and working current on microstructure and mechanical properties of 2219 aluminium alloy wire arc additive manufacturing parts. *J. Mater. Res. Technol.* **2022**, *21*, 781–797. [CrossRef]
106. Li, P.; Guo, S.; Yang, D.; Peng, Y.; Yan, D.; Li, D.; Wang, K. Ultrasonic vibration assisted arc augmentation Material manufacturing 2219 aluminum alloy structure and mechanical properties [J/OL]. *China J. Nonferrous Met.* **2022**, *1*, 1–12. Available online: https://kns.cnki.net/kcms/detail/43.1238.TG.20220919.1009.002.html (accessed on 19 September 2022).

MDPI AG
Grosspeteranlage 5
4052 Basel
Switzerland
Tel.: +41 61 683 77 34

Metals Editorial Office
E-mail: metals@mdpi.com
www.mdpi.com/journal/metals

www.ingramcontent.com/pod-product-compliance
Lightning Source LLC
Chambersburg PA
CBHW061545220926
51850CB00005B/251